Next-Gen Solutions for Sustainable Agronomy

Next-Gen Solutions for Sustainable Agronomy:

Application, Technologies, and Management

By

Wasswa Shafik

Cambridge Scholars Publishing

Next-Gen Solutions for Sustainable Agronomy:
Application, Technologies, and Management

By Wasswa Shafik

This book first published 2025

Cambridge Scholars Publishing

Lady Stephenson Library, Newcastle upon Tyne, NE6 2PA, UK

British Library Cataloguing in Publication Data
A catalogue record for this book is available from the British Library

ISBN: 978-1-0364-4965-0
ISBN (Ebook): 978-1-0364-4966-7

To my cherished family, whose unwavering support and love have been my constant source of strength.

To all the poor people across the world who fight to earn a living, who would have wished to make it in life, but the conditions failed, who was sentenced to death due to blackmail, and all the senior citizens, youth, the young generations, among the dead, the living and the unborn.

TABLE OF CONTENTS

PREFACE

The rise in the global population and the changing dietary habits have opened new frontiers in the agriculture sector. To meet the future demand and manage the resources for present and future generations, we must adapt our sustainable practices to become more maintainable and technology-driven. This book presents a critical insight into sustainable agronomy/agriculture in relation to next-generation (Next-Gen) computing, often overlooked in the spectrum of excellence in science, and puzzles many people because they don't know what it is. In fact, agronomy is a wide field where all scientific domains are represented in relation to plant production, but also with other monoculture systems, growing mushrooms, producing algae or microalgae blooms for omega-3 acids, and growing materials such as starch-based bioplastics, producing fruits or vegetables. Producing more food also raises concerns about the impact on nature. Climate change, increasingly impacting agriculture, is prompting debates on how plant-based systems, including livestock production, should be redesigned. Sustainable agriculture has often been at a loss to balance food and non-food goals, producing malnutrition on one hand and surpluses on the other. Addressing the root causes of malnutrition requires systemic changes focusing on sustainable rural development, but more importantly, an appreciation of the multifunctional roles of agriculture, not just in food security but also in preserving native ecosystems, especially biodiversity. We will look at the constraints and challenges facing the future of agronomy. Technology is increasingly thought of as a magic wand that can rescue farmers from predicaments like food shortages following extreme weather events and other unpredictable situations, whether societal or physical. The agronomists' perspective is that technology is indeed beneficial but also an available tool.

To extensively dissect and digest this trending paradigm, the book is structured into sixteen chapters and four parts. The initial part entails four chapters focusing on the introduction of sustainable agronomy and enabling technologies, presenting a comprehensive exploration of integrating sustainable agriculture and artificial intelligence, establishing foundational knowledge of sustainable farming principles and AI's transformative potential in the agricultural sector, and examining technologies such as precision farming, data analytics, and smart irrigation. Part two offers a scientifically rigorous and expert-level examination of the intersection between sustainable farming practices and the latest advancements in artificial intelligence, covering Chapters 5 through Chapter 8. It includes integrational of foundational agricultural principles with state-of-the-art AI technologies such as machine learning, robotics, and data-driven decision-making. It systematically explores how agricultural disruptive technologies can transform traditional farming practices, from optimizing precision agriculture techniques to leveraging big data for predictive analytics, thereby enhancing productivity and ecological sustainability.

Part three advances by delving into complex topics of sustainable agronomy like AI-enabled crop genetics, smart irrigation, and the use of autonomous systems for pest and disease management, providing a roadmap for implementing intelligent solutions across the agricultural value chain. It also emphasizes the scientific methodologies underpinning these innovations, offering a critical understanding of how computational models can address challenges in resource efficiency, climate resilience, and food security presented in Chapter 9 through Chapter 12. The integration of case studies in digital agriculture and supply chain optimization with cutting-edge research ensures that the material is not only theoretically sound but also grounded in practical, real-world applications. The final part of this book caters to a broad audience, including researchers, agronomists, and professionals in agricultural technology who seek to deepen their knowledge of AI's role in sustainable development by adding case studies and prospects of this field within Chapter 13 to the last one. It is also an essential resource for academic institutions and students in fields such as agricultural engineering, environmental science, and computer science, providing a comprehensive framework for research and coursework. Equipping readers with the tools to understand and harness the potential of AI in agriculture, this book positions itself at the forefront of scientific discourse on sustainable agricultural transformation. This book is intended to be a modern source of information on sustainable agronomy and agriculture, which could be useful to industry, students, academia, libraries, research institutes, laboratories, or other interested professionals in advancing agricultural sustainability and addressing global food security challenges.

INTRODUCTION

Agronomy - a term coming from the Greek words agrós, 'land,' and nómos, 'rule' - is the discipline involving scientific principles and the practice of field configuration. While it generally refers to crop science, dealing with field crops, it may also include other aspects of cultivation, such as genetics, livestock, economics, and the management of weeds, diseases, and pests. The practical side of agronomy involves the techniques used to improve crop production. While crop production can be influenced by genetics, disease, weather, and other factors, agronomical approaches to a sustainable yield are those that directly involve the management of the field, the influenced organisms, the resources, and the remaining market and policy influences. Over 8 billion people are currently inhabiting the earth, with the population expected to exceed 10 billion by 2050, regardless of the recent pandemic. Rapid population increases are putting extreme pressure on food production, land, and water resources. In addition, climate change further undermines food security across the globe. While changing the scenario is not an easy task, providing sustainable agronomy for food production is indeed profitable for thinking outside of agriculture. The challenge is how to transition to more efficient food systems, especially in developing countries, given that land and water scarcity is increasing at an alarming rate. In this context, unmanned aerial vehicles are an innovative solution that is increasingly being deployed for alternative farming approaches.

Sustainable agronomy has become a major issue over the last few decades, contributing to ecological balance and resource management using eco-friendly practices. Sustainable agronomy involves a collection of fundamental eco-friendly agricultural practices that can enhance agricultural productivity by minimizing the environmental impact. Crop rotation, organic farming, tactics to boost soil biodiversity, and proper weed, disease, and insect pest control are some of the underlying principles of sustainable agronomy. These principles are considered to be essential for producing high-quality and safe food at an affordable cost. Moreover, this also helps to develop and utilize agricultural land in such a manner that it minimizes damage to nature and other living things and has no impact on the social and economic development of regions and local communities. Continued implementation of these principles can ensure successful and sustainable agronomy by promoting and launching new-generation (Next-Gen) technology in agriculture.

Sustainable agronomy can also produce several other benefits, including increasing crop yields, coping with climate change, restoring soil quality, conserving land and water, protecting the environment, protecting natural resources and biodiversity, and lowering chemical residue in food. With a limited amount of pollutant environmental effects of crop protection products, they result in long-term, harmonious, and healthy plant protection. Sustainable agronomy primarily stresses the well-being of the people and supports the broader socioeconomic significance. Year on year, reasonable principles can offer a decent and long-term income for oilseed growers. By maximizing oilseed quality and price, cultivation within the principles can result in a greater return for a given harvest. In addition to generating the best revenues, the short term will undoubtedly result in a high yield on the program. Greater oil recovery from the oilseeds will likely add to the metal and pelleted food contribution, which might be marketed at a premium. Economic benefits are anticipated to be made. Regular *Globodera pallida* (PCN) and a large variety of weeds in soil weed regulation boost farmers' profitability. Agronomic solutions enhance the income and grow the revenues in oilseeds, resulting in leaders in profit. Apart from the broad sense, agronomic solutions embrace a range of suitable goods and practices. These goods and technologies might be applicable to any portion of the farming cycle and production. Agronomic plants, pests, and diseases can be used to range, depending on weather, or partly dependent on technologies.

Advancements in technology are transforming most of our daily activities, and the same is true for sustainable agronomy. It is widely recognized that efficiency and effectiveness play a vital role in farming. Several technological innovations to enhance decision-making and management in sustainable agronomy are in place. In precision agriculture, modern technologies gather data from various sources, allowing the farmer to make the best use of resources. It has the potential to reduce waste and save farmers money. It minimizes environmental impacts. Tools such as Global Positioning System (GPS), drones, and other sensor technologies collect data that ranges from the soil, weather, and crops. Their application areas include field variability, soil fertility, water quality and quantity, crop yield and quality, chemical, fertilizer, herbal, and organic substitution products, and machinery management. Remote sensing technologies such as multispectral, hyperspectral, and thermal, through drones and satellites, monitor plant conditions, stress factors, and vibration. These technologies have the potential to monitor fields quickly and can see large sections with one image. GNSS and Geographical Information Systems (GIS) are technologies used to map locations of interest and help with directions. Precision agriculture combines these technologies to monitor, respond, record data, and improve agricultural management in real-time. This results in many benefits to the farmers, including saving time and money.

Modern sensor technologies and IoT allow real-time data collection and electronic report management systems using digital applications. They monitor the crops and the soil and give the farmer real-time feedback. This session seeks to outline technological advancements that are moving sustainable agronomy practice from a traditional approach toward sustainability. It will help to inspect the resources completely, study the regional variability of resources, develop predictive services, and generate and synthesize new data to characterize and forecast the spatial and temporal variation of sustainability-related indicators. The potential applications of such progress in technology are both to evaluate different productivity and maximal input scenarios and environmental quality and crop yield to optimize the covariance equation in a classical loss function. The development of the information technology revolution in theoretical and practical econometric models of linking sustainable agronomy with farmers' assessments are the main achievements of research. Therefore, illustrating the scope, applicability, and limitations of each of the explored new technological improvements, as well as the provision of an empirical analysis corroborating each predicted estimate, is critical. Thus, this scientific work can help create awareness among the agronomy frontliners at regional levels. The heart of these technological advancements has been the very extensive development of sensors and sensor networks to collect data on the crops to produce high-resolution data with a large amount of information, thus improving statistical procedures around variable fresh but dirty data. In the past, fresh data were always rare. Thus, the development of many new sensors will allow a more accurate measurement of many properties, providing a clearer understanding of processes and saving costs in terms of both money and time spent in getting useful data.

As detailed within the book it highlights the importance of continued investment in research and training to produce the necessary tools and future scientists and extension agents required to support local communities in adopting and implementing sustainable agronomy. There is a need for a multi-stakeholder approach, including grassroots extension workers, to achieve a sustainable solution. It is through continuous investment in knowledge creation and massive investment in labor, technology, and resources that we can maximize the environmental, economic, and social benefits and enormous possibilities to drive sustainable agronomic systems. Policy reforms and incentives for the governmental, nongovernmental, and private sectors to consider and support sustainability concerns are essential.

It is through fostering broad public awareness of issues related to sustainable agronomy and food security and the commitment of the greater community to take the necessary actions to secure sustainable solutions. The final key to increasing the benefits of sustainable agronomy is through a substantially slow yet gradual consumer and community behavioral change that supports and promotes the consumption of sustainably produced food by becoming part of the growing consumer demand for safe local foods. Overall, there has been a shift in research focus in the area of sustainable agronomy to address some of these issues. With continued research, ongoing investment in agricultural research, education, and technological development, and closer collaboration, sustainable agronomy initiatives will play a greater role in the green future.

Description

In many parts of the globe, malnutrition, food insecurity, and hunger are the foremost encounters, mostly in low and middle-income countries. To cater to the rising global population growth as per the United Nations' development goals of Zero Hunger, climate change, and responsible consumption and production, this book examines the technological landscape that underpins the future of sustainable agriculture entailing agronomy, from precision farming and robotics to data analytics through crop genetics; this unveils the diverse tools contributing to more efficient and sustainable agricultural practices fostering the increase food production. Detailing an in-depth practical integration, smart irrigation systems, plant stress identification and detection through supply chain optimization to showcase the effective utilization of artificial intelligence for sustainable and resilient agricultural operations, making it well-suited for students, trainees, practitioners, and individuals seeking to deepen their understanding of how current agricultural technologies support sustaining farms and plantations. Moreover, the book serves as a valuable resource for educational purposes, providing insights for coursework, research projects, or training programs in fields related to agriculture, technology, and environmental sustainability. This comprehensive book equips readers with the knowledge needed to understand and address food shortages and technological integration concerns of the most pressing global challenges of this era.

PART I

INTRODUCTION TO SUSTAINABLE AGRONOMY AND ENABLING TECHNOLOGIES

CHAPTER ONE

THE CONCEPT AND PERCEPTION OF SUSTAINABLE AGRONOMY

Abstract

Sustaining global food demand with limited land and water resources calls for serious consideration of maintaining soil, water, and air quality. Worldwide agricultural practices over the last few decades have shown great emphasis on farm productivity, resulting in a significant increase in yield, but lesser attention was paid to environmental issues like soil erosion, loss of soil fertility, and heavy use of chemical inputs, leading to the deterioration of land, water, and soil. To meet global food demands, it is of utmost importance to maintain the sustainability of agricultural practices. It is a reality of immense complexity because it contemplates the essential aspects of the complex relationship between human beings, technologies, and the environment. With this setting, it is necessary to look at the relative need for precocity to understand that in relation to this dialogue and all the scientific aid of contemporary agriculture, sustainable agriculture has obtained little help. Scientific production is still reduced and limited, until now related mainly to specific problems and requirements that arise. At the university level, sustainable agriculture still receives little attention. Initiatives specifically aimed at the training of agronomists and others who consider the complex reality of Brazilian agriculture have not been as important, innovative, and promising as in other professional fields. The concept and perception of sustainable agronomy are identically limited, with an insufficient number of trained specialists to act effectively in the diverse tasks that await and predispose these professionals to confrontation, which will only happen if we foresee the innovative and multifunctional sustainable agriculture that is coming, which can and should involve academics and professionals. Preferences are very relative, and we respond. As in any field, the possibilities are always latent, and the conquests are consequences of intentions and actions. The possibility that we have had is the intention and the intelligent basis of the university.

Keywords: Agroecology, Agronomic Practices, Climate Adaptation, Crop Management, Environmental Resilience, Resource Efficiency, Soil Health, Sustainable Agriculture

1.1 Introduction

The term 'agronomy' has been used in a variety of contexts to refer to "deals with the principles and practices of production and management of field crops, that is, agronomic practices, or of the development and management of agronomic departments." It was in the first sense that the term was used in the first century when working with field crops. In fact, it was only from that century, and with the work of Varro, that the verb agriculture was expelled to make application and definition for the specialization agronomy (Nautiyal et al. 2024). Within agriculture, agronomy is traditionally situated at the interface between natural and social sciences or the environment and economy. Very often, agronomists are referred to as experts in the technical field and the human sciences of agriculture.

The term 'sustainability,' on the other hand, consists of many different words and synonyms, including sustainability, conservation, maintaining, and preserving, among others, and many different objects of conservation or elimination, including resources, services, products, development, supply, function, ecosystems, future, environment, and lifestyles. Nowadays, the concept of environment and sustainability is present independently in all areas of activity (Abdul-Majid et al. 2024). Due to the great global problems that have been arising, it became necessary to develop a new definition of agronomy that took into account not only the large volume of agricultural yield but also the conservation and preservation of the environment. Agronomy and sustainability of the environment are part of this new context and of an agro-technology capable of offering a technological architecture adequate to their needs. One characteristic of the occurrence presented is that it assumes several possible states, and in the repair process, the system will return to a given state, desired or not (Sharma et al. 2024).

Understanding the concept and profitable business of sustainable agronomy is essential for establishing an effective agricultural policy and stimulating the relevant practices at both farm and product-market levels. The way in which this concept is understood is described and discussed in the narrative literature. It is argued that investments and profits are much reduced in the credible, most familiar perceptions, while research-based incentives are preserved and expanded in their broader variation, containing, besides ecological, economic, social, and cultural principles. The main adverse effects of disregarding these broader principles point out the importance of this broader perception (Sandilya et al. 2024). Sustainable agronomy is operationalized in the most supported manner, where one concept of ecological endowment restrictions and sustainable ecological functions can be attached. The paradigm states that the social and ecological environment is converted via the economic environment into the economic structure based on human wants. This conversion leads to a biophysical reality and indicates where the sustenance of an economic system has to be ensured. It must be able to feed the population at all times. If it cannot do this, the envisaged societal, social, and cultural setup of an economic system is not sustainable (M. Singh et al. 2024). Only an economic system capable of doing this and of keeping the relationship between individual human beings, between people and organizations, and with respect to nature, such that the societal equation for externally unanswerable natural phenomena will not get disrupted, represents a quite restrictive sustainable content through revolutions, as presented in Fig. 1-1.

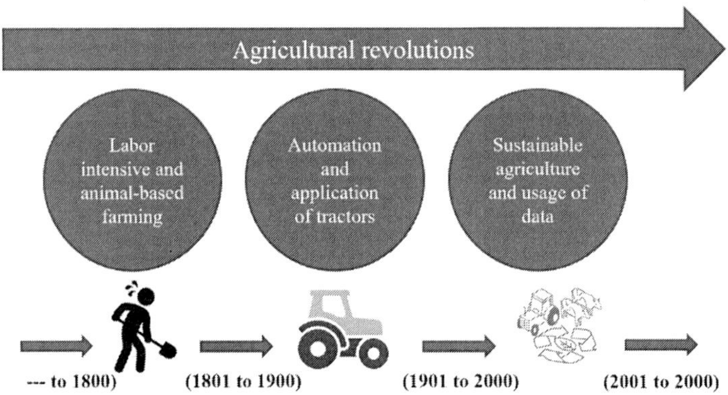

Fig. 1-1 Agricultural revolutions

1.2 Foundations of Sustainable Agronomy

Sustainable agriculture is associated with several concepts. This is primarily a commitment to justice in rural areas, hence the attitude of responsibility, because it is the responsibility of individuals to fully participate in shaping the common good and not allowing for the disintegration of the unity and sustainability of the environment. It is also a return to ecological origins, emphasizing the need for harmony between the requirements of economic development and the protection of the natural environment. In agriculture, the idea of sustainability is primarily a guarantee of food security, which reinforces economic security and, more broadly, political security. It also means preserving natural resources and the fertility of the earth not only for the present but also for future generations. In their perception, the function closest to the concept of sustainable agriculture is the agricultural function, particularly its educational role, understood as shaping citizens' attitudes toward sustainable development (Beig et al. 2024). Farmers' persistence and the conviction of almost unanimity that their work aims to shape the natural environment derive from their educational role as mediators of an ecologically sustainable rural environment. The rural environment is not only about the production of food and agricultural raw materials. Its role is understood universally; the function refers not only to material production but also to the health of citizens and the identity of their cultural and national heritage (Brower-Toland et al. 2024).

Agriculture encompasses all-time human activities: organization of risk, food production, landscape modifications, implementation of economic activities, and identification with tradition. Understanding agriculture in this aspect requires the use of a subsystem approach; it is impossible to grasp the essence of agriculture from merely a socio-economic point of view. The purpose of agronomy is not only to produce food and raw materials for the industry, as it has not only an exclusively investor-adjustment, socio-economic organizational function, or research function. It is aware of itself and the environment, its links with other sectors of the economy, and the wider farming functions performed in relation to the community, education, adaptation, and economics (Shafik, Tufail, De Silva Liyanage, et al. 2024). Providing citizens with diverse qualities while simultaneously being the manager of the natural environment, farmers and rural community operators critically shape the attitudes of citizens toward others, proving their civic potential and behaviors as citizens, providing them with a sense of belonging and integration, and filling their everyday lives with meaning. Consequently, the purpose of agronomy is to strive to meet the needs of future generations, hence the constant interest over several centuries in agricultural resources and their intelligent management in developing nations, for instance Sub-Saharan Africa (Shafik and Kalinaki 2024). If there is no such activity and no social and economic self-concept, it should be assumed that agronomy has not performed any function. This has good reasons. When these goals have been achieved, a function in the narrow and strictly determined sense may be discovered at one time. Then, is it justified, and what is the scope of activity in the area subject to analysis?

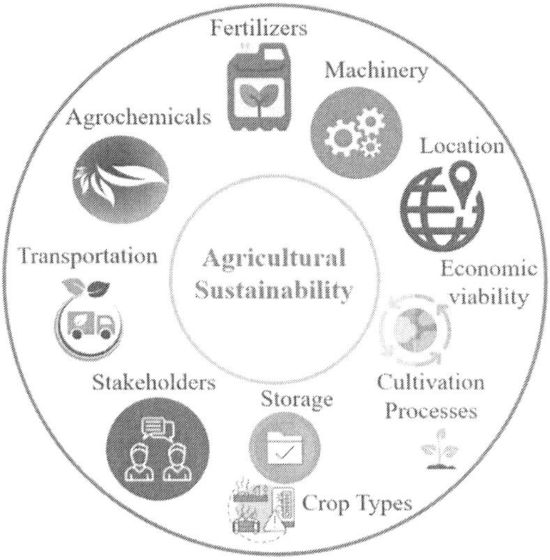

Fig. 1-2 Agricultural sustainability

1.2.1 Definition and Principles of Sustainable Agronomy

The word "sustainable" is often used but less frequently defined. It is mostly interpreted as a long-lasting process in which people manage and sustain themselves within the planet's resources. Sustainability serves as a measure of our human consumption and destruction of goods, limiting it to a sustainable level. In all such scenarios, both society and the planet exist sustainably. The term sustainability grew in the 1960s and 1970s from a sense of alarm regarding the detrimental impact of human actions on nature; the term sustainable development was coined in the 1980s to promote both objectives and approaches to achieve them (V. Singh 2024). Sustainable agriculture consists of the adoption of techniques that protect and conserve the basis of the agricultural environment. These practices can be implemented to avoid the suboptimal use of natural ecosystems and take advantage of them in an ecological and economically friendly manner. Promote healthy ecosystems, support current and future human generations, help increase the quality of life for current and future generations, and help produce sufficient and safe food. Keep humans at the center of development (Sharma et al. 2024; Moreno et al. 2024). Farmers should be able to cover their costs. Avoid external costs. Promote environmentally sustainable practices that support the improvement of soil quality using sustainable agricultural practices, as illustrated in Figure 1.2.

1.2.2 Historical Evolution

Written history records the origins of agronomical activities during the agricultural revolution in the fertile crescent, where the first civilizations were formed, and during the development of associations of horticulturists and crop growers in pre-Columbian Latin America (Shafik 2025a). In the first cases, horticultural activity is a transition activity, following very conservative times that will be called those gathered since the Neolithic. They represented the changes from primary systems based on predation to productive systems based on obtaining vegetable goods, partly in view of the conservation of the natural reserve and the territorial stability of the groups (Qu et al. 2024). Indigenous Latin America knows a symptom that receives a huge inflow from the tropical equatorial climate. Until 15,000 years ago, the environmental situation of a large part of the continents was more favorable, especially in the period between 10,000 and 7,000 years ago. Horticultural and cereal breakfasts are different, depending on the group of species that are generally distinguished in the primary crop systems, if the systems developed can be identified in a type of agronomy that recognizes three directions of agriculture: focused systems on open-fired food-collecting, woodland systems for agricultural application (Kumar et al. 2024; Muthumalai et al. 2024). Gardens appear as a hand tool and develop the process of domestication of plants useful for man; the main works are organized essentially around production processing and nutritional use. In pre-Columbian Latin America, evolution was parallel to domestication. The Incas were weakened by the discovery of South American starches, occupying 4th or 5th place, and above all, there was pacification and acceptance of cities that maintained the state structures for a long time, even after their victory over many competitions (Cisse et al. 2024).

1.3 Key Components of Sustainable Agronomy

There are three important blocks or aspects for any agronomic system or production system to be considered sustainable. These are social, economic, and environmental blocks or aspects. Social blocks include, firstly, the aspect of justice regarding the producers and laborers working in the areas to be cultivated; that is, how fair the wages paid are, how long the working hours are, whether child labor is used and whether they are obliged to do blood and sweat work, and what the social security levels are. An agronomic system in a country considered to be advanced in this sense may not be acceptable in another part of the country and may not meet the standard even in the world. Secondly, environmental and animal rights should be sought in the social aspect of any economic system (Moreno et al. 2024). Dignity, respect, rural exodus, democracy, life, food security, education, culture, body, and tradition are the other social aspects. The economic block includes the control over the producers and inputs of all systems/products within this chain, as well as the security problems. Increasing the quality of the natural environment and agricultural ecosystem within the economic block adds not only products but also innovative values.

Sustainable agriculture or agronomy should also allow the economic development of farmers and producers in addition to industrial and manufactured products (Nautiyal et al. 2024). Locally, healthy value creation with the lowest investment and expenditures is a part of the balanced economy of the region, as well as the creation of richer ecosystems favorable for the demands and variability of the climatic regions. It should not be overlooked that the purpose of the fight is for the existence of the platform with the observed countries rather than the complexity of the agronomic systems (Sandilya et al. 2024). Live correctly even in this world; that is, creating live products/services related to demand and country incomes is important for the sustainability of global food security since this is largely limited to three enabling factors: plants, including environment, host, and pathogens, as illustrated in Fig. 1.3.

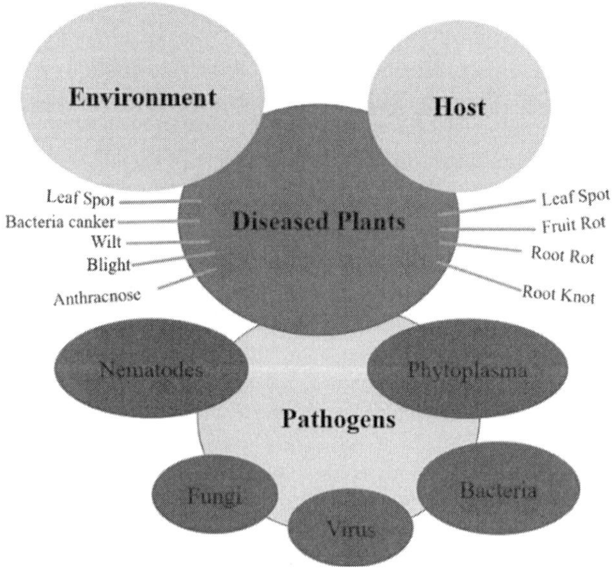

Fig. 1-3 Plant disease and the three enabling factors

1.3.1 Soil Management

One of the keys to sustainable agronomy is soil management in order to minimize degradation. Another benefit derived from good practices designed to improve the conservation of soils is the increase in agricultural productivity. The protection of soil from erosion is its most important characteristic because its deterioration is irreversible. Half of all the surface and all the subsoils around the world are already moderately to highly eroded. Using well-structured no-till techniques conceals seeds and protects the surface from the impact of rain, thus reducing the energy of the drops and causing them to become trickles (Naresh et al. 2024). In turn, it increases infiltration into the soil. The basic principles of minimum conservation tillage include minimizing soil disturbance, controlling weeds using mulches or chemicals, increasing soil organic matter by promoting the recirculation of nutrients and reducing runoff, and improving the physical structure of the soil depend the plant stresses as illustrated in Fig 1.4. The cultivation system most often used in tropical and subtropical climates is no-till because this type of climate is especially affected by soil erosion. In areas that are predominantly cultivated with wheat and other rotations well adapted to these kinds of conditions, minimum tillage techniques predominate as no-till techniques can lead to an increase in plant diseases if appropriate control techniques are not applied (Chirwa et al. 2024; Quayson et al. 2024).

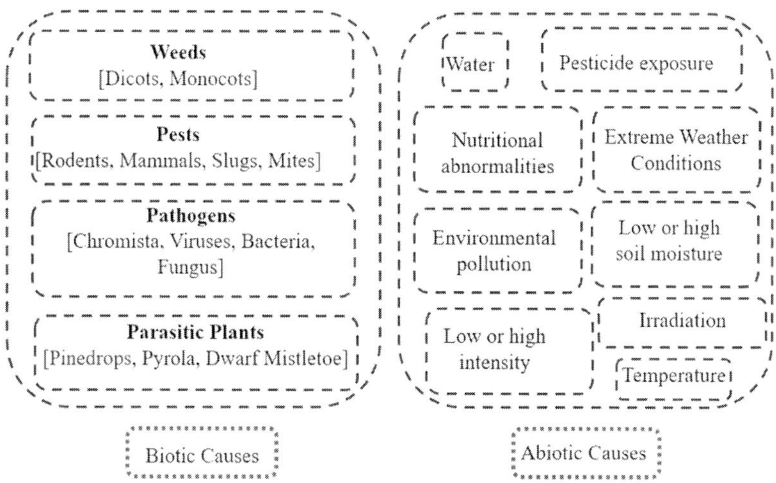

Fig.1-4 Plant biotic and abiotic causes

1.3.2 Crop Diversity and Rotation

Diverse cropping systems offer a larger portfolio of ecosystem services than mono-cropping. The many benefits of diversification derive from the improved temporal and spatial integration of farm activities that break the life cycle synchrony of pests, weeds, and pathogens and disrupt their structures, functions, and relationships necessary to survive and propagate. Favorable reciprocal relationships may also develop between some beneficial and harmful organisms, which would not occur if a uniform and single-crop environment prevailed (Moreno et al. 2024). Diverse cropping systems can also act as insurance against catastrophic crop failure and against unprecedented and unanticipated pest, pathogen, and weed pressures. In the case of the occurrence of new pests, pathogens, and weeds, farmers can use the new crop resources to breed crops that are resistant to the new pests and diseases. In regions where the knowledge of a crop system is highly developed, the consumed crop and other resources can be changed to minimize the negative impacts of pests and pathogens on yield. The recent resurgence of sustainable agriculture has refocused attention on traditional agronomic practices, the most significant of which is the diversification of the cropping system (Brower-Toland et al. 2024; Sandilya et al. 2024). While the degree of diversification is bounded by soil and environmental factors, diverse cropping systems offer better and more diverse sources of food, fodder, and fuel, increase agroecosystem stability and resilience, and ensure overall agronomic system sustainability. Crop rotation, intercropping, cover cropping, allelopathy, legume associations, and organic manuring are practical methods used to diversify cropping systems. A return to diverse crop systems is a major and necessary step to achieve environmentally and geopolitically sustainable agriculture (V. Singh 2024). Cultivating a diverse range of crops and crop varieties has long been recognized as a strategy to manage agricultural resources more efficiently and to reduce the risks associated with agricultural production. However, contemporary agriculture in most semi-arid, sub-humid, and humid tropical regions often includes only a few selected crops with limited vegetative, physiological, and genetic diversity (Chirwa et al. 2024).

1.3.3 Water Conservation

Water scarcity is one of the most challenging issues in the world. Rainwater harvesting can become a significant alternative to the increasing water crisis. The application of rainwater harvesting increases the economic benefits in the semiarid areas. Soils cover the majority of the Earth's surface area, and rainwater is the most important freshwater source for plants in non-humid land, but it is still not easily accepted as an adequate irrigation water source (Moreno et al. 2024). However, human societies have already used harvested rainwater for thousands of years within their water management systems; especially in non-regulated environments, rainwater harvesting is the predominant water supply option for many crop-growing systems. Scientists estimate that a significant portion of the total agricultural area is affected by water scarcity, and in these regions, mainly rainfed agricultural systems can be found (Qu et al. 2024).

1.3.4 Pest and Disease Management

One advantage of agroforestry is the inherent diversity of crops, which, combined with structured planting schemes including both long- and short-term planting, maximizes trees spatially and temporally and reduces the number of new pests that are likely to break out. There is a strong and significant reduction in disease. Variation in shaded establishments resulted in better management of diseases and pests. An increasing number of cases are also being developed, which help manage the risk of disease by diversifying land use through agroforestry management practices at the level of a farm (Shafik et al. 2023). However, because tree crops, instead of annual herbs, are a new crop and have specific logon for disease management, such management is part of the productive cycle since its determination is to be maintained to pests and diseases throughout the lifecycle of the crop, requiring farmer involvement. Moreover, standardized obstacles and knowledge systems face both formal research and ordinary farmers (Shafik, Tufail, De Silva, Haji, et al. 2025). Investment in research on food security environments favors poor exclusion, income generation, and self-sufficiency in the use of own consumption and farm products, where such production systems have no standard for adequate disease management, in which deep investment is realized because production is done for internal use (Naresh et al. 2024). Extending this understanding to tree crops makes a significant contribution to problems that remain somewhat unresolved due to the long external scenario of the crop with the rest of the crops, cultural management practices, and reproduction systems that vary, as summarized within Table 1.1.

Table 1-1 Key sustainable agronomy practices, environmental and crop impacts

Agronomic Practice	Brief Description	Environmental Impact	Impact on Crops	Crops and Regions	Reference
Soil Amendments (Biochar)	Application of biochar for soil health	Sequesters carbon, enhances soil fertility	Increases nutrient retention, water holding	Rice (Southeast Asia)	(Moreno et al. 2024)
Intercropping	Growing multiple crops simultaneously	Enhances biodiversity, better resource use	Reduces pest pressure, improves yield stability	Maize/Beans (Africa)	(Brower-Toland et al. 2024)
Controlled Traffic Farming	Precision traffic control on soil	Reduces compaction, improves aeration	Enhances root development, better yields	Cereals (Australia)	(Sandilya et al. 2024)
Green Manure	Incorporating cover crops into soil	Improves soil organic matter	Enhances nutrient availability	Sorghum (Sub-Saharan Africa)	(Sharma et al. 2024)
Biofertilizers	Use of microbial inoculants	Enhances nutrient cycling, reduces chemical fertilizers	Promotes root development, improves crop nutrition	Wheat (India)	(Chirwa et al. 2024)
No-Till Farming	No mechanical soil turning	Preserves soil structure, enhances organic matter	Reduces soil erosion, increases drought resilience	Soybean (Brazil)	(Quayson et al. 2024)
Precision Agriculture	Data-driven crop management	Optimizes resource use, reduces waste	Improves yield accuracy and predictability	Various (Global)	(Beig et al. 2024)

Drip Irrigation	Water delivered directly to the root zone	Reduces water wastage, minimizes runoff	Increases water-use efficiency	Tomato, Cotton (India)	(Shafik 2024a)
Integrated Pest Management (IPM)	Ecological pest control methods	Reduces pesticide use, enhances biodiversity	Reduces pest damage, improves yield quality	Rice (Asia)	(Shafik, Tufail, Apong, et al. 2024)
Agroforestry	Integration of trees and crops	Sequesters carbon, enhances biodiversity	Provides shade, increases crop protection	Cocoa (West Africa)	(Kalinaki et al. 2023)
Organic Farming	Use of organic inputs like compost	Reduces chemical pollution	Enhanced nutrient content in crops	Vegetables (Europe)	(Shafik and Kalinaki 2024)
Conservation Tillage	Minimal soil disturbance	Reduces soil erosion, improves water retention	Increased root growth, higher yields	Maize, Wheat (USA)	(Moreno et al. 2024)
Crop Rotation	Alternating different crops in succession	Breaks pest cycles, improves nutrient use	Enhances nutrient uptake, disease control	Corn, Soybean (USA)	(Qu et al. 2024)
Cover Cropping	Planting cover crops between main crops	Enhances soil fertility, prevents erosion	Improves soil moisture, reduces weeds	Legumes (Africa)	(Muthumalai et al. 2024)

1.4 Technological Innovations in Sustainable Agronomy

Achieving sustainability in crop production systems requires the development and adoption of innovative research results. This can be facilitated by the appropriate flow of information through extension and advisory services. This chapter has evaluated the concept and perception of agronomy on the environment, food safety, health, and the risks to people. It has also addressed the economic aspects and policymaking of sustainable agronomy. Reviewed concepts of agronomy for sustainability embrace practices that conserve the environment and manage weeds, pests, and diseases(Qu et al. 2024). In all agronomic practices, it is important to strike a balance between short-term and long-term benefits and costs and to recognize that risk management is integral to all economic and policy decisions. Over the past century, agronomic science has played an important role in significant yield increases of major crops and animal products, which in turn fueled population growth and improved the welfare of people globally. Currently, engineering innovative management programs for significant disease, pest, and weed problems using environmentally compatible, science-based solutions will be required (Naresh et al. 2024). This chapter represents an argument that new technological advances in agronomy are necessary to remove obstacles and prepare for new opportunities to develop sustainable agronomy. Corn, with its high resource use efficiency, high productivity, broad adaptation, and high net energy, makes it an ideal candidate for feedstock in biorefineries to produce bioethanol or other renewable biofuels (Moreno et al. 2024).

1.4.1 Precision Agriculture

The term "sustainable" is a fairly new and rather loosely used term. We can find the term "sustainable" in many combinations, such as sustainable economy, sustainable development, sustainable agronomy, sustainable agriculture, and sustainable tourism. Sustainable agronomy, using the principles of precision agriculture, is one of the appropriate ways to achieve better food production with a more rational consumption of resources (V. Singh 2024). Precision farming is an emerging concept, which in the very near future is expected to reformulate the aims of agronomy, leading to better and more rational food production, as well as sustainable agronomy. The search for high profits and better food quality, the necessity to reduce pollution, and the increasing consumer concern about the environment and health are factors favoring the changes associated with sustainable agronomy (Beig et al. 2024). The use of individual auto-steer machinery systems with precision seeding and planting, auto-system diagnosis, yield monitoring, and map-based unilateral or bilateral soil fertilization is highly demanded. These and other new techniques are eco-friendly, do not introduce a high mineral load, and grow increased agricultural yields through sustainability in the immediate future by means of conducting the needed stress analysis using the principles of sustainable agriculture (Kumar et al. 2024).

1.4.2 Biotechnology and Genetic Engineering

Genetic engineering and biotechnology represent a higher form of traditional plant-related technologies. From the government's perspective, maintaining a food balance is a pressing concern. This is because evidence has shown that, in times of chronic hunger and severely deficient malnutrition, many parents are themselves undernourished and physically weak (Shafik 2025b). The obvious way to break this vicious, intergenerational cycle is to help increase the productivity of staple crops that provide the potential for food-based approaches meant to address the complex problems of hidden hunger and essential nutrient deficiencies of the Indian population and other malnutrition-hit populations of the world (Adebunmi Okechukwu Adewusi et al. 2024). The developed world is largely reconciled to the benefits of genetically modified food crops. All manner of GM crops are marketed, not only in the fertile crescent of agricultural biotechnology but across Latin America, the EU itself, and, importantly, the developing world. In China, which is often characterized by the practiced ignorance of some healthy biotechnology critics as the unsuspecting potential victim of rapacious producers, analyses seldom acknowledge the scale and dynamism of the GM revolution that is reported to be taking place in multiple crops and locations. Furthermore, malnourished Indian children are not tokens or guinea pigs in the romantic war between agro-capitalism and anti-GMO activists (Cisse et al. 2024; Beig et al. 2024). The fact is that Chinese, Brazilian, and South African consumers are currently eating, often quite happily, a whole range of modified GM food supporting nutrient cycling, as presented in Table 1.2.

Table 1-2 Nutrient cycling in sustainable agronomy with environmental and plant impacts

Nutrient Cycle Component	Role in Sustainable Agronomy	Key Processes Involved	Environmental Impact	Impact on Plants	Benefits to Crop Yield	Reference
Silicon Cycle	Structural integrity	Soil dissolution, plant uptake	Reduces pest pressure	Strengthens plant cell walls	Higher resistance to stress	(Qu et al. 2024)
Molybdenum Cycle	Nitrate assimilation	Soil availability fluctuations	Reduces nitrate leaching	Enhances nitrogen utilization	Increased protein yield	(Kumar et al. 2024)
Boron Cycle	Cell wall and membrane synthesis	Soil sorption, desorption	Reduces boron toxicity	Improves flowering and fruit set	Better fruiting and seed set	(Muthumalai et al. 2024)
Manganese Cycle	Involved in photosynthesis	Oxidation-reduction reactions	Enhances redox buffering capacity	Promotes enzyme activation	Higher leaf quality	(Santoso et al. 2024)
Copper Cycle	Aids in lignin synthesis	Chelation, adsorption	Mitigates copper toxicity	Strengthens cell walls	Improved disease resistance	(Beig et al. 2024)
Zinc Cycle	Essential for enzyme function	Absorption, soil interactions	Limits micronutrient leaching	Improves enzymatic activity	Increases yield of zinc-rich crops	(Quayson et al. 2024)
Iron Cycle	Crucial for electron transport	Redox reactions	Mitigates soil and water iron deficiency	Prevents chlorosis, enhances vigor	Better overall plant health	(V. Singh 2024)

Magnesium Cycle	Central in chlorophyll molecule	Leaching, cation exchange	Reduces dependence on Mg fertilizers	Enhances photosynthetic efficiency	Increased vegetative growth	(Brower-Toland et al. 2024)
Calcium Cycle	Soil pH regulation	Weathering of calcium minerals	Prevents soil acidification	Improves cell wall stability	Enhanced fruit firmness	(Sandilya et al. 2024)
Potassium Cycle	Regulates plant water balance	Weathering, soil absorption	Reduces potassium runoff	Enhances drought resistance	Increased crop turgor pressure	(Muthumalai et al. 2024)
Sulfur Cycle	Essential for protein synthesis	Mineralization, oxidation	Reduces dependency on sulfur fertilizers	Increases protein content in crops	Improved seed quality	(Kumar et al. 2024)
Carbon Cycle	Soil carbon sequestration	Photosynthesis, decomposition	Reduces atmospheric CO_2	Improves soil moisture, nutrient content	Higher crop resilience	(Muthumalai et al. 2024)
Phosphorus Cycle	Availability of phosphorus to crops	Mineralization, Solubilization	Reduces phosphorus runoff	Improves root development	Enhanced fruit and grain yield	(Sharma et al. 2024)
Nitrogen Cycle	Nitrogen fixation and assimilation	Biological Nitrogen Fixation (BNF)	Reduces synthetic fertilizer use	Increases chlorophyll production	Enhanced biomass production	(Moreno et al. 2024)

1.5 Economic and Social Impacts of Sustainable Agronomy

Farming is the business of attaining the highest possible yields in order to make enough profit to ensure general prosperity. However, irrespective of improved technology in soil tillage, crop rotation, fertilization, and protection against pests, today's high yields resulting from a high level of agricultural growth are maintained either at the cost of severe damage to the soil or through heavy investments of human trust, partly in the form of financially burdensome measures (Shafik, Tufail, De Silva, and Apong 2025). This condition, in addition to the external economic effects pressing upon agriculture and the losses due to variations in global agricultural market prices, ultimately leads to income instability on the microeconomic level, which in turn fosters resistance against change and, thus, diversification and concentration (Sidik et al. 2024). The concept of comprehensive and sustainable agronomy encompasses not only changes in the production sector designed to stabilize and adjust agricultural income and revenues but also an ecological orientation in alignment with the principles of comprehensive and sustainable agronomy. In other words, the aim is to create a balanced relationship between society's demands and the production objectives of the agricultural community. The aim is to find a balance between competitive commercial production or commercial private agronomy and environmental goods of supra-economic value, common benefits that can only be generated from and secured on mutual grounds by the agri-ecosystems network, commercial farming, and non-entrepreneurial, non-private voluntary socioeconomic activities not oriented towards commercial incentives (Brower-Toland et al. 2024).

A socioeconomic dimension is consistently important in agronomy, with constant agronomic endeavors at the basis urging for the whole field to be occupied. Supplementary biotope functions providing societal services and integrating traditional activities are undoubtedly essential elements of agronomy land use, harmoniously integrated into the ecological surroundings. Such activities by no means compete with sustainable, productive agronomy to an attenuating degree. Much rather, the permanent involvement of people in productive agronomy combines with the sketched complementary servitude activities to create the necessary change (Naresh et al. 2024). Since, now as in the past, sustainable and comprehensive agronomy essentially forms a part of the infrastructure of regions interlinking the bio-ecological, commercial, and societal services sectors, contributions of this nature are prerequisites to the economic success of the whole region. In the altered processes that now emerge, this objective can only be achieved through organizational networks suitable for the respective economic, environmental, and regional conditions. In this sense, market economy requirements serve merely as a corrective measure facilitating normal interaction between the mechanisms of public administration, politics, and society (Santoso et al. 2024).

1.5.1 Economic Benefits

One of the most important short-term economic benefits is the potential reduction in production costs associated with reduced inputs. This requires less irrigation and reduced operation and maintenance costs. Less or absence of fertilization also leads to lower purchasing costs because no agricultural chemicals are required or fewer nutrients need to be applied. In general, the economic savings associated with lower input requirements are realized over time and not immediately. Nevertheless, the farmer is directly affected by additional economic benefits, such as price premiums for organic production or shorter production cycles (Sharma et al. 2024). A more immediate economic benefit is the opportunity to diversify production and, therefore, increase the long-term farm's economic resilience. Thus, it is not only about marketing products in the markets but also the ability of the unit to maintain or expand production continuously. There are important economic benefits associated with the optimization of the use of the resource base in the short, medium, and long term. In many cases, the greatest contribution of improved quality production to the search for new markets or niche markets is to attract price premiums that are necessary to compensate for lower production volume or to justify the cost of modernizing production techniques or farm operations (Sidik et al. 2024).

1.5.2 Social Equity and Inclusion

People who are involved with or impacted by sustainability, respect, justice, and fairness, as well as some issues to consider in relation to inclusive and equitable systems, are described. Promote social equity and inclusion according to people's needs and capacities and avoid weaknesses, oppression, and suffering. Organic systems may not be fully inclusive or equitable, but the systems, in general, provide greater space for integration with the least increases in scope and access disparities. All people must be included and find the use of current and new systems (Cisse et al. 2024). The new sustainable agronomy design is expected to create contexts for people's participation in decisions, to appreciate knowledge identities and insights of people, and to emphasize the importance of transparent and legitimate operations and flows. The key issue in this direction is the empowerment of people, which arises whenever conditions and resources can support the empowerment of local organic resources. The increase in power leads to the creation of dynamic space inside organic development and the organizations that contribute to the creation of such favorable contexts so that people can use them (Nautiyal et al. 2024).

1.6 Challenges and Barriers to Sustainable Agronomy

Sustainable agronomy is based on the principles of sustainable development and, more specifically, on the principles of the circular economy. On a more general note, these principles are based on the premise that economic, social, and environmental performance are interdependent and interrelated. Soils, water, and air are resources on which all of humanity depends and which we must protect. Sustainable development and achieving the overarching societal commitment to ensuring the welfare, health, and food security of present and future generations are inextricably linked to the physical, biological, and chemical properties of soils and their ecosystem services (Quayson et al. 2024). Agronomy is available to answer five basic questions: Who is to be fed? With what? In what quantities? With what investments and trade-offs? How best to satisfy food needs in both the short and the long term? However, the agronomical approach generally involves optimizing the use of one type of capital-economic capital-without properly addressing the other four types of capital: natural, social, human, and produced. Moreover, agronomy is not limited to the challenges of the present but must guide action to plan and adapt to future changes. Agriculture, as a sector, is fundamentally multi-functional: its role goes far beyond food production and is associated with non-fungible but crucial contributions to environmental protection and enhancement, as well as territorial structuring and vitality. The need to meet these multiple challenges through policies that combine different logic and tools is a gateway to an agroecological transition (Cisse et al. 2024; Santoso et al. 2024). What are the challenges and barriers involved in this agroecological transition?

1.6.1 Policy and Regulatory Issues

Policy instruments to promote sustainable agriculture should be targeted. If a sector receives favorable treatment in terms of subsidies, price support, or regulatory standards, it matters in management decisions. In some cases, the advice and support systems explain which direction to choose. A possible consequence is that land stewards become involved for the purpose that brings favorable treatment, not sitting idle in fear of more control. Adoption of supportive measures often requires a track record of regulatory compliance. This is an important landmark in that regulatory provisions have now to be multiplicative by market forces. Where the law is providing guidance, the mandatory character of the provision may be irrelevant (Kumar et al. 2024). Regulations may be more efficient if the government guides voluntary guides. These guides can exist at different levels, from very detailed guidelines to a more general form, with the degree of flexibility of the measures and the degree of uncertainty about the resulting environmental performance as increasing functions of the level. Individual decisions of farmers, influenced by their reference groups, are established in the framework of general legal requirements. The recent demonstration of the imperative environmental impact of widespread agricultural production has once again stimulated orientation perceptions of sustainable agricultural strategies (Chirwa et al. 2024). Encouraged by a boom of development help, these strategies try both to achieve more reversal of current agricultural development paths and to limit damage in those areas of the present ecological status of the country that have been missed in our current development history. While these perceptions seem appropriate and respond to a number of recent diagnoses and warnings on the timescales of our natural resources and landscape, the orientations themselves are, in fact, far from innovative, and there is little among the concrete strategies proposed to believe that agriculture will ever acquire a true obligation to biodiversity (Moreno et al. 2024).

1.6.2 Knowledge Gaps and Research Needs

In a communication published in this journal, an insightful analysis of the meaning of sustainable agronomy, the perception that agroecosystem stakeholders have of sustainable agriculture, and the research need to increase the presence of agronomy in this field is presented. Some of the proposals made require more theoretical foundation and, in any case, the review could contribute more authors and documents. It might also be useful to explore how sustainability is actually related to agronomy so that the action of its different aspects can be more clearly defined and measured (Quayson et al. 2024). As for the theoretical foundation, it is understood that a report is used to define sustainable agricultural production and supports the idea that sustainability is perceived differently by different stakeholders in or guardians of agroecosystems. In any case, the document is very well-referenced, which is suitable for the subject because it is interdisciplinary. Knowledge gaps that constrain the potential of agronomy to contribute to increased sustainability of the agroecosystem are identified, along with some recommendations for filling these gaps, and agronomic knowledge that is considered already useful to improve the perception of sustainable agronomy and the design of studies and practices, especially those related to plant breeding, is presented (Muthumalai et al. 2024). In this comment, the review is first contextualized, and then some observations on the concepts developed and the recommendations proposed are made. Likewise, knowledge useful to narrow the gaps in sustainable agriculture is proposed. It is also proposed to explore in greater detail the degree of sustainability that is carried through the action of agronomy to be aware of the positive contribution or the elements that require modification in the actions of its different aspects (Qu et al. 2024; Sharma et al. 2024).

1.7 Perception of Sustainable Agronomy

To make this section more insightful, we have to mention: \

a) Farming today is less integrated with society than it once was;
b) The rules of knowledge transmission between agricultural and non-agricultural systems have changed, also deeply conditioning the communication flows of agriculture.

Each form of agriculture has embodied in itself, both as an implicit perception and as a fundamental rule, a communicative system. In recent years, the issue of the sustainability of agricultural activity, as well as food, has met with the consensus of public opinion, which was sensitive to the over-mediatization of some issues. Despite the high consensus, we would like to know whether there is a correct perception of topics related to sustainability and agriculture. It is essential to understand what veracity of the concept of sustainable agriculture has been able to create a good degree of consensus, not only by farmers but also by consumers and public opinion in general, because, for many people, the good reputation of those who work for a topic related to the environment represents acknowledgment of the general interest for the "multiple functions" that agriculture would have to have in an economic system, a cultural system, a social system, to the production of nutritional well-being and quality of life, as well as to the protection of the environment and biodiversity (Cisse et al. 2024). Many surveys have been carried out on the perception of the consumer who is directed to the environmental attributes of the product—safety, absence of transgenic organisms and residues of chemical synthesis products, short and known supply chain. Curiously, however, the surveys on the perception of citizens regarding the factors that determine the general environmental sustainability of crops appear to be quite rare (Muthumalai et al. 2024).

The often sufficiency of agriculture and rural doctrines that argue the importance of public opinion does not appear to be accompanied by analytical, in-depth knowledge of the factors that make up the perception of sustainability, but rather scantily informed also from the cultural point of view. It cannot be said that public opinion is only formed under the lead of newspapers or in biased or conformist perception because these aspects only reflect one side of the story: while it is true that the press can often be accused of eco-spreading or over-emphasizing environmental news and issues, it is also true that sustainability and the biosafety of products tend, in any case, to be considered an item close to the people, something that, especially during some periods, has an extreme and captivating need for news (Beig et al. 2024). The educational and divulgative activity could, also in society, occur positively or negatively to the "popularization" of sustainability, making it become a municipal or social property. The recognition by the public, at large, of the peculiar features of agriculture, radically changed by the beginning of the third millennium, has become a need, as has the system of communication between agriculture and society. Raised by the reconsideration of farming in its ecological and multi-functional characteristics to the alternation of large phases of modernization and ecological conversion, it has entered fully into the garments of verbal-communicative, social, and cultural nature (Sharma et al. 2024).

1.7.1 Public perception

It has been widely recognized that the public can play an important role in reshaping the practices of agriculturalists and contribute to implementing agricultural policies. Several studies have shown that media messages, community outreach programs, and educational campaigns influence public awareness and attitudes about sustainable agronomy. Nevertheless, different interpretations coexist on the reasons behind a positive or negative public orientation. The acceptance of small-scale farming might be due to the fact that local or organic food is more flavorful or that it has been proven safer than industrially farmed produce (Brower-Toland et al. 2024). In some instances, however, participants in these studies have raised questions about the safety of farm products and even questioned the worth of small-scale farming. Indeed, conveying the impression that the quality of food is based on small-scale farming might give rise to the suspicion that small-scale farming is not up to those standards. The inclination of consumers to impress—not because of funds diversion, but because of the importance of their own or family's health—can be altered through information and public assistance measures (Moreno et al. 2024). Considering the contents of traditional means of communication, many studies have analyzed the opinions of various population groups on sustainable agriculture practices. In general, it appears that women are more critical, actively interacting with members of CSA farms; older adults who have lived through the phase conversion to intensive agriculture; rural people who experienced the changes in the rural environment firsthand; heavy consumers of organic or mega bioproducts. On the other hand, the most critical area is that of typical consumers whose knowledge can be based on various levels of distorted information (V. Singh 2024). From a different point of view, several others have ironically claimed that ecological blindness is responsible for the social unawareness to convey farm stress.

1.7.2 Farmer perspective

Apart from their time off work, as individuals and entrepreneurs, farmers have their ideas about what is sustainable agronomy and often use a definition that is less scientific and more experiential. Farmers are another economic player, with their professional activities, like others, being constrained by regulations. They produce, sell, and pay taxes and are therefore concerned with knowing what the conditions for their economic activities will be in the near or more distant future. It is in that capacity that they perceive the concept of sustainability in agriculture. They are therefore concerned with the future of grasslands, which are central to their businesses (Chirwa et al. 2024). They constantly seek to produce in quantities that allow them to survive while maintaining their way of working, which for them is sustainable and takes into account wider social and economic factors. To do so, they are curious and seek to be informed by all sorts of people who come and talk to them at meetings, who phone them, and also because they share their observations, concerns, and ways of doing things. These dialogues are important because they give us farmers' vision and knowledge (Beig et al. 2024).

Today, the future of agriculture also depends on the climate. In the future, our profession will depend on how it affects the climate and how we can mitigate this. For farmers, the climate also has an impact on the costs and revenue of their businesses. There are two key factors determining whether farmers choose to adopt sustainable practices. The first is financial: proposals need to be economically sound if there is to be support from professionals. The second is practical: it is not enough for research to say that a proposal is viable. Legislators and policymakers, whether regional, national, or European, often ignore farmers and their accounts of climate change: how can they plan to invest in the future when there is no money right now? They very much support all the innovative methods proposed and attend the meetings to discuss them so that ideas can be thrown out and new, potentially relevant ones can be brought in (Qu et al. 2024). They are very open to debate, but the proposals must attain a basic level of economic soundness. This is what the farmers also tell us. Information can also be a barrier, as presented in Table 1.3.

Table 1-3 Perception of sustainable agronomy across stakeholders

Stakeholder	**Perception**	**Benefits Identified**	**Challenges Perceived**	**Recommended Interventions**
Technology Providers	Development of sustainable agri-tech	Innovation in farming methods	High development cost	Partnerships with farmers
Media	Public awareness and knowledge dissemination	Increased visibility of sustainability	Sensationalism	Ethical Reporting
Urban Farmers	Integration of agriculture in urban areas	Local food production, reduced transport	Space constraints	Vertical farming technologies
Environmental Advocates	Support agroecological practices	Reduced carbon footprint	Policy resistance	Policy Advocacy
Financial Institutions	Investment in sustainable agriculture	Long-term ROI, ESG alignment	Risk of poor credit access	Risk mitigation schemes
Educational Institutions	Promote sustainable agronomy education	Skilled workforce development	Outdated curricula	Curriculum reform
Indigenous Communities	Traditional ecological knowledge	Conservation of biodiversity	Marginalization	Inclusion in decision-making
Consumers	Demand for sustainably grown products	Healthier food, ethical considerations	Price sensitivity	Awareness campaigns
Extension Agents	A bridge between research and practice	Knowledge dissemination	Limited resources	Capacity building

Agribusinesses	Long-term supply chain sustainability	Improved product quality	Short-term profit concerns	Incentive alignment
Agricultural NGOs	Advocate for sustainable practices	Environmental conservation	Lack of farmer participation	Community engagement
Researchers	Evidence-based agronomy practices	Data-driven decision-making	Research funding	Collaborative projects
Farmers	Economic viability, increased yield	Improved soil health, reduced input cost	High initial cost, knowledge gap	Training, subsidies
Policymakers	Policy integration, environmental regulation	Long-term sustainability	Political will, enforcement	Policy frameworks

1.8 Future Directions and Opportunities

The future for sustainability is rather broad and wide. There is also a broad understanding. It will come mainly to the shoulders of agronomy and soil science to lead the way in defining sustainable practices. There are opportunities for synthetic frameworks that promote soil as natural capital. What can we say about how soil natural capital is traded during the production years? What are the more efficient ways to do this? Integration of soil ecosystems leading to greater load or growing capacity looks like a promising way to lead the way forward. As the trading of soil capital is inevitably linked to the logistic function that leads it, one prediction would be a progression from land degradation to land sparing, a concept that links mortality at the extreme but transitions through a series of set saturation states (Sidik et al. 2024). And what set of tools would provide this outcome? Precision-based management tools utilize the intrinsic properties of soil optimally, matching plant requirements with soil resources and managing soil property heterogeneity. Specifically, the application of soil amendments by means of precision guidance techniques provides significant on-farm environmental benefits by increasing yields while at the same time reducing inputs (Moreno et al. 2024). Alternatively, there is also potential for further research and development to optimize the current precision methodology, increasing the effectiveness of reducing temporal and spatial resolution. Additionally, the optimization of the resource allocation strategy requires the establishment of representative and standardized research frameworks that have a clear goal from the outset, that being the promotion of precision management techniques as a tool capable of ensuring sustainable quality and quantity in the agricultural production chain (V. Singh 2024; Sharma et al. 2024).

1.8.1 Emerging Trends

The concern for agricultural development that is environmentally friendly has fostered a shift in the paradigms of agricultural development from one that is primarily technically oriented to agroecological parameters, from agricultural sectoral to rural community focus, and to political forces that are participatory and decentralized. These agricultural development models focus on sustainable development approaches that promote agricultural intensification along the principle of reducing the negative impact of environmental degradation beyond the focus on mere yield increment of food production. This is because they perceive this relationship as a consummative one, and the possibility exists of growing more on the same amount of land, which further reduces pressure on the environment. Such models include approaches to better land management practices, knowledge of where specific crops should be grown, determining systems of crop rotations, and devising tillage systems tailored for individual fields, all of which give additional competitive edges to this process (Moreno et al. 2024; Adebunmi Okechukwu Adewusi et al. 2024). Crop and animal husbandry are two dimensions of agriculture that are units of a wider agro-ecological system. For plant systems, the structure of the agroecosystem comprises the type of soil, the pattern of the crop type, the position of the crop in relation to the local environment, the tillage system, the selections of species and varieties, the nutrient and pest management practices, the irrigation system, and the cropping system. The concept of the suitability of crops to the local environmental niches constitutes the movement from mere productivity to sustainable agronomy that is involved with the mitigation of the impact of agricultural activity on climate change. Jurisdiction between grains grown under spring and winter seasons, planting historic planting dates, nitrogen rates, and landscape curvature from the yield layout, maximum yield, and partial economy optimum are economics and logistically relative (Brower-Toland et al. 2024). Crop modeling as an aid for farmers' productivity is more critical. Supplementary irrigation, carbon partitioning, and sequestration are among the issues that drive research in this direction.

1.8.2 Collaborative Partnerships

The success of sustainable agronomy initiatives depends on collaborative partnerships. A wide range of partners, including individuals, local, regional, and national governments, non-governmental organizations and foundations, the private sector, and universities, need to be engaged to ensure effective progress. Research responsive to specific country food security needs and national development goals can be supported by creative partnerships fostering the efficient, effective, and cooperative use of resources from governments, universities, donor organizations, and the private sector. University–private sector partnerships also contribute significantly to strengthening tertiary education in agronomy. Universities should be sensitive and accountable for meeting the specific needs of different groups, such as the rural sector and the private sector (Sharma et al. 2024). Private companies, either because they need to invest in human resources development or because they benefit professionally from offering advisory services, are very active as supporting members of this new approach in Technical Agronomy programs. These companies offer practical teachings in subjects of interest for their professional needs, support venues, and scholarships, and organize professional periods under the control of the faculty, who can closely follow the evolution of the students and intervene in case of deficiencies. Institutional closeness may stimulate invitations for short-term contracts during or immediately following graduation. The prospect has already mentioned the Professional Master's degree in Agronomy option (Quayson et al. 2024). Table 1.4 highlights how these innovations impact agriculture, the challenges they face, and actionable solutions for broader adoption.

Table 1-4 Future concepts and perspectives in sustainable agronomy

Concept/Trend	Core Idea/Description	Practical Impact on Agriculture	Key Challenges	Perspectives for Adoption
Urban and Peri-Urban Agronomy	Promoting farming in urban environments	Provides local food, reduces transport emissions	Limited space soil quality in urban areas	Vertical farms, rooftop gardens
Climate-Smart Certification	Certification for climate-resilient farming practices	Access to premium markets for certified products	Verification and monitoring costs	Partner with NGOs and certifying bodies
Biofertilizers and Biopesticides	Natural alternatives to chemical inputs	Improves soil health, reduces chemical dependence	Limited availability, inconsistent quality	Encourage local production, quality control
Circular Resource Use	Recycling waste (like crop residues, manure)	Enhances resource efficiency, reduces costs	Requires waste processing infrastructure	Community composting initiatives
Water Harvesting and Recycling	Capturing and reusing water for irrigation	Ensures water availability during dry periods	Infrastructure costs, water quality management	Low-cost water storage systems, training farmers
On-Farm Biodiversity Management	Promoting crop diversification and agroforestry	Increases ecosystem resilience	Cultural preference for monocultures	Demonstrate benefits through pilot projects
Participatory Breeding Programs	Involving farmers in crop development	Produces locally adapted, farmer-preferred crops	Coordination between scientists and farmers	Establish participatory research platforms

Renewable Energy in Agriculture	Use of solar and biogas for farm operations	Reduces carbon footprint, lowers energy costs	High initial costs	Financing schemes, government incentives
Integrated Pest and Disease Management (IPDM)	Combining biological, mechanical, and chemical methods	Reduces pesticide reliance	Requires farmer education and monitoring systems	Provide extension services and digital tools
Localized Nutrient Management	Precision application of nutrients based on need	Reduces fertilizer runoff, improves yields	Access to diagnostic tools	Mobile soil-testing services
Soil Health as Capital	Treating soil fertility as a long-term investment	Improves productivity and carbon sequestration	Short-term focus by some farmers	Incentivize soil-friendly practices with subsidies
Resilient Crop Varieties	Breeding crops resistant to drought and pests	Enhances food security in changing climates	Slow development cycles, regulatory approvals	Strengthen breeding programs, seed banks
Data-Driven Agronomy	Utilizing AI, sensors, and IoT for precision farming	Optimizes input usage (water, fertilizer)	Digital infrastructure and farmer access	Affordable tech solutions, rural connectivity
Agroecological Intensification	Enhancing productivity by mimicking natural ecosystems	Boosts soil fertility and biodiversity	Requires knowledge transfer and training	Localized knowledge sharing, agroecological hubs

1.9 Summary and Implications

The term "sustainable agriculture" was used for the first time during the 1980s following the recognition that farming models exclusively based on land development and exploitation concepts, mainly characterized by monocultures and extensive use of chemical inputs, were the primary cause of soil depletion, leaching of fertilizers and pesticides, destruction of ecosystem services and biodiversity, together with essential losses in the crop productivity, mainly related to organic matter poor content.

1.9.1 Summary of Key Points

The concept of sustainability has found numerous definitions, most of which were issued during political international conventions, probably the most significant conference in recent history for the promotion of environmental protection. Finally, we focused our attention in this chapter on the agronomic activity in the Mediterranean environment. We briefly analyzed the original vocations of these territories, whose identities were linked over the centuries to olive growing (V. Singh 2024). Then, in light of this historical experience, we stated at least two deep considerations. The first, referring to the concept of territory sustainability, is designed to return to the use of species adapted to local conditions, thus permanently solving the intrinsic problems of growing genetically inappropriate plant material (Shafik 2024b). Indeed, modern breeding programs aimed at improving overall phenotypic performance have led to the introduction of non-native and poorly adapted plant material. The enhancement of local genetic resources is designed to maintain genetic diversity, exploiting our biodiversity for the future challenge that we will necessarily have to face in a changing environmental and climatic context (Santoso et al. 2024).

1.9.2 Implications for Practice and Policy

Many of the implications for the "practice" of sustainable agronomy are a direct consequence of the discipline and philosophy espoused in this book. These implications also become the main policy messages. There is a need for stronger links between agronomists and ecologists. Conservation strategies, both in agriculture and in the development of natural resources, are more related than is generally perceived, and both, in the final analysis, are inextricably linked. The current popular trend for "integrated" approaches is a step in the right direction. But "integrated" has to be better than cursory; it should be "whole" rather than "a bit of," "joined-up" rather than "side-by-side." The specific challenge to plant and soil scientists is to first provide the technical support needed to deliver environmentally and economically sustainable farming (Qu et al. 2024). This needs to be practically and immediately useful, which probably includes the more routine component. More challenging is the need to develop the concepts of sustainable agronomy in the context of the social aspects of farming and the multiple demands on land, more closely with ecologists. The only "paradigm shift" I am prepared to consider without a great deal more question is the one paradoxically referred to as a "return to sustainable survival." The sort of thing we mean is captured in the following quotation from a paper on the economics of agroecosystems, “To plan for sustainability within any context, we need to depend on primarily ecological, but also sociological, definitions of agriculture.”

1.10 Conclusion

The chapter has availed the opinions and considerations concerning sustainable agronomy. This presentation concerns, among others, the issues of the definition of this relatively new scientific field, which aims to support crop production and influence the environment, supporting human development in the long term. It is worth considering the shaping of definitions of different levels of detail used to develop different types of universal and situational views. This is because the considerations stating that agronomy is a science dealing with the interrelations and governance of crop production and the environment predominate. It can be significant support for agronomists in different research and development tasks, especially if the broader consideration of this definition emphasizes the roles of agronomy among other agricultural sciences and presents other features and characteristics of this field. The concept presented forms the basis for the description of the perception of sustainable agronomy and specifies knowledge about this field. Most importantly, it may support the scientific recognition of its subject and scope, creating the possibility of more comprehensive and efficient development of this sustainable development field. Sowing progress begins with the concept and becomes a reality when the effectiveness of the decisions developed on the basis of this concept, including hypothesis testing and the development of practical technology, is confirmed. Some recognition of the concepts and doubts are also necessary components of scientific thinking, which does not necessarily have to indicate a lack of possibility for extending knowledge. For example, the presented insights of sustainable agronomy can be extended in the future. We hope that the scientists interested in this subject will be involved in further discussions.

References

Abdul-Majid, Mariani, Siti Aisyah Zahari, Norfaizah Othman, and Suhaila Nadzri. 2024. "Influence of Technology Adoption on Farmers' Well-Being: Systematic Literature Review and Bibliometric Analysis." *Heliyon*. https://doi.org/10.1016/j.heliyon.2024.e24316.

Adebunmi Okechukwu Adewusi, Onyeka Franca Asuzu, Temidayo Olorunsogo, Temidayo Olorunsogo, Ejuma Adaga, and Donald Obinna Daraojimba. 2024. "AI in Precision Agriculture: A Review of Technologies for Sustainable Farming Practices." *World Journal of Advanced Research and Reviews*, no. 1. https://doi.org/10.30574/wjarr.2024.21.1.0314.

Beig, Bilal, Muhammad Bilal Khan Niazi, Farooq Sher, Zaib Jahan, Munir Zia, Ghulam Abbas Shah, Ayman A. Ghfar, and Zahid Iqbal. 2024. "Development and Testing of Environment Friendly Nanohybrid Coatings for Sustainable Agriculture Technologies." *Environmental Research* 240. https://doi.org/10.1016/j.envres.2023.117546.

Brower-Toland, Brent, Julia L. Stevens, Lyle Ralston, Kevin Kosola, and Thomas L. Slewinski. 2024. "A Crucial Role for Technology in Sustainable Agriculture." *ACS Agricultural Science and Technology*. https://doi.org/10.1021/acsagscitech.3c00426.

Chirwa, William, Pan Li, Han Zhan, Yinyin Zhang, and Yanan Liu. 2024. "Application of Fine Bubble Technology toward Sustainable Agriculture and Fisheries." *Journal of Cleaner Production*. https://doi.org/10.1016/j.jclepro.2024.141629.

Cisse, Alioune, Ousmane Diallo, EL Hadji Malick Ndoye, Mamadou Sy, Ousseynou Sene, and Joel José Puga Coelho Rodrigues. 2024. "A Smart Farming Management System Based on IoT Technologies for Sustainable Agriculture." *Advances in Science, Technology and Engineering Systems Journal* 9 (1). https://doi.org/10.25046/aj090101.

Kalinaki, Kassim, Wasswa Shafik, Tar J. L. Gutu, and Owais Ahmed Malik. 2023. "Computer Vision and Machine Learning for Smart Farming and Agriculture Practices." In . https://doi.org/10.4018/978-1-6684-8516-3.ch005.

Kumar, Ashwini, Ayushi Trivedi, Nirjharnee Nandeha, Girish Patidar, Rishika Choudhary, and Debesh Singh. 2024. "A Comprehensive Analysis of Technology in Aeroponics: Presenting the Adoption and Integration of Technology in Sustainable Agriculture Practices." *International Journal of Environment and Climate Change* 14 (2). https://doi.org/10.9734/ijecc/2024/v14i24001.

Moreno, José C., Manuel Berenguel, Julián G. Donaire, Francisco Rodríguez, Jorge A. Sánchez-Molina, José Luis Guzmán, and Cynthia L. Giagnocavo. 2024. "A Pending Task for the Digitalisation of Agriculture: A General Framework for Technologies Classification in Agriculture." *Agricultural Systems*. https://doi.org/10.1016/j.agsy.2023.103794.

Muthumalai, K., N. Gokila, Yuvaraj Haldorai, and Ramasamy Thangavelu Rajendra Kumar. 2024. "Advanced Wearable Sensing Technologies for Sustainable Precision Agriculture – a Review on Chemical Sensors." *Advanced Sensor Research* 3 (3). https://doi.org/10.1002/adsr.202300107.

Naresh, Ram, N K Singh, Prashun Sachan, Lalita Kumar Mohanty, Sweta Sahoo, Shivam Kumar Pandey, and Barinderjit Singh. 2024. "Enhancing Sustainable Crop Production through Innovations in Precision Agriculture Technologies." *Journal of Scientific Research and Reports* 30 (3). https://doi.org/10.9734/jsrr/2024/v30i31861.

Nautiyal, Chanda Thapliyal, Pankaj Nautiyal, Gaurav Papnai, Harsh Mittal, Khusboo Agrawal, Shivani, Vishesh, and Raj Nandini. 2024. "Importance of Smart Agriculture and Use of Artificial Intelligence in Shaping the Future of Agriculture." *Journal of Scientific Research and Reports* 30 (3). https://doi.org/10.9734/jsrr/2024/v30i31864.

Qu, Bai, Zhenlei Xiao, Abhinav Upadhyay, and Yangchao Luo. 2024. "Perspectives on Sustainable Food Production System: Characteristics and Green Technologies." *Journal of Agriculture and Food Research* 15. https://doi.org/10.1016/j.jafr.2024.100988.

Quayson, Matthew, Chunguang Bai, Joseph Sarkis, and Md Altab Hossin. 2024. "Evaluating Barriers to Blockchain Technology for Sustainable Agricultural Supply Chain: A Fuzzy Hierarchical Group DEMATEL Approach." *Operations Management Research*. https://doi.org/10.1007/s12063-024-00443-x.

Sandilya, Darsana, Charlina Bharali, Angom Ringku, and Bobby Sharma. 2024. "Utilizing Greenhouse Technology Towards Sustainable Agriculture Using IoT 'TechFarm.'" In *Lecture Notes in Electrical Engineering*. Vol. 1061 LNEE. https://doi.org/10.1007/978-981-99-4362-3_34.

Santoso, H., M. A. Yusuf, S. Rahutomo, Madiyuanto, and Winarna. 2024. "A Cutting-Edge Precision Agriculture Technology to Support the Sustainable Oil Palm Industry." In *IOP Conference Series: Earth and Environmental Science*. Vol. 1308. https://doi.org/10.1088/1755-1315/1308/1/012053.

Shafik, Wasswa. 2024a. "Barriers to Implementing Computational Intelligence-Based Agriculture System." In *Computational Intelligence in Internet of Agricultural Things*, 193–219. Springer. https://doi.org/10.1007/978-3-031-67450-1_8.

Shafik, Wasswa. 2024b. "Biosensor-Based Drones Anomaly Detection Integration for Sustainable Agriculture Development." In *Achieving Food Security Through Sustainable Agriculture*, 283–316. IGI Global. https://doi.org/10.4018/979-8-3693-4240-4.ch012.

Shafik, Wasswa. 2025a. "Sustainable Agriculture and Diet for Urban Development." In *Sustainable Smart Cities and the Future of Urban Development*, 165–92. IGI Global Scientific Publishing.

Shafik, Wasswa. 2025b. "Sustainable Agriculture and Diet in the Metaverse Era: Sustainable Farming and Nutritious Eating in the Metaverse." In *Food in the Metaverse and Web 3.0 Era: Intersecting Food, Technology, and Culture*, 159–84. IGI Global Scientific Publishing.

Shafik, Wasswa, and Kassim Kalinaki. 2024. "Societal and Ethical Implications of Technology-Enhanced Agriculture and Healthcare: An African Context." In *2024 IST-Africa Conference (IST-Africa)*, 1–11. IEEE. https://doi.org/10.23919/IST-Africa63983.2024.10569306.

Shafik, Wasswa, Ali Tufail, Rosyzie Anna Awg Haji Mohd Apong, and Liyanage Chandratilak De Silva. 2024. "Internet of Things for Smart Agricultural Practices." In *Internet of Things Applications and Technology*, 190–217. Boca Raton: Auerbach Publications. https://doi.org/10.1201/9781003458401-13.

Shafik, Wasswa, Ali Tufail, Chandratilak De Silva Liyanage, and Rosyzie Anna Awg Haji Mohd Apong. 2023. "Using a Novel Convolutional Neural Network for Plant Pests Detection and Disease Classification." *Journal of the Science of Food and Agriculture*. https://doi.org/10.1002/jsfa.12700.

Shafik, Wasswa, Ali Tufail, Chandratilak Liyanage De Silva, Rosyzie Anna Awg Haji, and Mohd Apong. 2025. "The Role, Application, and Impact of Artificial Intelligence in the Agriculture Industry." In *Future Tech Startups and Innovation in the Age of AI*, 36–60. CRC Press.

Shafik, Wasswa, Ali Tufail, Chandratilak De Silva Liyanage, and Rosyzie Anna Awg Haji Mohd Apong. 2024. "Using Transfer Learning-Based Plant Disease Classification and Detection for Sustainable Agriculture." *BMC Plant Biology* 24 (1): 136. https://doi.org/10.1186/s12870-024-04825-y.

Shafik, Wasswa, Ali Tufail, Liyanage Chandratilak De Silva, and Rosyzie Anna Awg Haji Mohd Apong. 2025. "A Lightweight Deep Learning

Model for Multi-Plant Biotic Stress Classification and Detection for Sustainable Agriculture." *Scientific Reports* 15 (1): 12195.

Sharma, Ankit, Apoorv Prakash, Shiva Bhambota, and Santosh Kumar. 2024. "Investigations of Precision Agriculture Technologies with Application to Developing Countries." *Environment, Development and Sustainability*. https://doi.org/10.1007/s10668-024-04572-y.

Sidik, Muhamad Afifi Muhamad, Abdul Mutalib bin Leman, Dafit Feriyanto, Samir Sani Abdulmalik, and Supaat Zakaria. 2024. "GREEN TECHNOLOGY FOR SUSTAINABLE AGRICULTURE: BIO-FERTILIZER PRODUCTION FROM MUNICIPAL WASTE TO PRESERVE THE ENVIRONMENT." *International Journal of Innovation in Mechanical Engineering and Advanced Materials* 5 (3). https://doi.org/10.22441/ijimeam.v5i3.23743.

Singh, Manisha, Twinkle Kumar Sachchan, Prabhjot Kaur Sabharwal, and Ranjana Singh. 2024. "Smart and Sustainable Food Production Technologies." In *World Sustainability Series*. Vol. Part F1813. https://doi.org/10.1007/978-3-031-46046-3_1.

Singh, Vikram. 2024. "Advances in Precision Agriculture Technologies for Sustainable Crop Production." *Journal of Scientific Research and Reports* 30 (2). https://doi.org/10.9734/jsrr/2024/v30i21844.

CHAPTER TWO

DISSECTING THE ROLE OF ENABLING DISRUPTIVE TECHNOLOGIES IN SUSTAINABLE AGRONOMY

Abstract

A disruptive technology, or a disruptive innovation, distinguishes itself from other new technologies with the drastic change it brings to a particular field or industry by aiming to rewrite the rules of competition and affect the standards and habits of people in society. Adopting disruptive technologies helps increase the variety and performance of the products offered and serves the needs of people at lower costs. Terms such as exponential technologies or enabling technologies are commonly used to define technologies with similar impacts. Agricultural-related disruptive technologies are those focused on agriculture; however, the products or data obtained are used in other fields. These include, in particular, new-generation sensors, camera chips, big data applications, biotechnology, gene modeling techniques, robotics, miniaturized reactors used in the design of sensors and tractors, variable rate application technologies, precise agriculture-related software, and the Internet of Things (IoTs). Disruptive technologies play an important role in the economy of both agronomy and the world by creating products of high added value that can be commercialized with different combinations and taken to various platforms, renewable resources, re-equilibrated ecosystems, and, most importantly, new employment opportunities. Factors such as value-added product diversity, increasing employment opportunities, the agriculture industry's share in GDP, and the dynamic change of the agriculture industry with the economic strength it has achieved may change with disruptive technologies. The use of these technologies in agriculture and their inevitable interaction with nearby disciplines makes it a rapidly changing and evolving source of expertise known as sustainable agronomy. To increase the integration and adoption of these technologies, the role of stakeholders in the agri-food chain and the exploitation of these dynamics should be supported with appropriate investment in developments, including those needed in business models and governance arrangements.

Keywords: AgTech Innovations, Climate-Smart Solutions, Digital Transformation, Disruptive Technologies, Environmental Sustainability, Precision Agriculture, Renewable Energy Integration, Soil Health Management, Sustainable Agronomy

2.1 Introduction

Agronomy is the science and technology of using plants for sustainable and useful human purposes. It emerged around 7,000 years ago as a biological science dealing with crop and food production and has since developed with several additional facets such as the sustainable production of fiber, fuel, environment, ecosystem services, and non-food plants such as bioactive, medicine, flavors, dyes, and aroma chemicals—basically, any compound synthesized by plants. Throughout the long history of agronomy, farmers and farming systems worldwide have spanned a broad and complex spectrum of resource availability, technology, crop husbandry principles, sociocultural contexts, and environments (Zeng et al. 2024). The built-in symbiotic compatibility between humans and agronomy has enabled humans to develop as a species, surpassing all others in terms of population and the continuous quest for an enhanced quality of life. In the journey of agronomy from antiquity to the present time, several enabling and sometimes disruptive technologies have emerged from time to time. While some have been very well established in the agronomic mainstream, especially those developed since around 1970, others are in various stages of being piloted, matured, and successfully adopted, undergoing backlash due to societal resistance, moratoriums, and considerably delayed deployment (Das and Ghosh 2024). Disruptive technologies from the agronomic perspective may not really be technologies in the conventional sense but collectively refer to a range of scientific expertise, technical tools, methods, and shifts in thinking, all of which, when combined, can result in qualitatively new dimensions in agronomic objectives, organization, production processes, buyer and consumer awareness and expectations, and the role of the related policies in supporting and regulating these dimensions (Ma et al. 2024).

The global human population is set to reach nine billion people by the year 2050, which means that agricultural productivity will need to increase dramatically to feed and clothe the growing population. Therefore, more sustainable management of agricultural systems is essential to solve some major agronomic challenges. In fact, an increase in agronomic productivity is encouraged to meet this population demand while ensuring the conservation of the environment and the efficient use of natural resources. The initiative Agriculture 2030 aims to increase support for agronomic activities and the sustainability equation and to reduce the environmental impact and consequences (Kotelnikova et al. 2024). The challenges for sustainable agronomy are several and well known, such as the increase in pests and pathogens, losses in commodities, the emerging viral diseases that endanger the human species, and the production and management of superior quality crops and their production in the face of inefficient pest, weed, and nutrient management. Additionally, water use cannot exceed the constraints, and the water cycle must remain healthy; thus, the food security issue is in balance with environmental security (Zidan et al. 2024). Although no single system of sustainable agronomy is likely to be universally acceptable to all cultures or perfectly sustainable, the collection of agronomic approaches might become sustainable within a diversified cropping system and crop management. It is critical to fully assess the sustainability of all potential solutions, particularly to reduce the potential negative consequences of agronomic practices. Despite these heavy constraints and challenges, the future sustainability of agronomy strongly relies on enabling and disrupting technologies that will address and alleviate emerging agronomic challenges (Giannini and Marraccini 2023).

In conventional agronomy, a major challenge is to adapt crop structure and development by modulating the partitioning of carbon resources, particularly at the different stages of the reproductive phase. However, there is functional plasticity, known as the limitation imposed by the architecture of panicles or inflorescences and by the synchrony of stamen elongation, with activities determined by complex interactions on the seasonal or evolutionary scale. In addition, crop acclimation to the location, light reception, and possibly other mechanical considerations contribute to the late-stage development (Nowrose Jahan Lipi and Muhammad Maniruzzaman 2024). In order to stabilize and eventually increase biomass and yield in an unpredictable environment, it is necessary to fill in the gaps left in our knowledge. Moreover, sensitivity to climatic variations is due to the interactions mediated by time lag effects with the meristem age and development stage that facilitate the various agrometeorological constraints. Agrometeorological pressures impact agronomic performance through the responses of crop structure and development to a combination of environmental signals concerning the season, the light intercepted at the top of the canopy, and the local ultra-microclimate (Grace and Dawson 2024; Lai et al. 2024). For example, in order to perpetuate itself, maize has evolved a plastic system where staminate phase duration regulates the intercellular wavefront coordination with stamen elongation. The synchrony of meristem activity and stamen elongation, which operate through anisotropic growth and anther dehiscence through pollen shedding, meets the *morphofunctional* optimization of an upper limit of physiological efficiency in a variety of environmental conditions and ensures reproductive success. Functional and plastic coordination regulated through statures are the evolutionary adaptive strategies that enable the specification of the plant (Pranaswi et al. 2024). Considered the crucial trait of crop reproductivity, this pillar has been rarely documented and quantified.

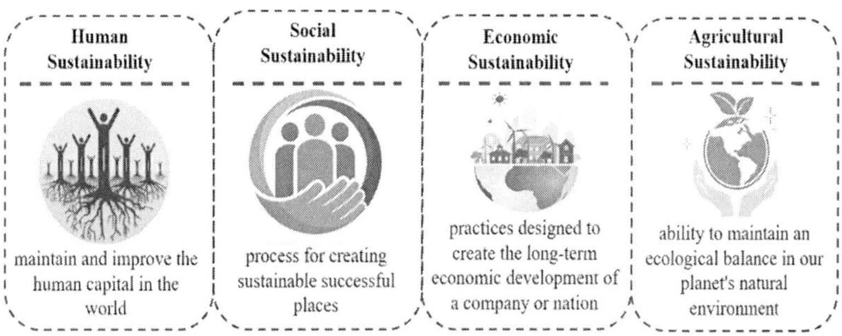

Fig. 2-1 Sustainable practices

Development and exploration of sustainable agronomy contribute to yield improvement using fewer resources. It is recognized that a sustainable agronomy system has less of a negative impact on the environment, requires fewer natural resources, has less vulnerability to climate change, and is more adaptable to dynamic natural conditions (Shafik 2025b). In the context of potential labor shortages, sustainable agronomy technology could lead to new farming methods, reduce the drudgery of repetitive tasks, minimize the uncertainty of the natural environment, and reduce resource inputs (Das and Ghosh 2024). While developing sustainable agronomy practices, it is important to ensure that food production goals are maintained to feed the 9.3 to 10.8 billion people expected to populate the Earth by the year 2050. Current sustainable agronomy practices tend to focus mostly on reducing the negative impact on the environment, with less emphasis on crop productivity and especially resistance management (Das and Ghosh 2024). The establishment of sustainable agronomy practices has a track record in yielding additional benefits, such as reducing unnecessary crop loss due to over-fertilization, increasing water use efficiency in agricultural production, setting additional crop yield records, and building resistance against harmful insects and weeds using sustainable practices, as illustrated in Fig. 2.1.

2.2 Intersection of Disruptive Technologies and Sustainable Agronomy

Disruptive technologies have already been implemented in many service sectors of economies and can offer pivotal contributions to relieve the inherent tension between ecological and economic objectives that agricultural systems have to adhere to in order to enable the delivery of societal benefits (Shafik 2025a). These technologies emanate from various empirical domains, such as the digital economy, communication, energy, mechatronics, etc., and are then combined with agricultural management, providing an opportunity to handle the complexity of biological systems and the practical demands placed on them efficiently (Giannini and Marraccini 2023). Examples of these technologies include information technology, sensor technology, geospatial technology, and biotechnology, promising to address these challenges directly through substitution, cost reduction, efficiency improvement, new product development, overcoming the inhibitory rules, legislations currently enforced, or reforms in the political context. This chapter reviews these technologies and their current and potential use to assess whether they can mitigate the multiple aspirations of industry and society without compromising the agronomic as well as environmental precepts (Das and Ghosh 2024; Zeng et al. 2024).

Disruptive technologies play a major role in sustainable agronomy, and many such technologies and methodologies have been established in various empirical domains. Disruptive technological innovations can provide pivotal intrinsic solutions to reconcile the inherent tension between ecology and economy in agricultural production systems. These technologies provide advanced functional tools for sustainable agricultural production, where the economy is achieved by simultaneously considering the well-being of the environment, the stability and efficiency of ecosystems, soil health and clean water, the safety and well-being of people, and the rights and economic status of individuals in society overall (Ma et al. 2024). Since these technologies provide renewed speed, can lead to cost reduction, more efficient information processing, and leverage innovation, they can offer an opportunity for a firm to create a dynamic competitive advantage. This chapter aims to assess the ability of technologies and methodologies to deliver sustainable agronomy by evaluating their functional architecture to be advocated at every stage in choosing and embodying disruptive technologies within the model for four principles of sustainable agronomy (Das and Ghosh 2024; Giannini and Marraccini 2023).

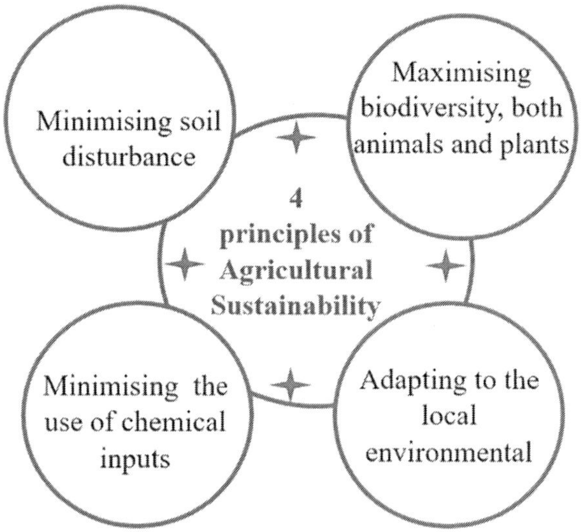

Fig. 2-2 The four principles of agricultural sustainability

2.2.1 Impact of Disruptive Technologies on Sustainable Practices

Disruptive technologies of the Fourth Industrial Revolution have transformed the entire landscape of the agri-food system with their unique selling propositions. They are not only fostering the incorporation of agricultural practices like conventional and precision farming but also addressing a wide range of challenges and transforming the sector at an exponential rate. In the era of contemporary agriculture, factors such as increased agricultural productivity, resource conservation, sustainability of farms, processing, and transportation activities, as well as the products derived from all these activities, are important (Ünal and Kızıldeniz 2022). The creation of farm-level value addition, continuous monitoring, aggregation of output, smart logistics, and expedited transport to industrial zones can be complemented by these technologies. In this chapter, we identify the most commonly reported and promising technologies to address agricultural challenges and transform agriculture for the future. Through a diversified range of strategies and implementations, these technologies have been identified in the current study, ranging from providing partial solutions to addressing a comprehensive range of agricultural and related problems (Philip and Suresh 2024). All people and systems in a food supply chain, from the production of food to consumption, can impact the consequences of the complete food supply chain on the natural environment. Moreover, estimating the overall sustainability of food items entails evaluating the environmental, social, and financial costs of generating food throughout the food supply chain by comparing their life cycle impact analysis. Developing the most favored technologies that turn product ideas of food start-ups and SMEs into reality, including those that bring scale, will boost the competitiveness of the agricultural sector (Lanki A. D. and Onwu C. A. 2024).

2.2.2 Case Studies and Success Stories

In 3.0, we provided a general picture of a number of EDTs with the potential to disrupt conventional agronomy practices in order to cope with the challenges faced now and beyond 2050. We are now providing a look at selected cases to illustrate how these general trends can be incorporated into real agricultural operations, revisiting technologies based on case history and successful implementations. Each case is reported in a separate paragraph where, after a brief introduction of each case, referring to either a start-up company, a consolidated industry, or a research institute involved in the development, we discuss which technologies are present, what the innovation is that is impacting agronomy, who the target user is, and what the current marketing strategies, present operations, and near-term prospects are (Catala-Roman et al. 2024). Each case is concluded with a brief overview of the lessons learned. As the data presented in each case come from different sources and generally refer to case history conducted in different years, which may include 2017 and 2018, during which case histories were not concluded yet to be generally shown in an associated paper, the number of the referred year is also always included in sustainable agriculture and rural development, as illustrated in Fig. 2.3. With a view to the dissemination of the information we report, following our analysis, we also include the site of the company or institute under study (Zhu and Wang 2024).

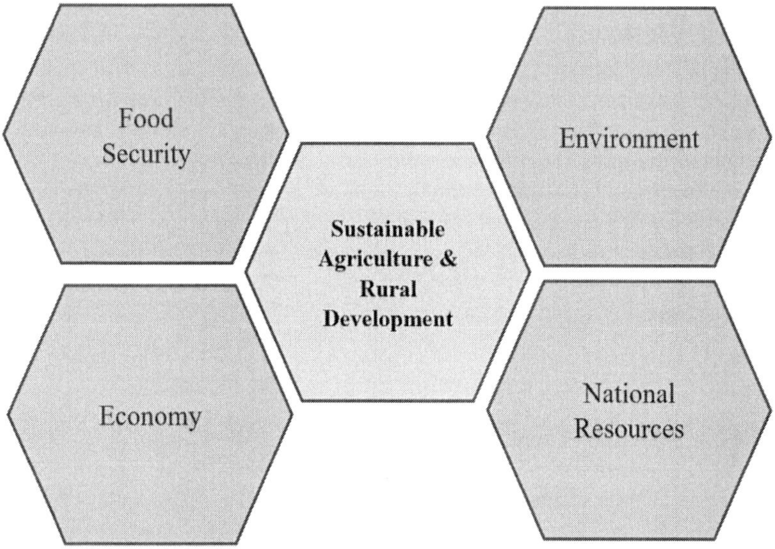

Fig. 2-3 Sustainable agriculture and rural development

2.3 Enabling Technologies for Sustainable Agronomy

A new agronomy, using digital technologies enabled by the ongoing data revolution, is not only timely but also presents the key pivot where all these three largely independent sensibilities are harmonized. It is timely because technologies are now available for developing and facilitating solutions that will help reduce global hunger and better manage our planet for the prosperity of all forms of life. Specific technology components derived from data sensing, mining, modeling, automation, and the realization of a low-cost, Internet of Things capable platform can obviate some of the resource barriers often associated with taking full advantage of advanced agronomy (Baviskar 2024). It shapes a sensibility of integrated science because connecting disciplines from the natural, social, and data sciences is needed to address sustainable development target challenges. It engenders a new developing world sensibility by focusing modernization processes not on the consumption capacity of developing countries but on those that drive, at reduced resource demands, their production capacity. A timely transformation, recognizing potential benefits from using data to tackle agronomic questions, is guided not just by the confluence of data technologies but by a realization that existing questions are familiar but new answers and strategies are required due to rapidly changing socioeconomic and physical environments (Anikina et al. 2024). The conclusion is not that sustainable agronomy is about learning to better manage scarcity, as that can create daunting barbed wire fencing around achievable goals of universal prosperity. Sustainable agronomy requires a more inclusive construct of modernity. Sustainable agronomy recognizes that our land use processes and priorities are pre-sustainable and confronts this ongoing path dependence with pragmatic, trait-specific interventions, with scope for further improvements, as our technological capacities increase and we develop an increasingly refined understanding of our natural and societal environments (Zeng et al. 2024).

2.3.1 Internet of Things (IoT) in Agriculture

The agriculture sector has evolved to become a hotbed of innovation in implementing and promoting enabling disruptive technologies such as the IoT, blockchain, and machine learning (Shafik and Kalinaki 2024). Adoption of these technologies in agricultural best practices is the key to transforming the agriculture sector and making it future-proof. Agricultural products, practices, and processes that are both supportive of economic viability and of the ongoing gains to the well-being of present and future generations meet the definition of sustainable (Das and Ghosh 2024). The management and use of agricultural resources to satisfy changing human needs and to ensure profitability for farmers while providing sufficient food, feed, fiber, and biomass for people and animals using non-renewable resources and causing minimum harm to the environment is termed sustainable agriculture. The IoT is a technological revolution that capitalizes on the global connectivity of various things that can be addressed to reach the internet for anytime communication. In simple terms, IoT is an intelligent network of things that communicate and relate to people (Kotelnikova et al. 2024; Ma et al. 2024). The IoT is, indeed, a complex technology ecosystem as it involves several elements comprising sensors, RFID/NFC tags, actuators, and devices connected to the internet for data exchange. The two other concepts often found to be linked to IoT are machine-to-machine (M2M) and the Internet of Everything (IoE). The Internet of Everything is the intelligent connection of people, processes, data, and things. The IoE value is generated by the things that collect and transmit data on their environment in order to be accessible to the web for further processing to anticipate and prevent challenges, such as the loss of rainfed crops or the management and maintenance of electrical distribution networks and traffic logistics (Zidan et al. 2024).

2.3.2 Artificial Intelligence and Machine Learning

Most recently, AI- and ML-driven methods are used to create high-resolution remote sensing products to provide immediate, actionable information to agronomists and landowners. This information is used to extract phenotypic traits and lends itself to a wide range of applications. AI- and ML-driven methods can determine and enable the discovery of complex relationships between driving forces, biotic and abiotic stresses on crop health, developmental strategies on agro-climate, and the requirements of different soil types under changing climates, which is key to directly quantifying benefits from the necessary choices around plant breeding, crop choice, tillage, weed and pest control, irrigation scheduling, and nutrient management on specific types of soils and crop management systems (Lai et al. 2024). By providing a constant and detailed quantitative measure of agronomic outcomes, effectively informing optimization models, AI- and ML-driven methods offer a pathway for benefits from the necessary choices related to agronomic practices and the locational placement of a wider range of new skills developed by agronomists in close collaboration with plant breeders, data analysts, computer scientists, sensor experts, and drone engineers. The social academy and industry should encourage their adoption and refinement (Pranaswi et al. 2024; Shafik 2024b). However, a word of caution: there is an essential need for microwave-specific education, avoiding the hype, learning from failures, and discussing how impactful, transparent, sustainable, and trustworthy codes and data analyses can be implemented, as this associated education is integrated and combined to communicate the potential value of AI and ML to the social academy and industry, and to help mitigate privacy and security concerns from the general public (Philip and Suresh 2024).

2.4 Data-Driven Decision-Making in Agronomy

The number of data-driven applications in both basic and applied agronomic research is increasing all over the world. Recent advances in sensor technology offer near-instantaneous in-field or on-field diagnostic capabilities. These include tools to monitor soil moisture, profile depth and fertility, and information from spectral sensing that can be used to diagnose crop health and predict yield variability. Simultaneously, a wide variety of precision and smart tools are available to manage a plethora of farm operations, from planting and replanting to harvesting, supply application, and maintenance (Demir, Dedeoğlu, and Başayiğit 2024). Many of these tools rely heavily upon and feed back into real-time information, enabling in-field decision-making. On the supply side, the wealth of remotely sensed spatial data, coupled with clear weather and accurately timestamped earth observations, fuels the ability to detect and predict threats to crops from diseases, insects, and crop lodging. Such capabilities and associated decision support systems are made possible by the increased availability, then reliability, and, finally, the efficiency and affordability of data management driven by recent advances in disruptive technologies (Naveen et al. 2024).

Many sensing tools and precision tools make use of small-scale information, which may well be unique at a scale meaningful for itself, making it a significant challenge to scale up use assumptions beyond the local scale/site, especially in environments where few, if any, pieces of support information exist to validate ad-hoc data-driven outcome models. The so-called integrity of such post-scaled-up exercises is often hindered by the underlying voluntary and exclusive behavior of the agricultural sector that perpetuates data silos. Oblique evidence also points to the increase in the use of cloud-enabled big data analytic agents and artificial intelligence algorithms to predict the unknown, a collection of measured exogenous variables that represent the domain's cognitive uncertainty space (Ma et al. 2024). Some have taken these clues as signs to democratize agriculture through the use of decision support systems that use intelligent behavioral computing, and that are dynamic, spatially explicit, and promote a learning-by-doing attitude (Shafik 2024d). Data literacy and easy access to agro-climatic indices might still be one of the prerequisites to adopting these intelligent tools. The key issue of making data-based decision support understandable, especially at the smallholder level, should direct research towards a greater understanding of learning preferences and—to some extent—also of social media and policies that might foster learning generally (Mohanty et al. 2024; Baviskar 2024). Be that as it may, the plethora of existing and potential applications of data-driven decision-making is enabling agronomy to operate on large spatial scales.

2.4.1 Importance of Data in Sustainable Practices

In a data-driven approach to optimization, we use data to compute the actionable opportunities, and then we map these actions to physical events that lead to better farm management. At each point in space and time, we can best utilize previously unavailable detailed engineered data sets. Swath-based high-resolution images informing about the precise state of the crops require the use of sensing technology throughout the growing season. In this context, machine learning algorithms are very useful to convert raw data into information. Hence, data-driven farming could be a boon to sustainable agronomy (Nowrose Jahan Lipi and Muhammad Maniruzzaman 2024). It promises to enhance our capacity to meet the ever-increasing demands of agricultural production while also having positive environmental and social impacts, referring to the three pillars of sustainable development. In the following, we discuss three specific cases of the role of data in boosting sustainable agriculture. The analysis is meant to provoke and help focus a broader discussion on the potential of sustainable practices enabled by data. It is by no means exhaustive but originates from our involvement with enabling technologies such as advanced sensing, data analytics, the Internet of Things, and unmanned aerial vehicles for environmental management. In particular, large volumes of imaging data are collected and processed using machine learning for identifying environmental stress conditions, people management, and remote sensing for qualitative and quantitative assessments of ecological indicators in agricultural landscapes (Baviskar 2024). All these enable increased contributions of agriculture to rural development by providing ecosystem services or by reversing rural exodus through local job creation, as illustrated in Fig. 2.4.

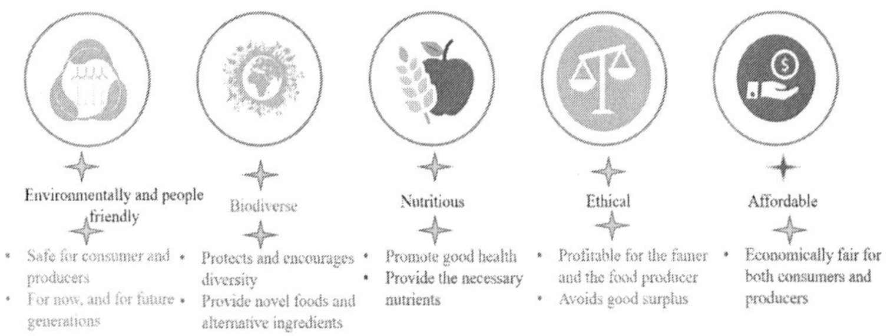

Fig. 2-4 Merits of sustainable agronomy

2.4.2 Analytics and Visualization Tools

Data is a critical element of decision-making in sustainable agronomy. The world is literally filled with data of untold dimensions, everything from local weather to available nutrients, from expected commodity prices to labor availability. Converting this data into knowledge and then into wisdom is at the heart of analytics. The growing field of data analytics blends statistics and a variety of software tools with signals from milliseconds to decadal. Suppose a given tool deems a signal as valuable (Shrestha and Wei 2024). In that case, it is expected to be reflected with minimum latency through visualization, either by fixing a current problem or by providing a view as to what the data may lead to within the realm of reason. Computer hardware is now at the point where considerable amounts of data can be quickly parsed and processed. The parallel-growing field of analytics packages allows for traditional statistical routines to be performed, although unsupervised and supervised machine learning routines and deep learning are becoming much more popular. There are a great many analytics packages available today, with a range of costs and training models to match every potential user. All require a data scientist or someone with similar expertise to interpret and communicate the results of the analytics back to the stakeholder group (Prajval V et al. 2024).

2.5 Precision Agriculture: Enhancing Efficiency and Sustainability

The environmental externalities associated with agronomic practices and the resulting economic imbalance are continuously raising the demand for new technologies capable of improving the overall performance of the agronomic supply chain to achieve better sustainability outcomes. This largely amounts to pursuing the most efficient approach to contrasting waste production, enhancing productivity, or accessing other mitigation options in order to cope with the environmental costs of agriculture while meeting the demand for food. This is what the enabling disruptive effect of site-specific technologies in agronomy is about, supporting the introduction of Precision Agriculture, a concept of agronomy that targets waste reduction and a halving of the environmental costs of the agronomic supply chain in terms of tension vis-à-vis sustainability objectives (Naveen et al. 2024). Precision Agriculture (PA) is an enabling technology for the irrigation, fertilization, spreading, and planting of maize capable of coping with high environmental pressures. The opportunity to intervene more precisely with a reduced dosage at the spatial and temporal levels while monitoring their effects on the physical, chemical, and economic variables within the boundaries of small plots, and to repeat the processes under the same variable or in modified conditions, is a rather consolidated statement in agronomic economics (Mohanty et al. 2024; Baviskar 2024). The real breakthrough moves from simple mentality—where the purposive action is that of investing more or less in terms of inputs while expecting to quantify the gain in production and change the cost structure—or rather the mechanization, standardization, and spatial homogenization of the treatment of the crops using the same dosages, machinery, technical references, units of measure, and execution times—or any notional characteristic of treatment scaled to the size of the field, which defines the difference between extensive specialization (Shafik et al. 2025).

2.5.1 Components of Precision Agriculture

The literature reports a multitude of combinations of equipment and software in PA systems, and several other integrations of disruptive technologies. The dGPS and RTK-GPS technologies remain fundamental, but they are augmented by IoT, robots, drones, artificial intelligence (AI), machine learning, virtual reality (VR), augmented reality (AR), additive manufacturing (AM), and automation and logistics (AL), seeking the best yield result. The integration of all these disruptive technologies in PA for a systemic compromise between multidisciplinary challenges, however, is still developing. The majority of systems usually have IoT and GPS; rarely do they also integrate robotics, automatic navigation, and drones (Ma et al. 2024). Machine learning is developing, yet within the available technologies, it is already achieving good results alone. When using machine learning, the PA systems are usually labeled as digital agriculture, but IoT, robots, unmanned aerial vehicles (UAVs), and AL, for example, may also integrate digital agriculture. A recent review classified and investigated various components of digital agriculture together with their cooperation. Information derived from numerous studies allowed us to propose the holistic theory of digital agriculture and define its eight principles, where only one of the principles is related directly to PA (Naveen et al. 2024). In order to refine the proposed theory of digital agriculture, the present chapter investigates the actual role of all of these enabling technologies with applications in precision agriculture and lists the existing and several future challenges besides the solutions inserted in the domains, each named national digital services. The most relevant challenges are those experienced, not so much as those potentially available (Ma et al. 2024).

2.5.2 Benefits and Applications

Innovative applications of artificial intelligence (AI) in agriculture, with the ability to deal with spatially distributed data for localized predictions and to focus on resource-preserving interventions, are becoming useful tools for farmer support. Autonomous transportation in agriculture can improve efficiency and reduce costs during field operations, leading to time and labor savings. The devices are now being applied in agriculture, collecting valuable data and connecting sensor networks to decision-support systems such as e-know-how, e-decision support, field zoning, and planning farm operation and management tools (Mohanty et al. 2024). The robust resilience of agricultural robotics and drones to increasingly uncertain climatic conditions and the consumption of natural resources must be regarded, in addition to potential sustainability challenges, as they are forcing the industry towards greater transparency, with regulators, food security stakeholders, and sustainability-concerned consumers around the world watching carefully. Consistent data can be one of the biggest problems facing those who wish to adopt precision agriculture. Scalable, multi-functional, lower-cost devices, connecting to global data collection and device management services, are perfectly designed to accelerate this deployment across agricultural robots and drone fleets (Dr. Mrinal Gaurav 2024). Such devices will further allow intelligent capabilities that lead to proactive and conditional data. Data will be transformed locally on the device to extract features independent of the cloud, process uninteresting data in real-time, or potentially in batch mode. This type of activity will not rely on constant data pipe connectivity, preserving privacy by utilizing customizable criteria-based smart rules. Furthermore, they will ensure that critical data valuable to the job, such as weather or geofencing data, will continue to be collected, and data will be available even when connections are poor or unavailable. With the autonomy and the business model of farming technology suppliers, there must be a smooth and sustainable path to deployment (Pranaswi et al. 2024).

2.6 Robotics and Automation in Agronomy

Certainly, the introduction of automation technologies should provide a step towards sustainable agronomy, but it is also recognized that agriculture is complex, and things are less clear when considering this in practice. The idea that sophisticated autonomous machines could work the land without human guidance provides an intriguing vision, but is it a realistic one as an endpoint, and to what extent should man be bypassed? After all, animal traction and human labor are key aspects of sustainability in traditional agroecology. This review aims to explore these issues. We start from the premise that, to make agriculture more sustainable, it is essential to reduce the number of damaging operations (Mohanty et al. 2024; Shafik, Tufail, Apong, et al. 2024). Although operations may be less frequent than on arable land due to reliance on the natural vegetation and its ecological and biological processes, jogging every 25 years will almost certainly damage them. We examine existing and projected automated systems and their likely impacts, but in looking forward, it should be appreciated that what is not possible today may be routine tomorrow. Many dry parts of the world rely upon non-native artificial water collection systems, yet for some years, large phlebotomists have been stinging the heel of those desert stones in possession of brains. Other agricultural automation is here now. Seldom a day can pass without an article about robots seen in the fields or robots that are soon to be seen in the fields; the robots are here now—idealized or not (Shrestha and Wei 2024). Before reviewing these developments, it may be worthwhile pondering what is meant by automation and the scale at which it may be useful.

2.6.1 Advancements in Agricultural Robotics

The early agricultural robotic systems, including automation of existing terrestrial vehicles, particularly large tractors and combine harvesters, robots, and smaller robotic devices for precision planting and weeding activities, were primarily developed for high-value specialty crops such as vegetables and fruits. Henceforth, input costs for the associated development and deployment of these technologies could be justified by the high returns received. The economics in other food and feed commodity types have variously enabled and limited advances. Although robotics for agriculture is a very active field, as many commercial products exist, several emerging innovations in agricultural robotics, including promising prototype robotic systems, will likely have significant roles in sustainable farming (Nowrose Jahan Lipi and Muhammad Maniruzzaman 2024). These include many highly specialized and versatile robots for numerous different tasks in various crop types throughout production stages, from ground surface preparation to harvest, and unmanned ground and aerial vehicles for various crop monitoring and other farm logistic activities. This paper concentrates on the specialized robotic devices. Among the influential factors in the progress of agricultural robotics are the adaptation of robotic system designs, advancements in sensing instrumentation, wireless communications, technology compatibility, partial cost performance designs, availability of ultra-low power off-the-shelf hard-to-obtain commercially available components, and possible uses of recycled components or raw materials. Two trends contribute to the progress of this field. The exponential escalation of microprocessor capacity allows incorporation of ever more intelligence and flexibility in robotic systems, and the elevated maturity of used robotics components results in fast adoption and hardened practice of the designed robot productions (Shrestha and Wei 2024; Catala-Roman et al. 2024). These trends tolerate accelerated deployment and scaling of robotic technology development.

2.6.2 Applications in Crop Management

The development of data-driven models that gauge water and nutrient uptake, as well as support decision-making for crop design management and food production, is imperative. Sensor applications include both image capture and wireless sensors to observe water availability, hormonal signaling, nutrient levels, and metabolite fluxes, both leaf- and root-centric. Such observations enable biosynthesis and developmental responses, stimulate growth and differentiation, and enhance photosynthesis through carbon allocation (Mohanty et al. 2024). Disruption has focused on the hydroponics niche, but positive disruptive impacts on crop management have yet to be realized. High-resolution scanning, utilizing magnetic resonance imaging or computed tomography, is being used. Interestingly, the use of a selective rhizotron tube and light field cameras incorporated within this technology can provide hybrid image capture capabilities directly within the root zone, further enabling direct bilateral observation of plant responses. To date, these models have not been plentiful and primarily describe water movement, such as that of soil moisture, and relate it to plant behavior (Tong et al. 2024).

2.7 Biotechnology Innovations in Sustainable Agriculture

In achieving the desirable goals of agronomy, biotechnology innovations are indispensable tools. By definition, biotechnology is the use of living systems and organisms to develop or make products or any technological application that uses biological systems, living organisms, or derivatives to make or modify products or processes for specific use. It involves DNA manipulation from one organism to another in order to obtain desirable traits, improved nutritional content, improved yield, or any other desirable characteristics. It is the effective and efficient use of bacterial, fungal, algal, or viral activities in releasing carbon, nitrogen, phosphate, and potassium nutrient elements from their bound states inside the dead organic matter of living soil to sustain efficient overall soil nutrient recycling programs (Nowrose Jahan Lipi and Muhammad Maniruzzaman 2024). The practical importance of biotechnology in sustainable agriculture for environmental safety, global food security, high yield, and poverty alleviation in developing countries is unparalleled. It is an inclusive science; the technologies to rationally alter and improve agricultural traits lie with many different scientific disciplines, such as plant breeding and genetics, molecular biology, genomics, proteomics, systems biology, plant production systems, soil science, and microbe-assisted production or trait improvement. Biotechnology has and continues to contribute to agronomy through various means. These include genetically modified organisms and gene editing, molecular breeding, microorganisms and microbial products, and breeding for the biofortification of staple crops (Philip and Suresh 2024).

2.7.1 Genetic Engineering and Crop Improvement

Genetic engineering was first developed in bacteria and viruses, and its use later expanded to fungi and plants. Unlike chemical and radiation-based mutagenesis, which usually leads to a cascade of unintended metabolic effects and significantly reduced crop fitness, which produces very little real improvement in crop genetic diversity, biological mutagenic methods used in conjunction with marker-assisted selection led to a significant increase in genetic variability that was further used as an abundant source of valuable new genetic variation for crop improvement by gene pyramiding. These new breeding techniques enabled the production of a variety of genetically engineered high-yielding crops that resist or tolerate natural pests and diseases, and herbicides, as well as extreme climatic episodes and soil stresses, while producing high-quality yields. They helped to reduce the usage of agrochemicals associated with conventional crop gene overdosing protocols and the negative environmental impact of excessive use of agrochemicals (Demir, Dedeoğlu, and Başayiğit 2024). That notwithstanding, such types of innovation scaling produced collateral mutations by affecting the morphology, genetic, genomic, and chromosome characteristics. *NovaerapT Mondrian durum* wheat with CRISPR-induced allele was semidwarf, with good resistance to pre-emergent herbicide and no fitness penalty.

2.7.2 Biological Pest Control

The use of living organisms, particularly arthropod predators or parasites, to control pests has been used in agriculture since ancient times. Modern biological control can be based on the use of chemicals to improve pest control or be associated with genetic advances in the production of resistant plants. Biological control has significantly contributed to the reduction of chemical use in agriculture. The genetic increase in plant resistance has enabled many times for the pest to be the one that ends up acting as a bioinsecticide, leading to a reduction in chemical use. Currently, the advancement of plant genetic techniques or the use of inorganic chemistry, which facilitates the improvement of plant defenses against pathogens, has received less attention (Naveen et al. 2024). These non-toxic or less toxic mechanisms must be the first option in the development of plants for the sustainability of agronomy. Biological control has the potential to fill the space left by chemical control. It must be based on risk management, rapid and visible results, information for producers, and precision in implementation. The promotion of biodiversity is key to this control. It has been shown that the presence of plants at the edges of crops with flowers and, therefore, large nectaries considerably increases the presence of natural enemies in the field. The application of Integrated Pest Management must apply to pesticides that can be biodegradable, to plant resistance, and to biological control. The increase in the quality and quantity of information about the agents involved is key to dictating the acceptance of the use of biological control in the field. It is, therefore, necessary to generalize and adapt to the particularities of the cases, the programs that promote the use of resistant crops, and the use and improvement of sterile insect techniques, natural enemies, and antagonistic microorganisms (Pranaswi et al. 2024).

2.8 Climate-Smart Agriculture: Adapting to Changing Conditions

The predicted impacts of climate change on agroecosystems, shifts in the distribution of potential for crop and livestock production, changes in crop yield potential, heat and drought stresses, water scarcity, and soil degradation highlight the need for scalable solutions to global agronomy problems. Genomics could facilitate such solutions through the development of crop varieties, knowledge, and breeding approaches for the control of plant responses to heat, radiation, and water stresses by improving the ability of crops to capture solar energy and carbon (Tong et al. 2024). Genomics has the potential to increase the production of crop biomass per year to effective heat use, which will alleviate food and carbon demand simultaneously. Agricultural production is sensitive to climatic conditions. The predicted impacts of climate change on agroecosystems include shifts in the distribution of potential for crop and livestock production, changes in crop yield potential, and heat and drought stresses. Climate change is forecast to be a slow-moving threat, with changes accruing over time as changing weather patterns emerge. Potential changes include altered distribution of bioclimatic limitations, changes in crop yield potential, heat and drought stresses, water scarcity, and land degradation (Prajval V et al. 2024).

2.8.1 Strategies for Climate Resilience

There are several agro-climatic zones in India, and each has its dominant seasonal pattern of occurrence of heat, rain, radiation, and humidity. Using the information contained from the successive days as well as from historical records, it is possible to deduce the critical timings for mature crops for heat, high rainfall, drought, delayed or early sowing or planting, hailstorms, frost, high radiation, and severe weather events. Critical timings for immature crops guide creating agro-advisories. The LPRs of future scenarios will be used as the choice of emission scenarios (Tong et al. 2024). Based on the LPR information from the future scenarios, the probable success or failure of cropping in the future can be assessed. The cropping-related postulations, which state the quantitative behavior of existing and future scenarios, can be utilized as enabling features for agronomy technologies. More recent work that uses GEV distribution over each grid box in India and the vegetative index appears very tempting. The critical period plots with data suggest that the future is very bleak if no policies are applied to adapt. With the scenarios, parameters are obtained for the future, and remodeled plots, along with measures of skewness, will be used for assessing the expected impact and the expected severity of the impact (Ma et al. 2024). Together, these features of future suitability and easy-to-use predefined cutoffs can be used as guidelines for creating the base for decision-making tools. The decision-making tools can also be used as potential advising tools not only for policymakers, and advocacy can be made to adapt to remove potential bottlenecks. Care will be taken to properly train these models and tools as conditions do change from hot, dry to warm, dry situations (Philip and Suresh 2024). We also analyze other climatic variables and indices that help in quantifying relationships.

2.8.2 Technologies for Climate Monitoring

Climate monitoring is needed to quantify current changes properly and to predict future patterns. The increasing frequency and magnitude of extreme events make it even more important to evaluate the effects of climate change. Agriculture is under direct risk, as it depends on a very narrow window of air temperatures compatible with the sowing and harvest of most major crops. Therefore, the development of both advanced technologies and novel approaches for climate monitoring over complex landscapes at a suitable resolution is essential in order to support climate impact understanding over crops and much better decision-making in agriculture. Among the most important remote sensing abilities for global change impact studies is the capacity to assess crop conditions at different stages of growth (Demir, Dedeoğlu, and Başayiğit 2024). Remote sensing is also a solution to maintain the growing population's food demands by reducing agricultural production losses from frequent extreme weather. The data are collected indirectly across agricultural fields at small spatio-temporal resolutions using rapidly advancing remote sensing technology that has the potential to improve agricultural prediction modeling. These remote sensing technologies are effective for disseminating farm management decisions and are also facilitating field monitoring for the farm digitization claimed. Factors limiting applications to the field level are slow processing of data, errors in thematic classification, and oversaturation of soil and crop properties. These limitations need to be overcome by remote sensing technologies to increase their ability to delineate field cropping system spatial patterns. Other critical bottlenecks include the lapses in spectral technology development, limited availability of data with low to delayed temporal resolution, and the trade-off between spatial and temporal resolutions in satellite sensors (Tong et al. 2024).

2.9 Policy and Regulatory Frameworks for Disruptive Technologies in Agronomy

Rapid developments in science, technology, and innovation, particularly with biotechnology and other enabling technologies, make it possible to achieve higher and sustainable yields faster and with less pressure on natural resources. However, these benefits are not free or without risks. It remains to be seen whether Europe, with its diverse socio-cultural factors and regulatory mechanisms, including more precautionary attitudes, can maintain and enhance global leadership in agriculture and agricultural technologies, including the advantageous position along the agrifood value chains. European agriculture, food security, and the related agrifood industries and their innovations are somewhat trapped between various trains of science, economy, ethics, and societal necessities on one side and lifestyles, health, environmental concerns, and conceptual values on the other (Lanki A. D. and Onwu C. A. 2024). The main disruptive agricultural-related technologies are enabled through or derived from biotechnologies, such as genetic engineering or genome editing. It is a regulatory challenge that the same organisms can also be produced by other means, including natural evolution. Additionally, the present regulatory systems and processes were established in the 1970s and often by the engineering practices of the day; they are currently incapable of dealing with current scientific hype and technology deluge (Mohanty et al. 2024). This chapter briefly describes some promises and factors associated with enabling disruptive technologies in agronomy and provides a policy reflection.

2.9.1 Government Initiatives and Incentives

Now more than ever, developments in agronomy are characterized by innovations in disruptive technologies, which interact with increasing challenges in the interfacing environment of the agrifood system. Sustainable intensification is a priority on many national research agendas. This chapter concerns the ways governments can best support, incentivize, and engage with various stakeholders involved in innovation to ensure such transitions are possible and desirable. "Enablement" includes not only direct funding inventories but also investments in infrastructure, service networks, and regulatory innovation (Ünal and Kızıldeniz 2022). Therefore, this analysis broadens the focus to the wider underpinning portfolio of government interventions for the agricultural and agrifood sectors and captures those efforts that tend to be neglected. Success depends on a national vision for the future of these sectors, cognizant of how it might contribute to other national goals, and the implementation plan that underpins these broader aspirations. Institutional coordination and strategic channeling of funding into critical enabler investments could shield transformative processes from path dependencies imposed by vested interests and favor a speedier realization of economic, social, and environmental opportunities and benefits (Baviskar 2024).

Since the high relevance of "Utilities" and "Competitive and Frugal Capital," their economic, social, and environmental goals, the intelligent design and use of these influential levers by governments are rarely a subject of intensive research. While general considerations of the role and impact that these organizations have on technological change patterns have been discussed in a generic and somewhat biased manner as technology push, pull, and infrastructure and investments, conceptually, these reflections are very loose and shallow when compared to the much denser literature on the other supportive contributions of government (Demir, Dedeoğlu, and Başayiğit 2024). Innovation systems and national systems of innovation concepts treat somewhat more precisely how general research, development, and innovation policies available to governments can influence their economies, their performers, and their sectors and the technologies these organizations develop. Across these policy frameworks, the supporting programs, instruments, or measures combined to achieve the desired impacts have advanced and diversified considerably in form and substance. Government support to research, development, and innovation is no longer the privilege of a few defense-related and large corporate companies (Prajval V et al. 2024).

The basic rationale for government intervention is macroeconomic, improving the efficiency and productivity of the agriculture and agrifood sectors that provide profound externalities of interest for the wider community. Its primary dualistic objectives are mainly focused on the creation of economic value and the satisfaction of social goals beyond the simple commercial enrichment of the private sector and the favored segments of the communities (Pranaswi et al. 2024). To strengthen research, development, and innovation and the enabling infrastructure, a more creative and empowering engagement of these sectors by the formal design of incentive programs that cater to the complexities of the actors, activities, lags, market imperfections, fragmented user groups, or neutral needs of the short and long-term markets is required. This is due as much to the market failures where private gains are deficiently associated with social benefits as to the market success and the risks associated with the long-term and high-risk character of agricultural technologies and practices, compared to the faster return on investment conditions being sought by the private sectors (Baviskar 2024; Shafik, Tufail, De Silva Liyanage, et al. 2024).

2.9.2 Ethical and Legal Considerations

The development and use of enabling, disruptive technologies in agronomy and agribusiness represent issues that need to be discussed and regulated before their mass implementation due to the concern for the magnitude of the impact they may have on society and the environment, as well as the complexity of predicting the degree of their implementation. Meanwhile, in contrast to strict land zoning in agriculture, which is implemented in many countries, the immediate application of these technologies as local initiatives is currently allowed. Even cultural heritage cannot legally stand in the way of technological revolutions (Pranaswi et al. 2024). For the purposes of limiting this mini-review, our intention is not to dive into the technical or policy aspects, such as security, privacy, or the protection of personal data of the citizens of agrarian areas, but rather to initiate the debate for local authorities or be an aid to anyone looking for answers on how to approach these issues. Furthermore, what all famous technological change theories have in common is that technologies are under the control of human decisions, that the dominant economic or political actors are leading the development, and that regulations are needed to limit the effects of innovations, which may be harmful, such as the destruction of equipment, city congestion, pollution, and excessive mortality (Philip and Suresh 2024).

2.10 Investment and Funding Trends in Agronomy Technologies

In the last few years, there has been notable activity in communication, connectivity, big data, and analytics, sensing, and robotics-related sectors such as Farmobile, CDI Lab, Rapid, Gamaya, Novozymes Biologicals, 30MHz, Rudron LLC DroneGrid, Climate Corporation, Netavo, Harvest CROO Robotics, Dronebase, Agricultural E-commerce CropsCo., granular.com, Precision Planting, ATA Robotics, and Blue River Technology. In contrast, investments in elemental agronomic technologies have often been in established and mature companies. For instance, the agrochemical conglomerates such as Dow Mexico and Syngenta in Mexico, Guatemala, and Honduras. The spreading diverse sectors of agronomy technology create opportunities for additional research and development, expansion, joint ventures, further business development, land economics, start-up momentum relative to large corporations, and strategic corporate venture capital activity. Since these opportunities relate to many countries, it may help extend networks of innovation and technology (Catala-Roman et al. 2024). Such developments may foster improved resilience and environmental outcomes. This paper can help the agronomy innovation community by providing a novel framework for classifying specific technology disruptions.

2.10.1 Venture Capital and Private Equity Investments

Foreign companies in search of sources of innovation that do not exist within their infrastructure, major agricultural companies identifying risks of unknown technologies, and venture capital interests that see agriculture and food as attractive investments all suggest a significant influx of financial resources to acquire or invest strategically in emergent agriculture and food technology opportunities. It appears that venture capital, and increasingly private equity money, is highly attracted to food and agriculture as a potential bonanza both in terms of constant demand for food and a complete view of a market opportunity. While recent evidence is enticing, history and a few intractable problems may offer pause for concern (Mohanty et al. 2024). But providing an attractive business model—particularly in comparison to information technology or medical technology where understanding the customer is problematic and the capital demands are excessive when competition becomes cutthroat. For the agriculture and food system, in contrast, the customer opportunity to gain relevant interest and validation is broad and can be peer-reviewed relatively easily. The basic theory of the venture capital model is most attractive to draw attention to agriculture. History indicates that many breakthrough technologies were developed not by large corporate research and development efforts, but instead by small teams of talented individuals, either in start-up companies or as small parts of large incumbents. By all accounts, the economics of successful agriculture ventures have been impressive, with resulting acquisitions or IPOs often in excess of 10 times the invested capital. In the agriculture and food industry worldwide, the compound annual growth of venture capital has been twice the average for all other capital transacted, putting the agriculture and food industry as an excellent option with its long product development timeline and reduced number of competitors (Shrestha and Wei 2024).

2.10.2 Public Funding Opportunities

Several institutions, both national and international, take a special interest in providing funds for areas of plight, such as food security and sustainable agriculture. The Sustainable Production and Prevention of Crop Disease initiative provides interested candidates the opportunity to come up with a high level of disruptive and unique ideas. This effort takes place in terms of a flexible, longer funding period of between 4 to 5 years to cover the length of time needed to answer all related questions and create necessary developments sufficiently. Another initiative engages in high-end disruptive and productive research in the competition that enables members of academia to create research in conjunction with industrial partners (Pranaswi et al. 2024). The duration is usually shorter, and decision-making is reached quicker than its counterpart. It also provides an early flow of knowledge exchange. The Global Food Security Programme funds research in terms of translational activities and cross-disciplinary research. This is very effective in piecing together information on highly complex problems. In addition, funding for innovative ideas is provided in the form of collaborative research and different types of supply chain projects. In addition to the list, it makes it a top priority to fund agriculture security and agri-tech via biotechnology in areas such as advanced food production technologies, advanced food processing technologies, agricultural equipment, field or harvest robotics, crop production technologies, and soil and crop management technologies (Nowrose Jahan Lipi and Muhammad Maniruzzaman 2024).

2.11 Collaborative Partnerships in Agronomy Innovation

Disruptive technologies, such as big data-driven tools, artificial intelligence, blockchain, or drones, are enablers of increased efficiency, accuracy, and scale in operations and bring advances in agricultural practices, soil science, and crop and animal breeding. There is a need for cross-disciplinary partnerships between sustainable agronomy experts and technology innovators that consider the wider perspective of food and environmental systems, recognizing local, national, and international needs and responding rapidly to changing external political and consumer agendas. These coordinated and inclusive approaches to innovation, coupled with participatory knowledge exchange, are essential for progress in data use and technology consumption

(Tong et al. 2024). Integration of theory-generated models and human expert decision-making, able to address stakeholders' missions, visions, challenges, and goals, can support evidence-based application of enabling technologies and stakeholder engagement. Development, testing, and deployment beyond the building blocks of technological innovation require ongoing interaction between technology-driven research and human-centered research agendas that aim to address big challenges and leverage multiple perspectives to foster creativity, creating new capabilities through collaboration between people and automated systems in a participatory environment, rather than creating technology that replaces human capabilities. Simple, clear, and user-focused interfaces can increase trustworthiness and use efficiency and effectiveness, disseminating technology results efficiently. Before running away with artificial intelligence-driven validation, there is a need to validate sensors, collect and annotate data carefully, use privacy-invading data and models ethically, and draw on in-depth scientific agronomy knowledge coined by human expert insights (Nowrose Jahan Lipi and Muhammad Maniruzzaman 2024).

2.11.1 Academic-Industry Collaborations

The fusion of ideas and complementary expertise is best accomplished through collaborations between academia and industry that enable the sustainable development of disruptive agronomic technologies. While academic basic research generates large amounts of biomolecular, omics, laboratory, and field data set and offers opportunities for understanding essential plant biology, many industry-led applied genomics, laboratory, and plant improvement activities will remain constrained in continuous access and application of the complete spectrum of basic research options necessary for addressing specific sustainability-related crop challenges. For a more focused alignment of academic-industry research activities to rapidly address current and direct future global agronomic challenges of joint interest, long-term research collaborations supported by plant-breeding companies and governing collaborations between university and industry scientists are necessary ingredients (Mohanty et al. 2024). A research and access model that provides a clear governance structure for both partners is outlined, and they can take full advantage of a broad range of academic research options. These should include novel discovery and gene editing capabilities, environmental performance studies, product development, and global cultivation of field trials on a variety of scales in accordance with local regulations. In these engagements, commercial entities commit to licensing, enabling innovation from university partners on reasonable terms.

With appropriate industry access, generous federal funding for basic research can accelerate private sector developments in seed biotechnology (Catala-Roman et al. 2024). These research collaborations exemplify a means to align the fundamental science with a shared plant improvement objective to accelerate innovative solutions and provide opportunities to generate substantial societal impacts in agriculture.

2.11.2 International Research Networks

Many of the enabling technologies raised in the previous sections are at an advanced state of development or feasibility and are seen as offering significant potential to contribute to the delivery of more sustainable systems of agronomy, agricultural ecosystems, and food security. However, the spectrum of these technologies is comprehensive, and the potential complexity involved in their effective combination so that they deliver real and usable benefits in real-world settings is often considerably underestimated. The organization of a session at a recent conference highlighted this dual challenge of providing innovative solutions and effectively combining them. It focused on the development of national, European, and international R&D networks and actions that could enable, support, and deliver joint, multi-stakeholder R&D programmes using UAV products for agriculture, addressing also the regulatory issues that needed to be developed (Shrestha and Wei 2024). Research can accelerate through robust networks facilitating the development of a range of translational projects that connect robotics with precision agriculture. Such projects could involve experts in core system design, computer vision, artificial intelligence, and agricultural science to seek solutions to the resource dilemmas faced by the agricultural industry in meeting its global environmental responsibilities. Pooling resources and expertise, including industry, after validating the feasibility of the core research outputs and their robustness in the face of variable field conditions and reserve margin for user-led implementation activities, can ensure scaling up the implications and wider societal impacts, enabling major improvements in farming systems to become a reality (Pranaswi et al. 2024).

2.12 Education and Skill Development in Agronomy Technologies

Disruptive technologies are those that displace an earlier technology, causing it to be obsolete, often in the market. Key enabling disruptive technologies for agronomy in the coming decades are presented and reviewed in relation to environmental and sustainable development in agronomy. Discussed technologies are big data, artificial intelligence, machine learning, robotics, the Internet of Things, blockchain, precision agriculture, gene editing, *biostimulation*, *photoselection*, photobioreactors, cellular agriculture, synthetic biology, and frugal innovation. The use of each of these enabling disruptive technologies in relation to major environmental issues—erosion, the greenhouse effect, air pollution, soil sustainability, preservation of water reserves, and food production—was analyzed (Tong et al. 2024). Education and skill development are another aspect of the problem. There are unmet needs in order to take full advantage of these new technologies. There is a lack of mentors and tutors, either online or onsite, that provide a face-to-face or at least a weekly opportunity for engagement and inspiration. There is also a lack of informed and sensitized personnel with the necessary skills to use the new disruptive technologies. It is important that we continually work to improve the interoperability of in-house developed databases, new public data repositories, and domain-specific, community-recognized open databases (Ünal and Kızıldeniz 2022). Domain scientists and bioinformaticians who manage big data resources need enhanced skills in bioinformatics, genetics, complex systems, and statistics. Commercial breeders of maize and wheat must now use mathematical models to evaluate and predict production ecophysiological traits, mainly related to drought resistance and photosynthetic mechanisms under different agro-ecological conditions. The collaboration between academia and industry researchers is critical for the development of new agronomy technologies and concepts. The traditional liaison role of extension is also critical since its function is precisely transferring information and connecting new technologies with the industry where these technologies will be adopted (Baviskar 2024).

2.12.1 Training Programs and Certifications

Training programs and certifications in enabling technologies are beginning. Some started on a volunteer basis, where an experienced professional has an apprentice or colleagues. As the demand grows, training centers and HR companies have also started designing certification programs. International standards for several enabling technologies are being created, while academia agrees on the role of economists, ethicists, and legal experts in mastering the potential of such technologies. In summary, it seems beneficial to create early the training center, professional roles, and certification programs. The need is urgent to speed up the development and maximize the competitiveness of the Republic of Croatia by accelerating governance and human resource readiness for the coming explosion of new technologies related to sustainable agronomy (Tong et al. 2024). Nevertheless, scholars should codify and share their experiences and results in specific technology training, enabling programs, and the ability for "learning to further progress." The strategy can support common competence certification systems and promote transitional, lifelong, and work-spanning education. Interdisciplinary and transdisciplinary approaches are necessary. In the process of designing programs, coherence among enabling competencies, national policies, and educational documents should be deployed to improve the fruition of common certification systems (Mohanty et al. 2024). Short courses on how to decrypt capabilities and on standards and certification processes could be requested. Research meeting policy and business agendas issues are necessary constituents of training and education programs, also for the unique knowledge support of decision-making, ethical standards, law, and policy development. This helps the development of common certification systems and inspires transitional, lifelong, and work-spanning education. Additionally, open-source and international collaboration policies for dedicated digital education resources will contribute to the success of all other policy measures supporting competence (Catala-Roman et al. 2024). They should facilitate citizens in future jobs and new employment opportunities.

2.12.2 Future Skills in Agronomy

The digital transformation occurring in agronomy, stimulated by the development and deployment of many enabling disruptive technologies, together with the unprecedented pace of change around the planet, is rapidly evolving a set of demands on the generation and management of innovative new career designs. These developments are creating the need for fundamental shifts in the way we define agronomist professionals, set goals and content in gaining agronomy qualifications, and implement strategies for ongoing and continuing professional development in achieving and maintaining a sustainable workforce. Recent innovations in the way development and extension have been carrying out their core functions are also addressing a highly disconcerting decline in demand for agricultural professionals and the consumer mismatch arising between publicly funded training outcomes in local human resource development and the ease in gaining access to skilled professionals in agronomy fields (Naveen et al. 2024). There has been a tendency when planning the use of new technologies to emphasize the technologies first and then engage users to apply the new capabilities. Many lessons have been learned from this flawed approach. The real challenge has always been the adoption process. It is only through users applying new capabilities that the disruptive technologies flip into truly enabling technologies, generating the endlessly fascinating creative uses of the technology built by the users to perform within innovative systems. The most striking features are how much the technologies change with usage, how the frontiers move away from the need for further pushing technological change in pursuit of opportunities at the intersection of big challenges, and how this cannot happen without the user partnerships which bring capacity to visualize and build attractive future states and strategic alignments that must create them (Lanki A. D. and Onwu C. A. 2024).

2.13 Barriers to Adoption and Implementation of Disruptive Technologies

Disruptive technologies are inevitably going to be met by barriers and limitations, from high cost and technical limitations in their current form to ethical and trust-based responses from the user communities. The two questions to be asked are 'What are they?' and 'How will these barriers be addressed?' Effective operation of markets should enable displacement of existing technologies and the further diffusion of disruptive ones. However, there are often significant obstacles that are to some extent tied up with the unique features that make the technologies disruptive. A potential barrier to the adoption of the use of technologies by the more progressive farmers looking to develop a sustainable farming system could be a lack of suitable information to recognize where the use of a disruptive technology could lead directly to a more sustainable farming system. Another issue is centered on the supply infrastructure (Philip and Suresh 2024). Both existing and emerging distribution infrastructure is set up to gain maximum advantage from bulk production and distribution. This concentration on large volumes creates areas that are generally referred to as cold spots in which low-level demand cannot encourage the development of any industry that could take the enabling technology and convert it into a product, leaving many potential users without the added value of the longer-term benefits or more effective technology that are already being incorporated into their business model. Frequently, a disruptive technology becomes disruptive at a time when the moral and ethical fabric of society, as well as the economies of business and countries, take significant strides (Prajval V et al. 2024).

2.13.1 Technological Barriers

We live in an era of acceleration-a time when technological, social, and environmental changes are often unprecedented in terms of development speed and impact. In the agricultural context, two ideas stand out as pervasive: the urgent need to supply sufficient, nutritious, and safe food to the booming and increasingly urban world population and the demand to do so in an environmentally and socially sustainable way. This makes agronomy one of the scientific disciplines poised to make a unique and critical contribution to 21st-century societies. Breakthroughs are needed in altering current agronomic practices to produce more with fewer inputs and reduced ecological impacts (Mohanty et al. 2024). However, the pathway to developing new and improved agronomic strategies often meets technological barriers related to the complexity of underlying biological phenomena. Emerging technologies, such as gene editing, RNA-guided regulation of gene expression, data management and analysis, and digital agriculture, have the potential to lead a technological revolution in agriculture by providing new tools and offering agronomic researchers opportunities to decipher complex aspects of plant biology that were previously unattainable (Catala-Roman et al. 2024). This capacity will allow the scientific community to generate more environmentally and socially friendly agronomic practices. In this chapter, we describe the following emerging technologies:

a) the major advances of gene editing for plant genome editing;
b) the RNA-based approaches inducing gene silencing or activating gene expression;
c) data approaches and digital agriculture as drivers for agronomy in the era of big and smart data; and
d) a multidisciplinary, translational approach to address biological questions and, consequently, to capture diverse and robust information that can ultimately be used to inform new and improved agronomic strategies.

2.13.2 Financial Constraints

It is commonly stated, to the point of truism, that offering hybrid or genetically improved crop varieties to farmers is a profitable activity given the very large returns to research. Once a successful variety is developed and a seed sufficient for commercial-scale production exists, a considerable net revenue can be generated. Moreover, the enterprise is attractive, especially to private companies often capable of promoting and distributing the hybrid or crop variety effectively. But it is also true that the development of productive crop hybrids, be they inbred line hybrids or hybrids coming from the three parental systems, requires large scale field trials. These have to be repeated for several seasons. A necessary stage is the production and conversion of land to hybrid seeds. Purification involves skilled labor and must also be accompanied by complex quality control procedures. All the associated costs have to be covered (Shrestha and Wei 2024). Once all costs are analyzed in relation to probabilities of success and benefits, we can conclude scientifically, rather than emotionally, that many low-income countries, probably most, cannot produce their improved hybrids. Small companies will be easily discouraged before they begin, and the large-scale ones, or the local subsidiaries of international companies, will be incapable of identifying and developing a market for virtually all the many and necessarily very wide-ranging conditions for which hybrids are required in the tropical zones. They may, in principle, be able to do better in other agroecological zones where conventional crop varieties can already be grown successfully without any hybrid or composite advantage. High levels of energy and capital availability also provide the staple crops of these latter regions (Pranaswi et al. 2024). The conclusion is that perhaps the only remedy available to low-income agriculture is collective action involving public or cooperative research.

2.14 Socio-Economic Impacts of Disruptive Technologies in Agriculture

It is also important to recognize and consider the possible negative impacts of the development and implementation of technologies affecting agriculture, namely, increasing levels of urbanization, changes in population dynamics, and the aging of active farmers, as well as destructive anthropogenic actions that may erode the potential for agriculture to respond to urban food demands. Socio-economic dynamics resulting from disruptive technologies have a profound impact on the social, economic, and environmental agricultural context. The implications of information-based technologies will relate to the reorganization of society in support of new industry forces. Data attributes from companies that exploit natural resources and transform them into products will be replaced by those that generate value around intangible data (Baviskar 2024). The modernization of agricultural production is only a small part of the impact, and the consequences of this technological transition, although unpredictable, will be an increased concentration of company assets and investments in data; thus, traditional businesses and staffing will be negatively affected. Broadly considered, data can interfere with all levels of activity of companies in any sector of the economy; hence, participatory platforms for creating or enhancing these technological support services are a necessity. By increasing global physical and social connectivity, digital-socio-technical systems linked to a heterogeneous network of actors affect the horizontal and vertical network coherence and the free access to and supply of digital tools and data democratization (Prajval V et al. 2024).

Radical changes based on re-learning of the rural environment will impact both rural and urban cultural natures and create new socio-economic issues, including the division of the unwanted side of the digital revolution by creating a significant difference between agricultural entrepreneurship, with or without extensive participation of technologies significantly involved in productivity gains or relationships with customers and the supply chain. The evolution of the model that is revising the links between entities belonging to the rural world is a particularly significant and disruptive technology that affects the importance of the arrangement of local economic structures. The rapid evolution of the growth curve of Internet technologies, which includes families of sensorized modular control devices and sophisticated artificial intelligence, is already homogeneous at all levels of organization of society, including entities typically characterized by secondary importance to the introduction of a modern digital approach, largely still based on traditional operational and decisional methods (Shafik 2024a). The forceful insertion of artificial intelligence algorithms aimed at decision-making for the management of companies is certainly characterized by great usefulness in the important continuity of the lifeblood of the organization and is used in data analysis and management, as well as in models or structures of innovative business initiatives. It will also involve farm and rural communities, creating scenarios that distract from the inadequate attention with which these changes have been observed. Providing feedback for initiatives that generate information for action represents an important task that must avoid the risks induced by poor service, thereby encouraging a unilateral relationship between citizens and companies aimed at creating data, otherwise creating conditions that could affect governance and generate new barriers related to the lack of transparency (Shrestha and Wei 2024).

2.14.1 Employment Effects

This paper analyzes the effects of using robotics and automation on employment, labor costs, the quality of employment, and location in the region in the major agri-food chain stage: production. Two long-term trends of enabling disruptive technologies have shaped human history: the demographic and the technological. The former refers to the increase of the population over time; the latter refers to the remarkable technological changes that allowed the development of new sectors, the reduction of costs through scale economies, and the transformation of social territory, such as the printing bankruptcy, which was partly caused by the entry of the printing press. These changes in the years 2025-2035 will be largely determined by two enabling disruptive technologies: robotics and automation. When these two technologies are combined with continuous technological improvements in sensors, computers, and machine learning, the tremendous potential to transform hardware and software into processes or products that, until now, had characteristics indelibly linked to human beings becomes very clear. The chain of major effects of such enabling technologies in production within the intensive agriculture segment, including plantations, represents a radical transformation of the sector (Tong et al. 2024). Since human workers will perform fewer repetitive tasks, the growth of the population in agri-food regions becomes possible, and the quality of employment and its location will increase. These theoretical effects are explored in this work.

2.14.2 Income Disparities

The Food and Agriculture Organization (FAO) of the United Nations defines small-scale farmers as those farming on fields no larger than two hectares. The percentage of farmers who meet this qualification varies greatly by country, from 94% in the Liberian Republic to less than 28% in Denmark. However, these farmers represent a large percentage of the global farming community, estimated between 70% and 80%, depending on whether the subsistence farmers and rural populations are included. These small-scale farmers are responsible for nearly 80% of the global food supply. Therefore, it is crucial to consider the efficacy of these enabling disruptive technologies for small-scale farmers to help address their contributing role in solving SDG goal number 2 ("Zero Hunger") (Shafik, Tufail, Namoun, et al. 2023). In a study with smallholder poultry farmers in the Philippines, enhancing the biophysical environment using free-range systems and paddy rice fields that incorporated resilient paddy rice fields, improved antibiotic alternatives, and reduced prevalence and shedding of infectious agents improved the income for 71% of the smallholder poultry farmers included. Similarly, a micro-flotation solution can be provided in a modular way and deployed in a small operation, reducing the minerals required for the production of animal protein and significantly reducing the feedstock cost compared to the status quo if the aggregation of ownership is also addressed (Shafik, Tufail, Liyanage, et al. 2023). Finally, the saving of resources due to the use of effective animal health management or inclusive feeding strategies creates not only a positive return for the animal production system, a positive return in terms of animal well-being, but also for the society around the production system.

2.15 Ethical Considerations in the Use of Technology in Agronomy

Ethical rationales and patterns are important. Major aspects are 'how is the technology used': Will it be used to prevent environmental problems from worsening and shifting in space by a more intensified but sustainable agricultural chain? Or is it used to solve problems that rely on the chain, as illustrated by events occurring and studies on the selection for restiveness behavior? Technologies are under development to make precision livestock farming one of the important technologies in animal production. However, since ethical considerations largely drive the fate of public acceptance, this section addresses the ethical considerations in the use of technology in agronomy. Ethical considerations regarding the use of new technology might play a role in the acceptance by stakeholders and, ultimately, the success of the technology (Kotelnikova et al. 2024). A plea for a holistic approach involving the ethical, cultural, and social aspects when deploying new technologies in agriculture is expressed. A close dialogue between research, technological development, and quantitatively studying the ethical implications is suggested. The options for technology application need to be developed with the main aim of serving humanity while respecting the other living beings with whom humans share natural and cultural contexts. Therefore, researchers should be prompted to conduct research beyond the material consequences of the technology and consider the concerns about human life, animal rights, biodiversity, solidarity, transparency, and fairness (Lai et al. 2024).

2.15.1 Data Privacy and Security

Farmers may be reticent to integrate disruptive technologies into their daily activities if they believe their farm data may be accessed and used in ways that are beyond their control. An important focus for the agronomy sector is to enable and encourage the widespread deployment of disruptive technologies, including big data applications, to enhance the adoption and impact of current and future approaches and technologies for sustainable agronomy, such as precision agriculture. We discuss the main aspects of data privacy and security regarding the various technologies. There are many concerns with data privacy and security at both the individual level of the farmer and the company or vendor level that supplies agricultural technology and data management and analytics platforms, as well as at the level of local, national, and international regulatory authorities and public policies (Beres et al. 2024). The implications of data-driven agricultural technologies are unprecedented, and to realize their incredible potential to support sustainable global food security, we must address the many challenges regarding data privacy and security as we also embrace ongoing technological progress. The development and use of autonomous, intelligent farm machinery and other advanced sensor and robotic technologies have been described. Data-driven AI models that rely on gigabytes, terabytes, or even petabytes of data require data to be stored, moved across networks, combined with other data, and shifted into the cloud to be analyzed by AI models or other types of data analytics tools, and to be translated back to put on an 'edge' device in an intelligible manner. While the pace of data generation speeds up, the specter of data privacy and security looms over the agricultural sector, potentially limiting adoption. There are many different perspectives on security: secrecy, authenticity, integrity, solidarity, utility, safety, and privacy of the data. Data must be kept secret from unauthorized access, use, and disclosure in order to find the right balance and make themselves invaluable to producers (Xiang et al. 2024). The data must be managed in such a way as to ensure it is not falsified or infected with malicious code that can lead to the execution of unauthorized activities.

2.15.2 Inclusive and Equitable Access

Prominent transformative technologies of the fourth industrial revolution have ushered in a largely private sector-driven approach, one that is exclusionary to unprivileged communities, which are instead the most needy and vulnerable and whose livelihoods have the greatest potential to be uplifted through access to these technologies. The sharing of the benefits of open-source technologies that are within the realms of public goods, derived from the commons by combining the knowledge that is dispersed among a global network of stakeholders, holds tremendous potential for enhancing the impact, inclusivity, and acceptability of these technologies and intensifying their use (Nowrose Jahan Lipi and Muhammad Maniruzzaman 2024). We have previously conducted in-field training programs disseminating socialist and participatory breeding and agronomy methods within the target rural community, and we have also recorded meaningful increases in crop safety and yield as outcomes. Subsidizing the costs of access to public goods through strategic public-private partnerships to align international seed, fertilizer, and crop protection agribusiness by innovatively construing access to these products as digital agriculture research or earmarked corporate social responsibility investments, leveraged by institutional publication and technology sharing of the crops it manages, can extend the availability of open source technologies to benefit underserved communities while also ensuring continued sustainable public access to advanced biotechnological tools (Lai et al. 2024). Table 2.1 dissects how disruptive technologies are transforming sustainable agronomy, detailing their core functionalities, environmental impacts, implementation challenges, and solutions for widespread adoption.

Table 2-1 The role of disruptive technologies in enabling sustainable agronomy

Technology	Core Functionality	Impact on Sustainable Agronomy	Environmental Benefits	Challenges to Implementation	Solutions for Adoption	Reference
Biopesticides and Biofertilizers	Natural alternatives to synthetic chemicals	Enhances soil health, reduces chemical dependency	Protects biodiversity, prevents soil degradation	Limited availability, inconsistent quality	Support local production, enforce quality standards	(Ünal and Kızıldeniz 2022)
Smart Irrigation Systems	Automated and adaptive water management systems	Ensures optimal water usage	Conserves water resources, prevents salinization	High installation costs, maintenance complexity	Low-cost irrigation innovations, training farmers	(Fürst-Preiß and Von Cossel 2023)
Automated Machinery and Robotics	Autonomous tractors, harvesters, and drones	Reduces labor costs, increases efficiency	Minimizes soil compaction, enhances precision	High cost, need for technical skills	Shared ownership models, training programs	(Philip and Suresh 2024)
Digital Platforms and AgTech Apps	Mobile applications for farm management and market access	Enhances knowledge dissemination and market linkage	Reduces post-harvest losses, empowers smallholders	Digital divide, connectivity issues	Improve rural connectivity, local content development	(Lanki A. D. and Onwu C. A. 2024)

Biotechnology and CRISPR	Gene editing for developing resilient crop varieties	Improves drought, pest, and disease resistance	Reduces the need for chemical inputs	Ethical concerns, regulatory hurdles	Transparent governance, public engagement	
Renewable Energy Solutions	Solar, wind, and bioenergy for powering farm operations	Reduces reliance on fossil fuels	Lowers carbon footprint, enhances energy resilience	High initial investment, maintenance needs	Subsidies, innovative financing models	(Catala-Roman et al. 2024)
Blockchain Technology	Securing data for supply chain traceability	Ensures transparency in agricultural value chains	Reduces food waste, supports fair trade	Lack of infrastructure, scalability concerns	Government and industry collaboration	(Shafik 2024c)
AI and Machine Learning	Predictive analytics for weather, crop yield, and pests	Enhances decision-making, boosts productivity	Minimizes crop losses, improves resource efficiency	Data privacy concerns, interpretability issues	Establish data-sharing frameworks, open-source AI	(Grace and Dawson 2024)
Precision Agriculture	Real-time monitoring using IoT, sensors, and drones	Optimizes input use (water, fertilizers, pesticides)	Reduces resource wastage, minimizes pollution	High cost of technology, limited digital literacy	Affordable solutions, capacity-building efforts	(Tong et al. 2024)

2.16 Emerging Technologies in Sustainable Agronomy

Sustainable agronomy can be defined as the combined application of concepts and technologies to produce sufficient crop production, meet economic and food security objectives, and, at the same time, reduce the environmental impact. Sustainable agriculture is about understanding, knowledge, and knowledge transfer and about changing farming practices. Multiple disruptive technologies can significantly help in addressing existing and forthcoming challenges in sustainable agronomy. The creation of highly productive and efficient agricultural systems capable of providing farmers with a secure and steady income, contributing to temperate and tropical food demands, while at the same time helping to restore natural, socioeconomic backgrounds, including biodiversity, along with research investment, is essential (Catala-Roman et al. 2024). The goals of sustainable agronomy include clean and spiritually satisfying agroecosystems. In order to achieve this, several enabling technologies will become available. They include sensors and other automated systems, engaging data analytics, artificial intelligence, machine learning, and wireless communication. These technologies should contribute simultaneously to producing more with less while reducing input use and environmental impact, preserving natural resources and reducing greenhouse gas emissions, restoring and enhancing biodiversity and ecosystem services, preserving and enhancing socio-economic benefits, trust, and comfort, and maintaining and promoting a favorable eco-compatibility balance and protection of human health (Albrecht and Albrecht 2024).

2.16.1 Predictive Analytics and Forecasting

Critical to enabling and accelerating the proliferation of many disruptive technologies used in plant sciences is the ability of many of these next-generation platforms to perform global system-wide quantification, abundance, or distribution, coupled with predictive analytics to understand and predict as an enabling technology. The ability to model and predict results based, in general, on pattern detection of large collections that represent highly diverse and contextually inclusive observations is transformative and perhaps one of the most critical areas of science to provide a firm predictive foundation for a system capable of understanding the complexity, uncertainty, and feedback mechanisms controlling the system (Kumar et al. 2024). Enhanced predictability of such emergent properties has been predicted as being pivotal to large agriculture-linked global economics and food security in an increasingly unpredictable, chaotic world.

2.16.2 Blockchain Technology in Supply Chains

Agricultural supply chains can be complex and fragmented, inspiring a wave of studies and programs aimed at better connecting them, thereby ensuring a more efficient and capable system in response to dynamic consumer needs. Disruptive technologies have been introduced with the aim of improving environmental externalities related to opportunities such as new value creation and reduction of public environmental costs. The proposed blockchain network application provides a platform where stakeholders can exchange information regarding international agricultural supply chains. This technology provides stakeholders with more direct effects of radical change through process optimization, innovation incentives, and environmental governance (Chao 2024). While the promise of these technologies should be highly emphasized, a major research effort focuses on understanding radical technological capacity. This work reviews recent transactions in supply chain management and organic food safety-related literature. After this work, a framework is developed to guide further responsible probing that focuses on the effects of blockchain technology throughout the supply chain and establishes a connection to related technologies, provideing open research results to stimulate further development of supplier connections in agronomy discussions (Sufaat and Juliandri 2024; Hugar and Waheed 2024).

2.17 Conclusion

It cannot be stressed enough that, while germline editing or the use of computer vision are benefiting from a surge in attention and investment, these potentially disruptive technologies remain difficult and expensive to develop and access, while their societal lifecycle benefits remain undervalued and little understood. The increasing urbanization is creating the most sustainable, safe environment in history, with the opportunity to focus on demanding a greener, healthier, and ingredient-rich diet, mostly produced by massive numbers of precision-engineering adapted crops tending to disappear from open fields, being replaced by cleverly designed agronomic vertical systems. A green economy transition, middle-income countries' growth, and the growing need for healthy and sustainable food for an increasing world population face a poorly developed agronomic sector, future agriculture, and agronomy jobs, increasing discouragement and uncertainty. Urgent policies, investments, and effective collaboration on national and global roadmaps are required to create an inspiring, inclusive, smart future with plenty of opportunities and ecological preservation. The relevance of emerging technologies should be seeded through university education, especially in agro-biosciences, including agronomy. The agro-food industry is highly knowledge- and employment-intensive, featuring many high-value-added and high-tech industries such as bioscience, robotics, and IoT for smart equipment, along with a national and global collection of differentiated and premium agri-food global brands. In brief, agronomy needs a disruptive model similar to how bioinformatics and computational biology represent a new breed of agronomists in the current era. Data-capturing advanced technologies with agronomy-focused experiments, layer by layer, are also enabling efficient bioinformatic processing within innovative and upgradable integrated big-data centers, starting to predict phenotyping data from plants' genotyping data. The opportunity might be real if plants could be grown and examined to discover the genes or gene networks that produce agronomically interesting phenotypes in the context of the stresses and nutrient levels typical of growing in real commercial fields.

References

Albrecht, Leandro P., and Alfredo Junior Paiola Albrecht. 2024. "Gadamerian Philosophical Hermeneutics and the Laudanian Perspective of Research Traditions in Agronomy." *Aoristo - International Journal of Phenomenology, Hermeneutics and Metaphysics* 7 (1). https://doi.org/10.48075/aoristo.v7i1.32325.

Anikina, A. V., E. K. Krutova, A. V. Klimova, and T. V. Smirnova. 2024. "Professional Standard «Agronomy» in the Context of Employer's Requests." *Professional Education in the Modern World* 13 (4). https://doi.org/10.20913/2618-7515-2023-4-12.

Baviskar, Dr. Prof. Pallavi. 2024. "Empowering Farmers through Technology: A Sustainable Approach to Agricultural Equipment Rental and Income Generation." *International Journal for Research in Applied Science and Engineering Technology* 12 (3). https://doi.org/10.22214/ijraset.2024.58825.

Beres, B. L., S. Simmill, W. Taylor, R. J. Dyck, and J. Hubert. 2024. "An Improved Design for a Zero-Tillage Experimental Plot Drill." *Canadian Journal of Plant Science* 104 (1). https://doi.org/10.1139/cjps-2023-0094.

Catala-Roman, Paula, Enrique A. Navarro, Jaume Segura-Garcia, and Miguel Garcia-Pineda. 2024. "Harnessing Digital Twins for Agriculture 5.0: A Comparative Analysis of 3D Point Cloud Tools." *Applied Sciences (Switzerland)* 14 (5). https://doi.org/10.3390/app14051709.

Chao, Kang. 2024. "Family Farming in Climate Change: Strategies for Resilient and Sustainable Food Systems." *Heliyon* 10 (7). https://doi.org/10.1016/j.heliyon.2024.e28599.

Das, Shaon Kumar, and Goutam Kumar Ghosh. 2024. "Soil Hydro-Physical Properties Affected by Biomass-Derived Biochar and Organic Manure: A Low-Cost Technology for Managing Acidic Mountain Sandy Soils of North Eastern Region of India." *Biomass Conversion and Biorefinery* 14 (5). https://doi.org/10.1007/s13399-022-03107-7.

Demir, Sinan, Mert Dedeoğlu, and Levent Başayiğit. 2024. "Yield Prediction Models of Organic Oil Rose Farming with Agricultural Unmanned Aerial Vehicles (UAVs) Images and Machine Learnaing Algorithms." *Remote Sensing Applications: Society and Environment* 33. https://doi.org/10.1016/j.rsase.2023.101131.

Dr. Mrinal Gaurav, Amit Kumar,. 2024. "Analysis of Cold Storage Capacity Utilization in Jharkhand: Special Reference to Agriculture Finance." *European Economic Letters (EEL)* 14 (1). https://doi.org/10.52783/eel.v14i1.1026.

Fürst-Preiß, Christine, and Moritz Von Cossel. 2023. "Biodiversity-Friendly Bioenergy—a Closer Look on Farmer's Experiences with Perennial Wild Plant Mixture Cultivation for Biogas Production." In *Biodiversity and Bioeconomy: Status Quo, Challenges, and Opportunities*. https://doi.org/10.1016/B978-0-323-95482-2.00017-1.

Giannini, Vittoria, and Elisa Marraccini. 2023. "On-Farm Experimentation in Agronomic Research: An Italian Perspective." *Italian Journal of Agronomy* 18 (4). https://doi.org/10.4081/ija.2023.2215.

Grace, Palli Susan, and Joy Dawson. 2024. "Influence of Nitrogen and Iron on Yield and Economics of Foxtail Millet (Setaria Italica L.)." *Environment and Ecology* 42 (1). https://doi.org/10.60151/envec/ysys8231.

Hugar, Sharanamma M., and Mohammed Abdul Waheed. 2024. "Using CNN to Identify NPK Deficiencies in Paddy Fields: An Advanced Detection Method." *International Journal of Intelligent Systems and Applications in Engineering* 12 (2).

Kotelnikova, A. D., T. I. Borisochkina, K. A. Kolchanova, M. A. Shishkin, F. S. Egorov, V. V. Okorkov, and O. B. Rogova. 2024. "Dataset on Elemental Composition of Soils and Plants under Long-Term Application of Mineral and Organic Fertilizers on Gray Forest Soils in Vladimir Region, Russia." *Data in Brief* 53. https://doi.org/10.1016/j.dib.2024.110057.

Kumar, Pravesh, Vineet Dheer, Pradeep Kumar, Jaykar Singh, Krishna Kumar Singh, Yogesh Kumar, and Anil Kumar Singh. 2024. "Effect of Irrigation Scheduling and Different Sowing Dates on Growth and Yield of Wheat (Triticum Aestivum L.)." *International Journal of Environment and Climate Change* 14 (1). https://doi.org/10.9734/ijecc/2024/v14i13820.

Lai, Qinghui, Yongjie Wang, Yu Tan, and Wenqiang Sun. 2024. "Design and Experiment of Panax Notoginseng Root Orientation Transplanting Device Based on YOLOv5s." *Frontiers in Plant Science* 15. https://doi.org/10.3389/fpls.2024.1325420.

Lanki A. D., and Onwu C. A. 2024. "Geographic Information System (GIS) Application in Soil Fertility Management: A Review." *Journal of Global Agriculture and Ecology* 16 (2). https://doi.org/10.56557/jogae/2024/v16i28592.

Ma, Ruijun, Yu Chen, Xiaohua Zhang, Toudeshki Arash, and Ehsani Reza. 2024. "Development Process, Mode and Technology Status of Apple Mechanized Harvesting." *Journal of Chinese Agricultural Mechanization* 45 (1). https://doi.org/10.13733/j.jcam.issn.2095-5553.2024.01.041.

Mohanty, Lalita Kumar, N K Singh, Pranav Raj, Aditya Prakash, Awanindra Kumar Tiwari, Vishal Singh, and Prashun Sachan. 2024. "Nurturing Crops, Enhancing Soil Health, and Sustaining Agricultural Prosperity Worldwide through Agronomy." *Journal of Experimental Agriculture International* 46 (2). https://doi.org/10.9734/jeai/2024/v46i22308.

Naveen, Naveen, Pritha Datta, Bhagirath Behera, and Dil Bahadur Rahut. 2024. "Climate-Smart Agriculture in South Asia: Exploring Practices, Determinants, and Contribution to Sustainable Development Goals." *Mitigation and Adaptation Strategies for Global Change* 29 (4). https://doi.org/10.1007/s11027-024-10126-4.

Nowrose Jahan Lipi, and Muhammad Maniruzzaman. 2024. "Effect of Different Planting Density of Sunflower Varieties on Yield and Yield Attributing Characters in Rice-Rice-Sunflower Cropping Pattern." *Research in Agriculture Livestock and Fisheries* 10 (3). https://doi.org/10.3329/ralf.v10i3.69592.

Philip, Bijin, and G. Suresh. 2024. "Cultivating Resilience: An In-Depth Exploration of Agricultural Insurance Initiatives and Their Transformative Role in India's Farming Landscape." *Universal Journal of Agricultural Research* 12 (1). https://doi.org/10.13189/ujar.2024.120109.

Prajval V, A. Krishnamoorthi, Riya Thakur, Ruchitha T, Manish Kapoor, Anushi, Abhishek Singh, and Karthik Chittibomma. 2024. "From Code to Crop: How Bioinformatics Is Transforming Crop Genomics in Modern Agriculture and Bettering Environment." *Journal of Advances in Biology & Biotechnology* 27 (4). https://doi.org/10.9734/jabb/2024/v27i4737.

Pranaswi, Dameruppula, Megha P. Jagtap, Gopal U. Shinde, Narendra Khatri, Stuthi Shetty, and Shravya Pare. 2024. "Analyzing the Synergistic Impact of UAV-Based Technology and Knapsack Sprayer on Weed Management, Yield-Contributing Traits, and Yield in Wheat (Triticum Aestivum L.) for Enhanced Agricultural Operations." *Computers and Electronics in Agriculture* 219. https://doi.org/10.1016/j.compag.2024.108796.

Shafik, Wasswa. 2024a. "Barriers to Implementing Computational Intelligence-Based Agriculture System." In *Computational Intelligence in Internet of Agricultural Things*, 193–219. Springer. https://doi.org/10.1007/978-3-031-67450-1_8.

Shafik, Wasswa. 2024b. "Biosensor-Based Drones Anomaly Detection Integration for Sustainable Agriculture Development." In *Achieving Food Security Through Sustainable Agriculture*, 283–316. IGI Global. https://doi.org/10.4018/979-8-3693-4240-4.ch012.

Shafik, Wasswa. 2024c. "Blockchain-Based Internet of Things (B-IoT): Challenges, Solutions, Opportunities, Open Research Questions, and Future Trends." In *Blockchain-Based Internet of Things*, 35–58. Boca Raton: Chapman and Hall/CRC. https://doi.org/10.1201/9781003407096-3.

Shafik, Wasswa. 2024d. "Generative AI for Social Good and Sustainable Development." In *Generative AI: Current Trends and Applications*, 185–217. Springer.

Shafik, Wasswa. 2025a. "Sustainable Agriculture and Diet for Urban Development." In *Sustainable Smart Cities and the Future of Urban Development*, 165–92. IGI Global Scientific Publishing.

Shafik, Wasswa. 2025b. "Sustainable Agriculture and Diet in the Metaverse Era: Sustainable Farming and Nutritious Eating in the Metaverse." In *Food in the Metaverse and Web 3.0 Era: Intersecting Food, Technology, and Culture*, 159–84. IGI Global Scientific Publishing.

Shafik, Wasswa, and Kassim Kalinaki. 2024. "Societal and Ethical Implications of Technology-Enhanced Agriculture and Healthcare: An African Context." In *2024 IST-Africa Conference (IST-Africa)*, 1–11. IEEE. https://doi.org/10.23919/IST-Africa63983.2024.10569306.

Shafik, Wasswa, Ali Tufail, Rosyzie Anna Awg Haji Mohd Apong, and Liyanage Chandratilak De Silva. 2024. "Internet of Things for Smart Agricultural Practices." In *Internet of Things Applications and Technology*, 190–217. Boca Raton: Auerbach Publications. https://doi.org/10.1201/9781003458401-13.

Shafik, Wasswa, Ali Tufail, Chandratilak De Silva Liyanage, and Rosyzie Anna Awg Haji Mohd Apong. 2023. "Using a Novel Convolutional

Neural Network for Plant Pests Detection and Disease Classification." *Journal of the Science of Food and Agriculture*. https://doi.org/10.1002/jsfa.12700.

Shafik, Wasswa, Ali Tufail, Abdallah Namoun, Liyanage Chandratilak De Silva, and Rosyzie Anna Awg Haji Mohd Apong. 2023. "A Systematic Literature Review on Plant Disease Detection: Motivations, Classification Techniques, Datasets, Challenges, and Future Trends." *IEEE Access* 11:59174–203. https://doi.org/10.1109/ACCESS.2023.3284760.

Shafik, Wasswa, Ali Tufail, Chandratilak Liyanage De Silva, Rosyzie Anna Awg Haji, and Mohd Apong. 2025. "The Role, Application, and Impact of Artificial Intelligence in the Agriculture Industry." In *Future Tech Startups and Innovation in the Age of AI*, 36–60. CRC Press.

Shafik, Wasswa, Ali Tufail, Chandratilak De Silva Liyanage, and Rosyzie Anna Awg Haji Mohd Apong. 2024. "Using Transfer Learning-Based Plant Disease Classification and Detection for Sustainable Agriculture." *BMC Plant Biology* 24 (1): 136. https://doi.org/10.1186/s12870-024-04825-y.

Shrestha, Manish Man, and Lin Wei. 2024. "Review—Perspectives on the Roles of Real Time Nitrogen Sensing and IoT Integration in Smart Agriculture." *Journal of The Electrochemical Society* 171 (2). https://doi.org/10.1149/1945-7111/ad22d8.

Sufaat, Imam, and Juliandri Juliandri. 2024. "IOT Rancang Bangun Alat Pengusir Hama Burung Pada Padi Sawah Petani Berbasis Internet of Things (IoT)." *Journal of Computer System and Informatics (JoSYC)* 5 (2). https://doi.org/10.47065/josyc.v5i2.4921.

Tong, Wenyu, Jianfei Zhang, Zhiyu Song, Guangqiao Cao, Yue Jin, and Xiaofeng Ning. 2024. "Research Status and Development Trend of Cabbage Mechanical Harvesting Equipment and Technology." *Journal of Chinese Agricultural Mechanization* 45 (1). https://doi.org/10.13733/j.jcam.issn.2095-5553.2024.01.044.

Ünal, Zeynep, and Tefide Kızıldeniz. 2022. "Smart Agriculture Practices in Potato Production." In *Potato Production Worldwide*. https://doi.org/10.1016/B978-0-12-822925-5.00010-4.

Xiang, Meiqi, Minghao Qu, Gang Wang, Zhongyang Ma, Xuegeng Chen, Zihao Zhou, Jiangtao Qi, Xiaomei Gao, Hailan Li, and Honglei Jia. 2024. "Crop Detection Technologies, Mechanical Weeding Executive Parts and Working Performance of Intelligent Mechanical Weeding: A Review." *Frontiers in Plant Science*. https://doi.org/10.3389/fpls.2024.1361002.

Zeng, Zhen, Huabo Cai, Jiaoyun Chen, Xiangli Liu, Yongjia Li, Yi Zhang, Jianwen Chen, Dehua Rao, and Wankuan Shen. 2024. "Improving Sugarcane Agronomy: Field Evidence for Ultrasonic Treatment Enhancing Yield, Growth, and Physiological and Biochemical Characteristics." *Industrial Crops and Products* 211. https://doi.org/10.1016/j.indcrop.2024.118276.

Zhu, Linjing, and Yuefen Wang. 2024. "Entrepreneurship and Carbon Footprints in Sub-Saharan Africa." *Problemy Ekorozwoju* 19 (1). https://doi.org/10.35784/preko.5759.

Zidan, Khadija, Laila Mandi, Abdessamad Hejjaj, Naaila Ouazzani, and Ali Assabbane. 2024. "Soil Fertility and Agro-Physiological Responses of Maize (Zea Mays) Irrigated by Treated Domestic Wastewater by Hybrid Multi-Soil-Layering Technology." *Journal of Environmental Management* 351. https://doi.org/10.1016/j.jenvman.2023.119802.

CHAPTER THREE

THE CONVERGENCE OF TECHNOLOGY AND SUSTAINABILITY IN MODERN AGRONOMY

Abstract

Agronomy is a branch of agricultural science and is one of its oldest. Its focus is on scientific crop production and soil management. Crop plants and their nutrition, protection from harmful pests, diseases, and environmental stress (like drought, salinity, temperature, among others), and their regeneration through sexual and asexual means are the principal objects of this branch of agriculture. Therefore, agronomy is critical for ensuring global food security. The agronomy of the past and present is and will remain multidisciplinary because it encompasses aspects of a few hard sciences of life, and every enduring phenomenon has a solid foundation of science. Soil science, horticulture, plant pathology, entomology, biotechnology, and even nanotechnology, which operate as interdisciplinary branches of science, are included in agronomy. Agronomy is a diverse field that plays a prime role in overcoming challenges like climate change and depleted natural resources for present and future generations. Moreover, it is the answer to doubling food production by 2050 in the wake of a burgeoning population. Inputs from the farming community, biologists, chemists, environmentalists, and bureaucrats are inevitable in this subject. Agronomy has helped us improve soil fertility, plant nutrition, and farming practices, which have established an excellent potential to strengthen rural economies and people's standard of living. With the raising and intensifying of agricultural systems, there has been an inevitable negative impact on the environment. Initially and still, the environmental impact of this management has been largely focused on the effects of land degradation through soil erosion and salinization. Stagnant growth in food production has led to food shortages, increasing prices, and disparity in food consumption. Clearly, there is an immediate need to increase crop productivity to provide food to a rapidly growing population. Thus, alternative strategies and innovative practices are required to bring about a new-age revolution in agriculture by increasing its productivity with minimal impact on nature.

These modern innovations in agriculture, which are meant to enhance yield further and safeguard natural resources for future generations, are expected to conform to the norms of sustainability.

Keywords: AgriTech Innovation, Climate-Smart Agriculture, Data-Driven Decision Making, Digital Agriculture Solutions, Green Agricultural Innovation, Precision Agriculture Technologies, Renewable Energy in Agriculture, Sustainable Farming Practices

3.1 Introduction

While the basic field of agronomy has been the cornerstone of agricultural activity for centuries, it has evolved with scientific developments and technological changes. Today, technology has greatly impacted agronomy by enhancing the monitoring of plant growth, leading to predictions about crop yields, and employing robotics for farm automation and management. Some key technologies that have revolutionized agronomy include the precision agriculture systems used to optimize water and chemical use through controlled irrigation, root and soil zone temperature monitoring, and vapor pressure settings. This approach also defines optimal sunlight exposure designed to improve vitamin and other phytochemical production, such as in raspberries and strawberries, in real-time using data from wireless networks of crop canopies (Singh et al. 2024). In the field, applications from remote sensing via drones and satellites are used to monitor plant health and productivity using such crop spectral indices. Robotic agriculture is another innovation in agronomy used to operate systems such as weeding using autonomously weeding tractors. Another example is the use of drones in agriculture for tasks in managing the farm, making drones a leading-edge equipment technology used for vertical farming, as sampled in Fig. 3.1. The integration of these advanced technologies has the potential to deliver reduced food waste, more efficient application of fertilizers, decreased water and chemical use, and increased crop yields. Technology is a 'game-changer' when coupled with decision-making that takes big data analysis into account. Agriculture that depends heavily on a production recipe and accumulation of best practices is likely to be used as a marketing niche—a tool for which buyers are prepared to pay more for an enhanced product (Rojas 2024).

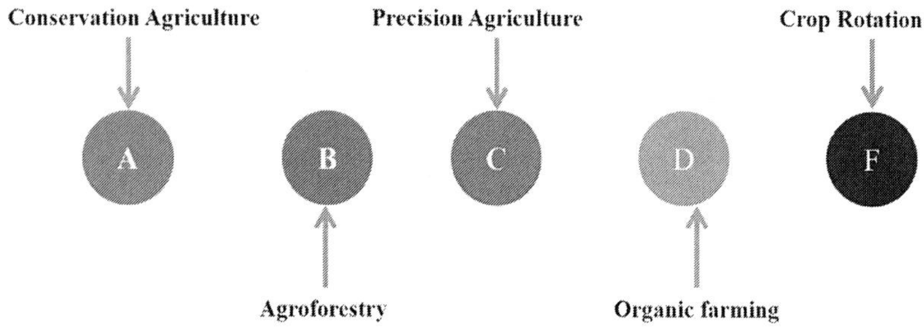

Fig. 3-1 Sampled sustainable agronomy practices

3.1.1 Precision Agriculture

Precision agriculture (PA) is an advanced technological approach that aims to make farming procedures more efficient. In this modern approach, precision instruments monitor and regulate the growth conditions, avoiding over- or under-fertilizing processes. The automatic and real-time conditions of crop growth could be adjusted. This technology has been developed rapidly along with several devices that could be used in precision agriculture. By using precision agriculture, optimal conditions of the field can be reached. Several techniques have been employed and developed in PA, such as soil mapping, spatial data interpolation and variogram estimation, prescription maps, and variable rate technology (VRT), which can be applied to lime and fertilizer (Verdi et al. 2024). The use of the best farming inputs based on field-specific needs could reduce the negative externalities or wastage on the environment, such as overused chemicals for those needed in plants. A more precise use of farming inputs could protect the environment through the use of pesticides and minimize nitrogen and other fertilizers, which could result in contaminating water resources (Etienne et al. 2024).

New technologies like the Global Positioning System (GPS) enable greater accuracy at planting and harvesting and allow farmers to know exactly where they planted a particular variety to the acre, which is particularly important for harvesting seed. The precision agriculture system relies on geographical information system technology, which requires a series of yield maps, farm boundary maps, plant population maps, and soil testing maps, and the GPS receiver has an inbuilt compact flash/data type media card to store the variable rate prescription (Singh et al. 2024). The UAVs fitted with multi-spectral sensors can detect crop conditions at the initial stage and warn based on the NDVI index or other inbuilt algorithms for taking timely corrective measures. All these technologies have given hope for converting subsistence farming into sustainable agriculture. It is envisaged that precision agriculture is the future of modern agronomy. It became possible only with the convergence of advanced information technologies in big data, automated cloud storage, real-time processes, artificial intelligence (AI), and the mobile revolution in communication for real-time field updates (Verdi et al. 2024).

3.1.2 Remote Sensing and GIS Applications

Remote sensing is the discernment of objects on Earth from images taken at a distance. For agronomy, remote sensing applications harness satellite and aerial images that show differences between crop species, growth stages, and changes in chlorophyll activity. Rendered in different, non-visible bands can indicate plant health and chlorophyll decline. Geographic Information Systems (GIS) are used to store, interpret, and manipulate data in the realms of location and space. Remote sensors on aircraft and in space platforms typically record many images. These images may track various conditions affecting plant operations over a range of sensors that provide the type and amount of data growers need to assess soils, crops, and field conditions. A significant development that accompanied the increased use of remote sensing in agronomy was the growth of GIS applications, which enabled agronomists to map and manage agricultural resources better (Singh et al. 2024). A GIS is built to collect, store, manipulate, analyze, and display spatially referenced information. GIS allows us to view patterns, trends, and relationships in data that are not apparent in tables or charts. This ability to visualize spatial relationships is an essential quality in agriculture, where much of the decision-making involves relationships among pieces of land, landowner objectives, and regulatory options. With the development of remote sensing and GIS, precision farming or site-specific management has been a fast-growing trend in agriculture (Reher et al. 2024). One of the key factors that enhance the utilization of GPS and remote sensing data in agriculture is the continuous development and improvisation of GIS technology. Remote sensing technologies have proven their potential in crop yield assessment and agricultural risk management. However, there are many challenges associated with data collection, processing, storage, and interpretation related to these technologies. Data accessibility is considered to be a significant concern in remote sensing through GIS technology use in the agricultural sector to make the findings useful for agronomic research, formulation, and implementation. The collaboration of agriculturalists and scientists must work hand in hand to develop remote sensing technology to be useful for the propagation of sustainable agriculture (Moreira et al. 2023).

3.1.3 Robotics and Automation in Farming

The use of robotics and automation is among the recent technological advances in the field of agriculture and has gained momentum in the last decade. The application of robots in agriculture is wide and varied, covering areas ranging from planting to pesticide spraying, cultivating, weeding, thinning, selective harvesting, and crop monitoring by means of ground robots, air robots, and hybrid robots. Automated harvesting, planting, thinning, weeding, drips, micro-sprinkling, and aerial seeding were practices to reduce dependency on human labor and labor costs. In addition to reducing human error and drudgery, automation improved the precision and timeliness of operations and lessened the physical impacts of agricultural work on human health (Valdés et al. 2024). There are critical reseasons for agricultural sustainability, as presented in Fig. 3.2.

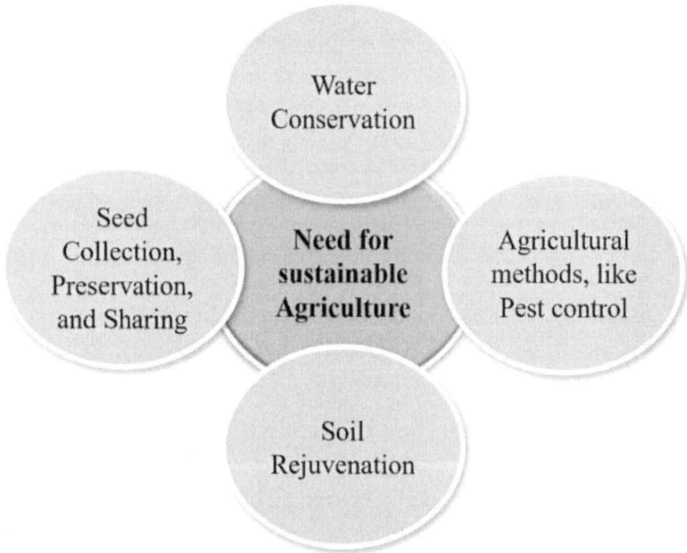

Fig. 3-2 Need for sustainable agriculture

Recent advances in environmental and field sensors, structural and model-based designs, machine learning algorithms, inexpensive equipment, and better environmental sensing capabilities provide robots with the capacity to decide in real time how best to operate based on the confluence of discrete environmental signals and learned or predicted relationships with agronomic production. They offer fast, round-the-clock, and labor-saving technologies with added benefits and cost assurance and are effective investments for farm sustainability. The alignment of this technology with the achievement of relevant goals helps to show the link between environmental preservation and human welfare in decreasing stress at work, increasing production outcomes, and high welfare demand (Sangaiah et al. 2024). Even deemed essential for the next farming actors, from large agricultural companies to small-sized farmers, under climate and environmental pressures, it improved the production of well-targeted aesthetics peculiar in segments of global markets; the driving next-generation technologies are presented in Table 3.1.

Table 3-1 Next-Gen Technologies Driving Sustainable Agronomy

Technology	Description	Application	Benefits	Challenges	Referene
Remote Sensing	Satellite and aerial imaging	Land use, crop health monitoring	Large-scale monitoring	Expensive satellite data	(Valdés et al. 2024)
Mobile Apps	Farmer advisory and information platforms	Market info, weather alerts	Increased access to information	Limited rural connectivity	(Moreira et al. 2023)
CRISPR Gene Editing	Precise genetic modification	Disease resistance, drought tolerance	Faster crop improvement	Ethical and legal concerns	(Reher et al. 2024)
Aquaponics	Combines aquaculture with hydroponics	Integrated fish and crop farming	Waste reduction, dual production	Balancing ecosystem	(Etienne et al. 2024)
Hydroponics	Soilless farming using nutrient solutions	Urban, controlled environments	Water efficiency, faster growth	Requires constant monitoring	(Verdi et al. 2024)
Vertical Farming	Multi-layer indoor agriculture	Urban agriculture, year-round farming	Space efficiency, reduced water use	High energy demands	(Rojas 2024)
Big Data Analytics	Large-scale data processing	Crop modeling, market trends	Informed decision-making	Data privacy concerns	(Singh et al. 2024)
Renewable Energy	Solar, wind, and bioenergy	Energy for farming operations	Reduced carbon footprint, energy savings	Initial infrastructure costs	(Moreira et al. 2023)

Biotech and GMOs	Genetically modified crops	Disease resistance, higher yield	Improved resilience, increased production	Public acceptance, regulation	(Shafik, Tufail, Namoun, et al. 2023)
Robotics	Automated machinery	Planting, weeding, harvesting	Labor savings, precision operations	High maintenance costs	(Shafik, Tufail, De Silva Liyanage, et al. 2024)
IoT Sensors	Smart sensors in fields	Soil moisture, weather monitoring	Precise input usage, reduced waste	Connectivity in rural areas	(Rojas 2024)
Blockchain	Decentralized ledger	Supply chain transparency	Enhanced traceability, reduced fraud	Adoption complexity	(Fontaine et al. 2024)
AI and Machine Learning	Predictive analytics	Pest prediction, yield forecasting	Data-driven decisions, early detection	Requires large datasets	(Damini, Ashwini, and Dipali 2024)
Drones	Unmanned aerial vehicles	Crop health, irrigation	Real-time data, efficient spraying	Regulatory restrictions	(Fernandez et al. 2024)
Precision Agriculture	Uses GPS, IoT, and sensors	Crop monitoring, soil mapping	Enhanced yield, reduced resource waste	High initial investment	(Chilaka et al. 2024)

3.2 Sustainable Agronomy Practices

To maintain economically viable and productive farming systems while protecting the environment, many approaches to sustainable agriculture have been developed. One common approach is organic farming, which has detailed standards for inputs and farm management. The guiding principles of organic agriculture include health, ecological balance, fairness, and care. This includes goals such as minimizing pollution, conserving water, and maintaining soil fertility. Other efforts include agroecology, which focuses on promoting ecological processes and the local knowledge of farmers in the development of an agricultural system for preservation over time (Xiao et al. 2024; Singh et al. 2024). They encourage the development of a series of sustainable farming systems with low use or more sustainable use of chemical inputs. IPM programs can control pests for the lowest cost both economically and environmentally.

Promoting practices for organically growing and producing food is a growing field of modern agricultural technologies and one of the most significant issues in agriculture. In addition to the positive relationship between the adoption of management practices promoting sustainability and the financial performance of farms, a wide selection of methods was employed to assess sustainability in terms of soil health, water conservation, and ecosystem services, as well as biodiversity. Agricultural practices of crop rotation are often used to promote soil fertility. Additionally, cover cropping also has several benefits for the environment by preventing soil erosion and increasing or adding nutrients to the soil (Etienne et al. 2024). The adoption of cover crops is also responsible for ecosystem services, such as conserving water, contributing to carbon capture, and resulting in no-till or less-till practices. One of the primary concerns of agriculture and its sustainability is dealing with environmental risks and the ways in which the agricultural industry is regulated. Despite a range of sustainable and organic practices being available, there are issues with adopting them. Supportive policy frameworks, including education and the provision of resources and advice to farmers, have been proposed as the best mechanism to ensure the adoption of an agriculturally sustainable mindset. It has been emphasized that the aim of a research program into the development of resilient farming systems must be to determine how systems can be designed to adjust continuously to changing pressures (Moreira et al. 2023). This is particularly relevant in the current era of rapid climate change, as illustrated in Fig. 3.3.

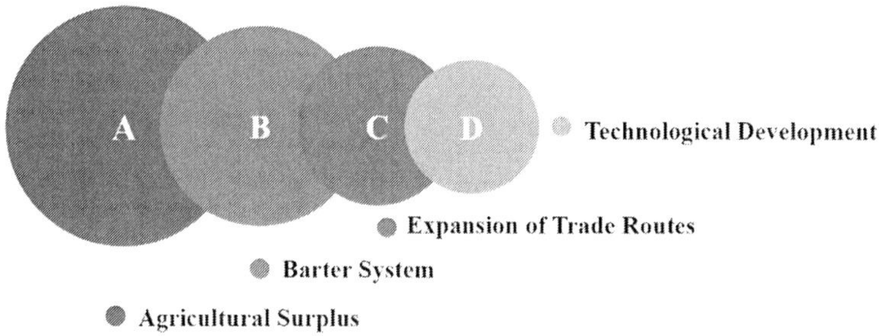

Fig. 3-3 Significant sustainable concerns

3.2.1 Organic Farming

Agroecological in action is a farming practice that fits within the goals of sustainability. Organic agriculture consists of farming systems that produce foodstuffs using methods that do not harm the environment, conserve soil health and fertility as well as non-renewable resources, follow the cycle of natural elements, and safeguard animal welfare. Farming techniques in organic agriculture depend on the health of local ecosystems, biodiversity, and self-reliance as far as natural inputs are concerned. These natural inputs can, in part, be produced on the farm but also sourced from elsewhere. The local ecological setting of farming, as well as the local social context of food production, are thus critical to the application of organic methods (Singh et al. 2024). Organic farming looks to sequester carbon, conserve water, limit erosion, and improve the health of its used soils. Inputs follow the precautionary principle in the sense that adverse effects on human health or the environment in the short-, medium-, or long-term lead to their exclusion. In terms of global food systems, organic may be seen as more of a home consumption or local selling commodity. The practice of certification, with many definitions and rules, has been aimed at providing integrity in the trade of organic products in meeting consumer expectations. However, certification can be a barrier to the practice of farming because of the costs involved (Moreira et al. 2023).

There is a growing acceptance of organic certification within the world food trade, something that can limit small-scale farmers. However, the approach can also benefit from reasonable regulatory hygiene measures to ensure the integrity of organic marketplaces. One key goal of organic farming is to reduce environmental impacts on natural biota. The focus on biodiversity has social and ecological dimensions, the latter being that of ecosystem services. The former revolves around a wider acceptance of global agro-biodiversity and includes aesthetic and cultural components (Shafik, Tufail, Liyanage, et al. 2023). Trends show that organic consumers are willing to pay premium prices for organic food, and this demand has a direct influence on the market prices for organic goods. A critique of the approach is that it can, in part, encourage trading and, therefore, does not fully stimulate resilient ex-situ agri-biodiversity systems. The focus on local markets can have many social benefits, including economic benefits for local economies. Organic, sometimes seen as a luxury, can have social benefits for communities dealing with economic exclusion. The rise in the development and practice of organic is often tied into wider concepts around food sovereignty or control over local food systems. The debate around the alternative of organic goes beyond this discussion, though it is an important consideration since alternative farming methods are key to developing sustainable human food systems (Shafik 2024a).

3.2.2 Agroecology

'Sustainability science is a transdisciplinary pursuit to contribute to enhancing global and local security for a wide range of environmental and social systems and services.' Agroecology can be described as the science of sustainable agriculture, and as such, it has necessary links with all parts of sustainability science as well as the wider concept of sustainable development. At its core, agroecology is a holistic paradigm, occupying the middle ground between competing reductionist 'hard' and social 'soft' approaches. The aim of a detailed understanding of the interactions of plants, animals, soils, etc., fits well with sustainability. The whole systems perspective and approaches to scientific research embedded in agroecology mirror the aims of the natural, human, and social sciences as part of sustainability science. Agroecology offers a detailed understanding of interacting ecological systems in relation to food and agriculture (Shafik, Tufail, De Silva, Haji, et al. 2025). This chapter sets out to illustrate how technology can be used in a way that encourages the seeking out and use of a diversity of ecological principles to build sustainable farming systems. We hold the view that other approaches will fail on these grounds. We have, therefore, drawn the argument from our understanding of what agroecologists are already doing at the individual or family farm scale (Garofalo et al. 2024). However, the value of a more structured and concerted use of agroecological techniques at the landscape level has also been noted, and agroecology could yet develop as a descriptor of practice at that wider level though impacted with chemical pesticides as illustrated in Fig. 3.4.

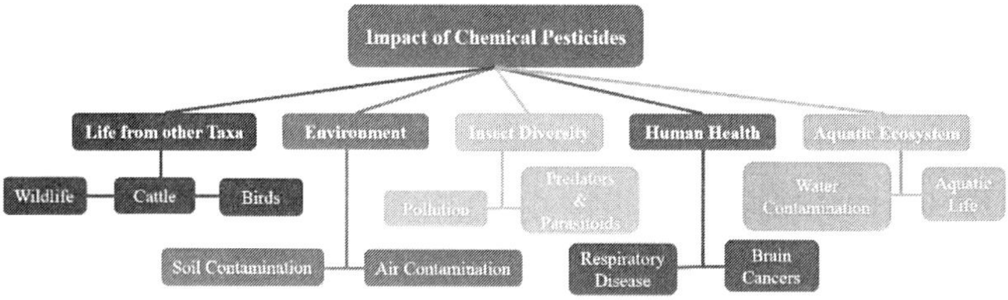

Fig. 3-4 The impact of chemical pesticides in sustainable agronomy

3.2.3 Integrated Pest Management (IPM)

A sustainable approach to controlling agricultural pests is based on the individual behavior of the pest and the monitoring of relevant factors in innovative crop production, which could also be called precision crop protection. IPM is aimed at balancing the need to protect crops from pests' adverse effects on the one hand and minimizing the risks of chemical pesticides to human health and the environment on the other, depending on the nature of the pathogen, as illustrated in Fig. 3.5. The classic definition of IPM states that it is an ecosystem-based strategy that focuses on the long-term prevention of pests or their damage through a combination of favorable biological, environmental, and chemical practices (Valdés et al. 2024).

To do so, IPM techniques are implemented using the most environmentally friendly technologies available and keeping an eye on the recent agronomical, social, and environmental shifts caused by the new global sustainability agenda. Special attention is placed on the strategies to monitor and reduce the use of toxic and harmful pesticides by choosing low-risk conventional, natural, biological control agents, or even physical measures, and relevant practices for urban and agriculture within a 'Zero Residues' perspective. Many successful examples show how the reliance on chemical pesticides can be reduced, and contaminants can be maintained under the quantifying limits while preserving a steady amount of yield. Indeed, this yield and production standardization can represent a crucial point in safeguarding the health of consumers, workers, and nearby residents (Verdi et al. 2024; Xiao et al. 2024). One of IPM's important aspects is that it focuses on the long-term prevention of pests; other sustainable practices are presented in Table 3.2.

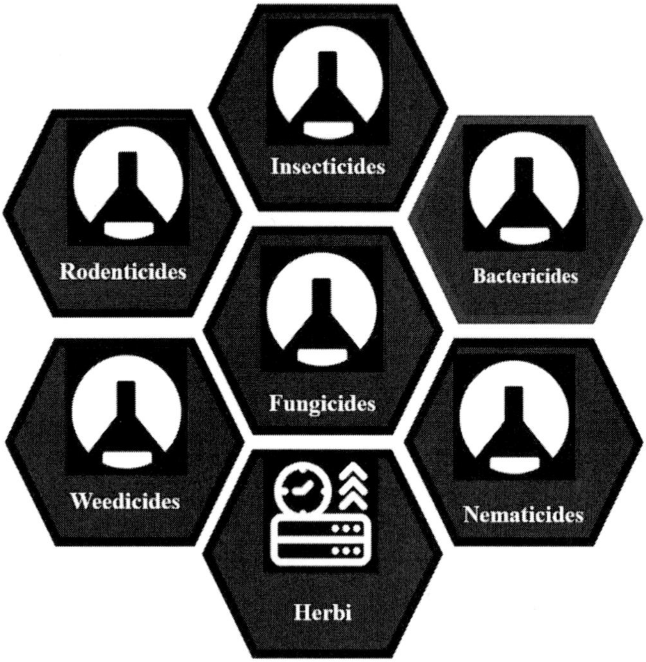

Fig. 3-5 Sampled agricultural stresses

Table 3-2 Sustainable Agronomy Practices and Their Technological Applications

Sustainable Practice	Description	Technological Support	Impact	Examples of Use	Reference
Digital Market Access	Online market platforms	Mobile apps, blockchain	Fair pricing, wider market reach	E-commerce for agriculture	(Carrillo Quiroga 2024)
Sustainable Livestock Management	Ethical and eco-friendly animal farming	Wearable sensors, robotics	Animal welfare, lower emissions	Automated milking systems	(Fontaine et al. 2024)
Regenerative Agriculture	Soil regeneration focus	Remote sensing, drones	Carbon capture, soil restoration	Carbon credit platforms	(Adeleke 2024)
Vertical Integration	Ownership of the supply chain	Blockchain, IoT	Increased profit margins	Traceable supply chains	(De Santis and Cammarano 2024)
Zero-Budget Farming	Low-cost input agriculture	Mobile advisory platforms	Cost reduction, increased self-reliance	Farmer cooperative platforms	(Reher et al. 2024)
Biodynamic Farming	Holistic farming methods	Soil and compost monitoring tools	Enhanced soil fertility	Biodynamic certification apps	(Sangaiah et al. 2024)
Crop Rotation	Rotating crops to improve soil health	Decision-support software	Nutrient replenishment	Crop planning apps	(Xiao et al. 2024)
Agroforestry	Integrating trees into farmland	GIS mapping, remote sensing	Carbon sequestration, soil protection	Satellite mapping	(Moreira et al. 2023)

Organic Farming	Avoidance of synthetic inputs	Digital monitoring tools	Healthier soil and ecosystems	Organic certification systems	(Fawzi et al. 2024)
Renewable Energy Use	Solar, wind, and bioenergy for farming	Solar panels, wind turbines	Lower carbon emissions	Solar-powered irrigation pumps	(Garofalo et al. 2024)
Sustainable Water Management	Efficient water use	IoT irrigation sensors	Reduced water wastage	Drip irrigation	(Valdés et al. 2024)
Integrated Pest Management	Ecological pest control	IoT sensors, predictive models	Reduced pesticide use	Pest detection apps	(Verdi et al. 2024)
Climate-Smart Agriculture	Adaptation to climate change	AI, weather forecasting systems	Resilience to climate variability	Early warning systems	(Carrillo Quiroga 2024)
Conservation Agriculture	Minimum tillage, cover cropping	Precision planting tools	Soil health improvement, erosion control	No-till seed drills	(Kalinaki et al. 2023)
Agroecology	Ecosystem-based farming	IoT, AI, drones	Biodiversity preservation, resilience	Crop-livestock integration	(Shafik, Tufail, Liyanage, et al. 2023)

3.3 Challenges and Opportunities in Integrating Technology and Sustainability

The integration of technology and sustainability in agriculture poses both challenges and opportunities. While upfront investment is needed for innovative technology and its application, high financial returns are expected on development funding. This potential shift in R&D investment is, however, challenged by the financialization of agrifood systems and the need to keep the return on capital high. An important question is how to improve technological advances for sustainable outcomes. It is, however, not clear whether these should conform to the direction of present technological trends. The question is also raised whether digital big data technologies and artificial intelligence can really produce more sustainable ways of production simply by looking for patterns in big data and building autonomous machines (Shafik, Tufail, Apong, et al. 2024; Garofalo et al. 2024).

A persistent question arises as to the political economy of this ecosystem, such as who will benefit from the use of such big data products and methods and at what cost? The social impact of the integration between the digital age and increasing sustainability comprises many dimensions, such as equity in who benefits from the new tools, facilitation of farmers' access to social and productive relationships, and the role of technology renewal in the evolution of labor dynamics. Policy and regulatory frameworks can help in overcoming the costs associated with the developmental cycle of sustainable technology adoption—production, learning, and management—and the uncertainties faced by society in the diffusion stage of technological deployment. The strategic alignment of sustainable innovation and digitalization hardly moves in the weather windows of mainstream economic policy tools, such as cluster competition policies (Valdés et al. 2024). Hence, to implement a supportive framework, which must actively guide the alignment effort, calls for policies and actions aligned to a transformative agenda.

The deployment of technology includes machinery, IT support, and computation/cloud data analysis, as well as a user-centered perspective that encompasses the transformation of information into knowledge and maintainable, knowledge-based decision-making and management more than just mere data crunching. This also means that the starting point of sustainable technology integration goes parallel to the production of a farming population owning the necessary skills to operate the necessary tools, as well as being comfortable in adopting, regularly updating, and upgrading the growing digital components of farming operations (Xiao et al. 2024). The integration needs may be a game changer for rural education, which must consider not only informing about the opportunities to farm better, healthily, and wealthier, but also offer practical tools. The conjoint consideration of the above issues leads to an innovation system perspective and to a roadmap of integrated action to align technology and sustainability. The strategic framework of reference moving from the SEEA-E framework to the degrowth synthesis goals, natural capital, human well-being, and distribution needs an operational framework because it deals with innovation, i.e., shaping the future, with tools and methods and an approach aimed at growth as well (Fawzi et al. 2024).

3.3.1 Economic Considerations

Many of the innovative technologies associated with the integration of sustainability into agronomic practices are expensive to take to market and may require significant financial investments. Innovative technologies to improve agronomic sustainability are primarily a product of research and development efforts and include such areas as breeding improved crops, soil microbiomes, and managing crop residues to promote carbon sequestration. The profitability and economic viability, in general, however, may not be known at the time of implementation of these technologies (Fawzi et al. 2024). Economic disincentives and costs to farmers at the time innovations are commercially available to implement are not generally known and must be, at least partially, forecast by evaluating the costs and benefits to farmers at the time of initial implementation. This initial economic evaluation offers potential economic costs and benefits. In other words, our current actions (or inactions) will directly impact the (un)willingness of farmers to adopt these necessary and innovative technologies. Market demand for sustainably produced goods will predominantly determine the implementation rate of newly commercially available agronomic technologies (Fontaine et al. 2024).

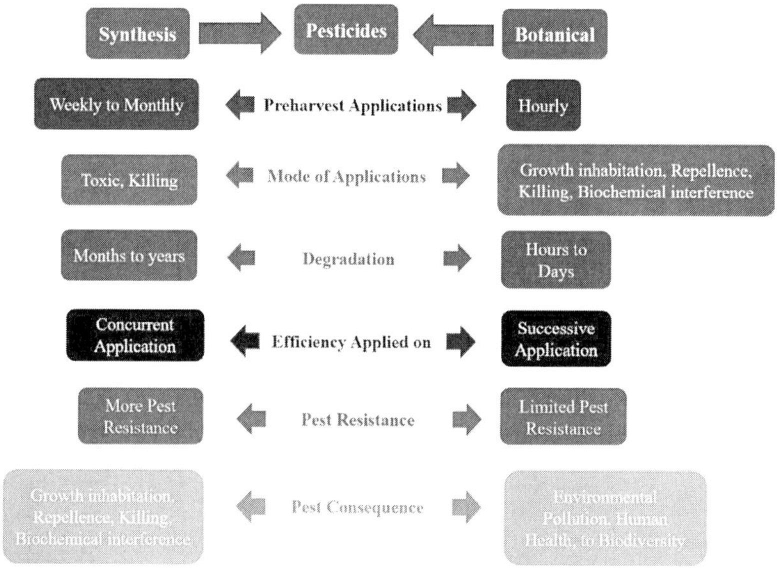

Fig. 3-6 Synthesis and botanical pesticides

With increasing global interest in net-zero value chains and corporate sustainability goals, early adopters of these technologies are likely to have a competitive advantage in the agriculture sector. Adopting healthy land management practices may also be supported by the needed financial support mechanisms. Many farmers may need access to credit or outside funding to adopt healthy land practices. The transition is likely to require bridge loans or inputs, for instance, and the repayment of these credit lines may be tied to land health improvement, which can be tracked on fields along the transition timeline using the same methods ideally used for incentive programs (De Santis and Cammarano 2024). Local cooperatives and partnerships may help broaden the availability of new technology like those underlying healthy land markets. Even developers without a normal profit motive may play important roles in investing profits into infrastructure and partnership development to gather or mobilize knowledge, monitoring, and verification across landscapes that can support healthy land program efforts (Carrillo Quiroga 2024). Addressing potential early adopter market imperfections can help broaden the availability and potential positive service of early technology on either synthesis or botanical pesticides, as illustrated in Fig. 3.6.

3.3.2 Social Impacts

Impacts may affect people through social equity, labor dynamics, agriculture, and social connectivity. Social equity in agriculture generally necessitates providing sufficient income and other resources to all farmers in order to produce the world's food supply sustainably. This requires equitable access to necessary resources and opportunities. All farmers must have access to sustainable agricultural technologies. Additionally, there are social implications of technology use in agriculture, including the displacement of traditional farming jobs and the growing need for new skills and labor dynamics. Alternatives to dairy and soybean-based products are often interpreted as increased connectivity for the agricultural community (Reher et al. 2024). In an attempt to maintain this social group balance, the development of the bioeconomy should play a special role in considering the position of the community, including the viewpoint of the farmer and grower.

Sustainable agricultural replications from the social point of view find it difficult for farmers to adopt agricultural systems, land use, and other management policies and technologies that focus mainly on increasing production and quality while considering environmental and social aspects. Technologies must be built with the participation of farmers and local communities so that they develop and are most suitable for an area with entirely environmental, economic, and social potential. New issues may arise if technical choices are determined at the local level, and there will be challenges to the capacity and potential to respond according to local resource capabilities (Xiao et al. 2024). Education and awareness programs with sustainable agricultural advantages may indicate repopulating and reusing part of their technology. However, social propensity among individuals can affect the performance of sustainable agricultural entities. Above all, people from rural areas can be discouraged from change due to rightful suspicion. User groups approved the current conventional systems and have been tested and working for a very long time. Thus, it is a good traditional approach. Technology assessment reports have a bias for the worst results and outcomes in order to steer the system into already established technologies and symposiums of the concerned category and interest (Fawzi et al. 2024; Shafik 2025a).

3.3.3 Policy and Regulation

Strong policy and regulatory frameworks are critical for the integration of technology and sustainability, including engagement with innovative farm-level and farm-gate processes and sustainable technology. Farmers are strongly incentivized by policy, with research demonstrating that they are receptive to further adoption guidance and expertise. Government policy also plays an important role in regulating and enforcing on-farm practices, waste management, biosecurity, weed control, and managing and protecting against pests and diseases. The rate of technology application is shaped by the regulatory environment, including the costs of compliance, demonstration of benefits, and other technology-specific requirements. In some cases, technologies may be developed to meet regulatory requirements (Verdi et al. 2024). Providing unambiguous advice about policy frameworks based on pragmatic and consistent evidence is critical for the integration of technology, innovation, and sustainability. Policymakers and regulators need to be aware of the potential impacts of new technologies and how they change the requirements for enforcement. Bureaucratic processes can introduce additional time and financial costs, but a good understanding of the applicable requirements will help reduce the complexities of negotiating the regulatory environment (Sangaiah et al. 2024).

There are emerging pressures for a governance system that can provide a licensing and permission framework that offers access to sustainable technologies while balancing these changes with the need to ensure environmental protection. Engagement and dialogue between regulators, farmers, technologists, and scientists are essential to develop best practices and foster an ongoing understanding of the emerging regulatory environment. Because of the need for technology transfer and community support, the adaptive co-management principles support the adoption of agricultural best technologies. Barriers to exchanging practices occur where local communities lack the capacity and opportunity to engage as equals in the co-production of knowledge (Carrillo Quiroga 2024). Policymakers supporting the implementation of sustainable agriculture need to be aware of the need for equity and inclusion of diverse enterprises and farming communities in the design of policy to enable their access to best practices. Policymakers need to be aware of potential perverse outcomes of regulation where the risks of new technologies do not enter their design. Exemption-based and outcome-based regulations will continue to be required in addition to innovations. Regulatory pathways need to strike a balance between the technological risks and impacts and the potential benefits of the technology (Adeleke 2024). Policymakers need to be open to ideas about how to finance alternatives. In some cases, the changes needed to facilitate the application of systems of technology and agronomy for sustainability require leadership to open regulatory pathways. Amendments to regulations are increasingly made based on the requirements of the relevant department and are not open to public scrutiny and comment. The changes are implemented largely because they assist no rule or gamification. This did not reflect adaptive co-management or the need for governance to engage with the community and affected participants; possible technological approaches are presented in Table 3.3.

Table 3-3 Summary of existing challenges, possible solutions, and impacts in tech-driven sustainable agronomy

Persistent Challenge (s)	Brief Description	Technological Solution	Potential Impact	Case Study
Financing Constraints	Lack of access to credit	Agri-fintech, mobile payments	Increased investment in agriculture	M-Pesa in Kenya
Energy Dependency	Fossil fuel reliance	Renewable energy tech	Reduced emissions	Solar-powered cold storage in Africa
Policy and Regulation	Inconsistent regulations	Policy analytics platforms	Harmonized agricultural policies	EU's Common Agricultural Policy reforms
Labor Shortages	Aging farmer populations	Robotics, automation	Increased efficiency	Automated harvesting systems
Fragmented Supply Chains	Lack of traceability	Blockchain, IoT	Better coordination, reduced waste	Blockchain for coffee traceability
Ethical Concerns	GMOs, AI in agriculture	CRISPR, ethical AI	Balanced technological advancement	Regulatory frameworks in EU
Environmental Degradation	Soil erosion, deforestation	GIS, remote sensing	Ecosystem restoration	Agroforestry projects in Kenya
Market Volatility	Fluctuating prices	Blockchain-based supply chains	Stable pricing and transparency	AgriLedger blockchain platforms
Pests and Diseases	Increasing pest outbreaks	AI-based pest detection	Reduced crop losses	PlantVillage's AI pest detection
Resource Scarcity	Limited water, land resources	Precision farming, smart irrigation	Efficient resource use	Israel's water-saving technologies

Knowledge Gaps	Lack of farmer tech literacy	Mobile training platforms	Improved skills and productivity	Digital Green's farmer training
Climate Variability	Unpredictable weather patterns	AI weather models, IoT sensors	Improved climate resilience	IBM’s weather forecasting for farmers
Limited Rural Connectivity	Poor internet access	Satellite internet, mobile networks	Wider tech access	Starlink connectivity for remote farms
Data Management Issues	Poor data quality, data silos	Cloud platforms, AI analytics	Improved decision-making	Agri-tech startups in Africa
High Initial Costs	Expensive tech investments	Government subsidies, fintech	Wider tech adoption	India’s solar pump subsidy

3.4 Case Studies in Technological Innovations for Sustainable Agronomy

A compilation of selected case studies that answer the technology intelligence question, "What case studies have there been of technological innovations successfully enabling sustainable agronomy?" Each of these case studies presents a technological innovation in the real-world practice of agronomy that demonstrates some level of integration of both technological and sustainability issues. The case studies look at the impact on resource use efficiency, environmental improvement, and contribution to the UN Sustainable Development Goals. It examines the "smart" application or development of a technology for agronomy, the conditions or other innovations that help it fulfill its intent for sustainability, and the evidence of sustainability outcomes (De Santis and Cammarano 2024). The selected case studies demonstrate a broad range of technological innovations in agronomy, from watering methods, supply chain tracking, and crop health evaluation to sea farming and vertical farming. They provide data on the increased efficiency and productivity of the farms, water and energy savings, maintenance of environmental concerns, increases in animal welfare, better food quality, and strong dissemination, promotion, and co-created innovation pathways. They attest to the need for tight collaboration and co-development of new technologies with agronomic practices and systems, and the crucial need for a user-centered approach in their development (Xiao et al. 2024). Importantly, seven of the case studies highlight that strong farmer involvement in the development of the technological innovation was a key factor for successful implementation on the ground and that the user-centered approach showed the real pathways to scale the application of the innovation in various agronomic and geographic contexts, ingrained in business and production operations. Two further case studies evaluate promising future opportunities for vertical farming, which, while not yet widely implemented, offer vast potential to contribute to the sustainability of food production worldwide. The scope of the selected case studies also examines the practicality and scalability of these technologies, demonstrating that even developed technologies can adapt to diverse climate, socio-economic, and market conditions (Garofalo et al. 2024).

3.4.1 Smart Irrigation Systems

One significant development in modern agronomy overcomes the challenge of water management. Smart irrigation systems measure environmental factors such as rainfall, humidity, soil salinity, and temperature, as well as plant factors such as crop water stress via sensors. They then use data analyses to alter the irrigation schedule in response to this data. These systems promise to reduce water wastage in agriculture, benefiting both plant yields and the natural environment. They also reduce operational costs since sensors determine the plant's needs in terms of moisture in the soil and alert the user when it is time to rinse. Smart irrigation systems aim to allow the water in the soil to be at an adequate level for the plant's needs, avoiding under-irrigation that can stress the plants and over-irrigation that leads to waste of water and soil leaching (Fontaine et al. 2024; Reher et al. 2024). This can become a major danger for the environment outside the farm, particularly in areas of intensive farming close to natural water basins. The use of sensors in agriculture for the management of water and its maintenance is an excellent practice for conserving the quality of water catchments outside farming operations (Shafik 2025b). This is a very good tool to assess when and how much water the crop needs efficiently, tailored to the specific stage of the plant growth cycle. Challenges surrounding the use of smart technologies for the practice of sustainable agriculture currently include issues relating to the cost of their implementation and operation in comparison to the resources available to smallholder farmers, as well as the general lack of existing infrastructure and skills to use and maintain these systems (Moreira et al. 2023). Therefore, large-scale education and the circulation of practices promoting the benefits for farmers, along with the reduction of pesticides, fertilizers, and economic resources for an increase in production and revenue, are required.

3.4.2 Blockchain in Supply Chain Management

Blockchain technology has already shown remarkable potential in tracking products from the farm to the table. Thanks to the transparent and immutable ledger of blockchain technology, users can confirm and verify the authenticity of their purchases. When applied in the agri-food sector, blockchain technology facilitates supply chain transparency and traceability of foodstuffs from primary production to consumption and can mold efficient cash transactions for farmers as it minimizes fraud and introduces farmers to expanded markets that have been difficult to access. One of the main characteristics of blockchain technology is the possibility for a customer to trace the entire path of the product, from the field to packaging, showing the date, time, and location of each handling event. Consumers can be informed about the geographical origin, the various steps of the production process, and the presence of a certification (Valdés et al. 2024; Garofalo et al. 2024).

Additionally, a blockchain hyper ledger offers the distributor complete protection against possible fraud along the way. In farmers' associations, blockchain may also be used as an efficient way to attribute incomes correctly to each farmer in the association according to the additional value they bring in terms of sustainable growing practices for either the certification process or for the direct sale of inputs (Shafik 2024b). However, the implementation of blockchain technologies in agriculture shows some challenges, such as standardization and maximized interoperability. The participation of multiple stakeholders in the entire chain, such as software providers and independent system integrators, as well as consortiums and governmental organizations, is of top importance for precision technologies in agronomy as well. In addition, the application of blockchain in the agrifood industry has strong potential in tracing the history and sustainability of the food supply chain, showing endogenous added value through efficient production (De Santis and Cammarano 2024).

3.4.3 Vertical Farming

A technique that optimizes space and resource use in agriculture and sees crops grown in vertical layers. Employing a controlled environment, often using hydroponics, these farms can grow a wide variety of produce and employ lighting systems that create artificial 'days' to maximize growth throughout the year. With space for crops growing upwards, many more plants can be grown in smaller areas. Research and development in vertical farming is on the rise and has been shown to be more sustainable with respect to land use, transportation emissions, and resource conservation such as water. However, setting up these technologies can be expensive as they require nutrient solutions, growing mediums, and lighting. LED lighting is often used because it is efficient and does not require large ventilation systems (Sangaiah et al. 2024). Some of the captured light spectrum can be converted into energy for the plants, making it even more efficient. One of the more energy-intensive components in hydroponic systems is related to the mixing and pumping of the media and nutrient solutions. Automating processes such as this with sensors and technology can further reduce energy use and enable 'smart' solutions to monitor and automate vertical farming systems. By using less land and resources, vertical farming is potentially more sustainable. It also has the advantage of being scalable, with smaller vertical farms capable of being set up in urban centers to provide fresh produce to city dwellers (Valdés et al. 2024). With urbanization on the rise, vertical farming is increasingly seen as a viable solution to enhancing food security in cities. Such methods are sometimes known as urban vertical farming since crops can be grown in high-rise buildings with controlled environment agriculture. Vertical farms come in a variety of shapes and sizes, from larger farms housed in warehouse-like structures to locally integrated units producing food in deserted underground stations and car parks. This allows for produce to be grown close to or inside buildings, thereby reducing the carbon emissions associated with transporting food. The bulk of research into controlled environment agriculture is focused on high-value crops such as herbs, salads, and baby leaves (Carrillo Quiroga 2024).

3.5 Future Trends and Innovations in Agronomy

The Future of Agronomy report details a number of areas of innovation in agronomy. Key trends include the use of artificial intelligence and machine learning for analysis and decision-making, as well as the potential applications of developments in biotechnology. The report details how genetic engineering might help in developing crop varieties that are more pest-resistant and resilient in the face of changing climates. It adds that climate-smart strategies rather than specific technologies are likely to increase agricultural production in the future and identifies digitalization as a key trend in agriculture (De Santis and Cammarano 2024). To harness the full potential of such developments, the report emphasizes that interdisciplinary cooperation will be key to unlocking new frontiers and addressing society's most pressing challenges, like population growth or the protection of natural resources to feed the growing human population. The convergence of digital with technological and natural sciences will lead to new opportunities in precision agriculture, both in terms of production and protecting the environment. At the same time, smart technologies have the potential to manage natural resources efficiently, ensuring economic and ecological sustainability. Adopting new technologies to further advancements in agronomy will no doubt have its impacts on different facets. One collective impact should, however, be in contributing positively to the management of primary resources efficiently and sustainably. A study of future trends in agriculture took an even broader view (Xiao et al. 2024; Moreira et al. 2023). They predicted a shift from conventional agricultural systems with low diversity and a high degree of technical inputs to agricultural systems that are essentially based on high-functional agro-biodiversity.

3.5.1 Artificial Intelligence and Machine Learning

Technological advancements in agronomy are driven mainly by developments in the field of artificial intelligence and machine learning. The success and higher reliability of AI tools arise when several complementary data sources assist them and are capable of analyzing large volumes of data based on supervised learning institutions. With AI, big data collected from various remote sensing devices on satellites, in drones, or sent by powered sensors are being processed and monitored, which allows for better predictability, more precise allocation of resources, and optimization in the fields of horticulture, expansive agriculture, forestry, livestock, and fishery (Garofalo et al. 2024). Software and advanced electronic applications based on AI now enable comprehensive monitoring of seasonal bio- and agro-processes of crop development, pest, and disease occurrence, hence decreasing the uncertainty concerning disease and crop outbreak-related losses. Still in the experimental stage, AI applications are used in models to predict crop yields and, on a smaller scale, precision farming due to significant costs. Incorporation of AI and machine learning systems in estimation models of everyday agricultural operations and food systems is expected to yield economic benefits, such as reduction of pesticide application in pest control, development of more targeted and effective medicine and crop protection chemicals, and increase in energy efficiency (Singh et al. 2024). AI-based models are useful supplementary tools in forecasting and sustainable management systems, including precision agriculture. AI enables the creation of an early pest and crop disease predictive system with virtual simulations or forecasting models. The ultimate achievable tool will serve to provide specific site-relevant strategic tactics for growing healthy crops with low crop protection application (Reher et al. 2024).

Prediction models covering pollination services can come into use based on AI. Another application of AI lies in agricultural machinery automation. Predictive analytics plays a significant role in optimizing the supply chain by equipping businesses with insights that can be used to make decisions that positively impact demand management, inventory optimization, supply network planning, and transportation and warehousing efficiencies. Data privacy and ethical issues are related to the growing information power of AI and big data handling. Researchers must ensure that AI algorithmic bias excludes any stakeholders' discrimination in the application of AI algorithms. Data privacy and certain algorithmic biases should also be continuously analyzed and assessed in the application of AI in agronomy (Moreira et al. 2023). The adaptation of farmers to the already existing IoT and future AI knowledge or skills might be a possible obstacle. According to some farmers, they are skeptical of cooperation with off-farm companies due to privacy protection or cyberattack concerns. Compliance with data protection regulation issues is a major concern for industries since they are based on the knowledge of individual field characteristics. The hardware or, better, software price could also be a limitation in terms of technology awareness. The advantage of previously conducted research and tools has to be adapted to those next-generation tools by a team dedicated to such toolbox updating. Ongoing research and an assumption regarding how we should tackle the drawbacks as mentioned above are essential for coping with the data mass in agronomy, then predicting and planning better future strategic actions to sustainably produce crops, livestock, and any forest tree or crop species, including under global policy frameworks (Fontaine et al. 2024).

3.5.2 Biotechnology and Genetic Engineering

Biotechnology, and in particular genetic engineering, are important and exciting areas of modern agronomy. With biotechnology, it is possible to develop plants that produce an increased amount of food for human or animal consumption. The same plants can have higher quality and be more resistant to environmental stresses or new pests and diseases. The field of applications for genetic engineering is becoming wider and wider (Shafik, Tufail, De Silva, and Apong 2025a). It is an interesting scientific approach and a powerful tool for agronomists to use when there is a need in plant breeding programs to introduce into a commercial crop a trait that is not normally present and could be an asset in coping with climate changes and the need for sustainable food production. Several other approaches or transgenic lines ensure an increased output in crop yield. In general, area-wide adoption of genetically engineered crops has increased the yield per hectare and has lowered management expenses and losses over conventional crops (Kumar et al. 2024). Genetic engineering could also help in compensating for the loss of crop variation in selected traits and newly bred cultivars adapted to intensive agriculture. It is imperative to note that these new crops must exhibit no off-target or side effects. Safety tests are mandatory. Testing must be conducted across seasons, hosts, and environmental conditions. It is necessary to assess the potential benefits of new biotechnology-derived crops because it is forbidden to manipulate human germline cells or to affect characteristics with the exclusion of severe diseases. An area that should be further developed using biotechnology is the restoration of contaminated or eroded soils (Verdi et al. 2024).

Despite the many potential benefits of genetically modified organisms, they will only be adopted fully if there is an improvement in consumer perception and stringent regulations. Many people are still skeptical about the benefits of genetic modification and make purchasing choices based not solely on physical attributes. International regulations for conventional food and feed commodities, plant varieties, and transgenic organisms are still very heterogeneous and create additional problems in trade management. Future biotechnological research must investigate the development of an innovative and sustainable approach. Genetic engineering, however, is a tool to be placed at the disposal of modern, environmentally conscious agriculture. The global population will be close to 10 billion by 2050; to feed this huge population, we must increase our current food production. Biologic technologies will play a major role in sustainably reaching this goal since there is already much research demonstrating the potential of biotechnology in agricultural and environmental systems. The adoption rate of GM crops in developing countries has been more than double the rate in developed countries in recent years. Farmers worldwide embraced this modern approach quickly as it proved to be a better option than their local alternatives (Sangaiah et al. 2024; Moreira et al. 2023). In particular, GM crops offer increased flexibility and security to small, resource-constrained farmers. The future potential of GM is quite promising, with the launching of new GM crops that will contribute to sustainable agriculture.

3.5.3 Climate-Smart Agriculture

The idea for this approach has been derived from an event with the aim to discuss and suggest ways for a sustainable future and the next steps when it comes to modern agriculture. This delineates the agenda for development, resilience, and sustainability, including SDG attainment (Shafik, Tufail, Liyanage De Silva, and Awg Haji Mohd Apong 2025). Moreover, the theme reflects the holistic approach of abiotic and biotic elements of agricultural systems and farming and identifies key features that modern farming action should take into consideration to maximize resilience without neglecting the importance of yield. Climate-smart agriculture (CSAg) is developing rapidly in the world as a framework and approach to increase the sustainability and adaptive capacity of farming systems coping with climate change impacts. CSAg integrates the social and economic dimensions of sustainable agriculture and food system development in addition to ecosystem management (Fontaine et al. 2024; Garofalo et al. 2024). True to the assessment of global change, CSAg focuses on achieving sufficient global food production combined with natural resource preservation and resilience to climate change . Altering the practices per se is a tested option for carbon sequestration and decreased emissions, and the vast majority of climate-smart practices can also contribute to yield assurance and even greater crop yield.

A wide array of farm-level and in-field practices, such as planting improved crop varieties, agroforestry and cover crops, diversification in crops, use of soil amendments and weather information, and implementation of water and soil managing techniques, are advancing as a climate-smart approach at a global scale and would increase resilience and yield if undertaken. A more sustainable and adaptive farming system consequently engenders higher incomes and capacity to endure climate-related shocks and fluctuations, thus signaling the necessity of a holistically oriented approach that enhances productivity and resilience with pertinence to land management practices in a context framed by policy, social aspects, and local economies (Fawzi et al. 2024). A farmer and community-oriented approach to the establishment of new links, power allies, and distribution of governance inclusive with them is the unique asset of CSA, which distinguishes it from taking too excessively broad an approach.

Expected challenges in embracement are poor accessibility to financial lending and expensive insurance, but with local capacity building to both access CSA technologies and training due to the inaccessibility of internet services delivery, apart from easier community mobilization, policy frameworks would integrate with the goals of forestry, agriculture, and the development community (Shafik, Tufail, De Silva, and Apong 2025b). A set of climate-smart practices for smallholders in Kenya, sub-Saharan Africa, has a positive return on investment and increases the likelihood of children being better nourished and less likely to suffer from malnutrition. In all other contexts, no report of negative effects has been exerted by the agroforestry synergies, including benefits like enhanced pasture, increased crop yield, and improved soil with the provision of windbreaks and evapotranspiration in addition to natural capital development (Fernandez et al. 2024; Carrillo Quiroga 2024). The amended pastures consequently increase the daily intake of poor, climate-vulnerable, low-income households in Kenya.

3.6 Conclusion and Key Takeaways

Technology is crucial in modern agronomy to develop and utilize new methods, models, and theories that support and promote more efficient and productive agricultural systems. At the same time, reaching higher sustainability in the agricultural sector is also a key driver. Sustainability concerns the needs of present generations in particular with regard to food security, energy security, mitigation of and adaptation to climate change, and more pressing environmental concerns. In modern societies, sustainability reports are often instrumental in fostering discussions about the future lines of research and innovation within agriculture and related activities. Agronomy and agroecology, as the mother disciplines of agricultural sciences, must sectorally develop and communicate their contents with the aim of fostering discussions about the future of plant, livestock, forest, and fish research, where a horizon scanning approach is possible. The main conclusions and key issues identified can be summarized as follows:

- The main challenge is to bridge technology and sustainability, as these two aspects are often found to be disjunctive. Bridging them by means of a collaborative commitment of scientists, stakeholders, farmers, and other actors is crucial.
- Sustainability is directly related to the production of food and other issues, such as climate change and disaster risk reduction, whose effect is becoming stronger; it is, however, also needed to improve territorial stability (managing better the Maximum Sustainable Yield through approaches such as Integrated Pest Management, agroecology, and permaculture).
- What is actually lacking is the desire to exploit the vast potential of agricultural development based on sustainability (diversity, precision, closed cycles, null waste, carbon sequestration, water savings, integrated and territorial empathy).
- The modern world is barely contaminated by sustainable agriculture and agronomy; the number of truly sustainable agricultural systems is actually limited, while many agricultural systems with high inputs are harmful to the environment as well as unprofitable for producers; therefore, they are also not sustainable. This is a gap that can only be filled through an operational and pragmatic focus.

References

Adeleke, Adeyemi Adegoke. 2024. "Technological Advancements in Cotton Agronomy: A Review and Prospects." *Technology in Agronomy* 4 (1). https://doi.org/10.48130/tia-0024-0005.

Carrillo Quiroga, Perla. 2024. "Perception of Agronomy Students on Agrochemical Pollution." *Revista Internacional de Contaminacion Ambiental* 40. https://doi.org/10.20937/RICA.54937.

Chilaka, Chibuzo, Alex J. Rinehart, Haoying Wang, and Frank A. Ward. 2024. "Sustaining Aquifers Hydrologically, Economically, and Institutionally: Policy Analysis of the Ogallala in New Mexico." *Science of the Total Environment* 921. https://doi.org/10.1016/j.scitotenv.2024.170727.

Damini, Prof. Miss Dhenge, Miss Kamble Ashwini, and Miss More Dipali. 2024. "Research Paper on What Is Artificial Intelligence and Its Applications." *INTERANTIONAL JOURNAL OF SCIENTIFIC RESEARCH IN ENGINEERING AND MANAGEMENT* 08 (02). https://doi.org/10.55041/ijsrem28575.

Etienne, Lucas, Pierre Franck, Adrien Rusch, and Claire Lavigne. 2024. "Apple Pest and Pathogen Reduction in Landscapes with Large Patch Size and Small Area of Orchards: A National-Scale Analysis." *Landscape Ecology* 39 (2). https://doi.org/10.1007/s10980-024-01830-2.

Fawzi, Nurul Ihsan, Basuki Sumawinata, Suwardi, Annisa Noyara Rahmasary, Ika Zahara Qurani, Raihan Garin Naufaldary, Ratu Nabillah, Heru Bagus Palunggono, and Budi Mulyanto. 2024. "Integrated Water Management Practice in Tropical Peatland Agriculture Has Low Carbon Emissions and Subsidence Rates." *Heliyon* 10 (5). https://doi.org/10.1016/j.heliyon.2024.e26661.

Fernandez, Romain, Loïc Le Cunff, Samuel Mérigeaud, Jean Luc Verdeil, Julie Perry, Philippe Larignon, Anne Sophie Spilmont, et al. 2024. "End-to-End Multimodal 3D Imaging and Machine Learning Workflow for Non-Destructive Phenotyping of Grapevine Trunk Internal Structure." *Scientific Reports* 14 (1). https://doi.org/10.1038/s41598-024-55186-3.

Fontaine, Sébastien, Luc Abbadie, Michaël Aubert, Sébastien Barot, Juliette M.G. Bloor, Delphine Derrien, Olivier Duchene, et al. 2024. "Plant–Soil Synchrony in Nutrient Cycles: Learning from Ecosystems to Design Sustainable Agrosystems." *Global Change Biology*. https://doi.org/10.1111/gcb.17034.

Garofalo, Pasquale, Marco Parlavecchia, Luisa Giglio, Ivana Campobasso, Alessandro Vittorio Vonella, Marco Botta, Tommaso Tadiello, et al. 2024. "ARMOSA Model Parametrization for Winter Durum Wheat Cultivation under Diverse Cropping Management Practices in a Mediterranean Environment." *Agronomy* 14 (1). https://doi.org/10.3390/agronomy14010164.

Kalinaki, Kassim, Wasswa Shafik, Tar J. L. Gutu, and Owais Ahmed Malik. 2023. "Computer Vision and Machine Learning for Smart Farming and Agriculture Practices." In . https://doi.org/10.4018/978-1-6684-8516-3.ch005.

Kumar, Rohit, Arvind Bhardwaj, Lakhwinder Pal Singh, and Gurraj Singh. 2024. "Environmental and Economical Assessment of Maize Cultivation in Northern India." *Process Integration and Optimization for Sustainability* 8 (1). https://doi.org/10.1007/s41660-023-00358-3.

Moreira, S. G., G. Hoogenboom, M. R. Nunes, and P. A. Sanchez. 2023. "Circular Agriculture Practices Enhance Phosphorus Recovery for Large-Scale Commercial Farms under Tropical Conditions." *Journal of Agricultural Science* 161 (6). https://doi.org/10.1017/S0021859624000042.

Reher, Thomas, Cas Lavaert, Brecht Willockx, Yasmin Huyghe, Jolien Bisschop, Johan A. Martens, Jan Diels, Jan Cappelle, and Bram Van de Poel. 2024. "Potential of Sugar Beet (Beta Vulgaris) and Wheat (Triticum Aestivum) Production in Vertical Bifacial, Tracked, or Elevated Agrivoltaic Systems in Belgium." *Applied Energy* 359. https://doi.org/10.1016/j.apenergy.2024.122679.

Rojas, Diana Alejandra Méndez. 2024. "Cytogenetics of Maize in Mexico as a Field of Transnational Exchange: The Case of Takeo Ángel Kato Yamakake." *Plants People Planet* 6 (5). https://doi.org/10.1002/ppp3.10483.

Sangaiah, Arun Kumar, Fan Nong Yu, Yi Bing Lin, Wan Chi Shen, and Akashdeep Sharma. 2024. "UAV T-YOLO-Rice: An Enhanced Tiny Yolo Networks for Rice Leaves Diseases Detection in Paddy Agronomy." *IEEE Transactions on Network Science and Engineering.* https://doi.org/10.1109/TNSE.2024.3350640.

Santis, Michele Andrea De, and Davide Cammarano. 2024. "Agronomic Management Factors Impacting Yield, Quality Stability, and Environmental Footprints of Barley in a Mediterranean Environment." *Field Crops Research* 309. https://doi.org/10.1016/j.fcr.2024.109334.

Shafik, Wasswa. 2024a. "Barriers to Implementing Computational Intelligence-Based Agriculture System." In *Computational Intelligence in Internet of Agricultural Things*, 193–219. Springer. https://doi.org/10.1007/978-3-031-67450-1_8.

Shafik, Wasswa. 2024b. "Biosensor-Based Drones Anomaly Detection Integration for Sustainable Agriculture Development." In *Achieving Food Security Through Sustainable Agriculture*, 283–316. IGI Global. https://doi.org/10.4018/979-8-3693-4240-4.ch012.

Shafik, Wasswa. 2025a. "Sustainable Agriculture and Diet for Urban Development." In *Sustainable Smart Cities and the Future of Urban Development*, 165–92. IGI Global Scientific Publishing.

Shafik, Wasswa. 2025b. "Sustainable Agriculture and Diet in the Metaverse Era: Sustainable Farming and Nutritious Eating in the Metaverse." In *Food in the Metaverse and Web 3.0 Era: Intersecting Food, Technology, and Culture*, 159–84. IGI Global Scientific Publishing.

Shafik, Wasswa, Ali Tufail, Rosyzie Anna Awg Haji Mohd Apong, and Liyanage Chandratilak De Silva. 2024. "Internet of Things for Smart

Agricultural Practices." In *Internet of Things Applications and Technology*, 190–217. Boca Raton: Auerbach Publications. https://doi.org/10.1201/9781003458401-13.

Shafik, Wasswa, Ali Tufail, Chandratilak De Silva Liyanage, and Rosyzie Anna Awg Haji Mohd Apong. 2023. "Using a Novel Convolutional Neural Network for Plant Pests Detection and Disease Classification." *Journal of the Science of Food and Agriculture*. https://doi.org/10.1002/jsfa.12700.

Shafik, Wasswa, Ali Tufail, Chandratilak Liyanage De Silva, and Rosyzie Anna Awg Haji Mohd Apong. 2025. "A Novel Hybrid Inception-Xception Convolutional Neural Network for Efficient Plant Disease Classification and Detection." *Scientific Reports* 15 (1): 3936.

Shafik, Wasswa, Ali Tufail, Abdallah Namoun, Liyanage Chandratilak De Silva, and Rosyzie Anna Awg Haji Mohd Apong. 2023. "A Systematic Literature Review on Plant Disease Detection: Motivations, Classification Techniques, Datasets, Challenges, and Future Trends." *IEEE Access* 11:59174–203. https://doi.org/10.1109/ACCESS.2023.3284760.

Shafik, Wasswa, Ali Tufail, Chandratilak Liyanage De Silva, Rosyzie Anna Awg Haji, and Mohd Apong. 2025. "The Role, Application, and Impact of Artificial Intelligence in the Agriculture Industry." In *Future Tech Startups and Innovation in the Age of AI*, 36–60. CRC Press.

Shafik, Wasswa, Ali Tufail, Chandratilak De Silva Liyanage, and Rosyzie Anna Awg Haji Mohd Apong. 2024. "Using Transfer Learning-Based Plant Disease Classification and Detection for Sustainable Agriculture." *BMC Plant Biology* 24 (1): 136. https://doi.org/10.1186/s12870-024-04825-y.

Shafik, Wasswa, Ali Tufail, Liyanage Chandratilak De Silva, and Rosyzie Anna Awg Haji Mohd Apong. 2025a. "A Lightweight Deep Learning Model for Multi-Plant Biotic Stress Classification and Detection for Sustainable Agriculture." *Scientific Reports* 15 (1): 12195.

Shafik, Wasswa, Ali Tufail, Liyanage Chandratilak De Silva, and Rosyzie Anna Haji Mohd Apong. 2025b. "An Enhanced Deep Convolutional Neural Network for Plant Disease Detection and Classification: Elevating Sustainable Agriculture." In *Artificial Intelligence and Data Science for Sustainability: Applications and Methods*, 297–322. IGI Global Scientific Publishing.

Singh, Kulpreet, Hardeep Singh, Peter Omara, Nkem J. Nwosu, Jennifer Bearden, Ravinder Singh, and Lakesh K. Sharma. 2024. "Potassium Deficiency in Cotton: Causes, Symptoms, and Considerations." *EDIS* 2024 (1). https://doi.org/10.32473/edis-ag475-2024.

Valdés, Cristian, Carlos M. Silva-Neto, Claudio Salas, María Teresa Muñoz-Quezada, Maribel Mamani, and Rodrigo O. Araujo. 2024. "What Can Insects Tell Us about the Use of Pesticides? The Detection and Quantification of Chlorpyrifos via QuEChERS and HEADSPACE-SPME Methods Using GC/MS/MS." *Agronomy* 14 (1). https://doi.org/10.3390/agronomy14010021.

Verdi, Leonardo, Anna Dalla Marta, Simone Orlandini, Anita Maienza, Silvia Baronti, and Francesco Primo Vaccari. 2024. "Evaluation of Biochar Addition to Digestate, Slurry, and Manure for Mitigating Carbon Emissions." *Agriculture (Switzerland)* 14 (1). https://doi.org/10.3390/agriculture14010162.

Xiao, Fengjin, Qiufeng Liu, Yun Qin, Dapeng Huang, and Yaoming Liao. 2024. "Agricultural Drought Research Knowledge Graph Reasoning by Using VOSviewer." *Heliyon* 10 (6). https://doi.org/10.1016/j.heliyon.2024.e27696.

CHAPTER FOUR

SUSTAINABLE AGRONOMY AND BOTANICAL PESTICIDES: PERSISTING AND NEW CHALLENGES AND PROSPECTS

Abstract

Botanical pesticides are attracting worldwide attention, both in traditional pest control and in crop protection programs, including integrated pest management for insects and diseases. Farmers around the world have discovered the significance of botanical pesticides, which have, to a great extent, relieved them of the hazardous effects of synthetic pesticides through their safe use in pest management. The magnitude of Dutch Elm Disease in Europe has accelerated research on botanical pesticides for the organic management of this disease. Allelochemicals are also exploited in the synthesis of new botanical pesticides that are biodegradable and environmentally friendly, with less or no residual effects. This harmonizes with the sustainability concept of modern agriculture. This has also attracted the attention of population groups in various parts of the world since the turn of the new millennium, marking their deep concern about the use of synthetic pesticides and their implications for human health and environmental protection. This has led to a growing trend to encourage the organic movement in all spheres of life. This aims to appraise the efficacy of existing chemical constituents and the possible *phytocompounds* for the plant protection process, constraints in adopting botanical pesticides, and their prospects. It critically evaluates the chemical classes of the *phytocompounds* that could contribute to plant health, offering protection against pests, diseases, and weeds, and presents potential applications. We describe the transition and evolution of traditional approaches in pest management based on plant extracts—the use of crude plant extracts, plant-derived pesticides, and semi-synthetic and synthetic pesticides, with specific reference to botanical insecticides. We present novel approaches in the extraction and isolation of *phytocompounds*, their in vitro and in vivo applications, and considerations for core plant health-enhancing technology and formulation technology. Its relevance in addressing modern agricultural challenges and the prospects and path forward, particularly in the context of the changes introduced after the COVID period, is further discussed. The essay stresses the need for finer resolution of pesticide risk assessment and to advance awareness of established merits as well as future application options at the landscape and ecosystem levels.

Keywords: Biopesticide Innovation, Botanical Pesticide Efficacy, Eco-Friendly Pesticides, Natural Pest Control, Pest Resistance Challenges, Plant-Derived Pesticides, Sustainable Pest Management, Toxicity and Environmental Impact

4.1 Introduction

The antecedents of botanical pesticides, or in other words, the use of natural plant compounds for pest control, are as old as agriculture. Diverse historical developments in various parts of the world during the last couple of millennia and up into the 19th century and their scientific analysis have been reported. Cultures such as the Greek, Roman, Chinese, German, Indian, and many others at various times have developed many pesticides from botanical raw materials. In the absence of rigorous sanitation and hygiene and of modern communication and transport facilities, crops protected against their indigenous pests using botanical pesticides evolved to be region- and country-specific and were influenced both by the kind of crops grown in each of the regions and by the cultural background, including the religion, of each of the peoples concerned (Yeshiwas et al. 2024). Traditionally, the development of plant protection materials was driven in each instance by demand or need due to the lack of availability of other means of control. Economic constraints and, in many instances, strong competition from botanical farm-made crop protection systems that had evolved over centuries had to be taken into account in the development of products and technologies. The historical overviews illustrated in these chapters clearly show how agricultural practices can remain unchanged over centuries, have little continuity between societies with different cultures, or dissipate within a compartmentalized civilization that does not communicate with others. While with the emergence of synthetic chemistry, the quest for an understanding of the mode of action became less and less important, the broad spectrum and quick activity of botanicals were eagerly sought, and in that much, interest in the development of the compounds and their mode of action ran parallel with that of the chemical compounds. The commercial and regulatory challenges for botanicals were equally as great in the past as they are now (Yates et al. 2024; Ankur et al. 2024).

Botanical pesticides are plant-derived pesticides categorized into distinct classes based on the target pests or vectors and uses. The general classification includes botanical insecticides, botanical fungicides, and botanical herbicides. Botanical insecticides target insects and related arthropods, including ticks and spiders, which are commonly pests of agricultural and horticultural crops and plantation crops (Shafik 2025a). They have shown a good inhibitory effect against vectors like deer ticks, mosquitoes, and cockroaches, demonstrating commercial potential for industrial and environmental applications. Common botanicals utilized as insecticides extracted from the plants are described. The details of the nature of action in synthetic pesticides and their toxicity may be referred to in related publications (Bi et al. 2024). Botanical fungicides kill or prevent the growth of fungi. These can be of interest for several reasons. The natural compounds may be utilized to prevent post-harvest spoilage of plant materials, wood, grain, etc., without the concern of human or animal toxicity. They also have a potential role in the management of fungal diseases of plants, including those caused by true fungi, oomycetes, and *fungi imperfecti*. Commonly used plant derivatives with good inhibitory potential are described. Botanical herbicides refer to unwanted plants that reduce the yield of agronomic or horticultural crops, deteriorate product quality, or block harvesting. According to the mode of action at the target plant tissues, these are classified into desiccants, growth retardants, chlorophyll inhibitors, and photosystem II inhibitors (Rzyska, Stuper-Szablewska, and Kurasiak-Popowska 2024). Some of these bio-herbicides with good potential also contain insecticidal properties and can kill or inhibit the development of arthropods.

4.1.1 Botanical Insecticides

The use of plants to fight or repel insects (or arthropods) is as old as agriculture; still, modern use of natural products from plants for insect management is regarded as negative. Insecticidal properties of plants, caused by a blend of active ingredients, some of which are toxic to insects (especially when they eat such plants), and others that are repellent, could be lethal when the concentration exceeds critical levels. Unlike synthetic insecticides, which are composed almost entirely of pure toxic chemicals that are lethal to insects, botanical insecticides usually contain a substantial proportion of compounds that are not toxic to insects but affect their physiology, including behavior, growth, development, metabolism, and resistance (Haruna S. G., Yahuza L., and Tijjani I. 2024). Many active ingredients of plants could serve as insecticides because they poison almost every enzyme and physiological system present in insects, as illustrated by common pesticides Fig 4.1.

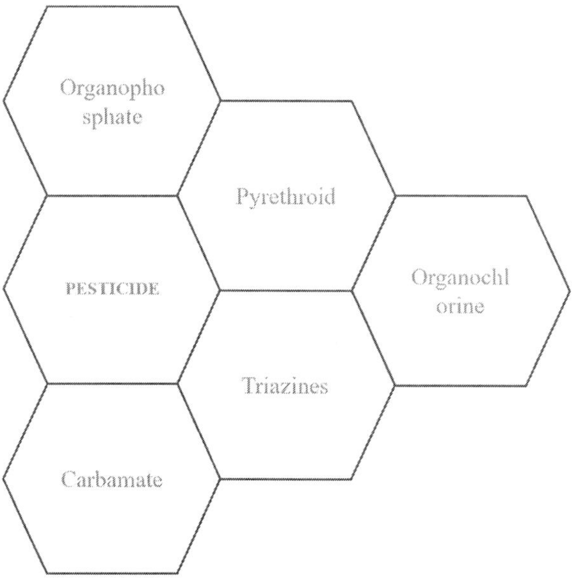

Fig. 4-1 Pesticides

A vast spectrum of plants known and used traditionally for pest control is suitable for preparing and commercializing insecticides based on neem oil and pyrethrins. The first could be toxic to parasitic wasps, whereas the fast detoxification of the second by aphids, whiteflies, thrips, etc., would lead to the eventual death of the unwelcome guest. Having diverse yet specific modes of action, particularly disruptors of key cellular systems and insect growth or development, these natural products have great potential for use in insect pest management. In real agricultural conditions, they have proven their worth as being effective in many cases. Botanical insecticides are suitable for them in biology and toxicology, as economic and even sociological considerations. Multiple applications of botanicals in tank mixes on the same field, under the premise of minimal risk use, may also underscore resistance management (Sa'adah et al. 2024). The future appears bright for botanical insecticides in general and those of neem and pyrethrin in particular. Although their potential as protectants of plants against possibly every pest is known and also as repellents against annoying insects seems only partly exploited, the predominant focus has been on toxicity against pests, especially the full range of ants, stored seed beetles, and other *intraductal*-cavity-feeding insects, root-feeding larvae, aphids, and thrips that have been fumigated. Such an expanded focus would also respond to progress in mechanization that brought with it on-farm storing for later sale under contracts and at market premiums (Arora, Choudhir, et al. 2024).

Plant products, such as essential oils, have known insecticidal properties and are effective in controlling pests in various aspects. Unlike many chemical pesticides, plant oils enrich the environment and have numerous other uses besides antifungal and insecticidal properties. Botanical insecticides are generally considered safe for humans and the environment, but they have a problem in being selective, and toxicity depends on, among other things, the ratio of toxic to non-toxic compounds present. The naturally occurring insecticides have a rapid decomposition rate and show low mammalian toxicity (Lansana Abou BANGOURA et al. 2024). Insects may develop resistance to botanical insecticides by reducing the techniques that minimize exposure. Various complexes in these insecticides consume insect metabolic enzymes and various co-factors. PH, organic solvents, temperature, light, etc influence the activity of these complexes. Insects can develop resistance to botanical insecticides by various mechanisms, of which target site insensitivity is by far the most frequently reported (Indermaur et al. 2024). Environmental factors play a major role in increasing the efficacy of plants as botanical insecticides.

4.1.2 Botanical Fungicides

An essential group of botanical pesticides since fungal plant diseases cause extensive yield and qualitative losses. Common antifungal metabolites produced by plants are secondary compounds, so the huge source of plant antifungal agents is enrichment and diversity for research and application of fungicides. Some popular plants and active plant-derived fungicides are garlic, essential oils, and plants such as chitosan, eucalyptus, litchi, clove, tea tree, neem, and hyssop. Both in vivo and in vitro modes of action are investigated for these botanical fungicides. Research and technology have elaborated the extraction, preprocessing, formulation, and application mechanism of botanical fungicides so effectively, which advances the effectiveness of controlling multi-pathogens and further steadily controls the disease at the field level (J. Ma et al. 2024). Formulation and application are already in advancement for botanical pesticides to upgrade their action nature, activity time, action range, and biosecurity. Botanical fungicides, including those active against bacteria and viruses, tend to have different levels of effects on pathogens from different life cycles in comparison to chemicals, including some that reduce the number of fungal spores, destroy *myxophytes*, and hyphae but do not inhibit spore germination, only sometimes stopping the incidence of the disease and leaving a long infection window. However, in many cases, a large application dose may have harmful or phytotoxic effects on natural beneficial microorganisms or macro-organisms (Modafferi et al. 2024).

Some plants have rapid evaporation, causing the remaining scented crop to be unsalable, and the harvest period should be considered. Some plants that are not studied with bioassay and field experiments can be alternatives for sustainable agriculture to use as bactericides (Shafik and Kalinaki 2024). Many plants should be optimized for future field use based on the different stages and requirements in the life cycle of the pathogens, such as bactericides that can destroy the cell wall, penetrate the spore wall, and simultaneously suppress the germination rate of spores. For reasons of lack of specificity and potential phytotoxicity, more acidic than chlorogenic acid should not be a good choice in plant extract and formulation, with low pH mostly including across-the-board effect agents such as weak acids, weak bases, and surface-active substances (Murugesan et al. 2024; Shafik 2024a).

The other challenge is that it currently takes some time for the success story of the botanical-based product to be released, so safe and long-lasting products should encourage the manufacturer to support the sustainable, cost-effective steps in re-registering and labeling them for different crop zones, which is in reverse deterred due to the cost itself until the proper guidelines are clear. The consumer has shown increased interest in biological control systems away from the chemical control system, so the market share of edible crops is mostly targeted for botanical fungicides, which have shown promise after a few years. When the concentration of public-private partnerships is parallel to the global market trend, the exploration of newly available botanical fungicides is expected to provide the best results towards niche or orphaned crops that are more affected by cost or yield. All of these have brought about a decrease in the cost of botanical-based products in the field, which is located in Europe and can achieve high productive capacity. Organic, herb, and extraction residues, as well as officer byproducts, can be used to create an integrated agro-scheme (Shafik, Tufail, De Silva, Haji, et al. 2025). Some samples are reported in the field for utilization in vineyards and pome fruit crops.

4.1.3 Botanical Herbicides

In farming, unwanted plant species are known as weeds. To control these, another group of classifiers called herbicides is used. Like synthetic or classic herbicides, botanical herbicides can also be used to control weeds. Clove oil, white or horticultural vinegar, and acetic acid are commonly used for this purpose, suggesting that an extract from the tissue of one of them can maintain allelopathic potential. Low solubility in water, rapid decomposition in the environment, and biodegradability during soil exposure are some of the key considerations in botanical herbicide use (Benson et al. 2024). Both of these criteria demonstrate the environmentally friendly nature of the botanical herbicide. Several factors, including extrinsic and economic concerns, demonstrate a major market shift from chemical to natural bio-based products. Although very little research has been published, it has demonstrated that the new total refractory metabolite mixture of crops has the potential to control weeds. The emergence of organic crop cultivation, where total production is locally marketed, does or will restrict the use of synthetic products (Bi et al. 2024).

It is important not to develop products that might prevent the possibility of increasing high-value food production because other products are not approved. In combination with other non-harmful control methods, natural herbicides are useful in the construction of sustainable farming. It is suggested that safer natural products in combination with resources are used in different levels of production for the totally unsuitable phase herbicide chemical reaction in organic farming (Yeshiwas et al. 2024). This area of work still needs further development. If conventional farming is no longer economically viable, there will be a move to organic production. The market for biopesticides is relatively stable, while the development of new synthetic biopesticides is on the rise. With recent advances in science, including genomics and structural biology, the development of the biopesticide market is changing. It is important to keep abreast or be ahead of current trends to take advantage of niche markets and new commercial opportunities; common Botanical Pesticides are presented in Table. 4.1.

Table 4-1 Common botanical pesticides and their sources

Botanical Pesticide	Botanical Pesticide	Active Ingredient	Active Ingredient	Mode of Action	Reference
Saponins	Soapwort, yucca	Saponins	Aphids, mites	Detergent action damages membranes	(Yeshiwas et al. 2024)
Menthol	Mint leaves	Menthol	Mosquitoes, aphids	Repellent and neurotoxic effects	(Yates et al. 2024)
Eugenol	Clove oil	Eugenol	Beetles, flies	Affects neurotransmission	(Ankur et al. 2024)
Linalool	Lavender and mint	Linalool	Linalool	Nervous system disruptor	(Bi et al. 2024)
Azadirachtin	Neem seeds	Azadirachtin	Locusts, aphids	Locusts, aphids	(Rzyska, Stuper-Szablewska, and Kurasiak-Popowska 2024)
Quassia	Quassia tree bark	Quassin	Caterpillars, aphids	Antifeedant and growth inhibitor	(Haruna S. G., Yahuza L., and Tijjani I. 2024)
Antifeedant and growth inhibitor	Various plants (e.g., mint, lavender)	Terpenes	Terpenes	Repellent, disrupts reproduction	(Sa'adah et al. 2024)
Citrus Oil	Citrus peel	Limonene	Ants, fleas, aphids	Ants, fleas, aphids	(Arora, Choudhir, et al. 2024)
Garlic Extract	Garlic (*Allium sativum*)	Sulfur compounds	Aphids, mites, whiteflies	Repels pests, inhibits feeding	(Lansana Abou BANGOURA et al. 2024)
Capsaicin	Chili peppers (*Capsicum spp.)*	Capsaicinoids	Mammals, certain insects	Irritates sensory neurons	(Indermaur et al. 2024)

Sabadilla	Sabadilla lily seeds	Veratridine	Caterpillars, beetles	Affects sodium ion channels	(Murugesan et al. 2024)
Nicotine	Tobacco (*Nicotiana spp.*)	Nicotine alkaloids	Aphids, thrips	Aphids, thrips	(Kadanakuppe Thammayya et al. 2024)
Rotenone	Derris and Lonchocarpus roots	Rotenone	Caterpillars, beetles	Inhibits mitochondrial respiration	(Eisen 2024)
Pyrethrum	Chrysanthemum flowers	Pyrethrins	Mosquitoes, flies, beetles	Disrupts nervous system	(Arora, Naaz, et al. 2024)
Neem Oil	Neem Tree (*Azadirachta indica*)	Azadirachtin	Aphids, whiteflies, mites	Aphids, whiteflies, mites	(Senewe et al. 2024)

4.2 Historical Overview of Botanical Pesticides

Given the significance of plant-borne compounds and health parameters, this paper draws attention to the persisting and new research challenges in the ecology of plant-derived biocides in general and a focus on biopesticides, in particular, that are being pursued in some pheromone chemistry for the modern agricultural economy. Throughout the ages, man has used plants to control pests in agricultural practices. Crops were mixed with strongly scented herbs, the floors of the barns and the stalls on which they were threshed were sprinkled with the same scented herbs, or herbs were burnt or fumed to fumigate the plant produce. Traditional knowledge of herbs and plant extracts with pesticidal properties passed down from generation to generation is seen to be the basis for botanical pesticides used these days (Arora, Choudhir, et al. 2024). The use of several plant and animal products, including pesticides, for the treatment of cattle, as well as pests like blood-sucking insects, *maggots*, ants, and locusts, has been documented. In traditional texts, we find many references to the use of bark, roots, leaves, and fruits of plants used as pesticides. During the 1940s and '50s, the hopes were high that the primary sources of plant extracts with pesticidal properties would provide lead compounds for the isolation of botanical pesticides of commercial interest (Lansana Abou BANGOURA et al. 2024). Many more plants with repellent, fumigant, and insecticidal properties, however, were tested between 1960 and 1980. It was pointed out that local and regional variations in ethnobotany, culture, and agriculture need cautious interpretation of test results. Cultural factors like traditional knowledge of plant extracts and the availability of wood to fire for rendering great quantities of extracts at a time influence the use of plant extracts. For example, the traditional use of certain branches and leaves sprayed along the boundaries of food storage buildings and traditional granaries is practiced by farmers in specific regions. Also, separating nuts from shells with a winnowing fan and seasoning green grams can be done easily if the stems and leaves of certain plants are burnt. Nuts remain intact, and the dry, brittle leaves do not rot, making it convenient to nibble grain as and when desired. Different ways of life are different and do not necessitate the use of such plant parts (Modafferi et al. 2024). How could we expect certain plant extracts tested against pests to be worked out under different intentions?

4.3 Classification of Botanical Pesticides

Botanical pesticides are mainly derived from plants and some organisms to protect them from harmful tissues. Based on the classification, botanical pesticides can be classified into insecticides, rodenticides, nematicides, miticides, fungicides, growth regulators, anti-feedants, attractants, herbicides, and molluscicides. In addition, they are grouped based on the plants they are derived from. Linking phytochemicals, a class of botanical pesticides, are classified according to their active constituents, such as pyrethrins, pyrethroids, neem oil, azadirachtin, essential oils, ammonia, and proteins. The first class of botanical pesticides is pyrethrins and pyrethroids, which are naturally derived from the flowers of Chrysanthemum sp. They have strong insecticidal properties with low mammalian toxicity, low environmental contamination, and degrade rapidly (Yeshiwas et al. 2024). Fog mixtures of active and dead insects exhibit fast knockdown and quick fly-back. The second class involves neem-based pesticides. They act against various pests in the target range of significant activity and are known to show broad-spectrum activity at higher concentrations. Chemically, neem comprises more than 100 active molecules, with azadirachtin being predominant. Azadirachtin inhibits steroid, ecdysone, and juvenile hormones and acts as an anti-feedant against chrysomelids, aphids, caterpillars, and some other pests. It is also antifungal against powdery mildews, downy mildews, root rot pathogens, and soil disease fungi. The third class includes rotenone and *ryania*, which have strong activities against various pests, including mites, sucking and chewing pests, and soil pests, both at contact and systemic ranges of activity. Rotenone is well known for its application in tea, coffee, and potatoes (Bi et al. 2024). Ryania is less known. The fourth class consists of pyrethrins and advanced generation pyrethroids from Kapani, which have a non-cumulative paralyzing action aimed at the insect nervous system caused by over-excitation of nerve cells, preventing normal responses or recovery. They have rapid knockdown with a long period of activity. They are used in agriculture and public health fields. The fifth class is classified as miscellaneous botanicals and includes *Abrus*, *Pongamia*, *Quassia*, pesticide ingredients, repellent soap, Malaysian Nephelium lappaceum, *Afzelia xylocarpa*, *Callicarpus* oil, edible oils, sodium chloride, Brugmansia, and Acorus. All provide acceptable control of selected pests at higher lethal doses, showing an individual target range of activity (Rzyska, Stuper-Szablewska, and Kurasiak-Popowska 2024).

4.3.1 Pyrethrins and Pyrethroids

This content, which contains pyrethrin 1 and pyrethrin 2, are natural products that are esters of monoterpenoid alcohol, *rethrolone*, and jasmonic acid from the genera of flowers of Chrysanthemum and have the R configuration. The pyrethroids are classified into two categories: type I pyrethroids and type II pyrethroids. Pyrethroids (type I and type II) represent synthetic derivatives; synthetic pyrethroids are developed by modifying natural compounds and have the potential to satisfy pest control efficacy requirements. These compounds are neurotoxins and act primarily on the nervous system of insects, especially on sodium channels. The mexiletine binding site between sodium channels and the receptor is the action target of pyrethrins/pyrethroids. Intervening two P-loop regions of domain IV and domain III, voltage-gated sodium channels are blocked. However, at low doses, deterrence rather than insecticidal effects has been observed in some insects (Haruna S. G., Yahuza L., and Tijjani I. 2024). The dose-response pattern of pyrethrins/pyrethroids suggests use mainly as insecticides rather than repellents. The pyrethrins are known for a rapid knockdown in houseflies and minimal residual toxicity. The effectiveness of permethrin products in agricultural settings has been shown in earlier field trials. Field trials also revealed relatively good efficacy of other pyrethroids, including beta-cyfluthrin, in cotton and corn. Nevertheless, adaptation through insect resistance is a known phenomenon in the case of resistance development to pyrethroids. Such considerations can nullify some of the advantages, albeit not necessarily in the area of public health, which pyrethrins/pyrethroids provide. Adverse environmental impacts of pyrethrins/pyrethroids on some non-target organisms, for example, bees, fish, hydrophytes, and microbes, have reduced their agricultural use. To minimize the chances of resistance of target pests to pyrethrins/pyrethroids, mixtures and multi-component semi chemicals that attract and stimulate feeding preferences can be used in traps, bait stations, and lures (Sa'adah et al. 2024). Regulatory considerations of natural pyrethrins have been largely unaffected by increasing contents of synthetic pyrethroids manufactured using allethrin. *Pyrethrin* regulations differ from pyrethroids in countries outside the United States. The application and usage of pyrethrins have been found to be safe, provided general standards of handling/loading safety are closely adhered to, and pre-bottled ready-to-use materials are chosen to avoid high exposures (Shafik 2025b). Concentrations range from 0.5–5% in aerosols and 0.05% in pet shampoos and a rose-petal rinse. Further work to limit human exposure has suggested additional formulations and encapsulated products. These changes could protect against some of the environmental detriments of current products: in some scenarios, encapsulated microspheres have approached release rates of the raw oleoresin compounds (J. Ma et al. 2024). To reduce reliance on these now fourth-generation synthetic pyrethroids, innovative new peppermint copper chitosan coatings could be used as synergists to pyrethrins, causing resistant head lice to die, as some effects of these pesticides are illustrated in Fig 4.2.

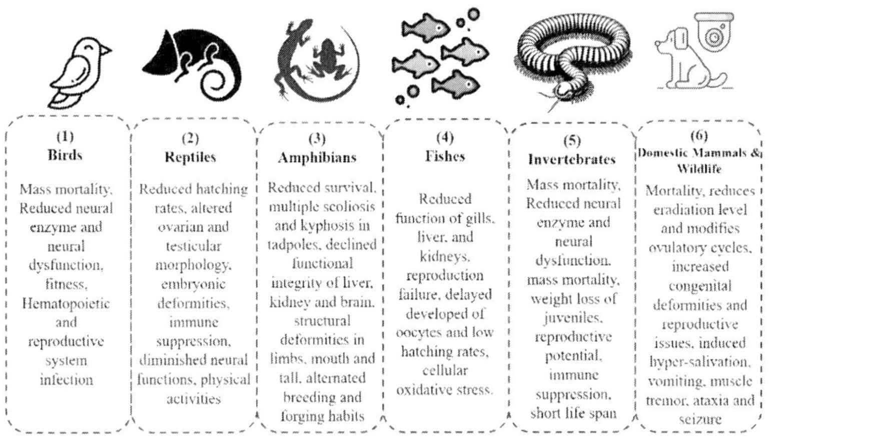

Fig. 4-2 Effects of pesticides

4.3.2 Neem-based Pesticides

Azadirachta indica, better known as neem, is a tropical evergreen tree in the mahogany family, Meliaceae. It is native to the Indian subcontinent and has been introduced to other continents. Pesticides based on neem seeds and leaves have been used in India, particularly against the desert locust and predatory grasshoppers, for over two millennia. The active ingredient of neem tree pesticides is now understood to be a complex onion-shaped molecule—*azadirachtin*—that has multiple modes of action, affecting feeding, molting, mating, and egg laying. At low doses, it interferes with pest growth and development; at medium doses, it acts as a "repellent," and at high doses, it leads to pest death. In traditional medicine, neem is known to have over 20 other pharmacologically active molecules (Kadanakuppe Thammayya et al. 2024). The concept of using these natural products of the neem tree with multiple modes of action in an integrated pest management system translates well to sustainable agriculture.

The development of neem-based pesticides presently focuses on the production of more active ingredients using biotechnology techniques. Over the past three decades, numerous research papers have been published revealing the variety of formulations and application methods that have been used to enhance the efficacy of neem-based pesticides on pests of horticultural crops. These include oil extracts and formulated neem emulsions, pastes, and granules. The major advantage is that neem products are generally much less toxic to nontarget and beneficial organisms like parasitoids, predators, pollinators, and soil organisms (Hassam et al. 2024). However, the main hurdles to the commercial development of these formulations are variations in efficacy due to environmental factors, such as rainfall, temperature, and humidity, as well as the development of resistance to neem molecules in some pest populations. The move toward more transparent regulatory frameworks for pesticide risk reduction helps prevent sublethal and long-term chronic effects in nontarget organisms. The registration status of neem pesticides and products is significantly different in various regions of the world, mainly as a consequence of sharply differing policies held by individual countries reluctant to relinquish the grip of the existing commercial and trade interests of the agrochemical sector (Arora, Naaz, et al. 2024).

4.3.3 Rotenone and Ryania

Rotenone, a crystalline compound that is classified as an isoflavone, is derived from soap nuts obtained from several plant species of the Polygonaceae family. Ryania, an amorphous powder consisting mainly of crystalline ryanodine as a single component, is extracted from *Ryania speciosa* and *Ryania mutabilis*, native to South America. In terms of efficacy, rotenone acts against a wide range of chewing and sucking insects, as well as some nematodes; its mode of application corresponds either to ingestion, resulting from contact, or is directly toxic by contact in insects. The compound has been used to renovate storage grains to minimize the development of indoor pests. Experiments in the field showed that rotenone has the same effectiveness as malathion for the control of onion thrips; in terms of safety, the organic product pyrethrin has the same activity (Ghosh 2024). Rotenone is generally not toxic to vertebrates except for some fish in water and in early life stages.

In insects, rotenone blocks complex I of the insect respiratory chain, directly influencing *adenosine triphosphate synthase* and producing reactive oxygen species in vitro within the gut. Because in tetany, Ryania *cruentata* insecticide paralyzes many insects associated with agriculture, gardens, and public health, and because the human and livestock consumption of tea made with dried *R. speciosa* leaves has an effect on the occurrence of *cruentata* insect populations, this finding reinforces the need to explore the potential of using crops of this native species of the Brazilian forest for biocidal purposes. Overall, both rotenone and *Ryania* are natural insecticides that have been used historically, the former for organic farming and the latter in commercial traditional formulations in mixtures with other synthetic active ingredients. However, in recent years, the use of these botanical compounds has declined. The use of botanical insecticides has not increased, remaining relatively small; nevertheless, the use of organic farming may increase their use once again (Istianto et al. 2024; Thompson et al. 2024). This is because synthetic chemical insecticides are not approved for organic farming approaches, challenged by the sampled issues as presented in Table 4.2.

Table. 4-2 Persisting and Emerging Challenges in Botanical Pesticide Use

Challenge	Description	Impact on Agriculture	Potential Solutions	Application	Referenc
Climate Impact	Weather affecting efficacy	Reduced field effectiveness	Development of weather-tolerant formulations	Development of weather-tolerant formulations	(Tesari, Leksono, and Mustafa 2024)
Lack of Public Awareness	Low consumer demand for organic produce	Limited market expansion	Awareness campaigns	Consumer indifference to biopesticides	(Yuan et al. 2024)
Extraction Environmental Impact	Resource-intensive extraction processes	Deforestation or habitat loss	Sustainable sourcing strategies	Overharvesting of Derris roots	(Li et al. 2024)
Complex Modes of Action	Complex Modes of Action	Delayed pest control results	Delayed pest control results	Neem requires multiple applications	(Rizvi et al. 2024)
Limited Shelf Life	Degradation during storage	Reduced usability	Improved storage technologies	Improved storage technologies	(Wulandari and Rahmadani 2024)
High Application Costs	More expensive than synthetics	The financial burden on farmers	Cost-sharing initiatives	Neem oil cost in smallholder farms	(de Jesus Oliveira et al. 2024)
Environmental Persistence	Environmental Persistence	Potential soil toxicity	Biodegradable formulation development	Rotenone persistence in waterways	(Gao et al. 2024)
Market Competition	Synthetic pesticides dominate	Synthetic pesticides dominate	Promotion of eco-friendly practices	Preference for chemical pesticides	(Pereira, Figueira, and Castilho 2024)

Knowledge Gaps	Farmer unfamiliarity with the use	Ineffective application	Training and Education	Limited understanding in rural areas	(Kampermann et al. 2024)
Regulatory Barriers	Strict approval processes	Delayed market entry	Harmonization of regulations	Varying EU vs. US biopesticide laws	(L. J. Ma et al. 2024)
Scalability Issues	Scalability Issues	High cost of application	High cost of application	Insufficient neem seed supply	(Scupinari et al. 2024)
Toxicity to Non-Target Species	Harm to beneficial insects or wildlife	Ecosystem imbalance	Target-specific formulations	Toxicity of rotenone to fish	(Rahma, Pramudi, and Liestiany 2024)
Variability in Composition	Plant extraction inconsistencies	Plant extraction inconsistencies	Standardization of extraction methods	Varying potency in neem extracts	(Khasanah et al. 2024)
Varying potency in neem extracts	Rapid degradation under UV/light	Rapid degradation under UV/light	Formulation improvements	Pyrethrins breakdown in sunlight	(Semerdjieva et al. 2024)
Pest Resistance	Pest Resistance	Reduced efficacy	Rotating pesticides, synergists	Resistance to neem in aphids	(Harte et al. 2024)

4.4 Mechanisms of Action of Botanical Pesticides

The activity of botanical pesticides is based on various biochemical and physiological mechanisms leading to toxic action, growth disturbance, and damage at the cellular and molecular levels of the target organism. Depending on their mode of action, these pesticides have been classified into the following groups:

a) Botanical insecticides;
b) Botanical nematicides;
c) Botanical molluscicides;
d) Mechanisms of action of botanical acaricides;
e) Mechanisms of action of botanical fungicides;
f) Mechanisms of action of botanical algaecides;
g) Mechanisms of action of herbicidal botanicals exerting growth inhibition.

The symptomatic study can be supported by the measurement of some metabolites when infestations occur, enabling any consequences of the infestations to be precisely identified. The general biochemical and physiological mechanisms by which botanicals act originally include neurotoxicity in insects and animals. Scans on the interaction of thymol and a formulation containing thymol with a varroa mite demonstrate how enzymes involved in redox regulation can play a role in the molecular effects related to the mechanisms of the interaction. The rapid development of molecular and omics biomolecular techniques provides fresh possibilities to perform analytical research focused on the mechanisms of herbal substances on different organism groups (Adimas, Adimas, and Abera 2024). It is evident that despite the 120-year-old finding, the mechanisms of action of different plant substances are not yet sufficiently known. Environmental factors and formulation aspects can substantially enhance effectiveness, and synergistic effects can also be utilized between different active ingredients in botanical agents. The reduction of crop damage to protect yield and improve food quality and safety is essential to meet the global challenge of a healthy world with a growing population (Karunakaran S. and Niranjana R.F. 2024). Integrating different strategies, including organic farming in which plant-based pesticides could be used as components, plays an important role. Despite several advantages and considerable developments, some challenges exist due to the lack of full evidence about the mechanisms of action of botanical pesticides. The main aims of this review are to provide details on the mechanisms of action of botanical active substances alone, showing some case studies and advantages of herbal agents as part of pest control, along with the difficulties that still need further investigation in basic and applied research in organic farming, as illustrated in Fig 4.3 (Zhao et al. 2024).

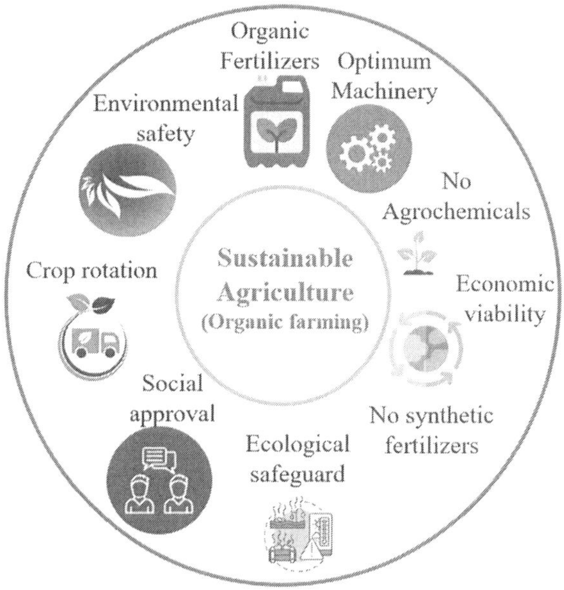

Fig. 4-3 Sustainable agriculture (organic farming)

4.5 Advantages and Disadvantages of Botanical Pesticides

Botanicals have shown that they can potentially be as lethal as synthetic pesticides. Therefore, just as there are pros and benefits, there are also downsides that need to be seriously considered. Another clear advantage of botanical product application is avoiding residues left on food crops since botanical pesticides can be more easily degraded and removed. This is especially important for export products, as many consumers are concerned about their health. In addition, botanicals have lower risks of pest resistance, as they contain not only the active ingredient but also many other components in the mixtures that can have different modes of action. However, botanical pesticides have some important weaknesses as well (Harte et al. 2024). As they originate from natural sources, they most often tend to be unstable and are of variable quality in the active ingredient or isolated form.

One important advantage of using naturally occurring and already used plant material is that its side effects are already known; thus, it is likely to have lower adverse effects on biotic non-target organisms than synthetic pesticides. Also, most likely, all of the properly used botanical pesticides are compatible with Integrated Pest Management. However, due to the dispersal of commercial whole plant botanical deposits at unauthorized sites from which the original plants originated, it should only be further tested thoroughly after making the toxicological tests for ecotoxicity of its commercial ingredient (Miloudi et al. 2024). A promising advantage of botanical pesticides is usually their smaller environmental impact, making them interesting for use in organic farming systems. Moreover, in Integrated Pest Management, experts do not work with small ecological sound black boxes but with open systems where the farmer and the farmer's ecosystem are considered as a whole. However, the major drawbacks of the commercialization of botanicals are the estimation of their possible legal issues and legislation, which may limit their use as multi-purpose products. The legislation and market approval for natural substances at different government levels as pesticides and food additives or supplementary food, must be established, which is not straightforward and involves long-term duration (Adimas, Adimas, and Abera 2024; Shafik 2024b).

Botanical pesticides share the same global regulatory framework as other conventional pesticides. Guidelines were put together to provide a means for evaluating the safety of residues present in or on food and to make recommendations concerning the acceptability of pesticide residues for international trade. Both safety and efficacy should be considered for the registration of botanical products (Shafik, Tufail, De Silva, and Apong 2025). Detailed safety assessment guidelines for manufacturing, agriculture, workers, consumers, and the environment, as well as guidance for pesticides having characteristics such as low mammalian toxicity, are provided. Safety and efficacy evaluations are normally aimed at national or international organizations. The approval process is very time- and labor-consuming, besides being expensive. Similar agencies or regulations are present in other countries, where the process might differ, for example, with respect to required data, tests, and approval timelines. While some botanical pesticides are compatible with organic farming, not all registered botanical pesticides of a country may be registered in other countries, and when registered, approval might cover other uses than for organic farming (Zenteno-Alegría et al. 2024).

In general, control is applied for all types of pesticides, including botanicals, and is based on the international guidelines for the registration of pesticides. Regulation differs between regions; in Europe, for instance, the use of a specific active compound in a plant protection product follows a two-step procedure: approval of the active substance and authorization of the plant protection product, both requiring a stringent risk assessment and involving various national regulatory agencies acting as evaluating regulatory agencies. The industry encounters challenges in obtaining and maintaining approval from many regions or countries on a product-by-product basis due to the complexity and resources required to generate data and information that are compliant with individual guidelines and assessment methodologies. This will, in part, explain the regional markets, including the prioritization of costs and lobbying that creates the perceived or actual content of the backlog, e.g., the discrepancies in approval times and levels of official raised concerns about residue analytical methods employed by many countries despite being in use and approved by some nations (Mohanty et al. 2024). The recent concerns expressed to registrants of botanical herbicides and subsequent withdrawal by one company were concentrated on the pesticide residue analytical method postulated. Convergence between the international harmonization of regulatory data requirements and guidelines performed by relevant experts to achieve international agreement with respect to both scientific literature and technology that a product is fit for purpose will pave the way to a smooth and economical process and outstanding trade of botanical pesticides. Capacity building and awareness will play a pivotal role in guiding the producers to acquire a basic understanding, prepare the dossier, and complete the level playing field with respect to the requested, substantially agreed upon necessary data and subsequent registration and free movement of the botanical pesticides (Popescu et al. 2023).

4.6 Current Trends in Botanical Pesticide Usage

Farmers increasingly appreciate the demand for sustainable pest management strategies due to the growing concern about environmental health and the demands of consumers for food and environmental safety. This has led to a paradigm shift in the mode of farming aimed at reducing the use of non-biodegradable compounds. The growth of organic farming is an indicator of the importance of botanicals in pest management. Because of sustainable pest management, most pesticide manufacturers are diversifying into the botanical market, leading to the growth of a homogeneous sector as well (Reher et al. 2024; Evizal and Prasmatiwi 2022). New formulations, adoption of precision application methods, and certification methods make the currently launched botanicals eco-friendly alternatives. Variations in the use of traditional botanicals in different states are due to the variation in the agricultural practices followed by farmers and the regulatory framework operational in each of the Indian states. These are some of the potential extension areas needing research input, development, and refinement to affect a sustainable botanical pest management system. In view of this trend in the international markets, a number of botanical pesticides poised in due course for international registration or approval are listed along with their status (D. Rizzo, Marraccini, and Lardon 2022; Ünal and Kızıldeniz 2022).

Although synthetic pesticides have been extensively used to control pests and diseases, emerging concerns about the undesirable impacts of chemicals have triggered the search for alternatives. Indeed, the use of plants with promise in pest management is becoming important in terms of the development of agro-policy (Shafik et al. 2023). Currently, various botanical pesticides are being introduced in three main agricultural sectors: plantation, field, and horticultural crops, with the highest growth, witnessed in botanical-based bio-pesticides in cereals and pulses due to better availability and application techniques in comparison to the minor sectors of tea and spices. Botanical pesticides are implemented either individually or in combination with other control methods. The main approaches are pest-predator compatibility, integrated pest management systems, farmer-focused methods, and integrated pest disease management (Baviskar 2024). There are several opportunities and challenges concerning the use of botanical pesticides, such as side effects on the environment, mixtures and formulations, genetic resistance, and the systemic effects on beneficial and off-target effects on wildlife, especially in organic farming. The evolution of improved formulations is another trend that potentiates the efficacy of an active ingredient alone or in combination. In some cases, a number of farms are integrating botanical pesticides in their fields to reduce the use of chemicals and move towards sustainable farming (Philip and Suresh 2024).

With their desired attributes and, more significantly, eco-friendly, botanical pesticides have often been considered prominent "future pesticides of choice." Despite the worldwide presence of hundreds of botanical pesticide products, it is surprising to note that they could find almost negligible market shares, let alone utilize these products as components of pest management toolboxes. The reliance of the industry on the development of synthetic active ingredient compounds, the registration-associated high costs, and the longer duration of registration processes all shed light on the current mindset of stakeholders. Addressing these connotations, this review discusses the challenges that bottleneck the adoption and implementation of botanical pesticides in agricultural systems (Demir, Dedeoğlu, and Başayiğit 2024; Popescu et al. 2022). The adoption enablers have also been discussed, and further directions are highlighted.

Resistance Management The evolution of resistance among pests and host plants is the major hurdle in integrated pest management. Therefore, to achieve a high level of pest control effectiveness and a highly sustainable cropping system, botanical pesticide options need to have adequate parameters that would support resistance management strategies. However, only a few of them in use have been found to be effective as a resistance management tool. Consequently, an increased frequency of applications (or application rates) of botanical pesticides in the field to attain the common level of control requirements results in a sudden rise in the cost of production (Pranaswi et al. 2024). The development of resistance of pests and pathogens to insecticides and fungicides can be a significant driving factor for the success of botanical pesticides. Resistance can develop through target site insensitivity, where the pest or pathogen has reduced sensitivity to the active ingredient or mode of action, or indirectly through the evolution of resistance to behavioral or physiological disturbances resulting from exposure to the active ingredient, such as interference with mating, feeding, vision, or olfaction. Resistance to many bioactive compounds of plant origin has so far developed at a slower pace than for synthetic chemicals, yet the concern around resistance has prompted many cautions and caveats around the use of botanical pesticides (Kampermann et al. 2024).

An effective tool for slowing down the evolution of resistance is the use of integrated pest management (IPM), which is based on the combination of multiple pest control strategies for reinforced control. Rotation and alternation of chemical pesticides are already recommended IPM practices and are based on the sequential use of pesticides, changing their target site and mode of action in order to overcome or avoid further selection pressure and, thus, the evolution of resistance. Similar principles could be applied to the use of botanicals to impede the likelihood of resistance evolution. Specific recommendations are that botanical pesticides with different modes of action should be used in rotation to suppress the buildup of resistance and that their combination with cultural practices to avoid resistance evolution by enhancing the effectiveness of the botanicals should be pursued. Continued monitoring of resistance to botanicals is essential within any agricultural system to inform which combinations and rotations of products are deemed successful at stalling resistance development (Yeshiwas et al. 2024; Adimas, Adimas, and Abera 2024). Both growers and consumers should be educated about the need for resistance management and the underlying mechanisms for it.

Formulation represents one of the more significant and challenging tasks for developing botanical pesticides. The active ingredients in botanical pesticides derive from plant sources; it is both time-consuming and expensive to achieve a consistent active ingredient for the manufacturing of botanical-based formulations. Highly variable bioactive substances present in source plant material make extracting, formulating, and ensuring the presence of the bioactive substance in the final product consistent. As almost all natural product-based compounds have poor physiochemical properties and are chemically unstable, advanced formulation technologies are required for enhancing stability during storage, distribution, and under field conditions. In addition to active ingredient variability, the bioactive must be only released in the field and post-application. Like synthetic products, botanicals also require a delivery system to release the active ingredient at the feeding site of the target pest (Kampermann et al. 2024). New micro- and nano-particles that are physically and chemically stable, contain highly active ingredients, and release them in a controlled manner have been developed. Innovative technologies such as micro- and nano-entrapped cells as bioactive ingredient delivery systems have been shown to enhance biological activity and promote specific delivery. The efficiency of bio-pesticides is hindered by factors such as rapid photodegradation, susceptibility to kinetic breakdown, ultraviolet degradation, sensitivity to heat, proteolysis, and biochemical decay. Industrial biopesticide products must have a longer shelf life to be able to compete on the market with traditional products, which is achieved by formulating a biopesticide in stable granulated form (Benson et al. 2024). Overall, biosafety regulations, microbiological safety, and legal practicality have also stimulated research into plant-based and bio-based pigments.

Satisfactory efficacy and persistence to field conditions are likely the two main parameters that corroborate the implementation of a botanical in integrated management. The pesticidal activity of each plant extract will depend on concentration, chemical profile, extraction method, and duration between extraction and handling. The pest-plant extract combination in which efficacy was evaluated may still suffer negative effects exerted by environmental constraints such as soil, temperature, moisture, and pH. For example, the activity of products containing neem is neutralized by soil properties and moisture (Tesari, Leksono, and Mustafa 2024; Arora, Naaz, et al. 2024). Studies are underway that aim to observe and quantify the behavior of these formulations containing the extracts when inserted into the soil. Research still in progress points out that although some commercial botanical formulations tried to convince users of similar efficiency to synthetic products and reported twice the efficacy, the mortality rate was lower than target expectations in the field when applied on the right schedule. This shows the necessity of tailoring the application of the botanical insecticide to maximize efficacy; at the same time, the application is based on scientific grounds to minimize environmental impact and influence in the integrated pest management of *S. frugiperda* (Benson et al. 2024).

Inorganic salts protect the insecticidal molecule from ultraviolet light by slow release of the active ingredient, conferring residual properties in a relatively safer way to the environment; many of the botanical insecticides contain this protective supporting formulation. Field comparisons of persistence between botanicals and comparable active ingredients are rare. Insecticidal IPM programs request a smooth performance of control agents, not only in the sense of efficacy but also persistence in the agent's action. Replacement of the active ingredient providing no persistence at all will reduce the benefits in pest management; therefore, it is necessary to monitor the active principal content in the botanical insecticide over time. It is also important to examine the long-term efficacy in order to observe the frequency with which the user would reapply the formulation to maintain control of the insect and also to monitor the content, acting as an indicator of formulation evaluation in the desired field trial area (Arora, Naaz, et al. 2024). Formulation innovation aims to reduce the persistent effect of conventional synthetic materials on the environment. The development of botanical insecticides is more costly than synthetic products because of production scale and greenhouse trials with the native insect. The main cost is the extraction of the biopesticide from the plant material or residues. Botanical pest management programs separately or in alternation/rotation with synthetic insecticides may have the potential to reduce damage and the frequency of pesticide use and enhance long-term pest control. The use of the botanicals should be optimized depending on disease risk, cultivation system, and storage parameters, and field management is key to profitable crop production (Bi et al. 2024).

4.7 Toward Commercialization and Public Acceptance of Technological Pesticidal Innovations

Innovative technologies are currently revolutionizing the field of botanicals by offering the replacement or improvement of the extraction processes of plants' chemicals, with the utilization of nanotechnology or enzymes and even genetic engineering to make the resulting botanical matrices more effective or easier to apply. Newly designed materials by nanoparticles can act as more efficient and effective supporters of botanical pesticide transport, leading to a decrease in dose and persistency as well as an enhancement of the effect in all kinds of applied agroecosystems. The present call for innovation is fueled by the increasing demand from farmers, consumers, and national food production ministries for effective new low-input techniques for sustainable and healthy crop production. (Yeshiwas et al. 2024) We review emerging technologies and approaches to develop fundamentally new or to improve existing botanical formulations and extraction/production technologies. We review the most recent methods to improve solubility, stability, and controlled release, such as diverse forms of solid nanoparticles. We also cover plant-produced antibodies and new kinds of stable recombined enzymes in the design of biopesticides that directly act on pests rather than as carriers for pest toxins. Furthermore, we cover new technologies that alter plant genomes to increase their resistance against pests, increase the plant's ability to produce and store active compounds or design the plant to act as a multi-action trap crop or auto-sterile to produce offspring with lower fitness. In all cases, we cover examples of developed technologies verified in the laboratory and field, as well as future needs before these technologies are applied on a large scale by agro companies.

4.7.1 Nanotechnology Applications

Nanotechnology has the potential to transform agricultural practices and the field of botanical pesticides. Botanicals contain an active compound derived from botanical material and are relatively low in toxicity but generally slow-acting, frequently volatile, and demonstrate poor water solubility. However, nanomaterials, engineered materials having at least one dimension in the nanoscale, can improve these properties while better protecting the pest-killing agents. Nanotechnology applications in botanicals are rapidly expanding, but the regulatory approval process presents barriers to their use in pest management. Furthermore, the general public remains dubious about the safety of using nanomaterials in pesticides despite their potential large environmental benefits with reduced chemical load and matrix specificity (Ankur et al. 2024). Various nanomaterials employed for botanical formulations include nanoparticles, *nanospheres, nanocapsules, nanopolymeric nanomaterials, nanoleonardite, nanofibers, nanogels, nanorobusts, nanosponge, nanoemulsions,* and nanoscale carriers. These nanomaterials can be conjugated to provide desirable properties such as targeted or controlled release. Several case studies have adapted their traditional application of botanicals, which involve plant parts or crude extracts that have been used since ancient times to control pests, into new formulations that incorporate a nanomaterial for maximum effect. As an example, zinc oxide nanoparticles synergistically improved diatomaceous earth, a traditional biopesticide, and exhibited a greater effect on adults in the flour beetle compared to their separate components at an equal or higher dose (Haruna S. G., Yahuza L., and Tijjani I. 2024).

The mixing strategy and minimal characterization of the nanoparticles make it likely that additives are already found on these particles in applications such as sunscreens. Most often, however, as nanotechnology evolves, garden variety nanoparticles will be used in food and will not likely have an aggressive effect on germination and plant growth within the high-end proposed rate. All non-polar surfactants are known to be carcinogenic, and the advice on their usage is only to be used within the label instructions. Many nanomaterials utilized in agriculture also have degradable and naturally derived types and control substantial user advantages, or they are generally regarded as safe (Lansana Abou BANGOURA et al. 2024). The scientific review in the next part is just about the use of nanoparticles and nanomaterials in ultralow volumes that have approvals temporarily and are used in very limited quantities because the use and degradation of nanomaterials readily lead to the manufacture of harmful arsenals or nanopolymers. Nevertheless, nanoparticle control is usually harsh or unreasonable, with nanoparticle products ranging from hundreds to thousands of dollars per hectare and several nanobenefits not yet well documented. There is current potential in the field of agriculture to investigate the environmental health of new nanoparticles or materials. Although high levels of nanoparticle manufacture have declined significantly, advances in the utilization of commercially available nanoparticles are progressively being controlled from the research laboratory to the field (Murugesan et al. 2024). This is particularly so if nanotechnologies are used widely and at higher levels than in the case of cultural goods, i.e., in applications requiring less physical input.

4.7.2 Biotechnological Innovations

Biotechnological innovations have and will continue to play a major role in the development and application processes of botanical pesticides. A number of recent advancements have also provided possibilities for the alleviation or solutions to these constraints. For example, the enhanced production of pest-resistant plants has involved creating plants with improved properties, particularly regarding the concentration of active compounds in botanical pesticides. Genetic engineering has facilitated new tools and methods for seed companies to offer resistance or reduced susceptibility to certain diseases via genetic engineering applications. Plant variety or cultivar development is one of the most promising areas for the application of biotechnology (Hassam et al. 2024; Eisen 2024). New varieties with tailored properties that permit cultivation in marginal areas and the efficient use of different types of cultivation regimes would be in high demand. Research on the genetic relationships of pests and disease resistance would be of major interest to the community. Relevant resistant and susceptible germplasm lines could be employed for gene expression analysis to help unravel the mechanisms that may be at play. These types of studies are fundamental for identifying, accelerating, and adapting to novel genetic resources and traits of interest that could subsequently be used in biotechnological applications such as the targeted generation of new varieties of crops. Biotechnology has opened up a number of possibilities in this research area. Biotechnology innovation offers the potential of identifying molecular markers for early detection of disease in living plants (Thompson et al. 2024).

Simultaneously, the application of biotechnology will allow the detection and early deployment of genetic resistance traits. Present developments in molecular biology and molecular markers make these markers suitable for use in molecular plant breeding. Despite these successes, several start-up issues must be addressed, ranging from public approval of genetically manipulated plants to the regulatory processes necessary to facilitate their full commercialization. The integration of biotechnological and environmental issues may provide fresh impetus to the profile of GM products. The probable use of GM with low environmental impact and advantages over existing synthetic chemicals provides an impulse further to exploit the possible synergy between biotechnology and botanical pesticides. The various applications of biotechnology are expected to enable further exploitation of botanical pesticides to meet some of these ongoing challenges that threaten their role as new prospects in the modern crop protection sphere. As a result of research and development in these areas, pesticides will be more environmentally sustainable. Moreover, there is an increased possibility of reviving or rediscovering a number of such pesticides based on indigenous knowledge or ethnomedicinal plant extracts. Cost is a major factor that restricts the development of new compounds. The application of biotechnological techniques for improving the botanicals industry is an important area of development (Miloudi et al. 2024). This aspect also needs to be the focus of future research into these pesticides.

4.7.3 Combination Formulations with Synthetic Pesticides

The translation of such theories to the field is realized so far by several examples. The first example is from a pre-mixed botanical fungicide incorporating a botanical with another synthetic on cucurbits. In southern Africa, mixtures and products such as neem- and tobacco-based products are being advanced. In the future, this idea and opportunity for commercialization will depend on demonstrating three conditions. The first is a significantly larger level of efficacy provided by the active ingredient. Synergistic and complementary actions have been demonstrated for some agents to rival some synthetic chemicals. However, potential mixtures must fulfill most of the contemporary guidelines for synthetic pesticide registration and include formulation products that have not yet been optimized for a combination to minimize any unpredictable or unmanageable effects (Karunakaran S. and Niranjana R.F. 2024). The gain is only fully realized in in vivo field trials, not in vitro or in silico experiments. The second factor is that the product must be acceptable in terms of human, livestock, and environmental safety, risks, and costs. Both of the above examples fall into this category, with botanical products being quite low in both of these areas, being minimally processed natural products based on traditional knowledge systems. The third criterion is that it is safe and meets global or national regulatory conformity and risk assessments of the bioactive species. Since the integration of synthetic and botanical pesticides raises new multifaceted concerns about toxicity, breakdown, and trade, the policy surrounding such complex and novel mixtures may need to be evaluated. Furthermore, if the putative benefits require the combination of ingredients to be used simultaneously, an additional level of knowledge of how this can be achieved in practice is required. It may in some cases be necessary to develop time-releasing delivery systems or temporal mixtures, i.e., to achieve temporally different degradations of ingredients or use patterns (Harte et al. 2024).

The commercial benchmarks aimed at testing and advancing synergistic actions are, as of yet, clouded by a lack of scientific and field-based knowledge and have only recently begun to be developed. With the above caveats and points to prove in mind, the future challenge for state-of-the-art integrated pest management will be to cooperate with the industry to produce compounds such as these. Suppose the botanicals show high overall natural recovery or branched metabolic routes in vivo. In that case, this presents not only novel synergy but also has the potential, in a situation where one path is disrupted by resistance, to enhance the efficacy of the alternative metabolic route via indirect bottlenecks or regulatory actions (Semerdjieva et al. 2024). Only then will we start to underline the basis for achieving synthetic-free farming goals with safe products that do not need to be disposed of, as yet unperfected before discharge, by farmers.

4.7.4 Nanotechnology in Delivery Systems

Diverse classes of phytochemicals have been widely used to control pests and diseases in crop plants. However, the drug-like activities of these phytochemicals also lead to the same challenges and failures as have been encountered in developing synthetic chemicals. A major deterrent to their practical use in agricultural applications is their inherent biodegradability and high toxicity characteristics. The growing requirement for environmentally friendly and degradable biopesticides has intensified research interest in both basic and applied fields on plant-based alternatives, including their commercial formulations (Kampermann et al. 2024). It has been found that the incorporation of nanotechnology in botanical pesticides has improved the bioavailability of the active ingredient, which is crucial for pest control, offering better stability, controlled release, and sustained pesticide effectiveness. Nanoemulsion, nanosuspension, and nanoencapsulation of botanical pesticides, have been prepared and loaded on different carrier materials such as liposomes, silica nanoparticles, and polymeric nanoparticles, followed by their validation to assess the sustained release of botanical derivatives. In one study, it was reported that azadirachtin nanoemulsion formulations caused improved bioefficacy in terms of greater larval mortality and pupation inhibition at about 1.0 ppm concentration, while the same toxicity was required at about 2500 ppm concentration in crude azadirachtin. Another study introduced GO-multilayer containing oxidative damascenone for the female attractant of Drosophila suzukii, a non-neurotoxic attract-and-kill biorational solution in a 10% lower dose than the commercial one. Plant essential oils like Mentha piperita, Symphytum officinalis, and Amygdalus communis oils have been encapsulated in denatured ethanol and microwave-initiated for molar ratio on ZnO nanorod for their release rate calculation to the targeted pest. Water-dispersible nanoformulations composed of silicon dioxide as support, spinosad, and vincaleukoblastine in sunflower against Spodoptera frugiperda have shown solid mortality rates of 92%, 65%, and 63%, respectively, after four days of application. Provided details about the differently prepared botanical nanoformulations for the control of several pests are described in this section (de Jesus Oliveira et al. 2024). Table 4.3 presents some insights for agriculturists considering the transition to more sustainable pest control methods.

Table 4-3 Comparative Analysis of Botanical and Synthetic Pesticides

Aspect	Botanical Pesticides	Synthetic Pesticides	Advantages of Botanical Pesticides	Challenges in Adoption	Impact on Sustainable Agronomy	Reference
Global Adoption Trends	Global Adoption Trends	Dominant in industrial agriculture	Supports global organic agriculture	Requires policy support and incentives	Strengthens global efforts toward sustainable farming	(Yeshiwas et al. 2024)
Research and Development	Increasing but underfunded	Increasing but underfunded	Increasing but underfunded	Limited R&D in some developing regions	Promotes innovation in sustainable agriculture	(Ankur et al. 2024)
Promotes innovation in sustainable agriculture	Promotes innovation in sustainable agriculture	Limited to chemical combinations	Limited to chemical combinations	Requires knowledge of interactions	Facilitates diversified, holistic pest control	(Bi et al. 2024)
Effectiveness in Extreme Conditions	Effectiveness in Extreme Conditions	Effectiveness in Extreme Conditions	Lower risk of secondary pest outbreaks	Performance variability in field use	Encourages adaptive pest management techniques	(Rzyska, Stuper-Szablewska, and Kurasiak-Popowska 2024)
Regulatory Requirements	Strict for biopesticides	Well-established for synthetic pesticides	Safer for regulatory compliance	Safer for regulatory compliance	Promotes the adoption of safer agricultural inputs	(Haruna S. G., Yahuza L., and Tijjani I. 2024)

Public Perception	Increasing demand for organic, eco-friendly options	Increasing demand for organic, eco-friendly options	Aligns with sustainability trends	Lack of awareness in certain markets	Drives the growth of sustainable consumer markets	(Arora, Choudhir, et al. 2024)
Shelf Life	Shorter, sensitive to environmental factors	Longer, stable under varied conditions	Encourages fresh production	Encourages fresh production	Encourages localized production and storage	(Lansana Abou BANGOURA et al. 2024)
Market Availability	Limited but growing	Limited but growing	Limited but growing	Limited but growing	Expands access to organic agricultural products	(Modafferi et al. 2024)
Application Frequency	Application Frequency	Longer residual activity	Encourages eco-friendly farming	Labor-intensive application process	Increases labor demand but promotes natural cycles	(Eisen 2024)
Effect on Non-Target Species	Generally safer for beneficial insects	Often harmful to beneficial organisms	Preserves pollinators and predators	Potential mild toxicity in some cases	Protects biodiversity and ecosystem balance	(Kavallieratos et al. 2024)
Environmental Impact	Biodegradable, minimal residues	Persistent, can cause soil/water contamination	Eco-friendly, reduces pollution	Shorter residual effect	Enhances soil and water quality	(R. Rizzo et al. 2024)
Source	Derived from natural plants (e.g., neem, garlic)	Chemically synthesized in laboratories	Renewable and sustainable	Resource availability and variability	Promotes the use of renewable plant-based inputs	(Harte et al. 2024)

4.8 Economic and Environmental Impacts of Botanical Pesticides

Several major constraints currently limit the use of botanical pesticides. Economics, both the upfront, though usually much less persistent, cost of botanical pesticide production and adoption of cultural practices difficult for many, especially smallholder farmers, to conduct, and adaptations to use the pesticidal plants by buyers and regulatory authorities need to go into the cost-benefit analysis for adopting botanical farming. As it pertains to growers, price fluctuation is equally related to food supply chain vicissitudes and official and political discourses on environmental neutrality and carbon taxation of cereals, as well as how and where they are grown after the recommendation of organic crops due to synergistic soil-to-fruit, nutrition and human health ties and the climatic, economic, and socio-political repercussions, may lead to huge economic losses in conventional as well as organic markets for families since many or most of either kind are managed not as businesses, but as lifestyles and legacies today (Scupinari et al. 2024). However, following the latest market trends and the shifting interests of rural families and lifestyle farmers, we may assume an increase in what sells under the organic and functional food labels, especially if manufactured in line with various circular, zero-kilometer, zero waste, and fair trade regimes. The environment is the main factor that has turned the transition toward this idea of a healthy and sustainable diet and greenscape. That is overshooting in global fisheries, pesticides, the concomitant chemical industry, and climate change (Pereira, Figueira, and Castilho 2024).

Botanical pesticides also exhibit their environmental impact in a very susceptible manner: increased crop and plot functional biodiversity, fostered plant and soil ecosystem services, reduced pesticide runoffs in heterogeneous soil and plot conditions, and increased farmer-society system resilience, all of them functionally dependent on climate, soil, and knowledge of local needs. Recently, the need for a fair balance in the description of the strategies developed to exploit the potential and to avoid caveats of botanical pesticides has been highlighted. This brief review aims to provide the best up-to-date overview of what has been published on the economic and environmental impacts of the use of botanicals as pesticides. The reviews, interviews, surveys, and experimental data on the potential for the use of commercial interest pesticides are based on lemongrass repellency that farmers may or may not need to use to prevent attacks by the rice blast and stem borer. The results show that evidence exists showing the negative and positive effects of pesticides on specific and general pests and lower tropics (Rizvi et al. 2024). However, the identification of the factors that determine the magnitude of these effects has been poorly studied. Therefore, the need exists for more advanced studies.

With growing consumer awareness and concern for food safety and environmental sustainability, there is a global trend of increasing demand for botanical pesticides as an alternative to synthetic chemicals. Currently, the key influencers driving market demand for botanical pesticides are consumer preference for pesticide residue-free food, the growing organic farming industry, increasing research and development activities in botanical pesticides, and the demand for environmentally secure and sustainable agriculture. Studies have been conducted on farmers' perceptions of botanical pesticides. Some farmers perceive botanical pesticides as individual values, while others do not. The studies show that pricing dynamics, including costs, become a major driving factor for acceptance among the majority of farmers across the world. The prices of botanical pesticides are comparatively higher than those of synthetic pesticides. This is due to their high production costs based on the ingredients and the smaller quantity produced at a time, manageable turnover for investment recovery by the manufacturer, sufficient capital to manufacture herbal pesticides, and the need to bring bioactive molecules to the lab and atmosphere (Li et al. 2024).

However, in order to improve the profitability of farmers applying botanical pesticides, there is a continuous need to rationalize the cost of herbs used for the production of pesticides. Some botanical manufacturers believe that if the real cost of production is not passed on to the consumer, it is difficult to promote botanicals in the market. In short, empirical evidence shows that agricultural economics is currently turning towards the acceptance of new practices, and agriculturalists need to look forward in order to meet social and environmental expectations. The above view is tested and validated in specific cases, and technical success stories are presented in the following part of this presentation. Historically, investment in botanical manufacturing units is considered low due to the misalignment of economic resources and knowledge, and it has faced serious threats due to various restrictions, hidden controls, and competition (Miloudi et al. 2024).

4.9 Impact of Climate Change on Botanical Pesticides

Climate change can directly impact pest management strategies in agriculture as well as in the public health sector. Botanical products can contribute to adaptive pest management strategies that allow building approaches to cope with the effects expected due to climate change. Changes in average and extreme temperatures are expected to enhance pest populations, increase the number of generations, and affect the distribution patterns of pests. In addition, alterations in precipitation regimes increases in CO_2 concentrations, and extreme weather conditions are expected to exacerbate the negative impacts on agroecosystems. Botanicals are also useful for making pest and disease management integral to production systems that harness the power of biodiversity (Ghosh 2024). Climate change can directly and indirectly influence pests as well as biological control agents. Expected alterations in migratory behavior or earlier arrival and quick dispersal associated with temperature increases make it important to develop strategies that spread control agents, botanicals, and biological controls. The insecticidal effects of botanicals used for pest control are altered by temperature and other climatic variables. This can lead to problems such as reduced efficacy, changing patterns of insecticidal or molting hormones, or resistance. Furthermore, not only are the control actions affected, but also the plants used to obtain the botanicals. Climate change affects their physiological status, phytochemical profile, growth, and yield. Moreover, non-target effects can increase due to the use of natural enemies. The habitat of these can be highly affected, so pest populations can become even more problematic. In the context of declining populations of some pollinators and natural enemies, such tools may also have a 'double effect' on the landscape, increasing management sustainability. Finally, although botanicals can have beneficial effects on human health by preventing the spread of diseases associated with pathogens, they are often toxic to non-target organisms. Therefore, higher doses or frequencies must be used to control pests, with the consequent increase in exposure to these natural enemies (Ragab et al. 2024). While the methods of adaptation to reduce pests in a changing climate for the production of food crops are of interest worldwide, it is important to note that the effects of increases in temperature and CO2 and declines in precipitation can vary widely. Moreover, global warming is not uniform, as the effects of a changing climate can differ in various ecosystems or regions. In many regions, for example, factors related to the movement of animals and the irrational use of land and water resources favor the stay and displacement of disease vectors as well as the decrease of biodiversity. Each year is characterized by variability due to the interaction between fluctuations in rain and temperature and regional factors that do not enable us to make future predictions. If we consider that different studies provide different results and the effects of climate change are difficult to predict and will vary in each growing area, there are limitations regarding the fastest and most profitable choices in reducing pests with species that use botanical preparations and biopesticides (Zhao et al. 2024; Harte et al. 2024).

4.10 Future Directions and Research Needs

To stimulate interest in the field of botanical pesticides and to make them virtually useful, they need to travel greater distances. Some areas where advancements are needed have been pinpointed above. In the present section, a multi-disciplinary approach tackling the research needs for making botanical pesticides useful has been discussed. The handling of botanical pesticides can gain direction if there is some systematic knowledge. There is more scope in the field for advancements, but it has been restricted due to a lack of proper approach. It has been observed from field visits and practical work that if a crop is infested with a pest, it is not always considered feasible to apply botanical pesticides due to non-availability or their failure. Several reasons, such as efficacy, status, formulation, production, shelf-life, stability, acceptability, and other concerns, need to be validated to extend the use of botanical pesticides (Rahma, Pramudi, and Liestiany 2024). Some of the directions and areas where research needs to be initiated and enhanced for developing botanical pesticides with greater acceptability have been pinpointed. Some of these areas are as follows:

Several missed opportunities, emerging research areas, and research opportunities yet to be explored in botanical research include developing a database and portal, wherein tools are developed to input and retrieve public data on the history of botanical invasive alien species including places, years, and reviewed products of phytochemical studies, in vitro studies of pathogens, and articles on efficacy, selectivity, and biodiversity impacts of these agents; accelerate understanding by faster and cheaper development for deploying new technologies such as gene editing in botanical research; a research topic on the importance of the endophytic microbiome or studying the microbiome before embarking on botanical field studies or fermentations; the significance of a comparative study in pesticide research which is a roadmap that sets out how to translate information garnered in the lab so that it informs in-field decision-making and can guide in planta and open field trials of botanical biopesticides among other topics (Kampermann et al. 2024).

There is also a need for a more interdisciplinary approach among chemists, biologists, and medical anthropologists to reconcile laboratory findings with ecological knowledge on a global basis. New opportunities have occurred recently where now the phytochemical literature can be accessed in large databanks and compared with lists of acceptable composition, toxicity, and residue data as the official reports before exploration of actual pesticidal properties. The working group has supported the clinical exploration of the mechanism of resistance and might now join in an exploration of pesticide evidence against malaria to create this proof of concept, particularly for tuberculosis sufferers (Gao et al. 2024). In regard to new technologies, including gene editing, the work on natural product chemical diversity and hybridization of the biopesticide with antibiotics tracks to discovery with so many novel genes still to explore. The preferred tools of choice are available. There are numerous research opportunities in both old and new directions, and the cannabis and essential oils market is again in *ferment*. we need passports where people have been exploring *antimycobacterial* for these species. Additionally, the pest-specific and relatively non-toxic lead to an opportunity to explore human, animal, and environmental safety, as well as action spectrum and photosensitivity. Long-term research would be necessary to find effective concentrations that control mandibulate pests and invertebrate parasites (Li et al. 2024; Yuan et al. 2024).

Bioprospecting and modifying underutilized but rich sources of biopesticides from the plant species that have not been tapped. Bioprospecting can also be done in reformatory reputedly minor weed and invasive plant species. Oils and extracts are two inquiries that deserve future research. In addition, research of underutilized crust and leaf cup medicine, and other products with good internal activity and selectivity. One area with almost no current research—due to the time required for such research—is the medium-term impacts on non-target pests and beneficial insects. In this case, the research could be done in parallel (but not sequentially) to quantify how sustained reductions of either invasive pests or beneficial insects affect the subsequent population stability and abundance of invertebrates at various trophic levels (Zhao et al. 2024). Outreach should also aim at engaging the public regarding the potential roles of botanical pesticide producers, academics, industry, regulators, policymakers, and conservationists and seek collaborators to integrate the work on efficacy and long-term ecological impacts. It is necessary to enhance the practical importance of addressing the long-term impacts of botanical pesticides on the environment. It is essential to strengthen efforts for enhanced practical relevance through close association among researchers, industry personnel, and policymakers. Given the radical differences in conceptual implications between 'chemical' and 'botanical' modes of action, an interdisciplinary approach to their design and implementation, considering all the expertise from human biology, chemistry, plant physiology, neurology, and agriculture, has revealed itself as the need of the hour (Scupinari et al. 2024). In summary, we believe that the idea of developing sustainable strategies in the biocontrol of plant pathogens based on predators and parasites, together with lower and highly selective eco-toxins such as 'botanical pesticides,' has great potential and has shown significant beneficial results. We believe, however, that the field has potential, which can only be realized through extensive field trials, as is the norm in pest biocontrol since the costs of failures are as great as the gains.

4.11 Conclusion

Botanical pesticides have enormous potential to contribute to sustainable agricultural production in the future. These mostly complement practices for the integrated management of pests. This essay does not recommend the rejection of all synthetic pesticides or the unconditional celebration of botanical pesticides. Instead, the former need to be regulated in terms of their non-disruptiveness of ecological processes and the hazards to both humans and non-target organisms from them, while sensibly used botanical pesticides can provide farmers with the flexibility to maintain the integrity of the multiple-pesticide governance approaches, often within the more research-based approach known as ecological engineering. Botanical pesticides can also be a useful part of a refugia approach within systems with internally regulating and particularly biological control agents. In the last several decades, there have been concerted efforts to minimize the negative environmental and health impacts of synthetic pesticides in agriculture and public health programs. This trend, in turn, has enhanced the importance and extensive usage of botanical pesticide research in integrated pest management strategies. The relevance of this outreach in the global context is well mirrored in an increase in published literature. The increased need for bio-products in agroecosystems calls for developing a robust plan of action. Technical developments have created an environment where it is increasingly appreciated that botanical products can play an important role in crop health and, hence, in food production, and this paper gives a fair snapshot of their current status. The case in favor is reinforced by the fact that the farmer can be the main stakeholder in new botanical pesticide technologies. Earlier negative effects of some synthetic pesticides have been blamed for the detailing of the specific mode of action of botanicals. The details of the alternatives are promising modes of action, but these should not be seen as restrictions for future endeavors. The report lists some of the obvious challenges for the faster upscaling of botanical pesticides as plant protection products, proposals for developing faster solutions to these challenges, and a few specific research areas that some of the African Union Member States would currently benefit from. In conclusion, given their proven potential, higher safety indices, opportunities for new markets for the valuation of botanical research products, and prospects of providing clean and pure alternatives from a broad spectrum of pests for both ordeal and safety, among others, make botanical alternatives an indispensable component in current farming technologies.

References

Adimas, Zemenu Tadesse, Mekuannt Alefe Adimas, and Biresaw Demelash Abera. 2024. "Plant-Based Bioactive Compounds for Grain Storage: A Comprehensive Review." *Cogent Food and Agriculture*. https://doi.org/10.1080/23311932.2024.2316152.

Ankur, Alka Gupta, Pragati Rawat, Mahak Singh, and Sanjiv Mullick. 2024. "Development and Characterization of Cymbopogon Winterianus (Jowitt) Essential Oil-Based Nano-Emulsion for Larvicidal and Antifeedant Activity Against Spodoptera Litura (Fab.) (Lepidoptera: Noctuidae)." *BioNanoScience*. https://doi.org/10.1007/s12668-024-01341-z.

Arora, Himanshu, Gourav Choudhir, Arunava Sengupta, Abhishek Sharma, and Satyawati Sharma. 2024. "Bioactive Metabolites of Licorice and Thyme as Potential Inhibitors of Cox1 Enzyme of Phytopathogens of Capsicum Annuum L.: In-Silico Approaches." *Journal of Biomolecular Structure and Dynamics*. https://doi.org/10.1080/07391102.2024.2303603.

Arora, Himanshu, Farah Naaz, Abhishek Sharma, Saurabh Dubey, Satyawati Sharma, and Gaurav Rajauria. 2024. "Thyme-Licorice Nanoemulsion for Anthracnose Management in Capsicum Annuum L. and Life Cycle Assessment of Its Production." *Biocatalysis and Agricultural Biotechnology* 56. https://doi.org/10.1016/j.bcab.2024.103029.

Baviskar, Dr. Prof. Pallavi. 2024. "Empowering Farmers through Technology: A Sustainable Approach to Agricultural Equipment Rental and Income Generation." *International Journal for Research in Applied Science and Engineering Technology* 12 (3). https://doi.org/10.22214/ijraset.2024.58825.

Benson, G. A. S., O. A. Oyetunde, A. O. Joda, A. O. Nofiu, and K. L. Balogun. 2024. "Particle Size Effect of Piper Guineense in the Control of the 'Sitophilus Zeamais Motchulski' on Maize." *Journal of Agriculture and Ecology Research International* 25 (2). https://doi.org/10.9734/jaeri/2024/v25i2585.

Bi, Xiaoyang, Meihong Lin, Yifeng Zhou, Dandan Li, Zuowei Xu, Lijuan Zhou, and Jiguang Huang. 2024. "Insecticidal Activity and Molecular Target by Morphological Analysis, RNAseq, and Molecular Docking of the Aryltetralin Lignan Lactone Helioxanthin, Isolated from Taiwania Flousiana Gaussen." *Journal of Agricultural and Food Chemistry* 72 (10). https://doi.org/10.1021/acs.jafc.3c06384.

Demir, Sinan, Mert Dedeoğlu, and Levent Başayiğit. 2024. "Yield Prediction Models of Organic Oil Rose Farming with Agricultural Unmanned

Aerial Vehicles (UAVs) Images and Machine Learnaing Algorithms." *Remote Sensing Applications: Society and Environment* 33. https://doi.org/10.1016/j.rsase.2023.101131.

Eisen, Lars. 2024. "Efficacy of Unregulated Minimum Risk Products to Kill and Repel Ticks." *Emerging Infectious Diseases* 30 (1). https://doi.org/10.3201/eid3001.230813.

Evizal, Rusdi, and Fembriarti Erry Prasmatiwi. 2022. "Gejala Produktivitas Rendah Dan Pertanian Degeneratif (Symptoms of Low Yield and Degenerative Agriculture)." *Jurnal Agrotropika* 21 (2).

Gao, Ying, Ethan Swiggart, Kaela Wolkiewicz, Prabha Liyanapathiranage, Fulya Baysal-Gurel, Farhat A. Avin, Eleanor F.P. Lopez, Rebecca T. Jordan, Joshua Kellogg, and Eric P. Burkhart. 2024. "Goldenseal (Hydrastis Canadensis L.) Extracts Inhibit the Growth of Fungal Isolates Associated with American Ginseng (Panax Quinquefolius L.)." *Molecules* 29 (3). https://doi.org/10.3390/molecules29030556.

Ghosh, Sunil Kumar. 2024. "Bio-Efficacy of Plants Based Formulations for the Management of Cowpea Aphid (Aphis Craccivora Koch.)." *Legume Research* 47 (2). https://doi.org/10.18805/LR-4636.

Harte, Steven J., Daniel P. Bray, Victoria Nash-Woolley, Philip C. Stevenson, and G. Mandela Fernández-Grandon. 2024. "Antagonistic and Additive Effect When Combining Biopesticides against the Fall Armyworm, Spodoptera Frugiperda." *Scientific Reports* 14 (1). https://doi.org/10.1038/s41598-024-56599-w.

Haruna S. G., Yahuza L., and Tijjani I. 2024. "Management of Fusarium Wilt of Tomato (Fusarium Oxysporum f. Sp. Lycopersici) and Related Soil-Borne Diseases Using Eco-Friendly Methods: A Review." *Asian Journal of Research in Crop Science* 9 (1). https://doi.org/10.9734/ajrcs/2024/v9i1257.

Hassam, U. A., A. Gulzar, B. Rasool, S. Zafar, T. Younis, M. Shakeel, D. Khan, et al. 2024. "Efficacy of Citrullus Colocynthis Seed Extract on Earias Vittella, Fabricius, (Lepidoptera: Noctuidae): Environment Sustainable Approach." *Brazilian Journal of Biology* 84. https://doi.org/10.1590/1519-6984.254479.

Indermaur, Elizabeth J., Charles T.C. Day, Amara R. Dunn-Silver, and Christine D. Smart. 2024. "Biorational Fungicides to Manage Cucurbit Powdery Mildew on Winter Squash in New York." *Plant Health Progress* 25 (3). https://doi.org/10.1094/PHP-11-23-0103-RS.

Istianto, Mizu, Rudi Cahyo Wicaksono, Otto Endarto, Susi Wuryantini, and Unun Triasih. 2024. "Preliminary Study: Effectiveness Test of Citronella Oil to Control Mealy Bug (Paracoccus Marginatus) on Papaya Fruit in the Field." In *AIP Conference Proceedings*. Vol. 2957.

https://doi.org/10.1063/5.0184046.

Jesus Oliveira, Joseane de, Eliana M. dos Passos, Suely M. Alves, Victor H.V. Sarmento, Thiago R. Bjerk, Juliana C. Cardoso, Cristina Blanco-Llamero, Eliana B. Souto, Patrícia Severino, and Marcelo da Costa Mendonça. 2024. "Microemulsion of Essential Oil of Citrus Aurantium Var. Dulcis for Control of Aleurocanthus Woglumi and Evaluation of Selectivity against Aschersonia Aleyrodis and Ceraeochrysa Cornuta." *Crop Protection* 178. https://doi.org/10.1016/j.cropro.2024.106586.

Kadanakuppe Thammayya, Shivakumara, Keerthi Manikyanahalli Chandrashekara, Akula Chinapolaiah, Ramya Ramesan Syamala, Shivakumar Kadukothanahalli Veerabhadraiah, Bhemanna Somanna Gotyal, Manjunatha Channappa, et al. 2024. "Comparative Effectiveness of Biorational Pesticides for Management of Phenacoccus Solenopsis Tinsley and Paracoccus Marginatus Williams & Granara de Willink in Gymnema Sylvestre (Retz.) R.Br. Ex Sm." *Heliyon* 10 (1). https://doi.org/10.1016/j.heliyon.2023.e23648.

Kampermann, Ivonne, David Bautze, Millicent Mapili, Martha Musyoka, Edward Karanja, Komi K.M. Fiaboe, Janet Irungu, and Noah Adamtey. 2024. "Insecticide Contamination in Organic Agriculture: Evidence from a Long-Term Farming Systems Comparisons Trial." *Crop Protection* 177. https://doi.org/10.1016/j.cropro.2023.106529.

Karunakaran S., and Niranjana R.F. 2024. "Phytochemical Analysis of Some Selected Indigenous Plants in Eastern Region of Sri Lanka Used for Minimizing Insect Damage on Stored Grains during Storage." *Proceedings of International Forestry and Environment Symposium* 27. https://doi.org/10.31357/fesympo.v27.7192.

Kavallieratos, Nickolas G., Nikoleta Eleftheriadou, Maria C. Boukouvala, Anna Skourti, Constantin S. Filintas, Demeter Lorentha S. Gidari, Filippo Maggi, et al. 2024. "Exploring the Efficacy of Four Apiaceae Essential Oils against Nine Stored-Product Pests in Wheat Protection." *Plants* 13 (4). https://doi.org/10.3390/plants13040533.

Khasanah, Nur, Mohammad Yunus, Dance Tangkesalu, Burhanuddin Haji Nasir, Lidya Evalinda Tauke, and Regita Devila Marzita. 2024. "Efficacy of Biduri Leaf Extract (Calotropis Gigantea L.) for Management of Fall Armyworm (Spodoptera Frugiperda J. E. Smith.) on Corn." *International Journal of Environment and Climate Change* 14 (1). https://doi.org/10.9734/ijecc/2024/v14i13875.

Lansana Abou BANGOURA, Aboubacar DIALLO, Adama Moussa SAKHO, Abdoulaye KEITA, Amadou Youssouf BAH, and Louncény TRAORE. 2024. "Ethnobotanical Investigations of Insecticidal Plants in Guinean Flora: Case of the Boke Region (Republic of Guinea)." *World Journal of Advanced Research and Reviews* 21 (1). https://doi.org/10.30574/wjarr.2024.21.1.0009.

Li, Mengmeng, Jingjing Li, Yanpu Meng, Yizhen Wang, Ming Gao, Jingao Dong, Zhiyan Cao, Lihui Zhang, and Shujie Ma. 2024. "Preparation of Nanoliposomes Containing the Extracts of Eleocharis Dulcis Corm-Peels and Ascertaining Their Aphidicidal Activity against Megoura Crassicauda and Acyrthosiphon Pisum." *Industrial Crops and Products* 207. https://doi.org/10.1016/j.indcrop.2023.117746.

Ma, Jiahui, Wei Liu, Xueyu Wang, Chang Lu, Zezhuang Hao, Ye Wang, Yuling Ding, and Yong Li. 2024. "Cnidium Officinale Makino: Phytology, Phytochemistry, Toxicology, Pharmacology and Prescriptions (1967–2023)." *Chemistry and Biodiversity*. https://doi.org/10.1002/cbdv.202301639.

Ma, Li Juan, Xiao Liu, Liwei Guo, Yuan Luo, Beibei Zhang, Xiaoxue Cui, Kuan Yang, et al. 2024. "Discovery of Plant Chemical Defence Mediated by a Two-Component System Involving β-Glucosidase in Panax Species." *Nature Communications* 15 (1). https://doi.org/10.1038/s41467-024-44854-7.

Miloudi, Soukaina, Imane Abbad, Bouchra Soulaimani, Abderrahim Ferradous, Abdelaziz Abbad, and El Hassan El Mouden. 2024. "Optimization of Herbicidal Activity of Essential Oil Mixtures from Satureja Alpina, Thymus Satureioides and Myrtus Communis on Seed Germination and Post-Emergence Growth of Amaranthus Retroflexus L." *Crop Protection*. https://doi.org/10.1016/j.cropro.2024.106642.

Modafferi, Antonino, Michele Ricupero, Giuseppe Mostacchio, Ilaria Latella, Lucia Zappalà, Vincenzo Palmeri, Stefania Garzoli, Giulia Giunti, and Orlando Campolo. 2024. "Bioactivity of Allium Sativum Essential Oil-Based Nano-Emulsion against Planococcus Citri and Its Predator Cryptolaemus Montrouzieri." *Industrial Crops and Products* 208. https://doi.org/10.1016/j.indcrop.2023.117837.

Mohanty, Lalita Kumar, N K Singh, Pranav Raj, Aditya Prakash, Awanindra Kumar Tiwari, Vishal Singh, and Prashun Sachan. 2024. "Nurturing Crops, Enhancing Soil Health, and Sustaining Agricultural Prosperity Worldwide through Agronomy." *Journal of Experimental Agriculture International* 46 (2). https://doi.org/10.9734/jeai/2024/v46i22308.

Murugesan, Pazhanisamy, Archunan Krishanan, Pasiyappazham Ramasamy, Kathirvelu Chandrasekaran, Jayasankar Rajendran, Sivagurunathan Paramasivam, Uma Chinnaiyan, Chellasamy Panneerselvam, Al

Thabiani Aziz, and Abdulrahman Alasmari. 2024. "Biorational Methods for Effective Pest Control Management in Stored Products for Agricultural Sustainability." *Entomological Research*. https://doi.org/10.1111/1748-5967.12697.

Pereira, Verónica, Onofre Figueira, and Paula C. Castilho. 2024. "Flavonoids as Insecticides in Crop Protection—A Review of Current Research and Future Prospects." *Plants*. https://doi.org/10.3390/plants13060776.

Philip, Bijin, and G. Suresh. 2024. "Cultivating Resilience: An In-Depth Exploration of Agricultural Insurance Initiatives and Their Transformative Role in India's Farming Landscape." *Universal Journal of Agricultural Research* 12 (1). https://doi.org/10.13189/ujar.2024.120109.

Popescu, Gheorghe Cristian, Monica Popescu, Moniruzzaman Khondker, David E. Clay, Silvia Pampana, and Mikihisa Umehara. 2022. "Agricultural Sciences and the Environment: Reviewing Recent Technologies and Innovations to Combat the Challenges of Climate Change, Environmental Protection, and Food Security." *Agronomy Journal* 114 (4). https://doi.org/10.1002/agj2.21164.

Popescu, Gheorghe Cristian, Monica Popescu, Silvia Pampana, Moniruzzaman Khondker, Mikihisa Umehara, Hisayoshi Hayashi, and Narong Touch. 2023. "Introduction: Sustainability as an Agroecological Strategy toward Resilience in Agricultural Systems." *Agronomy Journal* 115 (6). https://doi.org/10.1002/agj2.21483.

Pranaswi, Dameruppula, Megha P. Jagtap, Gopal U. Shinde, Narendra Khatri, Stuthi Shetty, and Shravya Pare. 2024. "Analyzing the Synergistic Impact of UAV-Based Technology and Knapsack Sprayer on Weed Management, Yield-Contributing Traits, and Yield in Wheat (Triticum Aestivum L.) for Enhanced Agricultural Operations." *Computers and Electronics in Agriculture* 219. https://doi.org/10.1016/j.compag.2024.108796.

Ragab, Amany, Mohamed A. Taher, Helmy H. El-Rafey, and Ahmed Ramadan El-Rokh. 2024. "Bioactive Compounds from Withania Somnifera Dun and Their Toxicity against Some Piercing Sucking Pests." *Applied Biological Chemistry* 67 (1). https://doi.org/10.1186/s13765-024-00880-z.

Rahma, Rahma, Muhammad Indar Pramudi, and Elly Liestiany. 2024. "Pengendalian Kutu Beras (Sitophilus Oryzae l.) Secara Organik." *JURNAL PROTEKSI TANAMAN TROPIKA* 7 (1). https://doi.org/10.20527/jptt.v7i1.2405.

Reher, Thomas, Cas Lavaert, Brecht Willockx, Yasmin Huyghe, Jolien Bisschop, Johan A. Martens, Jan Diels, Jan Cappelle, and Bram Van de

Poel. 2024. "Potential of Sugar Beet (Beta Vulgaris) and Wheat (Triticum Aestivum) Production in Vertical Bifacial, Tracked, or Elevated Agrivoltaic Systems in Belgium." *Applied Energy* 359. https://doi.org/10.1016/j.apenergy.2024.122679.

Rizvi, Syed Arif Hussain, Zhenhua Liu, Sandra Amarachi Ozuzu, Yasir Arafat, Caixia Han, Filippo Maggi, and Hua Shao. 2024. "Toxicity of Delphinium Brunonianum Royle Alkaloids against the Adults of Diaphorina Citri and Its Mechanism Study in Insect SF9 Cell Line." *Industrial Crops and Products* 208. https://doi.org/10.1016/j.indcrop.2023.117826.

Rizzo, Davide, Elisa Marraccini, and Sylvie Lardon. 2022. *Landscape Agronomy: Advances and Challenges of a Territorial Approach to Agricultural Issues. Landscape Agronomy: Advances and Challenges of a Territorial Approach to Agricultural Issues.* https://doi.org/10.1007/978-3-031-05263-7.

Rizzo, Roberto, Ernesto Ragusa, Giovanni Benelli, Gabriella Lo Verde, Valeria Zeni, Filippo Maggi, Riccardo Petrelli, et al. 2024. "Lethal and Sublethal Effects of Carlina Oxide on Tetranychus Urticae (Acari: Tetranychidae) and Neoseiulus Californicus (Acari: Phytoseiidae)." *Pest Management Science* 80 (3). https://doi.org/10.1002/ps.7827.

Rzyska, Katarzyna, Kinga Stuper-Szablewska, and Danuta Kurasiak-Popowska. 2024. "Diverse Approaches to Insect Control: Utilizing Brassica Carinata (A.) Braun and Camelina Sativa (L.) Crantz Oil as Modern Bioinsecticides." *Forests*. https://doi.org/10.3390/f15010105.

Sa'adah, Nadya Sofia Siti, Sukirno Sukirno, Tri Rini Nuringtyas, and Laurentius Hartanto Nugroho. 2024. "Toxicity of Ocimum Basilicum L. Leaf Extract against Spodoptera Exigua Hübner (Lepidoptera: Noctuidae)." *Plant Science Today* 11 (1). https://doi.org/10.14719/pst.3099.

Scupinari, Tamires, Helena Mannochio-Russo, Gustavo Galo Marcheafave, Vanderlan da Silva Bolzani, Estela de Oliveira Nunes, Waldir Pereira Dias, Clara Beatriz Hoffmann-Campo, and Maria Luiza Zeraik. 2024. "Effects of Crotalaria Spectabilis Extract in Meloidogyne Javanica and Heterodera Glycines Soybean Nematodes." *European Journal of Plant Pathology*. https://doi.org/10.1007/s10658-024-02821-4.

Semerdjieva, Ivanka, Valtcho D. Zheljazkov, Charles L. Cantrell, Lyubka Koleva-Valkova, Vasilina Maneva, Tzenka Radoukova, Tess Astatkie, et al. 2024. "Phytochemical Composition and Biopesticidal Potential of Pinus Mugo Turra Essential Oil." *Industrial Crops and Products* 209. https://doi.org/10.1016/j.indcrop.2023.118019.

Senewe, Rein E., J. V. Hasinu, M. Pesireron, and Ch Leiwakabessy. 2024. "Management of Clove Stem Borer Nothopeus Fasciatipennis Wat

(Coleoptera: Cerambycidae) on Clove Trees in Maluku." In *AIP Conference Proceedings*. Vol. 2957. https://doi.org/10.1063/5.0191213.

Shafik, Wasswa. 2024a. "Barriers to Implementing Computational Intelligence-Based Agriculture System." In *Computational Intelligence in Internet of Agricultural Things*, 193–219. Springer. https://doi.org/10.1007/978-3-031-67450-1_8.

Shafik, Wasswa. 2024b. "Biosensor-Based Drones Anomaly Detection Integration for Sustainable Agriculture Development." In *Achieving Food Security Through Sustainable Agriculture*, 283–316. IGI Global. https://doi.org/10.4018/979-8-3693-4240-4.ch012.

Shafik, Wasswa. 2025a. "Sustainable Agriculture and Diet for Urban Development." In *Sustainable Smart Cities and the Future of Urban Development*, 165–92. IGI Global Scientific Publishing.

Shafik, Wasswa. 2025b. "Sustainable Agriculture and Diet in the Metaverse Era: Sustainable Farming and Nutritious Eating in the Metaverse." In *Food in the Metaverse and Web 3.0 Era: Intersecting Food, Technology, and Culture*, 159–84. IGI Global Scientific Publishing.

Shafik, Wasswa, and Kassim Kalinaki. 2024. "Societal and Ethical Implications of Technology-Enhanced Agriculture and Healthcare: An African Context." In *2024 IST-Africa Conference (IST-Africa)*, 1–11. IEEE. https://doi.org/10.23919/IST-Africa63983.2024.10569306.

Shafik, Wasswa, Ali Tufail, Chandratilak De Silva Liyanage, and Rosyzie Anna Awg Haji Mohd Apong. 2023. "Using a Novel Convolutional Neural Network for Plant Pests Detection and Disease Classification." *Journal of the Science of Food and Agriculture*. https://doi.org/10.1002/jsfa.12700.

Shafik, Wasswa, Ali Tufail, Chandratilak Liyanage De Silva, Rosyzie Anna Awg Haji, and Mohd Apong. 2025. "The Role, Application, and Impact of Artificial Intelligence in the Agriculture Industry." In *Future Tech Startups and Innovation in the Age of AI*, 36–60. CRC Press.

Shafik, Wasswa, Ali Tufail, Liyanage Chandratilak De Silva, and Rosyzie Anna Awg Haji Mohd Apong. 2025. "A Lightweight Deep Learning Model for Multi-Plant Biotic Stress Classification and Detection for Sustainable Agriculture." *Scientific Reports* 15 (1): 12195.

Tesari, Tasafima, Amin Setyo Leksono, and Irfan Mustafa. 2024. "Effectiveness of Botanical Pesticide Combined with Beauveria Bassiana on Mortality, Nutritional Index and Fecundity of Spodoptera Litura L." *Cogent Food and Agriculture* 10 (1). https://doi.org/10.1080/23311932.2024.2320816.

Thompson, Courtney V., Steven D. Webb, Joseph A. Leedale, Peter E. Penson, Alicia Paini, David Ebbrell, and Judith C. Madden. 2024. "Using

Read-across to Build Physiologically-Based Kinetic Models: Part 1. Development of a KNIME Workflow to Assist Analogue Selection for PBK Modelling." *Computational Toxicology* 29. https://doi.org/10.1016/j.comtox.2023.100292.

Ünal, Zeynep, and Tefide Kızıldeniz. 2022. "Smart Agriculture Practices in Potato Production." In *Potato Production Worldwide*. https://doi.org/10.1016/B978-0-12-822925-5.00010-4.

Wulandari, Melisyah Putri, and St. Rahmadani. 2024. "Pengaruh Pemberian Pestisida Nabati Daun Tembakau Terhadap Mortalitas Keong Mas (Pomacea Canakuculaka)." *JUSTER : Jurnal Sains Dan Terapan* 3 (1). https://doi.org/10.57218/juster.v3i1.1039.

Yates, Ping, Juddy Janiol, Changbao Li, and Bao Hua Song. 2024. "Nematocidal Potential of Phenolic Acids: A Phytochemical Seed-Coating Approach to Soybean Cyst Nematode Management." *Plants* 13 (2). https://doi.org/10.3390/plants13020319.

Yeshiwas, Tadele, Melkamu Alemayehu, Fenta Assefa, and Selamawit Zelalem. 2024. "Compositions, Practices and Constraints of Home Garden: The Case of Bahir Dar City, Ethiopia." *Cogent Food and Agriculture* 10 (1). https://doi.org/10.1080/23311932.2023.2298025.

Yuan, Lisha, Tao Li, Yi Huang, Aoying Zhang, Shanchun Yan, and Dun Jiang. 2024. "Identification and Potential Application of Key Insecticidal Metabolites in Tilia Amurensis, a Low-Preference Host of Hyphantria Cunea." *Pesticide Biochemistry and Physiology* 199. https://doi.org/10.1016/j.pestbp.2024.105796.

Zenteno-Alegría, Claribel Orquídea, Luis Andrés Yarzábal Rodríguez, Jimmy Ciancas Jiménez, Peggy Elizabeth Álvarez Gutiérrez, Nina Gunde-Cimerman, and Ramón Alberto Batista-García. 2024. "Fungi beyond Limits: The Agricultural Promise of Extremophiles." *Microbial Biotechnology*. https://doi.org/10.1111/1751-7915.14439.

Zhao, Qiuyu, Guoxing Wu, Pu Yang, Yuanchong Shi, Zuoyi Fu, Haifeng Mo, Chunlan Shi, and Shuhui Yu. 2024. "Metabolomic and Transcriptomic Analyses Reveal the Molecular Mechanism Underlying the Massive Accumulation of Secondary Metabolites in Fenugreek (Trigonella Foenum-Graecum L.) Seeds." *Genes* 15 (3). https://doi.org/10.3390/genes15030343.

PART II

ENABLING TECHNOLOGIES IN SUSTAINABLE AGRONOMY

CHAPTER FIVE

PRECISION AGRICULTURE AND ENABLING TECHNOLOGIES FOR SUSTAINABLE AGRONOMY

Abstract

Precision agriculture, often known as site-specific crop management or precision farming, refers to a holistic approach to farming management in which decisions related to agriculture are made using information from multiple sources. These include previous and present prices of inputs and products, knowledge about specific crop management practices, and site-specific information obtained from digital data sources, such as field maps, satellite images, aerial photographs, lab analysis, and livestock performance maps. The concept of precision agriculture evolved initially with the widespread use of new technology, which meant growers were shifting to mechanical plowing and sowing. This was the start of a move to precision sowing, and while it wasn't identified as such, it marked the beginning of precision agriculture. The basic need for precision agriculture was to feed the increasing population of the world in the face of a decreasing cultivable area and increasing pressure on natural resources. This was the main philosophy that moved scientists and researchers to come up with the technological advancements that became known as precision agriculture. The development of global information systems in agriculture has occurred in the wave of technological revolutions from the 1970s to the present. Parallel events occurring during this time have advanced personal computers, the World Wide Web, cheap web camera technology, and powerful geographic information systems. Unlike other agricultural revolutions, technology is driving the change, and new technology will drive the industry. Technology has made it cheaper, faster, and more accurate to map environmental variables, including soil electrical conductivity, terrain maps, spectral radiance, multispectral imagery, hyperspectral imagery, and radar maps. With large quantities of local data, it is not hard to imagine where this could lead.

Keywords: Agricultural IoTs, Agricultural Robotics, Artificial Intelligence, Big Data Analytics, Sustainable Farming, Drones for Crop Monitoring, GPS-Guided Machinery, Machine Learning, Precision Agriculture, Remote Sensing Technologies, Variable Rate Technology

5.1 Introduction

Precision agriculture, in its most precise definition, is a management strategy that uses information technology to ensure that the crops and soil receive exactly what they need for optimum health and productivity. It's being practiced throughout the world in all types of crops and animal farming. Precision agriculture seeks to integrate technology and data analytics to optimize agricultural production and environmental quality. According to this definition, precision agriculture can include multiple methodologies and technologies, which can be used individually or jointly (Kobe et al. 2024). For example, the transversal use of positioning technologies in agriculture is an important part of precision agriculture. However, we can go further and define precision agriculture as an integral and multi-professional approach, which seeks to incorporate technological advances in the knowledge of the biophysical relationships of agricultural productions, integrating different data sources in such a way as to know in detail the variability observed in them (soil-plant-atmosphere systems), so that producers can make efficient decisions to manage it in the interest of increased profits and optimizing the use of natural resources like water and agrochemicals (Justina and Thenmozhi 2024).

Moreover, precision agriculture includes automatic data acquisition in a real-time system and its use in spatial characterization and management and resource conservation, such as water and seeds. Precision agriculture is also considered a completely new way for farmers to monitor, manage, and optimize their yields based on the spatial observation of the soil and crops, and others view precision agriculture as soft computing for an agrotechnological problem-solving approach with precision scale resolution (Babos et al. 2024). The area of the use of precision methods varies from machine guidance and control to planting, seeding, water management, soil management, agro-environmental and climate studies, etc. Precision agriculture encompasses all technologies used in modern farming that make practices more precise. Instead of making general, broad recommendations, precision agriculture aims to target specific crop needs in specific spots. These targeted spots can be areas that lack proper nutrition, suffer from soil erosion, show heavy pests, or don't hold up as well as others. Precision agriculture begins its intervention during the preparation phase of planting and continues these practices in each phase of the long, complicated growing process of a crop (Anusha, Anandan, and Venkata Krishna 2024). There are eight noticeable precision agriculture trends, as pointed out in Fig. 5.1.

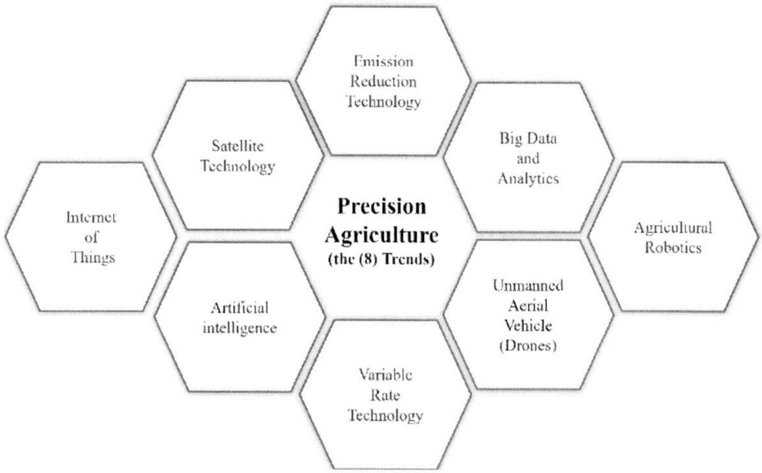

Fig. 5-1 The eight precision agricultural trends

By using precision agriculture methods, a variety of benefits can be brought that make agricultural production more sustainable. Weed management, for example, can be done using far less intervention, such as fertilizing and spraying. This can lead to a 20 percent increase in crop yields while decreasing input costs anywhere from 20 to 50 percent. These practice changes in water retention and efficient use of resources can drastically reduce nutrient and chemical run-off, implementing these practices not only financially beneficial for farmers but also a responsible recommendation for the health of surrounding ecosystems. Smallholder farmers can see these benefits more readily, as the use of chemical and nutrient applications will increase profitability per crop. This increase in profitability per land unit can, in most cases, mean the difference between feeding a family or not for many smallholders (Kobe et al. 2024). Increased food security can then help bolster a community from the bottom up, increasing community resilience and individual livelihoods.

Spatial and temporal crop variability is crucial in the decision-making process of precision agriculture. Understanding, predicting, and managing this variability represent key challenges that can be addressed by means of different sciences, such as remote sensing and wireless sensor networks. Continuous soil and crop monitoring significantly contributes to understanding plant needs and their interactions with the new cultural practices. In order to enable fast and intelligent decision support for the adoption of these precision cultural practices, this study proposes the use of hybrid intelligent systems that will use data mining to process data from sensor networks (Babos et al. 2024). Furthermore, these techniques should be able to handle the high spatial and temporal grain of the input data available from wireless sensor networks. In this context, this study presents major contributions to the specification and design of an architecture that supports continuous and adaptive mining of the knowledge extracted from the data collected by wireless sensor networks in precision agriculture. This architecture is supported by a knowledge management approach that uses data warehousing concepts for data acquisition, data quality, and its characteristics preservation for dimensionality reduction, and a classical data mining phase for the end-user extraction of the acquired knowledge, as illustrated in Table 5.1 with the enabling next-gen technologies.

Table 5-1 Significant Next-Gen Technologies in Precision Agriculture

Technology	Description	Application	Benefits	Challenges	Reference
Mobile Applications	Smartphone apps for farm management	Crop monitoring, data input	Easy access to real-time data	Limited in areas with poor connectivity	(Monteleone et al. 2024)
Renewable Energy Integration	Solar and wind power for operations	Powering farm equipment	Reduces operational costs	Dependence on weather conditions	(Romano et al. 2024)
Autonomous Vehicles	Self-driving tractors and harvesters	Planting, spraying, and harvesting	Increases efficiency	High cost and technical complexity	(Pauline A. Ongadi 2024)
Blockchain	Secure data management and transparency	Traceability of agricultural products	Enhanced data security	Adoption barriers in rural areas	(Garcia et al. 2024)
Geographic Information Systems (GIS)	Spatial data analysis	Mapping field variability	Enhanced land use planning	Requires specialized expertise	(Corceiro et al. 2024)
Climate-Smart Technologies	Tools adapting to changing climate	Drought-resistant farming practices	Improved resilience	High cost of technology	(Eron et al. 2024)
Soil Sensors	Embedded sensors for soil properties	Nutrient level monitoring	Precision in soil management	Installation and maintenance costs	(Gopi and Karthikeyan 2024)

Big Data Analytics	Processing large datasets	Weather forecasting, market trends	Improved risk management	Complex data management	(Kim and Park 2024)
Drones	Unmanned aerial vehicles	Field scouting, imaging	Fast, detailed data acquisition	Regulatory and battery limitations	(Govi et al. 2024)
Robotics	Automated machines for farming	Weeding, harvesting	Reduces labor costs	High cost of development	(Devi and Rajan 2024)
Artificial intelligence	Data analysis and decision support	Yield prediction, disease detection	Enhanced decision-making	Data quality and interpretability issues	(Nguyen et al. 2024)
Variable Rate Technology (VRT)	Adjusts input rates based on data	Fertilizer and pesticide application	Optimized resource use	Requires high-quality data	(Morales-García, Terroso-Sáenz, and Cecilia 2024)
Remote Sensing	Satellite and drone imagery	Crop health and stress detection	Large-area monitoring	Resolution limitations	(Silva et al. 2024)
Global Positioning System (GPS)	Satellite-based location tracking	Precision planting and mapping	Accurate field positioning	High initial setup cost	(Gauci et al. 2024)
Internet of Things	Sensors connected via networks	Soil moisture, temperature monitoring	Real-time data collection	Data integration and security	(Bütüner et al. 2024)

5.2 Sensing Technologies

Precision agriculture relies on diverse technologies for data collection. This new agri-food system of production and management requires the collective use of several complementary sensors that can be described in two main categories. Remote sensing consists of technologies deployed high above the earth's surface, and proximal sensing includes sensors that are placed at low altitudes above the target of observation or in close contact with targets of observation. Remote sensing technologies offer the capacity for wide and dense coverage of plots and can provide a deep and accurate understanding of crop health and soil conditions (Kobe et al. 2024). From a technical point of view, this type of sensing detects and processes radiation reflected and emitted by the target of interest. The term "target" refers to the typical object of observation, in our case, land and crops, while "radiation" encompasses a wide range of spectrums, including visible and invisible spectrums. The closer the spectrums are to the sun, the more they are the source of all energy on earth and the driver of biological activity in living cells, converted into plant biomass and influencing chlorophyll levels (Babos et al. 2024).

Increasing plant and chlorophyll levels are usually associated with high leaf nitrogen content. It is for the same reason that the ratio between red and near-infrared spectrums allows accurate prediction of vegetation index: NDVI index = (NIR - R) / (NIR + R). It reflects plant chlorophyll and biomass levels and the ability of the crop to absorb sunlight and grow. Vegetation indexes are used to detect the health of the crop in different phenological stages and are used to ensure timely recovery from defects or diseases that may affect the final yield. The farming-related and agri-environmental solution with IoT is also data-driven but refers to very precise and systemic environmental monitoring at the farm scale by some well-selected and multisensory technology placements in pilot farms. They essentially integrate imagery and sensing technology with automatic well-debarked positional data supporting type of analysis and scanning and hyperspectral analysis to support the high-resolution agricultural field mapping algorithm further (Balayan et al. 2024). Precise soil and environmental information are expected to result in more precise nitrogen fertilization.

5.2.1 Remote Sensing

The term remote sensing includes a number of technologies critical to the development of precision agriculture. In particular, remote sensing is an excellent way to monitor how a crop is growing in a large area. By using a variety of methods, including satellites, drones, and planes, remote sensing can help to quickly assess how a crop is doing and identify areas or plants that may be under stress. There are many uses of remote sensing in precision agriculture, including monitoring crop development, assessing the health of plants, identifying weeds, diseases, and insect problems, tracking soil characteristics, and mapping management zones within a field. Aerial photography has historically been used to monitor crops on a regular basis. Since the first satellites began orbiting the Earth, remote sensing of crops has moved to this more futuristic platform (Al-Ammary and Ghanem 2024). Recent advancements in sensor and satellite technologies now provide spectral and spatial resolutions more detailed than ever before. Analysis indicates that the data from these satellites can be used to identify and differentiate between crops such as corn, milo, and soybeans. Additionally, hyperspectral imagery can be used to monitor plant health through changes in chlorophyll content and photosynthetic rates. In addition to being relatively inexpensive and capable, small satellites in the hands of private companies can cover a broad array of fixed and urgent-use applications for agriculture with essentially real-time observation (Gauci et al. 2024). There are, however, challenges to using satellite data, especially when it comes to the use of this data in decision-making using a standard precision agriculture cycle, as shown in Fig 5.2.

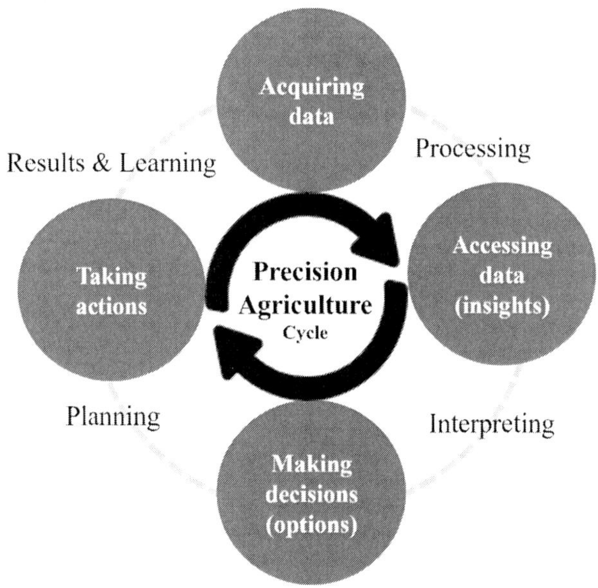

Fig. 5-2 Precision Agriculture Cycle

5.2.2 Proximal Sensing

Proximal sensing is a crucial technology in precision agriculture. Due to its low cost, availability, and speed of data collection, different types of tools are emerging in this sector. Most common among the African options are soil probes, handheld sensors, and mobile applications. These tools primarily provide localized or spatially specific information about the farm fields. The data collected from these sensors enable farmers to make informed decisions for farm management, with near real-time operations for input application and irrigation scheduling, managing resources for improved production, and cost-effective solutions for any area of interest (Babos et al. 2024). The sensors and technologies range from the assessment of soil moisture, soil moisture tension/deficit, electrical conductivity, soil type/map, nutrient deficiency, and leaf wetness. Most of them are used for in-situ soil assessment. With the possibility of integrating collected data with GIS and remote sensing, the use of some of the tools could be improved with more functionalities removed from mobile and put on the desktop to integrate with other agricultural technologies to promote data storage and efficient decision-making (Kobe et al. 2024).

The integration capabilities include sending reports and history of data collection and trends while allowing inputs through the interface, evaluation of completed input based on farm needs in order to recommend certain actions to the farmer, and displaying favorite areas to compare data and evaluation on input efficiency and effectiveness in comparison to other favorite areas of interest. These tools currently work well. However, consideration should be made so that much of the interface between the farmer, sensor tool, and final analysis and recommendation can take place independently from farmer decisions, possibly at the Specialized Extension Unit or computer services (Anusha, Anandan, and Venkata Krishna 2024). This can guarantee an unbiased viewpoint and make the best use of available resources and conditions not only on the farmer's farm but in other areas and soil types that are of interest. The system or these tools may compete favorably with other remote sensing techniques, which tend to take much longer to provide feedback, depending on satellite passes over target sites. Nationally, these technologies are being used by a few project personnel with a sound educational background, some farmers, and student interns (Romano et al. 2024).

5.3 Data Analytics and Artificial Intelligence

An immense quantity of data is collected using a combination of sensing technologies in the field. Processing and analyzing such a large amount of data for decision-making forms the first step toward introducing precision agriculture. A simple descriptive statistics analysis can help in offering some understanding and answering the "What has happened?" question. These statistics may provide information related to averages, medians, modes, and variance for a given set of historical data. A more formal approach for gaining insights from historical data is the use of statistical analysis (Monteleone et al. 2024). Statistical techniques offer a wide range of methods to examine long-term trends and to study the correlations. Predictive modeling is able to answer the "What is likely to happen?" question. In predictive modeling, historical data are used to make predictions about future events. Popular tools for predictive modeling in agriculture are machine learning algorithms. Machine learning is a method of data analysis that automates analytical model building. A number of machine learning algorithms are available, including linear regression, decision trees, and support vector machines. Machine learning models can be trained in an automated way to identify patterns and make decisions without being explicitly programmed to do so (Ortatas et al. 2024).

The application of machine learning can be advantageous for managing agricultural processes with multiple stages. Machine learning can help in maximizing profit, managing resources in a precise manner, and also in environmental protection. A combination of predictive analytics, machine learning, and sensing technologies can also pave the way for closed, fully automated farming systems. The future of data analytics is in the use of artificial intelligence or, specifically, machine learning algorithms that can develop prediction models on data, self-learn from past mistakes, and make decisions based on input data (Silva et al. 2024). Deep learning, a kind of machine learning, is able to process complex data, such as images captured by remote sensing from the field. While classical machine learning must be supplied with data that has been hand-selected and fed to the algorithm by a user, deep learning can automatically and hierarchically learn from the data. AI-empowered data analytics enable proactive management decisions through timely interventions to improve yield and resource efficiency (Kim and Park 2024).

5.3.1 Machine Learning Algorithms

Machine learning is the study and construction of computer algorithms that improve their performance on some tasks with experience. Three basic types of machine learning algorithms exist. Supervised learning algorithms predict an output variable from input data with a collection of labeled training data. Unsupervised learning algorithms partition input data into clusters and try to discover hidden structures using only input data. Reinforcement learning algorithms learn to interact with an environment by performing actions and receiving feedback from the environment in the form of rewards. Data scarcity is a major challenge in agriculture owing to the crop growth cycle, land heterogeneity, weather dynamics, and variation in microclimatic conditions (Shafik et al. 2024; Shafik, Tufail, Liyanage, et al. 2023). However, machine learning offers a possible solution to developing predictive crop models that can help farmers and stakeholders in risk assessment as well as yield optimization.

The adoption of machine learning techniques allows for the prediction of future outcomes based on historical data. For example, in agriculture, learning algorithms can predict future crop yields based on historical data, evaluate the disease risk of a crop, and notify a suitable irrigation schedule based on environmental conditions. There are supervised, semi-supervised, unsupervised, and reinforcement machine learning techniques. Each of them can be implemented using a wide range of methods and software packages (Devi and Rajan 2024). In precision agriculture, the outputs of machine learning techniques are used to drive decision support systems in order to allow farmers to schedule their farm activities in an accurate and well-informed manner. Data quality is another important consideration when using machine learning models. The selection of appropriate attributes and the availability of historical data are critical to model preparedness for real-life decision support systems. Continuous model accuracy evaluation and model retraining are required since machine learning models need to adapt on a regular basis in real-life scenarios (Monteleone et al. 2024).

5.3.2 Deep Learning Techniques

Recent years have witnessed an increased interest in the scientific community in employing deep learning techniques for precision agriculture. Deep learning, a subset of machine learning, is concerned with modeling high-level abstractions in data using algorithms called artificial neural networks. These are composed of several layers of interconnected nodes, which enable the analysis of data in complex patterns or structures. The main advantage of deep learning over other approaches is to automate the feature extraction from raw data, which is important for interpreting large and complex datasets. The training and validation of the deep learning models are typically computationally intensive and require a very large set of high-quality data, which is a challenging issue for many real-time applications, including those in agriculture (Romano et al. 2024).

Deep learning has been extensively used in precision agriculture as an emerging technology to automatically recognize and interpret the raw information obtained from agricultural data, which can be in the form of images, sounds, or text. In what follows, we provide a brief survey of the state-of-the-art applications of deep learning in precision agriculture. Several studies have shown the potential of using deep learning and neural networks for crop monitoring based on images. Deep learning techniques have been extensively employed for automating the detection and diagnosis of plant diseases. In addition to plant diseases, insect and pest attacks are also extensively monitored (Devi and Rajan 2024). A deep learning technique was presented to develop an advanced economic threshold forecasting system to predict the need for pest control measures. Although deep learning was applied in greenhouse conditions using images of the tobacco budworm, promising conclusions can be generalized to modern agriculture. Sensors based on deep learning can be used to count the number of fruits and the percentage of fruit coming from the whole plant. In another study, the deep neural network model was used to predict the head lettuce yield values (Gopi and Karthikeyan 2024). Random Forest and deep learning techniques were used to predict the number of cherries in Japanese cherry trees.

5.3.3 Crop Monitoring and Yield Prediction

Monitoring of land utilization at a spatial scale and rate technically feasible for precision agriculture demands the application of advanced technologies, including ground- and air-borne remote sensing. Sensor technologies, including hyperspectral sensors, synthetic aperture radar, and LIDAR, are used or have the potential for use in crop monitoring. These instruments are capable of remotely measuring canopy reflectance and backscatter at spectral and temporal resolutions that are sensitive to microscale crop canopy variability. High technology logistics requirements limit the utility of these sensors for crop monitoring by non-specialist end users. Ground-based systems utilizing ground vehicles, ground robots, and proximal sensing are used for crop monitoring (Pauline A. Ongadi 2024). This range of ground-based systems is gradually expanding, and current designs are finding increasing use for research, demonstration, and in some cases, commercial applications. The worldwide scale application in support of precision agriculture remains problematic due to the cost and logistical requirements.

However, with the increasing intensity of horticulture and specialty cropping for which precision agriculture will be most beneficial, there may be a stronger incentive to develop and deploy a diverse range of technologies in the near future. For crop monitoring, yield prediction, and other crop management and customer relation activities, a plant environment monitoring system is required. The proposed system, which has potential applications for other crops and fields, monitors both plant growth and growing conditions such as illumination, nutrition, and water supply (Morales-García, Terroso-Sáenz, and Cecilia 2024). Through quantitatively correlating the effects of plant growth on the growing conditions, fertilization and watering can be precisely controlled, which in turn improves the quality and quantity of cherry tomatoes. In addition, by monitoring the number of various indicators, such as chlorophyll and the amount of absorbed light, the system predicts the fruit volume and growth speed about one month in advance and gives the prediction to the pre-marketing customers, who can then plan their resellers accordingly using appropriate application processes, as depicted in Fig. 5.3 (Kim and Park 2024).

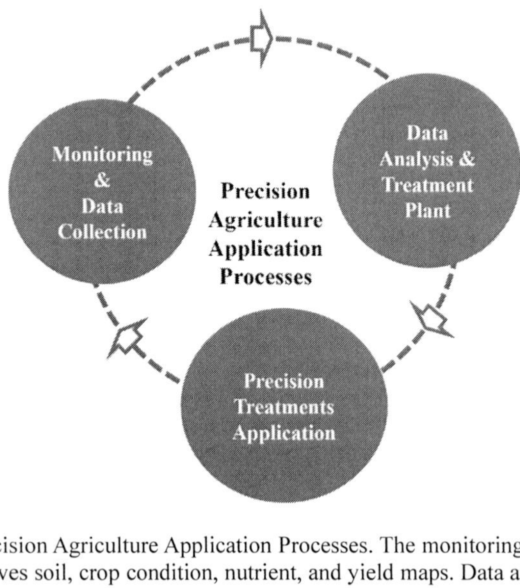

Fig. 5-3 Precision Agriculture Application Processes. The monitoring and data collection involves soil, crop condition, nutrient, and yield maps. Data analysis entails post-processing, modeling, and prediction management zones and the precision treatment application involving soil preparation, crop management, and harvesting.

5.3.4 Weed and Pest Detection

One of the most persistent problems of modern agriculture is the control of weeds. Weeds compete with the main crop for light, nutrients, and water.

Even worse, they can alter the microclimate in the surroundings of the cultivated plants. Weeds act as hosts for several pathogens and pests and can serve as a reservoir for many of the latter. Commonly used tactics to control the presence of weeds are mechanical or chemical methods. The latter is the most widely used, yet it represents a growing concern because it has been linked to several environmental issues, specifically the spreading of toxic substances, the increasing development of herbicide-resistant weeds, and the reduction of biodiversity. With the potential exhaustion of arable land and the growing global population, the issue of feeding the population is becoming more intricate. This issue involves the capability of controlling weeds and the increasing demand for herbicide spot treatment to reduce the environmental impact of their use (Corceiro et al. 2024). Decision support on when to apply herbicide automatically will also reduce the risk of spreading herbicide-resistant weeds.

Despite the fact that it is the most commonly used method, herbicide applications are often really ineffective and excessive. The former happens because the level of weed infestation is unknown most of the time. The latter is due to the fact that the application is usually performed on preset schedules, neglecting the spatial variability of weed emergence. Alternatively, the use of robots equipped with machine vision and machine intelligence techniques can be used to selectively apply herbicides in the spots where the weeds are located. A general network of functioning includes the following steps: Acquisition of aerial images by drones (Romano et al. 2024). Construction of weed and plant indices. Application of segmentation algorithms to locate individual plants and weeds. Chemometric analysis to infer the biochemical contents of the plants and weeds from the acquisition. Data processing through online machine learning to adapt the number, type, and location of the treatments. Several satellite and airborne remote sensing systems have been successfully used for weed detection. Traditionally, these methods are designed to distinguish two or three main ground land use types approximately within one week after the crop emergence. However, specific methods are required to deal with pest detection within the crop rows, particularly for weed detection (Monteleone et al. 2024).

5.4 Automation and Robotics

The rise of the machines: automation and robotics in precision agriculture
Autonomous vehicles, in particular drones and tractors, are rapidly changing the agricultural landscape. Autonomous vehicles, with guidance from GNSS and on-board sensors, are capable of fully autonomous flight paths without any human intervention. Tractor-driving robots have the potential

to plant seeds, apply fertilizer, manage pests, and harvest with high precision. Robots such as drones have found extensive applications in agriculture for selective and variable rate pesticide application, spray operations, detection of plant stress, greenhouse monitoring, livestock tracking, managing water and pesticides, and monitoring crop diseases (Kobe et al. 2024). Field trials have shown the potential for fruit picking, both above and on the ground, with automated systems offering the capability of collecting data daily for better harvest management. The capability for drones to carry out assigned tasks, such as planting and fertilizing operations, with minimal human intervention is transforming the process of farming. Drones used in precision agriculture can capture images from high resolutions, providing farmers with real-time and repeated snapshots of crop health to make educated decisions to improve margins and reduce waste. These measurements are used to guide, in a completely autonomous manner, soil cultivation, planting, fertilizing, watering, weed control, harvesting, monitoring crop yields, and monitoring for spoilage. Since drones carry out petty fogging, a decreased environmental impact of pesticide sprays and a lesser load of inert pesticides used on the environment is expected (Justina and Thenmozhi 2024). Robots can reduce the occurrence of human errors related to pesticide applications, such as missed areas due to manual spray paths in fields.

A robotic system designed for precision spraying and forwarding autonomous mobile robots exhibited significant improvements in both efficiency and accuracy over manual methods. In comparison to manual spraying done by humans using backpack sprayers, farmers could save between 21 and 42% in the cost of removing weeds in apple orchards when using the robotic ground system, which could lead to a reduction in orchard operations cost and a positive return on investment (Shafik 2024b). Additionally, various multinational corporations have already produced prototype autonomous motor vehicles that are specifically designed for seeding, planting, and harvesting operations. Automation in agriculture can enhance productivity by conducting planting, fertilizing, weed-killing, and harvesting operations with precision. However, automation still presents a number of significant technical, legal, and social issues before its widespread implementation (Anusha, Anandan, and Venkata Krishna 2024; Alves and Cruvinel 2024). These include the cost of automation, regulatory and health and safety issues arising from the use of uncrewed aerial vehicles for chemical spraying, and the possible need for training of a more technically skilled workforce as some manual tasks are made redundant by automation. The worldwide agriculture sector will require an extra two billion workers by 2050 to meet growing populations and rates of agricultural production. However, with the labor force already saturated and long working hours causing tension, young individuals who are considering the agriculture sector as a potential working opportunity are becoming less numerous. The development of technologies for manipulating field environmental data and precision in all farming activities, including planting, fertilizing, pest control, and reaping, is one potential solution to this challenge. Automation can slightly advance in every sector of the value chain, from planting to pest control, to improve productivity and lessen the necessity for a large workforce while allowing farmers to make data-driven decisions based on field environmental data accurately (Al-Ammary and Ghanem 2024).

5.4.1 Autonomous Vehicles

Applications some of the autonomous vehicles used in precision agriculture include unmanned aerial vehicles or drones, which are also called unmanned aircraft systems for safety purposes. Others include ground vehicles such as tractors, harvesters, driverless robots, and uncrewed ships. These vehicles could be equipped with various sensors like global positioning systems, satellite imagery, robotics, and low-altitude remote sensing in the case of drones to enable the collection of real-time data related to soil and crops (Shafik 2025). Replacing human farmers employed in repetitive field-based activities such as crop and livestock monitoring, seeding, and weeding allows for a reduction in labor-associated costs. In addition, these autonomous vehicles perform repetitive tasks around the clock with similar precision and speed and do not require regular health and safety breaks as humans (Silva et al. 2024). They could also execute precision-based farm operations that may not be accurately achieved using human labor. For instance, the operation of some agricultural tools, such as drones with single rotors, is complex to be operated by human hands. Environmental Impact Similarly, these vehicles have positive impacts on the environment and human health. Overall, their operations decrease the discharge of chemical inputs, including pesticides, fertilizers, and irrigation water. Most of the current agricultural robotic applications involve precision-based operations, potentially reducing chemical inputs. Also, there is evidence of decreased fuel consumption and soil compaction for some of the autonomous vehicles in agriculture like driverless tractors (Govi et al. 2024). These ground tractors can obviate the manual labor of tire wheel deepening that increases soil compaction. Dual-layered or multi-layer cloud computing architectures can assist in overcoming battery life limitations by commanding the ground vehicles from the electrically driven quadcopters or drones overhead. In particular, the majority of the drones would utilize multirotor or fixed-wing packages powered by batteries that can operate for about twenty to seventy minutes. These advanced technologies and architectures could enable different categories of drones to cooperate and optimize the total routing time to satisfy real-time scheduling goals autonomously (Kobe et al. 2024).

5.4.2 Robotic Systems

These are some of the enabling technologies for precision agriculture due to their potential to work 24/7, perform repetitive tasks, and work in harsh and adverse weather conditions. Several robots are developed for planting, weeding, spraying, monitoring, thinning, and harvesting. These robots are able to perform crop-specific treatments based on crop, soil, and plant properties. In comparison to manual operations and precision agriculture, robotic systems show various operational advantages, including high accuracy, on-time intervention, enabling faster operations, and performance in adverse weather conditions. Robotic systems are not able to replace human labor in all kinds of agricultural tasks; however, they can collaborate with human labor on various agricultural tasks (Anusha, Anandan, and Venkata Krishna 2024). In this context, human workers can supervise and guide robots' operations. Although robotic systems have several operational advantages, they are not widely used by end-users in agriculture due to various reasons. The high initial investment, low adaptability of the technology, a lack of maintenance and repair services, and the productivity of the robotic systems compared to manual workers are some challenges in the market that slow down the adoption of the technology. Furthermore, the precision of the robotic systems can change over days due to variable environmental conditions such as high and low temperatures, humidity, airflow, and nighttime. To overcome these challenges, researchers focus on AI and machine learning algorithms for modeling environmental and weather conditions and for their impact on the robots' operations (Yuan 2024). There are five ways of approaching precision agronomy, as presented in Fig. 5.4.

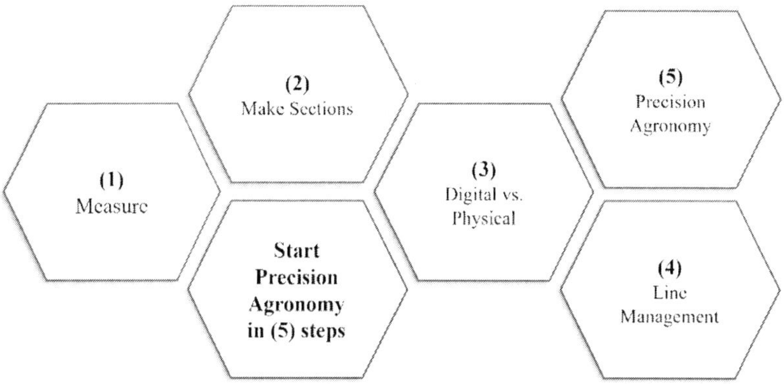

Fig. 5-4 The five steps of starting precision agronomy. It commences with measurement, making sections depending on the requirement, testing the approach, whether digital or physical, line management and finally, precision agronomy.

5.5 Geographic Information Systems (GIS) in Agriculture

GIS technologies have emerged as essential tools for the field of precision agriculture. GIS is a computerized information management system used to capture, store, manage, analyze, describe, and output different types of spatial and geographical information or phenomena. In agriculture, GIS is used to provide farmers with the capability to visualize, manipulate, analyze, interpret, map, and interact with their farm's spatial data. GIS can piece together various types of data, including field topography, soil quality, crop yields, tillage activities, animal movement, water runoff patterns, weather systems, government regulations, and other information related to the farm. The maps generated can be used to help gain an understanding of spatial trends, patterns, and interactions in the fields. Combining data obtained from satellite imagery or GPS is helpful for farmers and ranchers to properly assess the needs of their land and efficiently manage their resources (Ortatas et al. 2024). Being able to accurately and quickly assess the above factors is the key to responding effectively to weather events and reducing the likelihood of environmental degradation, such as soil erosion.

The main functions of GIS in agriculture are concerned with data spatialization and different forms of objective data viewing and valuation produced in agriculture. For instance, these systems are useful in the realization of soil use and occupation studies, distribution of the soil potential to crop delineation and environmental fragility. It also helps in the analysis of climate and crop interaction, in computer-aided teaching, in the making of the

Organic Geographic Registration, and in Blockchain technology. GIS provides the ability to analyze a large amount of field-related data. With the introduction of spatial analysis into GIS, it is possible to go a step further and analyze the spatial variability of a set of related parameters such as temperature, wind speed, humidity, and rainfall within a geographical study area. These systems enable the identification of certain parameters' spatial patterns, the aim being to analyze and identify the factors explaining these patterns (Gauci et al. 2024). Once the visual interpretation and analysis of GIS-generated maps are done, the interpretation of this data becomes useful to optimize resource allocation, assimilate factors in the landscape, and provide a link between different factors under an integrated environment. As such, GIS techniques form part of PA techniques in this field. Having environmental data in a numerical format is not that useful when it cannot be visualized for the purpose of integrating environmental information as part of agriculture development projects (Devi and Rajan 2024).

Geospatial technologies are based on the use of GIS tools, which help us to manage and analyze large volumes of spatial data generated by remote sensors and stored in a database. Thus, a geographical information system integrates computer software, hardware, as well as digital image processing and remote sensing techniques, GPS, and thus the internet to store, display, and manipulate information about places on the earth's surface. In addition, Geographic Information System tools provide support for scientific data analysis and modeling of spatially distributed processes, aiding decision-making processes by the visualization of input data as thematic maps (Kim and Park 2024). Specific GIS applications permit the juxtaposition or overlay of these thematic maps to allow the analysis of possible relationships between the variables. A Geographic Information System integrates hardware, software, and data for capturing, managing, analyzing, and displaying all forms of geographically referenced information. GIS allows us to view, understand, question, interpret, and visualize data in many ways that reveal relationships, patterns, and trends in the form of maps, globes, reports, and charts (Pauline A. Ongadi 2024).

5.5.1 Mapping and Spatial Analysis

What is the most important component of precision agriculture? Many researchers agree that mapping is one of the most important components because seeing the spatial relationships that are present in fields and between fields is crucial for management purposes. Tools for mapping and performing spatial analysis are available in many Geographic Information Systems packages. Each of these packages can map a field and then analyze different aspects of that field in a spatial context. This section will give an overview of the mapping and spatial analysis tools available to help make information-based decisions in agriculture. Mapping in GIS applications can involve a wide range of uses in agriculture (Bütüner et al. 2024). These include crop mapping, pest risk assessment, soil mapping, protection of indigenous lands, and forestry applications. Crop mapping is useful for a variety of reasons, but in precision farming, it can be used to estimate crop area and assess the severity of crop damage or pest infestation. In much the same way, soil mapping can be used to identify soil types that are present in each field, and weed mapping can identify stressed areas. Thematic mapping techniques incorporate traditional image processing techniques with ground-based data to assign specific attributes to various zones (Balayan et al. 2024).

One of the most important principles applied in precision agriculture is that of site-specific management practices. Advancements in the development of Geographic Information Systems have made spatial analysis tools more accessible to the general producer interested in implementing site-specific management concepts. These new tools have concentrated primarily on the use of crop or yield maps to identify high and low-yielding areas of a field. This can be subsequently correlated with soil types and management

zones throughout the field. An important use of Geographic Information Systems in precision agriculture is to aid in the development of management zones in a field, which helps to optimize input applications in the field. Spatial data has been used to help in developing prognosis models for areas affected by natural disasters and pests. More directly related to precision agriculture, mapping and GIS have been used to solve three problems of importance: dealing with current growing season variability, minimizing the effects of topographical heterogeneity, and minimizing yield variability (Ortatas et al. 2024; Bütüner et al. 2024). Many studies have shown that integrating GIS and PA techniques into farming practices leads to increases in resource-use efficiency. However, while good integration tools and concepts are known, there are technological challenges in integrating various applications and interpreting the results of the analyses. Surveys of current applications and perspectives in the area of advanced detection technologies in agriculture observed that mapping, integrating sources of information for deciding on a crop, spatial management support, and decision-making were the main areas where intelligent and automated tools were wanted, and stakeholders saw the greatest opportunities. During many precision agriculture workshops, the need for root cause analyses (mapping and GIS-type applications) was the most mentioned need for special types of knowledge that could be used during proactive strategic planning for the various interested parties. In the great majority of the contributions to the conferences on precision agriculture, the need for root maps, i.e., cause-effect analyses using maps and spatial characteristics, has been mentioned. This means that, for the involved parties, having maps with information on spatial relations is essential for taking strategic measures (Morales-García, Terroso-Sáenz, and Cecilia 2024). Thus, mapping and analysis of spatial relationships are important parts of the use of technology for making well-founded decisions.

5.5.2 Decision Support Systems

GIS can be used to generate information for precision agriculture. Recent applications of GIS in agriculture often make use of Decision Support Systems (DSS). This new generation of information systems can be used to integrate various types of data, including data from GIS, to aid in decision-making. This, in turn, will help to improve agricultural output by informing the grower of past, current, and future events. A DSS can be just one part of an operational system. It can also be the front end of an agro-informatics or precision agriculture system. A DSS can be employed in many agricultural venues, including farm management, in-season management of precision agriculture, tissue or petiole analysis for nutrient use in c, irrigation scheduling, site-specific nitrogen requirements, refuge seed planting for Bt technology, Sudden Death Syndrome (SDS) product trials, and planning zones for yield monitor calibration. Many decision support systems have been developed for agriculture. Some of the most popular involve cash crops. Decision support systems have focused on in-season management of crop production (Kim and Park 2024; Gauci et al. 2024). These systems have been developed using prior and current knowledge that the user would have brought to computer software.

Decision support systems were used to identify specific issues, and these DSS helped end-users make management decisions. In so doing, they saved the farmer time and money. In essence, the computer system served as an assistant to help the farmer with time-management practices. Developing models and maps that are accessed via the Internet is, in effect, a part of precision agriculture. Users enter data into the computer models, and the computer runs programs to develop a specific result for that cartographic area. This interactive technology is often called a Decision Support System. Some farmers are better able to cope with and manage this new precision agriculture information. The clientele for these new, easier interfaces is usually younger farmers, but they span the age groups. The capital-rich producer can afford the new technologies and does not seem to mind the time needed to learn how to use them. Research studies show that big producers with much land use GIS to delineate and collect field data (Garcia et al. 2024). They also raise more cash crops and have more reason to farm semi-logically to make a profit, among other benefits.

5.6 Challenges and Future Directions in Precision Agriculture

The growing need for data has significantly expanded the range of agricultural technologies and services used by stakeholders in the precision cultivation network. Companies have very different strategies to offer precision agricultural technologies and services: large companies are involved in promoting their networks of connected technologies dedicated to smart, sustainable farming, centered around advanced machinery and the use of sensors, while small and medium-sized companies focus on heterogeneous solutions, promoting the digitization of agriculturally specific models and using local, innovative solutions that employ more open-ended, integrated, interdisciplinary technology. However, the potential implementation of the technological framework introduces several challenges in precision farming. Data privacy issues need to be resolved to establish compliance between the data collected in the technological network of precision farming and European legislation. Another key research direction for the future is the integration of new enabling technologies (Shafik 2024a). More cutting-edge techniques have been applied in state-of-the-art precision agriculture systems, such as machine learning to build dynamic models in order to predict crop yields, integrating imagery data from multiple sensors, and physical data to develop deep-learning-based interpretability models.

There is a need to identify the existing bottlenecks in the adoption of new technologies in agriculture at a farm level and in the wider context of the agricultural products supply chain, and to improve and promote social innovations. The adoption of new digital technologies in agriculture is complex, and there is extensive territorial diversity. There is an identified need to promote social innovations in agriculture that should be triggered by designed and run by digital technology users. This needs the development of educational and training tools that will work directly with precision agriculture stakeholders from varying farm scales and sectors, including farmers and agronomists, and will assist in the transition of the precision agriculture paradigm downstream across the wider agricultural network (Babos et al. 2024). The results of the report highlight some of the future research priorities related to the implementation of the Technology Framework and map out some potential directions for the future. To untap the societal and economic potential of the digital transformation of the agricultural industry, it is necessary to ensure that the existing challenges and obstacles are addressed. Here we aim to develop stakeholder collaborations that are able to develop successful integration paths. It is envisaged that these collaborations will be able to deliver excellence while being truly democratic and inclusive, representing the different stakeholders, cultures, and practices in Europe across primary and secondary production in various agricultural sectors, geographical locations, and sizes of operation (Anusha, Anandan, and Venkata Krishna 2024).

5.6.1 Data Privacy and Security

Data privacy and security are of great importance in precision agriculture because the involved technologies are capable of monitoring fields and livestock with great detail and spatial resolution. Due to the greater availability of technologies that enable further data acquisition, privacy becomes an important issue as the possibilities of monitoring are triggered and objective norms of the data acquired and derived are established. At the same time, the possibility of earning money with sensitive data is used by the providers of data analytics in agribusiness as a preference factor for their services. Despite the importance, a uniform and comprehensive data protection regulation for the member countries of the European Union has not yet been established. However, the collection of individual consent for data use and access, as well as defects in cybersecurity, are the major sources of threats to data privacy and security (Ortatas et al. 2024). With the expansion of the IoT community in precision agriculture, solutions must be identified to guarantee the security of these systems and their users. Data must be available for authorized users only, and it must not be corrupted.

Food and agriculture were not susceptible to cyberattacks in the past, but this assumption was recently proved wrong. By breaching the supply chain, hackers could access information like pesticide and fertilizer information, geographic location, and possibly even the type of crops grown by the user or the type of livestock and other private data. About half a decade ago, dairy cows were fed contaminated feed (Silva et al. 2024). The question remains: who owns the data, and who is authorized to share it? Also, what are the obligations of those who handle data, the technology providers? It is, therefore, important that existing data handling and sharing structures in agriculture are reviewed. Data privacy and security are essential requirements for further development in precision agriculture. Data sharing requires a detailed review of the further development of the specialty and associated regulations. It would also include a review of the obligations of technology providers who handle a significant amount of user data (Shafik, Tufail, Namoun, et al. 2023).

5.6.2 Interoperability and Standardization

In contemporary agricultural practices, several technologies and devices, such as sensors, precision variable rate applicators, remote sensing, decision support systems, equipment telematics, and animal tracking, are used for different management operations. In order to be beneficial, these devices must provide better data accuracy, reduce costs, increase efficiency, enable site-specific management applications, and cover a wide area, among others. However, the lack of uniform protocols and standards among these devices and systems is an obstacle to technology interoperability. Interoperability enables different devices, systems, and applications to exchange, interpret, and apply data from integrated systems in a seamless way to achieve the desired outcome (Devi and Rajan 2024). It is needed to manage and interpret operational data from multiple equipment and software providing farm maps of yield, fertilizer application, grain moisture, and seeding maps, for example, commonly used for decision making. Several studies agreed on the significance of interoperability for precision agriculture solution adoption and suggest that it may help farmers, research organizations, and technology providers. Interoperability problems may result in loss of efficiency, impact delayed business, and poor customer satisfaction, while continuously expanding data silos may lead to inefficiency for farmers and may block confidence and use by both farmers and their advisors (Yuan 2024).

A lack of communication standards in protocols and data formats is a critical issue. For best productivity and compatibility, it is crucial to have agreed standards at both the communication and data levels. This realization may stimulate the accumulation of installations to develop an efficient recommendation system that can help farmers apply solutions to their problems. It is recognized that the overarching aim is to remove differences and uncertainty to stimulate adoption over a range of collaborating groups. Ultimately, true integration must mean connecting all the many components required for the delivery of precision agriculture into one easy-to-manage, easy-to-assemble system (Silva et al. 2024). We envisage that the combination of innovative technologies will increase the value created. The reality is, however, that farm solutions created from siloed data, even at the technology provider or user level, have not yet unlocked the value every financier envisaged. The data is not just siloed at the farm level; it's siloed above that and at a more robust level across the technology service providers, right up to their processes and practices. The tremendous efficiency gains and subsequent savings won't come until all service activities can be integrated, and that will require a mass of data standards (Govi et al. 2024).

5.6.3 Socio-economic and Environmental Impacts

Agriculture is a vitally important part of the world economy. It contributes to food security and employs large numbers of people, even though its relative importance has been declining over time, largely because of the more rapid growth of other sectors that supply a range of goods and services that often have priority in an economy in transition. Furthermore, economic growth accelerates more rapidly in countries where the structural composition of the economy moves from being predominantly agricultural to becoming more diverse. This shift in the structure of the economy is indicative of a natural process of economic development at the national scale: when food demand growth created by income growth slows down, the weight of agricultural production activities starts to diminish. It is this structural shift that has allowed accelerated growth to take place in other vital sectors of the economy, generating additional demand for products in the agricultural sector (Eron et al. 2024). Agriculture is also important from a social and environmental viewpoint. It provides the means for people to satisfy essential life-sustaining needs, generates a sense of social cohesion by providing families with the capacity for self-sufficiency and mutual support, grounds people by connecting them with the place they live, and forms part of that existential bond between people and nature, which is a vital and defining characteristic. From an ecological perspective, agriculture depends on a stable natural environment for its ultimate survival, and it is in agriculture's best interests to maintain the quality of the natural resources it uses and occupies, as summarized in Table 5.3 (Pauline A. Ongadi 2024).

Table 5-3 Challenges and Solutions in Precision Agriculture Adoption

Challenge	Description	Impact	Proposed Solution	Outcome	Reference
Cultural Resistance	Resistance to change in farming practices	Slows adoption in traditional farming	Awareness campaigns and success stories	Increases willingness to adopt	(Justina and Thenmozhi 2024)
Limited Funding for R&D	Insufficient investment in innovation	Slows technological advancements	Public-private partnerships	Accelerates R&D progress	(Kobe et al. 2024)
Market Uncertainty	Fluctuations in input and output prices	Financial risk for farmers	Market analytics and forecasting tools	Reduces market risks	(Saxena et al. 2024)
Environmental Impact	Potential e-waste from devices	Environmental sustainability concerns	Recycling and reuse programs	Minimizes e-waste	(Alves and Cruvinel 2024)
Short Lifespan of Technology	Rapid obsolescence of equipment	Frequent need for upgrades	Modular, upgradeable systems	Prolongs equipment lifespan	(Latino, Menegoli, and Corallo 2024)
Lack of Standardization	Inconsistent protocols for data sharing	Data interoperability issues	Development of global standards	Facilitates data integration	(Yuan 2024)
Limited Local Availability	Technologies not available in all regions	Limits access for rural farmers	Local manufacturing and distribution	Enhances regional availability	(Ortatas et al. 2024)

Energy Dependence	Reliance on electricity for devices	Increases energy costs	Integration of renewable energy sources	Lowers energy expenses	(Gauci et al. 2024)
Maintenance Costs	Ongoing expenses for repair and updates	Adds to long-term operational costs	Development of low-maintenance systems	Reduces total cost of ownership	(Morales-García, Terroso-Sáenz, and Cecilia 2024)
Technical Expertise	Farmers may lack technical knowledge	Slows adoption and usage	Training programs and extension services	Enhances farmer capability	(Nguyen et al. 2024)
Data Privacy Concerns	Farmers' data security at risk	Trust issues with technology providers	Use of blockchain for data security	Builds farmer trust	(Devi and Rajan 2024)
Regulatory Barriers	Complex regulatory approvals	Slows adoption of new technologies	Streamlined regulatory frameworks	Accelerates innovation adoption	(Gopi and Karthikeyan 2024)
Data Management Complexity	Large volume of data needs processing	Difficult to manage and interpret	Cloud-based platforms, AI tools	Simplifies data processing	(Corceiro et al. 2024)
Connectivity Issues	Poor internet in rural areas	Limits real-time data access	Use of low-power wide-area networks	Ensures data accessibility	(Garcia et al. 2024)
High Initial Costs	Expensive equipment and installation	Limits adoption for small-scale farmers	Subsidies, financial incentives	Increases accessibility	(Monteleone et al. 2024)

5.6.4 Technological Challenges

The sustainability of agricultural production, in particular in Europe, is faced with many technological, economic, socio-economic, environmental, and policy challenges, which must be addressed to ensure consistency with the scenario "Agriculture 2020". Modern information and communication technologies have an important role. Precision agriculture and organic farming are two different directions of modern agriculture, which may help transform and enhance the current European agricultural production systems. This paper does not discuss the signals for agricultural policies that can help promote these changes. Precision agriculture uses technologies such as Global Positioning Systems, Satellite Remote Sensing, Geographic Information Systems, Variable Rate Technologies, Automatic Guidance Systems, Soil Electro-Magnetic Induction, Yield Mapping and Monitoring, and Expert Systems to enable farmers to apply natural resources in the best possible way (Gauci et al. 2024). The results of implementing precision agricultural practices, such as yield increase, input cost reduction, reduced soil erosion, and decreased environmental pollution, ensure a sustainable use of available natural resources.

An enormous amount of information is generated. Presentation, integration, and management of that information are extremely challenging. Systems integration is a major challenge. Technologies are available, but there are still many technological challenges that must be addressed, including creative and innovative research in telecom, electronics, mechatronics, computer science, and systems engineering. The wide diffusion of technologies is still hampered by the fact that 'early' adopters of these strategies have to bear the additional risk of unproven technologies. The economic sustainability of the organic processing industry, which is transforming organic farm products into consumer-ready products, may be upset by inadequate equipment and systems capable of exploiting the technologies of precision agriculture in the organic crop farm (Devi and Rajan 2024). Needed are telecommunication and automatic guidance systems capable of communicating in real time, at a marketplace speed and precision, the exact position and name of organic and non-organic farm harvested plots to the corresponding trailers of different grain qualities.

Any technological breakthrough that may empower organic farm production capability would be considered a leap forward. A similar argument could be conceived for precision agriculture, where the role of original equipment manufacturers seems even more significant to ensure effective farm implementation. To address these challenges, our vision is to design and develop a multimedia-enabled toolbox known as prototype-based knowledge-rich precision agriculture techniques. The key enabling component needed for such a toolbox is a problem-solving compass. We envision a prototype problem-solving-based approach to link existing tools and technologies lacking cooperative capabilities (Garcia et al. 2024; Corceiro et al. 2024). The concept lies in the one-to-many relationship between the source of problems and the problem-solving requirements, thus utilizing a general-purpose problem-solving model. To realize this vision, we need to prepare a top-down method that:

a) combines techniques for descriptive and predictive models;
b) incorporates scientific procedures for decision analysis to build and validate a persuasive case;
c) develops tactics and strategies by integrating non-quantitative bases and
d) attains outcomes through deductive and inductive reasoning.

5.7 Conclusion

The prototype problem-solving compass should be rich in knowledge tools and apply prototype problem-solving characteristics in the form of high-level requests achieved from the mission, which defines the requirements of the high-level problem-solving features needed in the precision agriculture mission. Representational learning of unsupervised data in deep learning creates meaningful patterns that can be used for classification, clustering, or regression. These algorithms have the potential to remodel or re-represent complex data such as plant image analysis. There is a vast literature that deals with how AI has been used to predict different diseases in various applications. Among the disease predictions, our focus is on AI-mediated automated prediction, which assists all patients and doctors and has been validated using an agricultural sensor. There are competing methods that enable the prediction of diseases based on the data from a mobile or wearable device. The major challenge is the communication, decision-making, and smart data integration between data, patients, and doctors. Patients are generally unable to interpret sensor data in real-world applications. Discrete wavelet transform along with support vector machine and multi-layer perceptron algorithms have been used to predict falls in ubiquitously collected sensor data. Many other researchers have succeeded in achieving their goals using other techniques. Thus, it is also predicted here that diseases or problems can be identified using the features calculated from the total possible data. Many researchers have discussed factors such as the accuracy of diagnosis, reliability of prediction, monitoring of patients in real-time, and reduction of healthcare costs through monitoring. For this, various patents have been analyzed according to their relevance and explained their limitations side by side.

References

Al-Ammary, Jaflah Hassan, and Mohammed Essam Ghanem. 2024. "Information and Communication Technology in Agriculture: Awareness, Readiness and Adoption in the Kingdom of Bahrain." *Arab Gulf Journal of Scientific Research* 42 (1). https://doi.org/10.1108/AGJSR-07-2022-0113.

Alves, Gabriel M., and Paulo E. Cruvinel. 2024. "Parallel and Distributed Processing for High Resolution Agricultural Tomography Based on Big Data." *Multimedia Tools and Applications* 83 (4). https://doi.org/10.1007/s11042-023-15686-2.

Anusha, D. J., R. Anandan, and P. Venkata Krishna. 2024. "A Novel Deep Learning and Internet of Things (IoT) Enabled Precision Agricultural Framework for Crop Yield Production." *Journal of Autonomous Intelligence* 7 (4). https://doi.org/10.32629/jai.v7i4.1218.

Babos, Diego V., Amanda M. Tadini, Carla P. De Morais, Bianca B. Barreto, Mayara A.R. Carvalho, Alberto C.C. Bernardi, Patricia P.A. Oliveira, José R.M. Pezzopane, Débora M.B.P. Milori, and Ladislau Martin-Neto. 2024. "Laser-Induced Breakdown Spectroscopy (LIBS) as an Analytical Tool in Precision Agriculture: Evaluation of Spatial Variability of Soil Fertility in Integrated Agricultural Production Systems." *Catena* 239. https://doi.org/10.1016/j.catena.2024.107914.

Balayan, Anurag, Rajnish Mallick, Stuti Dwivedi, Sahaj Saxena, Bisheshwar Haorongbam, and Anshul Sharma. 2024. "Optimal Design of Quadcopter Chassis Using Generative Design and Lightweight Materials to Advance Precision Agriculture." *Machines* 12 (3). https://doi.org/10.3390/machines12030187.

Bütüner, Alperen Kaan, Yavuz Selim Şahin, Atilla Erdinç, Hilal Erdoğan, and Edwin Lewis. 2024. "Enhancing Pest Detection: Assessing Tuta Absoluta (Lepidoptera: Gelechiidae) Damage Intensity in Field Images through Advanced Machine Learning." *Tarim Bilimleri Dergisi* 30 (1). https://doi.org/10.15832/ankutbd.1308406.

Corceiro, Ana, Nuno Pereira, Khadijeh Alibabaei, and Pedro D. Gaspar. 2024. "Leveraging Machine Learning for Weed Management and Crop Enhancement: Vineyard Flora Classification." *Algorithms* 17 (1). https://doi.org/10.3390/a17010019.

Devi, P. Saranya, and A. Senthil Rajan. 2024. "An Inquiry of Image Processing in Agriculture to Perceive the Infirmity of Plants Using Machine Learning." *Multimedia Tools and Applications*. https://doi.org/10.1007/s11042-023-18052-4.

Eron, Francisco, Muhammad Noman, Raphael Ricon de Oliveira, and Antonio Chalfun-Junior. 2024. "Computer Vision-Aided Intelligent Monitoring of Coffee: Towards Sustainable Coffee Production." *Scientia Horticulturae* 327. https://doi.org/10.1016/j.scienta.2024.112847.

Garcia, Deny Carolina, Leonardo Soares Cangirana, Ricardo Guimarães de Queiroz, and Régio Marcio Toesca Gimenes. 2024. "DETERMINING FACTORS IN THE ADOPTION OF PRECISION AGRICULTURE TECHNOLOGIES BY GRAIN PRODUCERS." *Informe GEPEC* 28 (1). https://doi.org/10.48075/igepec.v28i1.32166.

Gauci, A. A., J. P. Fulton, A. Lindsey, S. A. Shearer, D. Barker, and E. M. Hawkins. 2024. "Precision of Grain Yield Monitors for Use in On-Farm Research Strip Trials." *Precision Agriculture* 25 (2). https://doi.org/10.1007/s11119-023-10092-y.

Gopi, P. S.S., and M. Karthikeyan. 2024. "Red Fox Optimization with Ensemble Recurrent Neural Network for Crop Recommendation and Yield Prediction Model." *Multimedia Tools and Applications* 83 (5). https://doi.org/10.1007/s11042-023-16113-2.

Govi, David, Salvatore Eugenio Pappalardo, Massimo De Marchi, and Franco Meggio. 2024. "From Space to Field: Combining Satellite, UAV and Agronomic Data in an Open-Source Methodology for the Validation of NDVI Maps in Precision Viticulture." *Remote Sensing* 16 (5). https://doi.org/10.3390/rs16050735.

Justina, Michael J., and M. Thenmozhi. 2024. "SorghumWeedDataset_Classification and SorghumWeedDataset_Segmentation Datasets for Classification, Detection, and Segmentation in Deep Learning." *Data in Brief* 52. https://doi.org/10.1016/j.dib.2023.109935.

Kim, Han Na, and Jin Hee Park. 2024. "Monitoring of Soil EC for the Prediction of Soil Nutrient Regime under Different Soil Water and Organic Matter Contents." *Applied Biological Chemistry* 67 (1). https://doi.org/10.1186/s13765-023-00849-4.

Kobe, Martin, Melanie Elias, Ines Merbach, Martin Schädler, Jan Bumberger, Marion Pause, and Hannes Mollenhauer. 2024. "Automated Workflow for High-Resolution 4D Vegetation Monitoring Using Stereo Vision." *Remote Sensing* 16 (3). https://doi.org/10.3390/rs16030541.

Latino, Maria Elena, Marta Menegoli, and Angelo Corallo. 2024. "Agriculture Digitalization: A Global Examination Based on Bibliometric Analysis." *IEEE Transactions on Engineering Management* 71.

https://doi.org/10.1109/TEM.2022.3154841.

Monteleone, Sergio, Edmilson Alves de Moraes, Roberto Max Protil, Brenno Tondato de Faria, and Rodrigo Filev Maia. 2024. "Proposal of a Model of Irrigation Operations Management for Exploring the Factors That Can Affect the Adoption of Precision Agriculture in the Context of Agriculture 4.0." *Agriculture (Switzerland)* 14 (1). https://doi.org/10.3390/agriculture14010134.

Morales-García, Juan, Fernando Terroso-Sáenz, and José M. Cecilia. 2024. "A Multi-Model Deep Learning Approach to Address Prediction Imbalances in Smart Greenhouses." *Computers and Electronics in Agriculture* 216. https://doi.org/10.1016/j.compag.2023.108537.

Nguyen, Trung Viet, Lam Son Le, Syed Attique Shah, Sufian Hameed, and Dirk Draheim. 2024. "PenChain: A Blockchain-Based Platform for Penalty-Aware Service Provisioning." *IEEE Access* 12. https://doi.org/10.1109/ACCESS.2023.3344038.

Ortatas, Fatma Nur, Umut Ozkaya, Muhammet Emin Sahin, and Hasan Ulutas. 2024. "Sugar Beet Farming Goes High-Tech: A Method for Automated Weed Detection Using Machine Learning and Deep Learning in Precision Agriculture." *Neural Computing and Applications* 36 (9). https://doi.org/10.1007/s00521-023-09320-3.

Pauline A. Ongadi. 2024. "A Comprehensive Examination of Security and Privacy in Precision Agriculture Technologies." *GSC Advanced Research and Reviews* 18 (1). https://doi.org/10.30574/gscarr.2024.18.1.0026.

Romano, Elio, Andrea Bragaglio, Carlo Bisaglia, Alberto Assirelli, Elia Premoli, and Simone Bergonzoli. 2024. "Case Study on the Economic and Environmental Impact of the Introduction of the Variable-Rate Distribution of Fertilizer in Wheat (Triticum Aestivum L.) Cultivation." *Sustainability (Switzerland)* 16 (4). https://doi.org/10.3390/su16041612.

Saxena, Vinay, Mala Singh, Parul Saxena, Mukesh Singh, Arun Pratap Srivastava, Navneet Kumar, A. Deepak, and Anurag Shrivastava. 2024. "Utilizing Support Vector Machines for Early Detection of Crop Diseases in Precision Agriculture a Data Mining Perspective." *International Journal of Intelligent Systems and Applications in Engineering* 12 (16s).

Shafik, Wasswa. 2024a. "Barriers to Implementing Computational Intelligence-Based Agriculture System." In *Computational Intelligence in Internet of Agricultural Things*, 193–219. Springer. https://doi.org/10.1007/978-3-031-67450-1_8.

Shafik, Wasswa. 2024b. "Biosensor-Based Drones Anomaly Detection Integration for Sustainable Agriculture Development." In *Achieving Food Security Through Sustainable Agriculture*, 283–316. IGI Global. https://doi.org/10.4018/979-8-3693-4240-4.ch012.

Shafik, Wasswa. 2025. "Sustainable Agriculture and Diet in the Metaverse Era: Sustainable Farming and Nutritious Eating in the Metaverse." In *Food in the Metaverse and Web 3.0 Era: Intersecting Food, Technology, and Culture*, 159–84. IGI Global Scientific Publishing.

Shafik, Wasswa, Ali Tufail, Chandratilak De Silva Liyanage, and Rosyzie Anna Awg Haji Mohd Apong. 2023. "Using a Novel Convolutional Neural Network for Plant Pests Detection and Disease Classification." *Journal of the Science of Food and Agriculture*. https://doi.org/10.1002/jsfa.12700.

Shafik, Wasswa, Ali Tufail, Abdallah Namoun, Liyanage Chandratilak De Silva, and Rosyzie Anna Awg Haji Mohd Apong. 2023. "A Systematic Literature Review on Plant Disease Detection: Motivations, Classification Techniques, Datasets, Challenges, and Future Trends." *IEEE Access* 11:59174–203. https://doi.org/10.1109/ACCESS.2023.3284760.

Shafik, Wasswa, Ali Tufail, Chandratilak De Silva Liyanage, and Rosyzie Anna Awg Haji Mohd Apong. 2024. "Using Transfer Learning-Based Plant Disease Classification and Detection for Sustainable Agriculture." *BMC Plant Biology* 24 (1): 136. https://doi.org/10.1186/s12870-024-04825-y.

Silva, César de Oliveira Ferreira, Célia Regina Grego, Rodrigo Lilla Manzione, Stanley Robson De Medeiros Oliveira, Gustavo Costa Rodrigues, Cristina Aparecida Gonçalves Rodrigues, Eduardo Antonio Speranza, Ariovaldo Luchiari, and Luciano Vieira Koenigkan. 2024. "Summarizing Soil Chemical Variables into Homogeneous Management Zones – Case Study in a Specialty Coffee Crop." *Smart Agricultural Technology* 7. https://doi.org/10.1016/j.atech.2024.100418.

Yuan, Wenan. 2024. "AriAplBud: An Aerial Multi-Growth Stage Apple Flower Bud Dataset for Agricultural Object Detection Benchmarking." *Data* 9 (2). https://doi.org/10.3390/data9020036.

CHAPTER SIX

ROBOTICS AND AUTOMATION IN SUSTAINABLE AGRICULTURE

6.1 Abstract

Food is essential for the survival of living organisms. Because of the fast-growing world population, offering abundant and safe food supplies will be put on the agenda. The importance of agriculture, which is the main resource of the food supply, will be much more revealed. However, in the fast-evolving modern world, social, ecological, and economic sustainability have been discussed a lot in relation to agriculture. This is a concern about consuming fewer natural resources and emitting fewer environmental pollutants to lower the impact on the environment while reducing costs at the same time. Generally, these approaches are included in sustainable agriculture. But when it comes to achieving sustainability in agricultural systems, it does not matter which method is being discussed; the important thing is that we achieve both environmentally and economically balanced systems. There will be no condition under which we use the agricultural areas, which are rapidly decreasing, both in the right place and without ruining the natural balance. With mechanization, the adoption of agricultural machines was possible and it has even contributed to making agriculture more efficient, but such machinery is physically rigid and not capable of dealing with unforeseen events. Robotics allows intelligent machines to respond to stimuli received from the environment. It is capable of working in a production chain, both inside and outside drop areas, and it is also possible to make alterations in their work through the use of software. The software is capable of implementing different types of tools, instruments, shapes, fingers, or specific sensors for different applications. Robotics reduces the need to suppress the environment and also reduces occupational hazards, especially when work is performed in inadequate and dangerous places. In the future, it will not be easy to allocate more fertile areas for agricultural fields to produce food. New ideas and methods are needed to preserve and constantly produce food in available agricultural areas for the next generations. To sustainably manage our vital resources such as soil, water, and air, new advances in technology that allow humans to understand better and manage these resources are necessary. With this motivation, the role of robotics and automation in sustainable agriculture is discussed in this chapter.

Keywords: Agricultural Robotics, Autonomous Farming Machinery, Precision Automation, Sustainable Crop Management, Smart Harvesting Systems, Automated Weed Control, AI-Powered Farm Robots, Labor-Saving Technologies

6.2 Introduction

The agricultural sector is one of the most important sectors in many developing countries. Nevertheless, this sector faces several challenges, particularly as pressure to achieve sustainable development intensifies, and concern about climate change, water use, and pesticide residues increases (Shafik, Tufail, De Silva, and Apong 2025). One of the major challenges that face the agricultural sector is how it can adapt to the increasing demand for food imposed by a growing global population and, at the same time, minimize this impact on the environment and socio-economic development (Carolan 2022). The deployment of robotic and automation systems in agriculture can contribute to alleviating these challenges by enabling efficient operations that can alleviate a shortage of labor and supplement the use of increasingly toxic agrochemicals with more efficient and selective application methods. Furthermore, the development of more advanced technologies that can work without the need for human support can address the problems of abandoned fields due to the migration of young rural people to cities. Sustainable agriculture is a major issue addressed by numerous organizations worldwide and encompasses multidisciplinary approaches combining technological and socioeconomic aspects (Ullo and Sinha 2021).

On the one hand, sustainable agriculture is driven by the need to face current societal challenges like poverty, climate change, the increase in world population, and malnutrition by ensuring economic stability for farmers and food security for all. On the other hand, it is necessary to consider more general perspectives like the local and global effects of agricultural practices on ecosystems and societies. Four main problems that have notable implications for the sustainable development of the agricultural sector have been highlighted: food supply, labor shortage, climate change, and efficient and selective use of agrochemicals. Robotics and automation technologies are definitely part of the solution to these problems. The presence of agricultural technology in developed countries is quite conclusive and is present during all stages of the agricultural activity production chain (Yigitcanlar et al. 2020). In underdeveloped countries, there is less participation in technology, and that is why agricultural activities are more extensive labor activities. Robotics applied to agriculture is still scarce, but it is capable of functioning in domestic environments. Thus, it is possible to observe that it has established, positioned, and profiled seeds following the lines of seedling modules in planting operations, in the distribution of all the crops' nutritional inputs, identified and utilized all the crops' weeds and pest plants growing in that field, evaluated the plants by counting the number of spots or diseases that a certain number of crops presents, and is capable of predicting the production yield (Barnes et al. 2021).

6.3 Evolution of Robotics in Agriculture

There have been several projects on robotic systems in the context of agriculture. The earliest recorded work seems to be that of John Wright, who invented the first agricultural machine, known as the "Beverston Plough," in 1730. The objective of the project was to mechanize the backbreaking and labor-intensive job of digging, hoeing, and raking in maintained beds in the kitchen garden lawn of the Earl of Scudamore, thereby allowing the Earl to harvest the maximum number of crops (Carolan 2022). According to the inventories of the Earl's Gardens, this historical robotic system was equipped with a four-wheeled chassis, a side-operated machine taper, which acted as the lateral guidance of the device, and a guiding apparatus that was equipped with a single vertical and two horizontal handles for height adjustment, thereby allowing the Earl's gardeners to manipulate the tool during the various operations. These simple guiding devices allowed the Earl's gardeners to rapidly drag the Beverston Plough along a straight line while inserting the cutting tool in the ground to minimize the inclusion of undesired herbs in the maintained garden beds (Barnes et al. 2021). There are several types of agricultural robotics, as presented in Table 6.1, with their impact on sustainability.

Table 6-1 Robotics and Automation Types in Sustainable Agricultural Practices

Type of Robot/Automation	**Description**	**Primary Function**	**Primary Function**	**Impact on Sustainability**	**Reference**
Pest Detection Robots	Identifies pests in early stages	Identifies pests in early stages	Reduces crop losses	Limits pesticide application	(Yi et al. 2024)
Seeding Robots	Precision seed placement	Ensures optimal seed spacing	Enhances crop uniformity	Enhances crop uniformity	(Geckeler et al. 2024)
Autonomous Greenhouses	Fully automated indoor farming	Fully automated indoor farming	Year-round production	Reduces energy and water usage	(Guldenring, Andersen, and Nalpantidis 2024)
Packing and Sorting Robots	Packing and Sorting Robots	Packing and Sorting Robots	Increases processing efficiency	Increases processing efficiency	(Fountas et al. 2024)
Fertilizer Robots	Increases processing efficiency	Artificial pollinators for crops	Artificial pollinators for crops	Ensures consistent pollination	(Parai, Gupta, and Islam 2024)
Soil Sampling Robots	Collects and analyzes soil samples	Soil health assessment	Informed soil management	Informed soil management	(Khadatkar et al. 2024)
Autonomous Sprayers	Targeted spraying of pesticides	Targeted spraying of pesticides	Targeted spraying of pesticides	Targeted spraying of pesticides	(Carolan 2024)
Targeted spraying of pesticides	Targeted spraying of pesticides	Targeted spraying of pesticides	Increases crop yield	Increases crop yield	(Narsale et al. 2024)
Irrigation Automation	Irrigation Automation	Precision irrigation	Precision irrigation	Conserves water resources	(Akhter et al. 2024)

Livestock Monitoring Robots	Reduces livestock losses	Reduces livestock losses	Faster, more uniform planting	Faster, more uniform planting	(Halstead, Zimmer, and McCool 2024)
Weeding Robots	Eliminates weeds without chemicals	Eliminates weeds without chemicals	Decreases herbicide use	Protects soil and water health	(Zhang et al. 2024)
Harvesting Robots	Robots designed for picking crops	Harvesting fruits, vegetables	Harvesting fruits, vegetables	Reduces post-harvest losses	(Lincoln and Elliott 2023)

6.3.1 Historical Overview

Although the inspiration to develop robotic systems for agricultural applications can be traced back to the 16th and 17th centuries for both field and horticultural tasks, it was only in the 1980s that the scientific and technological approach to agricultural robotics became more focused (Shafik 2025a). Research has shown that in most countries, the key driver for research has been the lack of availability of manual labor at the time of planting, weeding, thinning, and harvesting, but also the interest in being able to identify the exact position of a plant in the field. In the case of intensive farming of horticultural crops such as tomatoes, there is a marked interest in increasing the economic characteristics of this system. The development of the Fruit Piano was the first reference to an autonomous fruit-picking robot that travels on a rail system in a typical orchard (Zhang et al. 2024). This project led to the development of a stand-alone robotic system that can pick singles or bunches of grapes and place them in a small bin located on the main support module; examples of existing agricultural robotics are demonstrated in Fig 6.1.

Fig. 6-1 Sampled examples of agricultural robots

The BoniRob project further developed the concept of a stand-alone system and showed the feasibility and potential of removing grasses and other weeds in the existing crop without the need for the previous development of a digital map. Researchers have shown that the implementation of a suitable docking system for the electric and compressed air supply can reduce the total weight of the robot to a reasonable value. In recent years, governments have shown a strong interest in this topic, and Europe funded a project as part of a program. This program financed the development of four scenarios for the horticultural world and is testing a robot designed to collect lettuce and water pumps with each drop, improve the information on color changes using technology, and place the collected vegetables on a motion module (Parai, Gupta, and Islam 2024). After the development of the technology, a detailed economic assessment of each robot will be made, showing the potential for commercial adoption in the short to medium term.

6.3.2 Current State of Robotics and Automation

Today, we now have unmanned aerial vehicles with payload capacities ranging from small to extra-large payloads. While some UAVs, like small micro-UAVs and multi-rotor UAVs, are equipped with only one sensing device, such as RGB color, hyperspectral, or thermal cameras, others, like traditional fixed-wing airplanes and gas-powered helicopters, are able to carry full environmental monitoring payloads or dedicated agricultural sensor clusters for spectral, fluorescence, and thermal-based applications. Fixed-wing UAVs are well suited for large-area remote sensing applications, while swarm-based or car charger UAVs are great for closer inspection of smaller subareas within the flight path. In general, fixed-wing UAVs are faster and cover a larger surface area but have weaker maneuverability and longer take-off and landing times (Geckeler et al. 2024). Fixed-wing UAVs, however, are also capable of ultra-long-range missions when they are equipped with multiple batteries.

The autonomous control of modern-day agricultural machines using embedded electronic control systems is perhaps the most advanced type of robotic technology currently deployed in modern agriculture. Nowadays, modern farmers utilize advanced, high-tech machines, like tractors combine harvesters, and potato planting and harvesting machines, that are capable of fully autonomously managing different agricultural processes. For instance, small robot tractor systems, independent platforms, and autonomous tractors are standard tractors that have been fitted with GPS systems, cameras, and other tools for long-line cultivation and production and can generate significant cost savings and support large-scale precision agriculture in the area. New designs for UAV-based crop scouting that employ a robotic pyramid system for accurate crop management of maize fields have also been developed (Yi et al. 2024; Parai, Gupta, and Islam 2024). Work has also been done on developing mobile systems for ground robot-based crop scouting that are capable of providing autonomous information on plant structure at the field level. A robot harvester designed from a user point of view for the mass harvesting of growing potatoes has been developed to respect the root area of the plant and has demonstrated success in field labor. Coordination of the prototype harvester with the robotic manipulator arm has to be tested in the future (Geckeler et al. 2024). Thus far, despite the promising results, relatively few companies are actively involved in the development and supply of agricultural robotic equipment using AI technologies, as illustrated in Fig 6.2.

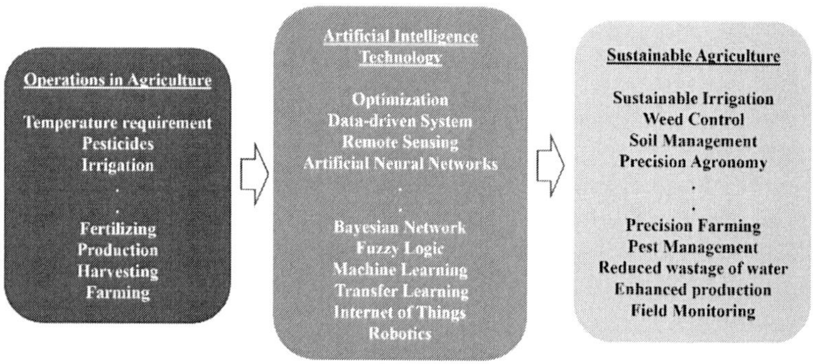

Fig. 6-2 Transforming sustainable agriculture using artificial intelligence

6.4 Benefits of Robotics and Automation in Agriculture

With the shift in focus from quantity to quality in recent years, most industrial operations have adopted automation in their work practices, including chemical, food processing, and automobile industries. Benefits of employing robots in the industrial sectors include reduced operational costs, optimization of production, increased product uniformity, job preservation, reduction of toxic fumes in the working area, improved worker safety, and increased productivity. The same beneficial effects are expected when automation and robots are brought into farm environments. The agricultural industry, with its ever-growing need for products, is known to offer more potential for robotics and automation, with today's robot use below 1% (Zhang et al. 2024). Benefits of employing robots in agricultural activities include operational efficiency in large agricultural areas, high productivity from controlled activities, as well as the ability of robots to perform multivariate tasks simultaneously, the ability to perform labor with high risks to human safety, reduction of workforce during labor peaks, and working in unsuitable weather conditions. There will also be environmental benefits through reduced chemicals, fuels, emissions, fertilizer use, and the release of biological crop agents and organic products. Remote-controlled sensors and robots can be used for site-specific crop management by recognizing crop stresses and field patterns associated with soil variability, predicting yield, and assessing post-crop harvest quality, allowing for efficient monitoring of water, timely water, and fertilizer use, as well as disease control. This helps to reduce the quantity of pesticides and fertilizers used, which would otherwise be over-applied, causing a negative effect on the entire ecosystem (Akhter et al. 2024). The use of robots and automation in agricultural activities may result in higher productivity than inter-row post-crop crops.

6.4.1 Increased Efficiency and Productivity

The increase in efficiency will generally result from the removal of some of the constraints that limit the productivity of current systems. For almost all activities, tasks are not performed at the maximum levels of efficiency and reliability of performance capable in terms of achieving the desired objectives. The inherent variability of the soil and crop systems means that performance is often less than desired. Increased technical abilities, the ability to work 24 hours a day every day, and the ability to reach performance levels at times considered physically impossible will mean that the precise rather than the best will become both possible and feasible. The other side of increased efficiency is likely to be increased output and loss minimization (Khadatkar et al. 2024). Huge volumes of waste are produced in all systems and can only be reduced by improving the efficiency of the process. The losses of marketable products are of great concern across the world, and their reduction will be actively pursued. Also, where the significant added value in the process is from the carrying, transplanting, planting, and post-harvest handling, the value in the increased efficiency, therefore, lies in the capability to control the environment to produce uniform plants that automatic arrangements can handle (Yi et al. 2024). Both of these factors combine in favor of automated handling for most tasks.

6.4.2 Resource Conservation and Environmental Sustainability

Brooks argued that scaling farming operations to the point where robotic automation increases the performance of the entire system to commercially viable levels is inherently wasteful of resources. While on the one hand, growing food crops in controlled environments like greenhouses reduce the need for chemical application and also provides better ecological sustainability, the very materials used in constructing these greenhouses, in combination with the need for illuminated growth, often use up lots of extra energy, making what grows within more expensive (Zhang et al. 2024). The high cost of constructing these systems is such that they are viable in only a fraction of a percentage of climatic zones on Earth. Similarly, fully mixed micro-illumination is the holy grail of plant growth in such environments, for when we need to shine a laser in order to deliver a tailored spectrum of sunlight to a plant, we also employ hundreds of times per photons expended in the old ultraviolet-and-mineral approach. Even then, we have not overcome the native incompatibility that plant growth is green: in a properly tuned photosynthesis environment, most photons are absorbed, and all others are an eyesore green that is easily captured by *photodiodes,* and photosynthetic efficiency falls (Carolan 2024).

An elegant alternative strategy to such direct-to-photosynthesis solutions is to take advantage of how a plant grows: initially small, round, and slow, plants are easy to tend by robots deployed in a crowded urban environment from relatively inexpensive, supposedly recyclable land nearer the equator. Initial growth is supported by a widely available mixture of sunlight, humidity, crop nutrients, and rainwater; that run-up helps a growing plant to be larger and stronger, enabling it to endure better the challenging, continuously changing microenvironments characteristic of outdoor plant life (Halstead, Zimmer, and McCool 2024). These labor-light solutions are efficiently scalable — anyone can purchase an inexpensive sensor-based drip irrigation system; a national chain can more effectively handle developing, scaling, and deploying commercial-quality control and decision-making software, and larger still can afford banks of growth lamps to produce high-quality seedlings for outdoor transplantation, as summarized in Table 6.2.

Table 6-2 Robotics and Automation Benefits in Sustainable Agriculture

Benefit Category	Description	Technology Application	Agricultural Impact	Sustainability Contribution	Reference
Innovation and Technology Adoption	Drives technological advancements	AI-Powered Robots	Encourages modernization	Strengthens sustainable practices	(Ullo and Sinha 2021)
Enhanced Marketability	Enhanced Marketability	Enhanced Marketability	Blockchain Integration in Robots	Blockchain Integration in Robots	(Belforte et al. 2006)
Waste Reduction	Limits agricultural waste generation	Sorting Robots	Less waste in the supply chain	Supports circular economy	(Lowenberg-DeBoer et al. 2020)
Supports circular economy	Supports circular economy	Autonomous Greenhouses	Steady supply of crops	Reduces seasonal food shortages	(Langemeier and Boehlje 2021)
Reduces seasonal food shortages	Minimizes use of pesticides and herbicides	Minimizes use of pesticides and herbicides	Improves crop safety	Protects ecosystems	(Ma et al. 2017)
Animal Welfare Improvement	Monitors and enhances animal care	Monitors and enhances animal care	Ensures better health outcomes	Reduces environmental impact of livestock	(Marinoudi et al. 2019)

Food Quality Assurance	Sorting and Packing Robots	Sorting and Packing Robots	Consistent quality standards	Reduces food waste	(Eastwood et al. 2022)
Energy Efficiency	Uses renewable or optimized energy sources	Uses renewable or optimized energy sources	Reduces reliance on fossil fuels	Reduces reliance on fossil fuels	(Ahmad et al. 2022)
Improved Soil Management	Enhances soil health and fertility	Soil Sampling Robots	Informed decision-making	Informed decision-making	(Lincoln and Elliott 2023)
Climate Resilience	Climate Resilience	Autonomous Greenhouses	Maintains stable production	Enhances resilience to climate change	(Halstead, Zimmer, and McCool 2024)
Enhances resilience to climate change	Identifies issues before they escalate	Identifies issues before they escalate	Prevents major crop losses	Prevents major crop losses	(Khadatkar et al. 2024)
Precision Agriculture	Accurate application of inputs	Autonomous Sprayers	Minimizes resource use	Reduces environmental footprint	(Geckeler et al. 2024)
Resource Efficiency	Optimizes use of water, fertilizers	Irrigation Automation	Reduces input waste	Conserves natural resources	(Shafik et al. 2023)
Conserves natural resources	Enhances yield and quality of crops	Enhances yield and quality of crops	Maximizes output	Supports food security	(Shafik, Tufail, De Silva, Haji, et al. 2025)

6.5 Key Technologies in Agricultural Robotics

The trends and technologies in agriculture-related AI, neuro-fuzzy, generative and transfer learning, machine vision and data fusion, hyper- and multispectral, and crowd-sourced imaging systems, mobile and social networking, cloud, fog, and edge computing, and IoT in precision agriculture are heavily discussed and researched. Major agritech companies consolidate to attack the digital agriculture market, encompassing AI, machine learning, smart farming, robotics, and IoT. While the market is projected to grow, many of the technology and commercial value expectations have not been met. That space has been dominated by precision agriculture robots, offering expanded and commoditized existing capabilities (Ma et al. 2017). There are only a few commercially available platforms that offer autonomous field data collection robot services, such as field monitoring, short-range scouting, autonomous bench-carting, precision pesticide spraying, robotic harvesting and fieldwork, weed control, laser weeding, and phenotyping.

Reliable precision agriculture robots require three fundamental subsystems: robust perception systems responsible for sensing and dynamic mapping of flora and fauna across vast and varied terrains, diagnostics systems maintaining the health and welfare of the robot machinery and plant life, and robot action and control systems executing repair and care operations on the robot machinery and the crops when and where they are required to boost or stabilize productivity and quality(Shafik 2025b). Together, these subsystems form the basic foundation for autonomous robot-based digital farms operating across broad geographical areas and from the inside of greenhouses or multi-tier vertical farms. The key technologies and associated state of the art of current agricultural robots are covered. These include precision agriculture platforms, unmanned helicopters, micro-nano drones, and VTOL systems providing short-range aerial scouting and servo-controlled drones for follow-and-selfie actions (Langemeier and Boehlje 2021).

6.5.1 Drones and UAVs

Drones and unmanned aerial vehicles have emerged as advanced tools for various applications in agriculture. Farmers have started to implement drones for uses related to crop monitoring, including establishing crop health, crop stress detection, and yield estimation. Drones are now used with multispectral cameras, hyperspectral cameras, and LiDAR sensors. Normal color cameras are used to take pictures of the crops, which can monitor visible injuries when the biomass of the crop is too high. Multispectral and hyperspectral cameras are equipped with additional filters that offer better imagery, enabling the detection of the health of the plants (Barnes et al. 2021; Ullo and Sinha 2021). Optical wavelength sensors within the multi-spectral cameras can detect shifts in light reflected by the plants that indicate stress from pests, nutritional deficiencies, salinity, water stress, or diseases. The LiDAR sensor records the distances to a target in the vicinity of millions of pulses. It has become a useful tool in agriculture, with applications available for plant height measurements, crop level measurements, surface reconstruction, and crop volume, improving sustainability, productivity, and quality, as demonstrated in Fig 6.3.

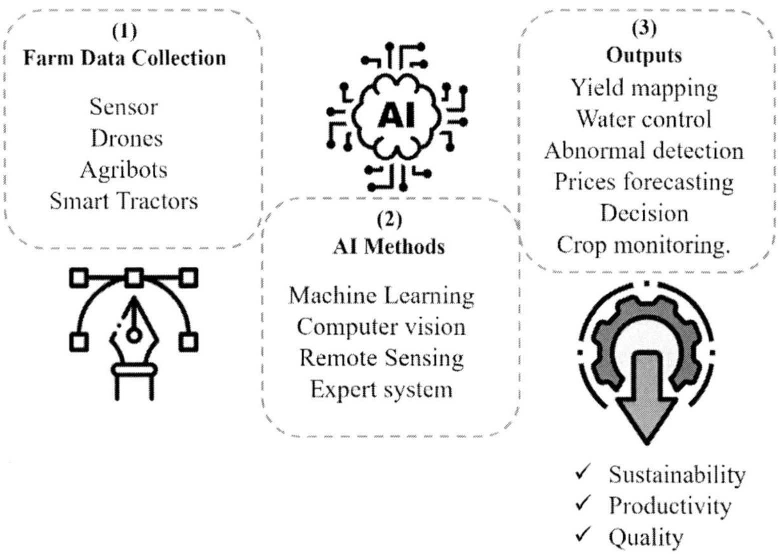

Fig. 6-3 Artificial intelligence approaches in improving sustainability, productivity, and quality.

The possibility to fly with the drone automatically over the field using GPS and autopilot technology is one of the most important current and futuristic applications of UAVs in agriculture. This has made remote sensing simple by saving time and reducing the risk associated with commuting over large fields. When the drone scans the area, imaging sensors in these devices with high spatial and temporal resolutions can analyze a wide variety of field characteristics based on visible and non-visible spectral data. Currently, due to the orthogonal visibility of the ground and the objective and subjective environmental conditions, UAVs equipped with off-the-shelf photographic and multispectral sensors suffer from weaknesses in obtaining high-quality images (Ma et al. 2017). Moreover, this results in different resolutions in different parts of the images and in acquiring images with many occlusions, shadows, and light variations. An image captured mostly using low-altitude UAVs, which has neighboring plants partly shadowing each other, is another source of error in spectral data caused by varying light conditions. However, research studies are being conducted to ascertain that merging UAV-shot, visible-range, slow-speed images with computer vision techniques, shadow and occlusion reduction, and the generation of orthophotos with corrected radiometries and light conditions enable the generation of high-quality multispectral image mosaics (Ismail et al. 2023).

6.5.2 Autonomous Tractors

Autonomous tractors combining robotics with automation and artificial intelligence to replace human operators have been under development for some fifty years. The technology is advanced, and some seven companies already have products on the market. The companies sell what is termed 'small' and 'medium' scale machines, which are essentially programmable steering assistance and variable work rate systems fitted to traditional small petrol engine tractors and tractors based on small commercial vehicles. Additionally, a university is currently developing and testing a small, true driverless robotic tractor. The key enabling technology for the larger tractors is electronic differentiation. Autonomous tractors enable improvements in energy efficiency, productivity, and economics, and reductions in soil loss and carbon emissions (Halstead, Zimmer, and McCool 2024; Kan et al. 2021).

The development facilitates more extended operations, greater work rates, fewer tilling operations, matches vehicle power to requirements, enables precision farming, reduces the number and weight of machines, including reduced use of the pivotal and powerful ploughing machine, smaller tractors with autonomous haulage robots for lighter roles, machines with lower power consumption and embedded emissions, machines of optimized performance with less maintenance, fuel and lubricant requirements, reduced use of plastics, rubber, aluminum, steel, and other construction materials, and reduced cost of ownership. These are all sustainability outcomes. It seems that real-world sustainability and particularly ecological issues can be achieved only when farm machinery is enabled by information technology to make local post-processing decisions unique to each square meter, as the tractor operates autonomously across the machine (Marinoudi et al. 2019; Ahmad et al. 2022).

6.5.3 Robotic Harvesting Systems

Robotic systems for fruit and vegetable harvesting are the most common type of robotic system and have been attracting research interest since the 1980s. The high value of fruits and vegetables makes the cost of labor for harvesting using human workers a much smaller fraction of the cost of the produce compared with many other agricultural tasks, making robotic harvesting a rational step for mechanization. In addition, fruit tree aesthetics and yield per unit land demand a large amount of precise manipulation and would benefit from robotic systems. The other main type of robotic harvesting robot is shearing-type devices used in pruning. Initially, pruning by shearing was demonstrated in the 1980s using commercial systems, and the first robotic system that pruned a plant was developed in the 1990s (Ismail et al. 2023; Deo, Joshi, and Brahmanakar 2022). Research on robotic plant pruning has continued to be active, spurred initially by an example and bolstered more recently due to advances in LED lighting and vertical farming systems, which allow year-round growth using LEDs as opposed to sunlight. It needs to be mentioned that despite long-term research on pruning and harvesting robots, relatively few commercial systems have been developed. This display of failure of agricultural robotics should not be frowned upon; after all, huge amounts of engineering and research accomplish nothing for various problems that are currently lacking satisfactory solutions. However, as automated systems for harvesting and pruning are increasingly often requested due to unreasonable demographic necessities for low-paying jobs, interest in these systems has continued. The other problem is that since working with agricultural robotic systems never ceases to be hard, designing and building such robotic systems are interesting also for primarily non-commercial reasons. Indeed, this chapter on robotic apple picking is one result of that (Barnes et al. 2021; Ullo and Sinha 2021).

6.6 Applications of Robotics and Automation in Different Farming Practices

Agricultural practices vary greatly around the world according to available technology, naturally occurring climates and seasons, and relative socioeconomic and demographic contexts. Thus, in evaluating the effect and the dynamics of technological innovation in agriculture, a technology-oriented classification is not enough. This addresses the issue of the impact of robotics and automation in different farming practices, considering also qualitative aspects. According to a technological criterion, the main applications are as follows: robotics and automation for field production of plantation crops; horticulture intensive agriculture; protected cultivation; animal production in confined conditions; precision stockmanship; weed management and pesticide application; harvesting; and fruit and vegetable transport in plantations (Carolan 2024).

6.6.1 Precision Agriculture

Precision agriculture relies on high-tech ICT for data gathering, processing, interpreting, and using it in order to optimize decision-making for improved crop and livestock operations. The major components of precision agriculture currently include global positioning systems for vegetation monitoring, soil mapping, yield monitoring, weather information management, and field operations; variable rate technology systems; geographical information systems; sensors and measurement systems to map soil volumetric properties; infrared thermometers; simulation modeling of crop dynamics; and machine vision for weed detection and other applications. Several GIS, GPS, and sensors can be mobile or wearable robotics, imaging sensor systems, remote sensing, autonomous and advanced manned aircraft, and drones (Langemeier and Boehlje 2021).

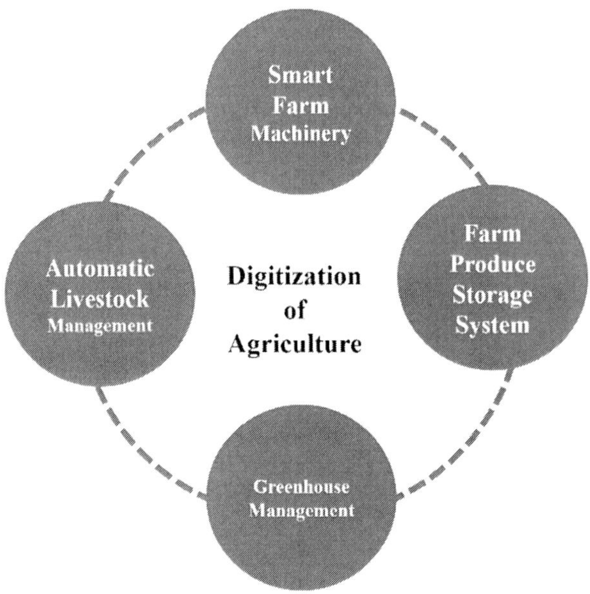

Fig. 6-4 Digitalizing Agriculture

The main benefits of precision agriculture are improved productivity and environmental impact. Reducing the use of water, seeds, fertilizer, herbicides, and pesticides will improve crop and farm operations. Cloud computing and data analysis allow the entire pipeline to be outsourced to experts. Furthermore, it is also possible to outsource crop management; a farmer can visit the field just a few times during a ripening period. Robot-assisted precision farming increases efficiency, reduces soil erosion and compaction, and reduces pollution due to spills or improper handling of mineral fertilizers and pesticides (Barnes et al. 2021; Langemeier and Boehlje 2021; Narsale et al. 2024). The benefits are much larger on larger farms than on small farms. The market for farm management software and services reached considerable growth. Compared to drivers, autonomous tractors save a considerable amount of time that is wasted in long field distances, like turning 180 degrees at row ends. It is also possible to work during the night when temperatures are lower, and there is more humidity, which reduces stress in plants. To end, improved crop management can allow the sale of premium quality products at higher prices through digitalizing farming, as illustrated in Fig. 6.4 (Halstead, Zimmer, and McCool 2024).

6.6.2 Livestock Management

For a couple of decades, farming and livestock management have already been automated by utilizing reusable magnetic marks for pet identification, automatic steering of tractors, and applications of fertilizers. Farm robots for harvesting low-hanging fruit have also been well-established for a long time. Sheep and cattle farming in Australia frequently use livestock-shepherding robots and other livestock-roaming robots to regulate herds in often remote locations. Parasites and contagious diseases create a danger to animal health and efficiency, with several yet unknown dangers to human health or food safety (Deo, Joshi, and Brahmanakar 2022). Some parasites and contagious conditions are species-specific, and consequently, animal welfare assurance methods can include farming segregation from different varieties in combination farming, as well as in transportation. Many contagious conditions in farm animals can turn into an epidemic. Illness may usually spread from patient to patient and consequently from production unit to production unit. Additionally, farm animals are often subjected to contagious diseases or parasites during transportation. In the traveling process, farm animals receive varying levels of protection from the parasitic or contagious elements, get stressed and strained, and require rest and food to recover and be nursed back to an everyday routine (Ahmad et al. 2022). Not addressing these needs poses health risks to the farm animals.

Despite the species involved, disease and parasite treatments are often a set of mass cures when one animal is attacked. Their vulnerabilities are affected by various reasons, such as farmers striving for maximum production output per year, planters aiming for maximum production output per year, those wanting to ensure the health and welfare of their production companies within their environmental management strategy, and veterinary interventions, which appear to be a recurrent but often costly threat to production businesses (Raj, Karthiban, and Cse 2022). Farmers often look for ways to improve the health and welfare of livestock, which creates opportunities for preventative measures and improvements in health and welfare. Advances in preventive health for animal welfare may have large-scale outcomes and will evolve, offsetting herd and flock potential drug use and thereby helping farmers effectively change their practices to higher levels of human well-being (Narsale et al. 2024).

6.6.3 Weed Control and Crop Monitoring

Mechanical weed control within a crop row has been successfully achieved using just preventive control while exploiting repetitive shapes and machine vision. Automatic crop row vision for the chemical in-row weeding treatment of the assembled machine. Passive and active reflectance plants interact to illuminate the crop rows. UV light was in the spectral range pulled in to set the ground true. The present approaches integrate both location and crop plant area for in-row spraying. A color imaging system is used for monitoring data in the red region by using a method that identifies a part of the red color and the location and width, leaving passive reflectance as a result. Controlled lighting can be used to form high-contrast images of the red or the least reflection during adverse conditions or to improve sunlit conditions with patterned UV light (Marinoudi et al. 2019). In the application, the approach is promising but still relies on RGB. Automatic row finding is object spectrally three areas are discriminated, a soil/non-crop biomass. A system has been developed for chip camera-based real-time weed detection. Reflectance systems may be successfully integrated with the control system by detecting weeds within the row on both sides. An RGB image analysis method based on differences in spectral bands and multidimensional thresholds using a simplified shape recognition model achieved similar results. An automatic image recognition system identified the crop row based on the position and histogram of the color points present in the images. This has already been tested in the sun if an additional UV light is installed parallel to the row, even in the presence of heavy solar conditions, with an automatic control system for weeding with the same photo responses (Lincoln and Elliott 2023; Narsale et al. 2024).

6.6.4 Harvesting and Sorting

Harvesting is the second most critical phase in the agricultural production cycle. Progress is made in developing robotic systems for harvesting that pertain to activities such as fruit picking, weeding, pruning, and plant cutting. Sorting is a processing system used primarily in the grape and citrus sector. It generates added value to production chains based on the use of machine vision for the automatic sorting of products according to their size and quality. Integrated machine vision has allowed improvements in various technological processes and in the performance of expert systems for the differentiation of perishable products based on their macro and micro properties (Fountas et al. 2024). These systems can combine sensing hardware based on different physical principles to yield results that are more accurate than using only a single sensor. Robotic systems for the sorting of products integrate multiple sensors such as color, x-ray, and spectroscopy, among others, to classify products based on multiple properties to enhance production efficiencies by eliminating human labor, reducing rejection of products, or increasing the marketability of products. The newly developed sorting technology for various local and export markets has been described (Ullo and Sinha 2021; Lowenberg-DeBoer et al. 2020). For this study, some noted case studies in sorting systems have been discussed, and they are predominantly citrus and grape harvesting, using IoTs, as shown in Fig 6.5.

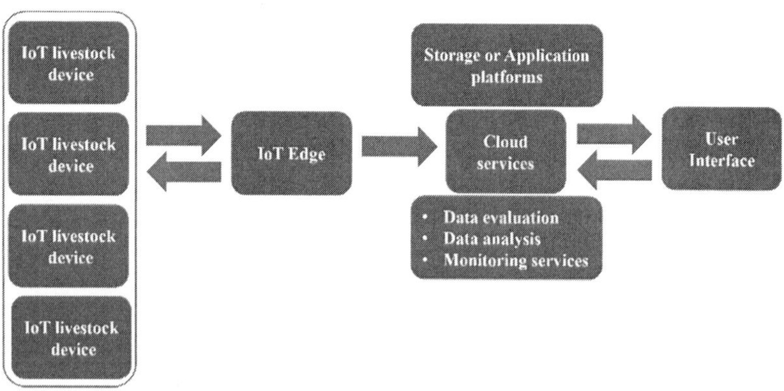

Fig. 6-5 Application of internet of thing and cloud-based systems for livestock monitoring

6.7 Challenges and Limitations of Robotics in Agriculture

Robotic and intelligent systems are used in many different agricultural applications, including livestock management, crop assessment, harvesting, and the implementation of precision agriculture strategies. However, the field is relatively new and can only partially contribute to sustainable agricultural development at present. The challenges imposed by agricultural operations are strong and require the development of innovative perception, decision-making, and manipulation capabilities that were not considered in traditional robotic development. The present transition phase necessitates time-consuming developments of various self-developing or learning capabilities based on multisensory data analysis (Lincoln and Elliott 2023). Various robotic and intelligent systems are involved in everyday agricultural life. Robotic and intelligent systems for sustainable agriculture are large, often distributed, and difficult to evaluate and develop. Presently, the development of such systems is not supported by former operational studies, leading to the requirement of strong experimental effort, which limits the scope and applicability of intelligent automation solutions. The successful deployment of robotic and intelligent systems in agriculture calls for integrating knowledge and methodologies from quite different technological fields, combining biosystems engineering, data mining and big data analysis, machine learning, as well as ethics and rural sociology (Ahmad et al. 2022).

6.7.1 High Initial Costs

The initial cost of purchasing and installing robots and automation machinery is a significant barrier to entry. The machinery and technology are prohibitively expensive for small to medium-sized agricultural operations, and even larger corporate operations will not implement the technology until they can demonstrate a significant return on investment. Moreover, many U.S. farms are family operations, and farm families often have low personal incomes that do not allow direct purchase of such equipment. Since family farms often have lower creditworthiness than for-profit corporations, it can become difficult for them to finance the purchase of expensive machinery and technologies (Narsale et al. 2024). That is why large corporations and manufacturers will need to offer agribusiness-specific financing models to these small and medium-sized farms. Farmers with access to financing options or high profits can benefit from increased yields and low labor costs from robots and other automation technologies. However, relatively high initial costs prevent much of the use of the technologies and equipment. The price of commercial robots is approximately $2,000 per acre, and the cost is close to $1,300 per acre for small robots that may also be used for commercial agricultural operations. Because many farming systems in the United States are already structured and producing, early adopters must separate purchasing land and becoming owner-operators from purchasing commercial farm robots. This also holds for crop-based enterprises seeking to expand. Financing is curtailed as potential buyer versatility is undermined. This problem creates a market for improved credit options and for agricultural lenders who understand potential borrowers and who are far more willing to finance the type of technology that is being purchased. Ferreting them out will not be simple, and the recognition of all the current risks and returns involved will be additional work (Ahmad et al. 2022).

6.7.2 Complexity of Agricultural Environments

Robotic systems designed for use in agriculture are defined as systems that have passed a test of robustness, whether that system has been used in field applications or designed using the stored knowledge of experienced users of the potential host facility. Unlike industrial settings, farmlands represent complex, unstructured, and dynamic environments with changing and unpredictable lighting conditions, a variety of terrain, including different levels of roughness, three-dimensional work, flexibility and fragility of plants, and the presence of interacting living organisms, such as pests or beneficial livestock. These conditions necessitate different systems apart from vision and sensor standards used in factories and address the need for new processing logic to apply to the big data background. In fact, vision and sensor fusion in aligning and adapting to hardware problems are also crucial steps involved in developing these systems (Langemeier and Boehlje 2021). Environmental perception is also considered essential in data collection, decision-making, robotic arm planning, and end-effector-plant interactions. Visual and sensor standards associated with industrial settings can be applied to field applications using modern techniques comprising hyperspectral imaging, photonic systems, and 3D imaging solutions. However, the move from these modern developments to in-situ applications is challenging in terms of data transmission, energy consumption, robustness to different light conditions or sensor radiation, and the availability of appropriate harvesting equipment. Indeed, the main challenge is to apply these robotic innovations to the development and deployment of field systems capable of making decisions on the best shape in multiple directions towards different fields and ready for harvesting many crops, from the shoot to the root, with great plasticity of capacity by virtue of the addition or subtraction of functions when necessary. Collecting and processing accurate data is essential for a system to diagnose and make the right decisions (Barnes et al. 2021; Ullo and Sinha 2021). The right decisions are either to help the farmer better manage the field or have the robot act on vegetation with the least cost of energy or wear and tear on the hardware.

6.7.3 Technical Complexity and Integration Challenges

One of the key issues in the ease of adopting advanced robotic systems for farmers is the technical complexity of the entire farming automation systems, which calls for qualified farm workers. Often, what is needed to run a specific technology is outside the skill set of traditional farmers. The efficacy of automated systems depends on the dynamics of the integration of different machines performing different operations. This has grave implications for equipment vendors providing different machines for different operations, aiming to enhance the affordability, flexibility, and usability of the technologies. Furthermore, the extra effort required by the operator to integrate these machines may compromise the improved usability of the machines. Therefore, operators may not be able to simplify path planning, coordination, and fleet management despite the use of autonomous machines designed for such increased capability (Ahmad et al. 2022). Agri-robots and related tools and equipment may not be directly transferable to local conditions with no additional software tools to help customization. Poorly integrated systems may be more difficult to operate, repair, and maintain due to the interface between parts. The complexity of the automated farm system also results in complicated operation and maintenance. User- or maintenance-friendly designs can certainly reduce the demands on the operator of an automated system, but their capabilities to operate and maintain automated systems need to be expanded and supported by a variety of actors. Also, field operators need to develop a continuous learning process to keep pace with improvements in automated systems, resulting in changing hardware and changing user approaches (Fountas et al. 2024). The technical issues must be addressed before robotic technologies can go mainstream.

6.7.4 Adaptation to Diverse Agricultural Environments

Adaptation of robotic and automation solutions to various agricultural conditions is a significant challenge. Agriculture is inherently an endeavor to manipulate the environment in order to obtain a desired response – either through applied inputs, like water or nutrients, or through avoiding conditions that are harmful to crop development and production. Widespread variability in the natural conditions of every field, which results from localized environmental factors, can create an enormous diversity in conditions within and across different farms. For example, large farms tend to span multiple soil types, climatic zones, and other factors that affect crop development and often produce multiple different crops or varieties (Ullo and Sinha 2021). Because of this, a solution that has an optimal design or parameter setting for one context may not perform effectively or efficiently in another. Collaboration from the outset among the researchers who develop these new technologies, the farmers who will use them, and industry members who would commercialize them is therefore critical for success. This is what we mean when we describe the development of robotic solutions or testing of existing technology across a range of agricultural contexts. Historically, this is a major area in which robotics have failed to deliver on their full promise. Although agricultural mechanization is widespread in some areas and being developed in others, the complexity of agricultural environments – notably the variability in the conditions of each field within a particular farm and across a region or country – creates unique demands for mechanization. Despite this, there are many positive examples where researchers have undertaken this extra work to adapt their solutions to local needs and have ultimately shown benefits across a broad range of settings. Understanding local conditions and local expertise is crucial for arriving at successful answers. Although there are not many robotic examples of work across a wide range of systems and settings, in the area of crop health management, examples do exist (Carolan 2024; Ahmad et al. 2022). In general, variability-adapted solutions described here display great potential to contribute towards sustainable improvements, including restrictions on chemical usage and best practice management of the soil and crop, as presented in Table 6.3.

Table 6-3 Summary of Challenges and Resolutions in Implementing Robotics in Agriculture

Issues	Brief Description	Impact on Agriculture	Practical Resolution	Outcome for Sustainability	Reference
Ethical Concerns	Displacement of traditional labor	Social and economic challenges	Reskilling programs for displaced workers	Promotes inclusive development	(Lowenberg-DeBoer et al. 2020)
Energy Storage Issues	Battery life limitations	Battery life limitations	Advanced battery technology	Prolongs operational hours	(Geckeler et al. 2024)
Limited Adaptability	Limited Adaptability	Restricts broader usage	Modular and adaptable robot designs	Enhances flexibility	(Fountas et al. 2024)
Data Overload	Excessive data generation	Difficulty in analysis and decision-making	AI-driven data analytics	Streamlines data interpretation	(Khadatkar et al. 2024)
Weather Dependency	Adverse weather may affect robotics	Operational downtime	Weather-resistant robotics	Weather-resistant robotics	(Akhter et al. 2024)
Cultural Resistance	Farmers resistant to adopting new methods	Slows integration into traditional farms	Awareness campaigns and success stories	Encourages wider acceptance	(Zhang et al. 2024)
Market Access Limitations	Market Access Limitations	Slows down adoption	Local manufacturing and distribution	Expands technology reach	(Ismail et al. 2023)
Energy Dependence	High energy consumption of robots	Increases operational costs	Integration with renewable energy	Integration with renewable energy	(Marinoudi et al. 2019)

Environmental Concerns	Potential e-waste from old machinery	Potential e-waste from old machinery	Potential e-waste from old machinery	Reduces environmental impact	(Shafik and Kalinaki 2024)
Maintenance and Repairs	Ongoing need for servicing	Ongoing need for servicing	Ongoing need for servicing	Ensures continuous operation	(Ma et al. 2017)
Regulatory Barriers	Complex regulations for robotics use	Complex regulations for robotics use	Streamlined policy frameworks	Faster adoption	(Barnes et al. 2021)
High Initial Investment	Expensive to acquire and implement	Limits adoption by small-scale farmers	Government subsidies, loans	Broadens access to technology	(Kan et al. 2021)
Data Privacy Concerns	Reduces trust in automation	Reduces trust in automation	Secure data management systems	Faster adoption	(Langemeier and Boehlje 2021)
Technical Expertise Required	Farmers may lack the skills to operate robots	Hinders effective use	Training and extension programs	Builds local capacity	(Marinoudi et al. 2019)

6.8 Future Trends and Innovations in Agricultural Robotics

Innovations in robotic systems for agriculture are not limited to the commonly discussed elements of low cost or improved perception for robotics in the near future. There are still numerous technical challenges that need to be addressed with regard to robot navigation, robot perception, and robotic ability. Addressing these challenges will lead to new technologies and increased interest in applications, such as agricultural robotic systems, which can help address the critical issue of global food security. These factors combined will ensure that agricultural robotics remain a thriving field of research, development, and innovation for many years to come. Developments in sensing and computational technology can equip robotic systems with the ability to adapt to the specific needs of the crop in a way that is difficult to achieve with generic systems (Langemeier and Boehlje 2021). As different regions have different crops and farming methods, developing the methodology and equipment that is adapted to the local context can improve the practicality of the integration with innovative robotic systems. As this technology becomes more widely adopted by agriculture, it can help contribute systematically to the identification and exploitation of increasingly diverse patterns and provide both innovative systems as well as promote the future development of a new agricultural growth model, which new technological breakthroughs would drive. This increased collaboration provides an invaluable service by further increasing the visibility of agricultural robotics within the wider Agri-Tech community (Ullo and Sinha 2021; Lowenberg-DeBoer et al. 2020).

6.8.1 Artificial Intelligence and Machine Learning

Recent advancements in artificial intelligence (AI) and machine learning are predicted to drive the next wave of digital agricultural solutions. In addition to LSTM or similar models already in use, other developments in machine learning, such as deep convolutional neural network (CNN) models for crop classification and object detection, have become mainstream with the implementation of remote sensing in agriculture and the advent of powerful machine and deep learning GPU servers. Today, researchers have begun deploying reinforcement learning models that allow their pre-trained dexterous robots to also teach other less skilled robots in the network. The most compelling research in artificial intelligence for field robotics comes from a team led by José Antonio López-Espí of the Automation and Machines Robotics Research Laboratory (Deo, Joshi, and Brahmanakar 2022). As per their work, the robotic observation station is trained as a stand-alone automated bean leaf walker in a supervised manner, as well as trained to follow a different bean leaf rolling robot on a reversing or resetting policy to avoid obstacles or accumulate redundancy in its dataset. Their main contribution stems from being able to provide time-synchronous manual tomato-picking actions, significantly capturing both human and robot egocentric perspectives. Their work will be able to reduce the sampling bias and increase both production efficiency and deployment of their model due to the unfair human instructions. Their work also opens up new areas of interdisciplinary research in task automation, including, but not limited to, synthetic data creation and model supervision (Ma et al. 2017).

6.8.2 Robot Swarms and Collaborative Systems

Robots have little capacity to understand the working environment. There has been very little formal work on multi-robot systems applied to real-world problems. Collaborative systems have the potential to improve the performance of robot implementations of tasks. Each robot could specialize in one of the necessary sub-tasks, making simpler and more reliable robots feasible. This will allow the development of modular robots that assemble themselves when performance guarantees are required. Even modular and self-reconfigurable robots are based on very unreliable technology; the physical connections that allow them to change shape are the most likely to break. Swarm robotics is inspired by the observation that simple animals can often solve complex problems in groups far better and faster than a single individual (Ullo and Sinha 2021). The outcome of such self-organization can be far more efficient and able to adapt to varying environmental conditions, as opposed to the often-fragile cooperative systems developed through traditional robotics approaches. Swarm robotics has shown that the search efficiency of tracking-locating tasks increases in programmed swarms as a function of large numbers. However, relying on large numbers also increases the probability of encounters with obstacles for each robot and increases the likelihood of robots being involved in situations that are potentially harmful to their safety. The implementation of programmed swarms creates a map of solutions, which will be used in one or more particular applications by triggering the mission execution of the different robots involved according to the requirements (Lockhart, Marvin, and While 2023).

6.8.3 Sustainable Energy Sources for Robotic Systems

The choice of renewable energy for powering robotic systems will contribute to the sustainability of agriculture by reducing the environmental footprint. These robotic systems, mainly used in pollination and autonomous farm operations, require constant access to energy sources. Several energy systems, such as solar, wind, wave, geothermal, hydraulic, and bioenergy, are currently used to power robotic systems, but in the field of agriculture, researchers have mainly focused on solar energy for charging battery-based systems for land-based as well as aerial robots. Moreover, these energy harvesters can be integrated without any significant additional weight in most designs. Exploiting energy harvesting from solar irradiation has the additional advantage of making the robot independent from power cords, increasing its operational range and usability (Quaglia et al. 2022). The utilization of sustainable energy sources would reduce the operational costs of machines and help to address the continual increase in energy prices. The resulting independent farming approach could also serve as an occasion to change the current reliance on fossil fuels.

Unmanned electric vehicles, such as drones and robots powered by batteries or energy harvesting technologies, need efficient ways of generating, storing, and managing the power to sustain farming operations. By consuming less energy, robotics could be directly consistent with the principles of sustainability and environmental preservation. Integrated solar panels are technically feasible for a highly autonomous system. Aerial vehicles could be designed to have relatively lower power consumption than ground vehicles. However, trade-offs in design must carefully balance size, robustness, affordability, and energy budget (Chong et al. 2023). These considerations are not unique to aerial drones but apply to all robotic systems. Apart from the technical aspect, flying robots are also an agile and quick alternative to traditional ground vehicles for monitoring large farmland. Solar power is a credible, sustainable energy choice. Hence, one trend in the scope of future development of autonomous systems is generating energy from accessible, sustainable sources to provide power for autonomous farming systems. In addition, optimization of overall energy and energy storage based on the consumption specifics of robots is another focus area and represents a fundamental research direction. Energy sources that are sustainable and environmentally friendly are likely to make a profound impact on the economic and ecological sustainability of farming (Ishibashi et al. 2023).

6.9 Case Studies and Success Stories

As in any other dynamic field, there are many successful and innovative applications. Reporting some of them. Some are still at the prototype stage. Others are already in regular use. Some of these are still considered pre-market. But they are clearly addressing some challenging and costly aspects. In this respect, we may take the experience of large teams dealing with rice and maize. These crops represent a potential volume of business and opportunities. Other groups of researchers have pioneered their efforts in addressing the highly price-sensitive issue of picking fruits and vegetables. It covers a wide span of different fruit types, different ways the fruits are arranged in the vine or plant, different levels of maturity, and variability for each parameter. Robots developed to address these tasks have been tested in experimental settings with some degree of success (Ismail et al. 2023). Medium-term challenges here reside in keeping track of the fruits that grow with the plant, conserving all the information, and handling all the different varieties that have evolved by means of selection according to different criteria. Various interesting ideas have been proposed and prototyped for different types of fruits. A key, common, and challenging aspect is the identification of the fruit in space and in time, e.g., while it is still growing. It is reasonable to expect that these approaches will be used in large greenhouses and high-tech farms where the quality of the information on the status of all fruits is fundamental for business intelligence activities (Geckeler et al. 2024; Guldenring, Andersen, and Nalpantidis 2024). Other economic sectors might also profit from these ideas and applications.

There are many examples of commercially available as well as research-stage prototypes for the robotic and automation implementations of agricultural tasks. Below, we provide a brief overview of some of the major examples for each aspect of crop farming, including machine perception, intelligent control, and dexterity. The most successful and widely adopted example of the application of robotics and automation in agriculture is heavy field machines such as tractors, combine harvesters, and ploughing and sowing robots. In addition, among the traditional stationary installations of modern competitive farms, one could find the whole spectrum of electro-mechanical devices of different levels of complexity and autonomy, starting from automated relatively static devices for milking cows and finishing with fully integrated systems for cleaning, feeding, and growing animals (Lincoln and Elliott 2023; Zhang et al. 2024). These machines come in different shapes and sizes, have different underlying implementations, and are intended for vastly different tasks related to crops and animals.

Working in the field with such precision-equipped machines started several decades ago and uses diverse sensors and actuators for optimal performance of different seeding, watering, and harvesting tasks. In recent years, the field of precision agriculture has exploded in the number of research papers, as well as industrial startups and prototypes of commercial robots and autonomous systems. Estimates suggest that in 2018, there were more than 200 different companies around the world developing agricultural robotics and automation solutions in tech-driven sectors of farming, numbering more than 1 to 4 automated solutions for typical problems in planting, irrigation, crop scouting, and health assessment, weed and pest control, and harvesting crops (Chong et al. 2023).

6.10 Ethical and Social Implications of Agricultural Robotics

At the same time as the number of scientific reports on the effectiveness of robotics and automation for agriculture is multiplying, recognition of the need for a broader perspective on the overall effects of introducing high technology in agriculture beyond economic impacts and the consideration of ethical, social, and political implications arising from their wide adoption is growing in parallel. Discussions about the ethical and social impact of implementing ICT in agriculture should also consequently include robotics and automation. The pervasive development of advanced technologies in agriculture is labeled "digital agricultural transformation." Robotization and automation of agriculture require solving scientific, technological, policy, and acceptance challenges (Ismail et al. 2023). Without facing such complex issues and resolving concomitant challenges, broader societal opposition or even ethical dilemmas may become insurmountable barriers to the design, development, testing, deployment, and use of service robots in agriculture. Reasoning on the ethical and societal progression of autonomous robots in agriculture will likely produce such "responsible innovation" or "responsive regulation." A reason for this is that the introduction of robots in all possible professional activities can worsen the condition of the already fragile relationship between humans and the environment. It may also increase threats to animal welfare standards if animal tasks like herd or flock management, health monitoring, and treatment cannot be improved (Marinoudi et al. 2019).

These threats could be reduced only if prosperous robot systems can fully replace labor on farms with a subsequent reduction in educational costs for future generations. The need to assess the ethical and social effects associated with service robotic technology innovation is increasingly recognized by the research community as one of the central issues deserving attention. Interviews conducted through the course of development prove that farmers take ethical, political, legal, social, and environmental issues into account while evaluating products for their farms. Providing transparency of robotic technologies may help mitigate resistance from operators (Halstead, Zimmer, and McCool 2024). In a more evidence-based approach, pilot studies, action research, and even outdoor field trials with service robots in actual professional contexts are considered methods to support the identification of the most relevant ethical, regulatory, acceptance-related, political, environmental, and social concerns. Responses from public and organizational actors to these challenges provide plenty of indicators and favor the ongoing digital transformation of agriculture. The document should be systematized in order to provide interested public and private stakeholders with clear indications for the ethical assessment of key decision-making processes on the development and use of farm service robotics and precision farming technologies (Barnes et al. 2021).

6.10.1 Impact on Rural Communities

There is an urgent need to address the declining availability of human labor for manual agricultural tasks, especially those that require high precision. An increasing number of robotic and automated systems are being developed and deployed to address the needs of specialty crop production. As it turns out, most of these crops are produced in rural areas that have little access to advanced technologies and, at the same time, are facing a technology exodus due to the absence of enterprises requiring advanced expertise in cutting-edge engineering solutions. If the needs of the growers can be satisfied with locally developed and even open-source software and hardware technologies, such as robotics and automation systems, not only can the economies of rural communities be revitalized, but the resilience of the agricultural sector can be achieved overall (Raj, Karthiban, and Cse 2022).

Robotic and automation technologies not only help with the production of crops, such as planting, germination, weeding, thinning, harvesting, conveyance, and sorting, but, most importantly, contribute to the dignity of the workers who will be doing more suitable jobs with high-value responsibilities using such technologies with at least the same current harvest-parity effort because of the high-tech training programs associated with these tools. Indeed, these new-generation growers will have an edge over their counterparts in the industry, as they will, in fact, be the greatest brand marketers in the future with their crop-to-gourmet cuisine prepared using their organic agricultural produce touted at price premiums (Chong et al. 2023). In anticipation of the next 30-year chapter until 2050, we are laying out a roadmap and a strategic plan for an Agricultural Renaissance centered on Robotic and Automation Systems for our nation powered by rural communities.

6.10.2 Job Displacement and Reskilling

Given that increased automation often displaces manual and low-skilled labor with machines that are highly efficient at specific tasks, the immediate question that comes to mind is what to do with the workers who are displaced. A great simple analogy can be found in the Industrial Revolution period where an increase in mechanization caused poverty and displacement for many, driving these displaced workers to destroy the automated machines. It is important to understand and emphasize that people are our most important and valuable resource. In economics, the principal costs of automation and mechanization are in terms of their effects on human capabilities and the well-being of workers and their families. One obvious situation that might first come to one's mind would be an entire displacement of agricultural workers (Langemeier and Boehlje 2021). This would only be likely in history if the entire agricultural system were mechanized by a mechanism that could do everything simultaneously and on a large scale. This kind of mechanization or automation is not feasible at present or in the immediate foreseeable future over a period to ensure an adequate food supply to provide everyone with enough to eat. However, it will continue to evolve with different stages of technology combined, initially within separate crop systems, in agronomic, breeding, and engineering advances to develop autonomous systems for a spectrum of field crops and activities that form the backbone of the food system (Zhang et al. 2024; Lincoln and Elliott 2023).

6.11 Conclusion

In a short span, robotics and automation have dramatically changed the way farming is done. The present contribution aimed to address the issue of increasing food production while not neglecting the importance of sustainable agriculture. Two major themes that cutting-edge research was divided into were plant phenotyping for sustainable agriculture and field robots as a key technology for the sustainability of modern agriculture. Topics of discussion included individual strengths and weaknesses of alternative robotic strategies, sensors, and platforms for field phenotyping. This young and rapidly maturing discipline of robotic agriculture is setting the scene for the next generation in terms of how it should be implemented. The adoption of robotics and automation is set to revolutionize agriculture. However, the optimal solutions are yet to be found, and several related research challenges are yet to be addressed. To a great extent, these will depend on the disciplinary area and practical context in which the research is to be applied. In this chapter, two main application areas (open field and protected) have been treated. The principal challenge is that of innovation, as research on technology is essential to developing these innovations to a high enough standard to be a commercial reality. Moreover, the development of adaptive policies has been set as one of the future steps. The rapid development of this technology signifies that policies need to be sufficiently adaptive to accommodate further developments. Another challenging step is to draw target groups, technologists, farmers, and policymakers into the development and evaluation of the tools. This will increase the likelihood that farmers will accept and use these technologies. At the same time, technologies need to be developed in a way that requires no special skills or training on the part of users; this applies to farmers, technologists, and policymakers. Moreover, prospects will range from product development to pilot tests and eventually to demonstration farms and follow-up activities, depending on technology availability and readiness. In general, awareness of these technologies will need to be heightened via comprehensive communication strategies. The integration of these tools into practical adoptions of scientific experimentation at various sites is anticipated to take form. The journey in robotic tools and precision agriculture research aimed at future-oriented development for application in sustainable agriculture has begun. The eventual success and assimilation of the technologies into the mainstream will depend in no small part on proactive policy and decision-making aimed at creating an enabling environment for such technologies to be developed, tested, refined, and eventually adopted. At present, the greater challenge appears to be the development of policy instruments that stimulate and guide the research and deployment of such technologies while protecting the interests of the broad spectrum of stakeholders within the sector.

References

Ahmad, Mohd Najib, Mohamad Izzuddin Anuar, Nordiana Abd Aziz, Mohd Azwan Mohd Bakri, Zulkifli Hashim, and Idris Abu Seman. 2022. "Addressing Functionalities of Agricultural Robotic (Agribots) and Automation in the Agriculture Practices: What's next?" *Advances in Agricultural and Food Research Journal* 4 (1). https://doi.org/10.36877/aafrj.a0000298.

Akhter, Afroza, Ambreen Nabi, Sumati Narayan, Sabia Akhter, Bilal Ahmad Lone, Vaseem Yousuf, Bhinish Shakeel, et al. 2024. "Digital Technology: A Game Changer in Vegetable Cultivation." *Annual Research & Review in Biology* 39 (2). https://doi.org/10.9734/arrb/2024/v39i230631.

Barnes, Edward, Gaylon Morgan, Kater Hake, Jon Devine, Ryan Kurtz, Gregory Ibendahl, Ajay Sharda, et al. 2021. "Opportunities for Robotic Systems and Automation in Cotton Production." *AgriEngineering* 3 (2). https://doi.org/10.3390/agriengineering3020023.

Belforte, G., R. Deboli, P. Gay, P. Piccarolo, and D. Ricauda Aimonino. 2006. "Robot Design and Testing for Greenhouse Applications." *Biosystems Engineering* 95 (3). https://doi.org/10.1016/j.biosystemseng.2006.07.004.

Carolan, Michael. 2022. "Digitization as Politics: Smart Farming through the Lens of Weak and Strong Data." *Journal of Rural Studies* 91. https://doi.org/10.1016/j.jrurstud.2020.10.040.

Carolan, Michael. 2024. "Who and What Gets Recognized in Digital Agriculture: Agriculture 4.0 at the Intersectionality of (Dis)Ableism, Labor, and Recognition Justice." *Agriculture and Human Values*. https://doi.org/10.1007/s10460-024-10560-9.

Chong, Yue Linn, Jan Weyler, Philipp Lottes, Jens Behley, and Cyrill Stachniss. 2023. "Unsupervised Generation of Labeled Training Images for Crop-Weed Segmentation in New Fields and on Different Robotic Platforms." *IEEE Robotics and Automation Letters* 8 (8). https://doi.org/10.1109/LRA.2023.3293356.

Deo, Kaushik, Prasad Joshi, and Rohit Brahmanakar. 2022. "Design and Fabrication of a Multipurpose Agriculture Robot." *International Journal for Research in Applied Science and Engineering Technology* 10 (5). https://doi.org/10.22214/ijraset.2022.43276.

Eastwood, C. R., B. Dela Rue, J. P. Edwards, and J. Jago. 2022. "Responsible Robotics Design–A Systems Approach to Developing Design Guides for Robotics in Pasture-Grazed Dairy Farming." *Frontiers in Robotics and AI* 9. https://doi.org/10.3389/frobt.2022.914850.

Fountas, Spyros, Borja Espejo-Garcia, Aikaterini Kasimati, Marilena Gemtou, Hercules Panoutsopoulos, and Evangelos Anastasiou. 2024. "Agriculture 5.0: Cutting-Edge Technologies, Trends, and Challenges." *IT Professional* 26 (1). https://doi.org/10.1109/MITP.2024.3358972.

Geckeler, Christian, Emanuele Aucone, Yannick Schnider, Andri Simeon, Jan Philipp Von Bassewitz, Yunying Zhu, and Stefano Mintchev. 2024. "Learning Occluded Branch Depth Maps in Forest Environments Using

RGB-D Images." *IEEE Robotics and Automation Letters* 9 (3). https://doi.org/10.1109/LRA.2024.3355632.

Guldenring, Ronja, Rasmus Eckholdt Andersen, and Lazaros Nalpantidis. 2024. "Zoom in on the Plant: Fine-Grained Analysis of Leaf, Stem, and Vein Instances." *IEEE Robotics and Automation Letters* 9 (2). https://doi.org/10.1109/LRA.2023.3346807.

Halstead, Michael, Patrick Zimmer, and Chris McCool. 2024. "A Cross-Domain Challenge with Panoptic Segmentation in Agriculture." *International Journal of Robotics Research* 43 (8). https://doi.org/10.1177/02783649241227448.

Ishibashi, Kyosuke, Mitsuo Komagata, Hiroki Ishikawa, Osamu Azami, and Ko Yamamoto. 2023. "Compact Water Pump and Its Application to Self-Contained Soft Robot Hand for Vegetable Factory." *Advanced Robotics* 37 (15). https://doi.org/10.1080/01691864.2023.2238800.

Ismail, Bukhary Ikhwan, Muhammad Nurmahir Mohamad Sehmi, Hishamadie Ahmad, Shahrol Hisham Baharom, and Mohammad Fairus Khalid. 2023. "Robotic Research Platform for Agricultural Environment: Unmanned Ground Vehicle for Oil Palm Plantation." *Journal of Cases on Information Technology* 25 (1). https://doi.org/10.4018/JCIT.328579.

Kan, Xinyue, Thomas C. Thayer, Stefano Carpin, and Konstantinos Karydis. 2021. "Task Planning on Stochastic Aisle Graphs for Precision Agriculture." *IEEE Robotics and Automation Letters* 6 (2). https://doi.org/10.1109/LRA.2021.3062337.

Khadatkar, Abhijit, A. P. Magar, C. P. Sawant, and R. U. Modi. 2024. "Development and Testing of Automatic Seedling Extractor in Robotic Transplanter Using Mechatronics for Nursery Seedlings." *Discover Applied Sciences* 6 (2). https://doi.org/10.1007/s42452-024-05670-2.

Langemeier, Michael, and Michael Boehlje. 2021. "Automation and Robotics in Production Agriculture." *Farmdoc Daily* 11 (57).

Lincoln, Jennifer M., and K. C. Elliott. 2023. "Emerging Technology in Agriculture: Opportunities and Considerations for Occupational Safety and Health Researchers." *Journal of Safety Research* 86. https://doi.org/10.1016/j.jsr.2023.06.001.

Lockhart, Andrew, Simon Marvin, and Aidan While. 2023. "Towards New Ecologies of Automation: Robotics and the Re-Engineering of Nature." *Geoforum* 145. https://doi.org/10.1016/j.geoforum.2023.103825.

Lowenberg-DeBoer, James, Iona Yuelu Huang, Vasileios Grigoriadis, and Simon Blackmore. 2020. "Economics of Robots and Automation in Field Crop Production." *Precision Agriculture* 21 (2). https://doi.org/10.1007/s11119-019-09667-5.

Ma, Rui, Koon Hoo Teo, Shintaro Shinjo, Koji Yamanaka, and Peter M. Asbeck. 2017. "A GaN PA for 4G LTE-Advanced and 5G: Meeting the Telecommunication Needs of Various Vertical Sectors Including Automobiles, Robotics, Health Care, Factory Automation, Agriculture, Education, and More." *IEEE Microwave Magazine* 18 (7). https://doi.org/10.1109/MMM.2017.2738498.

Marinoudi, Vasso, Claus G. Sørensen, Simon Pearson, and Dionysis Bochtis. 2019. "Robotics and Labour in Agriculture. A Context Consideration." *Biosystems Engineering* 184. https://doi.org/10.1016/j.biosystemseng.2019.06.013.

Narsale, Swapnil Ananda, Patekar Prakash, Hari Prasad Mohale, Ravi Baraiya, Samad Sheikh, Parmar Bindiya Kirtikumar, Chovatia Ravikumar Mansukhbhai, Rishikesh Venkatrao Kadam, and Indulata Tekam. 2024. "Precision Aquaculture: A Way Forward for Sustainable Agriculture." *Journal of Experimental Agriculture International* 46 (5). https://doi.org/10.9734/jeai/2024/v46i52360.

Parai, Krittibas, Daya Sagar Gupta, and Sk Hafizul Islam. 2024. "IoT-ID3PAKA: Efficient and Robust ID-3PAKA Protocol for Resource-Constrained IoT Devices." *IEEE Internet of Things Journal* 11 (6). https://doi.org/10.1109/JIOT.2023.3325583.

Quaglia, Giuseppe, Luigi Tagliavini, Giovanni Colucci, Ardit Vorfi, Andrea Botta, and Lorenzo Baglieri. 2022. "Design and Prototyping of an Interchangeable and Underactuated Tool for Automatic Harvesting." *Robotics* 11 (6). https://doi.org/10.3390/robotics11060145.

Raj, Starlin Daniel, Mr Karthiban, and M E Cse. 2022. "Artificial Intelligence in Agriculture: A Literature Survey." *International Journal of Creative Research Thoughts* 10 (6).

Shafik, Wasswa. 2025a. "Ecological Health Impact in the Marine World." In *Radiation Status in the Marine World*, 307–35. Springer.

Shafik, Wasswa. 2025b. "Sustainable Agriculture and Diet for Urban Development." In *Sustainable Smart Cities and the Future of Urban Development*, 165–92. IGI Global Scientific Publishing.

Shafik, Wasswa, and Kassim Kalinaki. 2024. "Societal and Ethical Implications of Technology-Enhanced Agriculture and Healthcare: An African Context." In *2024 IST-Africa Conference (IST-Africa)*, 1–11. IEEE. https://doi.org/10.23919/IST-Africa63983.2024.10569306.

Shafik, Wasswa, Ali Tufail, Chandratilak De Silva Liyanage, and Rosyzie Anna Awg Haji Mohd Apong. 2023. "Using a Novel Convolutional Neural Network for Plant Pests Detection and Disease Classification." *Journal of the Science of Food and Agriculture*. https://doi.org/10.1002/jsfa.12700.

Shafik, Wasswa, Ali Tufail, Chandratilak Liyanage De Silva, Rosyzie Anna Awg Haji, and Mohd Apong. 2025. "The Role, Application, and Impact of Artificial Intelligence in the Agriculture Industry." In *Future Tech Startups and Innovation in the Age of AI*, 36–60. CRC Press.

Shafik, Wasswa, Ali Tufail, Liyanage Chandratilak De Silva, and Rosyzie Anna Haji Mohd Apong. 2025. "An Enhanced Deep Convolutional Neural Network for Plant Disease Detection and Classification: Elevating Sustainable Agriculture." In *Artificial Intelligence and Data Science for Sustainability: Applications and Methods*, 297–322. IGI Global Scientific Publishing.

Ullo, Silvia Liberata, and G. R. Sinha. 2021. "Advances in IoT and Smart Sensors for Remote Sensing and Agriculture Applications." *Remote Sensing* 13 (13). https://doi.org/10.3390/rs13132585.

Yi, Tao, Dongbo Zhang, Lufeng Luo, and Jiangtao Luo. 2024. "View Planning for Grape Harvesting Based on Active Vision Strategy under Occlusion." *IEEE Robotics and Automation Letters* 9 (3). https://doi.org/10.1109/LRA.2024.3357397.

Yigitcanlar, Tan, Nayomi Kankanamge, Massimo Regona, Andres Ruiz Maldonado, Bridget Rowan, Alex Ryu, Kevin C. Desouza, Juan M. Corchado, Rashid Mehmood, and Rita Yi Man Li. 2020. "Artificial Intelligence Technologies and Related Urban Planning and Development Concepts: How Are They Perceived and Utilized in Australia?" *Journal of Open Innovation: Technology, Market, and Complexity* 6 (4). https://doi.org/10.3390/joitmc6040187.

Zhang, Jun, Ningbo Kang, Qianjin Qu, Lianghuan Zhou, and Hongbo Zhang. 2024. "Automatic Fruit Picking Technology: A Comprehensive Review of Research Advances." *Artificial Intelligence Review* 57 (3). https://doi.org/10.1007/s10462-023-10674-2.

CHAPTER SEVEN

AGRICULTURAL DATA ANALYTICS FOR FARM MANAGEMENT INFORMATION SYSTEMS

Abstract

Farming all over the world has been changing due to various advancements in agriculture. Traditional farming practices have been replaced by advanced technology, which is coupled with advanced knowledge of how the plant grows and how to maintain it. Agriculture has become a precision science with the aim of maximizing profit and minimizing loss. To maximize profitability in farming, the decision-making system should be knowledge-driven rather than intuition-based or experience-based. By gaining knowledge from the field, such as soil fertility, weather conditions, crop management decisions, etc., it is possible to reduce uncertainty during the optimization of the farm. This can be achieved through the process of data analytics in the agricultural domain. Data analytics has become one of the main streams of analytics in the 21st century, while the domain of agriculture plays a notable role in each country's economic development. The main role of data analytics in agriculture is that it provides managerial and pragmatic support, enabling managers and farm administrators to make comprehensive decisions. The domain of agriculture can be defined as a major sphere in which big data is being used in a widespread manner. In sustainable agriculture, advanced data analytics are used for improved decision-making. The use of advanced data analytics and methods enables us to improve crop or plant productivity, raising the national and international economic environment. By using suitable data analytics techniques, it is possible to obtain present and future technical information from farming data that is derived from or related to information technology. This practical information can be utilized to improve the performance and effectiveness of the future decision-making system.

Keywords: Agricultural Data Insights, Crop Monitoring Tools, Data-Driven Decision Making, Farm Management Systems, Farm Productivity

Optimization, IoT in Agriculture, Precision Farming Analytics, Real-Time Data Integration, Smart Agriculture Solutions

7.1 Introduction

The growth of technology in the agricultural domain has gained huge attention in university and college research departments due to the promise of advancements in agricultural practices and the benefits sought from research projects. Agriculture companies are focused on using technical advantages in decision-making to improve the performance and optimization of the farm as it tends more toward the economic perspective. The decision-making model is not statistically based; thus, to resolve this issue, research in the field of data analytics-integrated farming would be helpful (de Souza Filho et al. 2023). Farmers are interested in knowing the current condition of their farm business via the concise derived information provided from the historical data of the farm and other related data. It is possible, using data analytics, to design strategies accurate to the national and international markets. Data analytics are becoming increasingly popular and important in agriculture due to their wide range of benefits. Improved decisions are paramount for any agribusiness to remain competitive and profitable. By receiving actionable insights, farm managers can make quicker and more informed decisions and act on them more easily (Carrer et al. 2015).

The potential benefits provided by improved decision-making through data-informed insights are extensive, including higher yields, cost reductions or savings, improved and smarter resource management, asset or input usage efficiencies, and sustainability benefits by optimizing inputs for production and minimizing damages and waste. Reliability and speed of data are important factors in providing any of these benefits, either on their own or in concert, and significant data lag can reduce both the magnitude of the benefit and the applicability of data analytics solutions to those decisions. The examples described demonstrate that practical farm use of data analytics can translate into significant value realized on farms (Carrer, de Souza Filho, and Batalha 2017). The value of these outputs is derived not from the outputs themselves but from the resulting efficiency and gains that are made possible using data. Some of the benefits of adopting data analytics in agriculture that a data analytics-integrated business system should target are multiple, including the ability to interpret data for meaningful insights and improved and efficient decision-making. The need for grower capacity for comprehensive data usage is exemplified in the adoption of data-intensive farming; ensuring timely and reliable data to drive such data analytics applications is fundamental (Husemann and Novkovic 2014).

Conversely, some important challenges and limitations have also been highlighted regarding the adoption and implementation of data-driven farm management information systems. One of the critical issues is related to data collection, which might be a cumbersome and error-prone task. Moreover, farmers may use multiple data systems on their farms, and the information may be locked in inaccessible islands of automation. As a result, silos of data have been created, and the systems from which the data originate are unable to share the information in interoperable or other useful ways. For farmers, investing in expensive technology and data analysis tools is challenging because of the variable, high-risk, and often low-profit margin agriculture business, which can make meeting operational expenses difficult (Fountas et al. 2015). In principle, large amounts of information produced could be useful for decision support, but in practice, this potential is rarely fully exploited.

Moreover, the digital divide and low digital skill levels among farmers and agriculture stakeholders are thought to be a barrier to the adoption of data-driven systems embracing farm and food system management. Accordingly, a steep learning curve in data analytics utilization might act as a limitation for uptake, regardless of potential benefits as perceived by many stakeholders. Therefore, a preference for cheaper and easy-to-interpret and act-upon data analytical approaches is consistently suggested as a requirement for practical applications (Kassahun et al. 2022). Therefore, there is also the issue of privacy and control over farmers' and other stakeholders' information, which might substantially hinder the general uptake of digital data systems. Despite the many possibilities and advantages of agricultural data analytics, very few farmers are currently applying the concept at their farms because of several impediments besides the roles of sustainable agriculture, as illustrated in Fig 7.1.

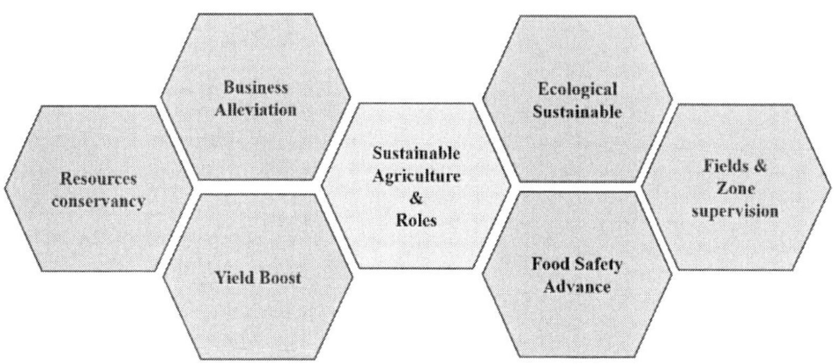

Fig. 7-1 Sustainable Agriculture and its associated roles

7.2 Fundamentals of Farm Management Information Systems

A Farm Management Information System (FMIS) is typically the first step toward the digital farm and acts as the central hub for collecting, storing, and analyzing different sets of data across domain-specific farming tasks. In principle, an FMIS lets various agricultural data sources and collection technologies 'speak' to each other in order to streamline certain agricultural operations. FMIS can be described as a system used in farms with data that comes from the field to the FMIS, where all the data are stored, and local analyses are made to allow the farmer to make better decisions about farming.' A typical FMIS has components, including software, hardware, and communication infrastructure. The software usually comprises an FMIS software platform that features a user-friendly interface to provide easy and intuitive access to the wealth of information stored on the FMIS. FMIS relies on the input of a range of different data types. A typical FMIS needs to support data storage and retrieval capabilities as well as a farm management decision engine alongside some visualization tools. In general, FMIS is predominantly focused at the management and operational level, supporting the interconnectedness of farm tasks associated with the implementation of farming strategies (Schulze Schwering, Bergmann, and Isabel Sonntag 2022). These systems are aimed at providing appropriate farm management recommendations to increase farm productivity and reduce the cost of production, sustainability, and profitability.

Existing FMIS gathers data throughout a farm's operational cycle from various data sources and collection tools. In that sense, FMIS are innovative, as they compile available data and provide a means for making data-driven decisions at the farm level. In general, data collected on the farm might be grouped into three classes: strategic, tactical, and operational data. A detailed description of data collection tools is outside the scope of this chapter. However, providing some examples in terms of data sources can help appreciate the diversity of data used in FMIS. Many different sensor systems are used to obtain information from the farm. These could include satellite, LIDAR, and GPS systems. Additionally, weather stations may be used to record various agricultural relevant conditions outside the field (Lewis 1998). However, not only is the collection of data itself relevant in FMIS, but also the ability to provide user-friendly interfaces to support the farmer's engagement with and analysis of the data gathered through data boost, as illustrated in Fig 7.2.

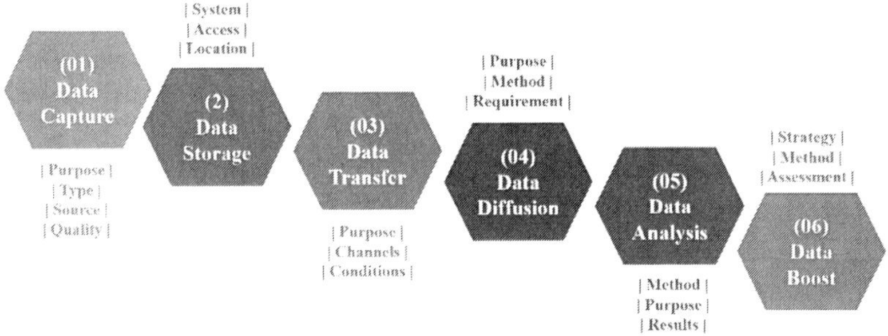

Fig. 7-2 Six Data processing steps

7.2.1 Components and Architecture

FMIS are modern systems providing comprehensive farm management tools based on combining information from different sources like farm equipment, data loggers, geo-information systems, and weather forecasts. The modern FMIS is a complex group of technological elements that handle different types of data, such as text, multimedia, and sensor result data. This requires suitable hardware, software, databases, and network infrastructure. These devices must be scalable for future expansion and flexible to allow different configurations for various farming operations. The actual FMIS has components like computer hardware and peripherals, communication devices, software for farm management, data storage, and security systems, providing data administration and a convenient, user-friendly interface for the applications (Nugawela and Sedera 2020). The whole FMIS architecture can be divided into separate but linked functionalities, expressed as hardware or software components. These elements can be combined in different ways to allow the implementation of forecast systems for small to large farming operations. An essential issue for today's FMIS is the application of cloud computing technology to enable the integration of IoT devices from different manufacturers. This prediction needs to address robust, secure, and reliable clouds serving data and applications on a short- and long-term basis, combined with FMIS operators for many applications and customers. In practice, FMIS can be depicted with different architectures, from specific for a separate group of devices and systems, up to universal, the cloud-based model combined with complex decision rules and machine learning approaches (Tummers, Kassahun, and Tekinerdogan 2019).

The FMIS needs to use the Internet of Things to manage data created by the combination of diverse devices and software. In practice, this means the possible connection with data from several sensors into one IoT device, which helps to control and maintain the connectivity from the device to the sources of IoT. The IoT device can be placed in various consecutive locations because they are user-oriented; it can be in the cloud, fog, or edge. Devices managed by FMIS can exchange information such as data, measurement results, processed data, applications, and even their configurations and consequently share information with others. Forecasts can be generated by system management with the information bases and technology used (Salami and Ahmadi 2010). The FMIS is a wide system surrounded by a lot of technology and is needed to guarantee the protection of people and information. This requires several groups of knowledge and consideration. The main topic in this part of the chapter is to describe the informal architecture and structural support in FMIS development, as well as the required technology oriented towards IoT and cloud computing, as summarized in Table 7.1.

Table 7-1 Components of Agricultural Data Analytics in Farm Management Information Systems

Component	Description	Data Sources	Key Technologies	Impact on Farm Management	Reference
Compliance and Reporting	Ensures adherence to regulations	Farm records, government databases	Regulatory compliance software	Regulatory compliance software	(de Souza Filho et al. 2023)
Waste Management	Tracks and minimizes farm waste	Waste sensors, production data	IoT, analytics platforms	Promotes sustainability	(Carrer et al. 2015)
Energy Management	Monitors energy use and efficiency	Energy meters, IoT devices	Energy analytics software	Energy analytics software	(Carrer, de Souza Filho, and Batalha 2017)
Market Analysis	Tracks market trends and prices	Market reports, real-time data	Data visualization tools	Informs pricing and sales strategies	(Husemann and Novkovic 2014)
Irrigation Management	Automates water usage based on soil moisture	Moisture sensors, IoT devices	IoT, automation software	Conserves water resources	(Fountas et al. 2015)
Yield Prediction	Forecasts crop output based on data trends	Historical data, climate data	Predictive models	Enhances production planning	(Kassahun et al. 2022)
Livestock Monitoring	Tracks health and productivity of animals	Wearable sensors, cameras	IoT, data analytics	Improves animal health	(Schulze Schwering, Bergmann, and Isabel Sonntag 2022)

Labor Management	Manages workforce schedules and productivity	Workforce data, mobile apps	HR management systems	Increases labor efficiency	(Bermeo-Andrade and González-Bañales 2022)
Financial Analytics	Tracks costs, revenues, and profitability	Financial records, market data	Predictive analytics	Improves financial planning	(Melzer, Bellingrath-Kimura, and Gandorfer 2023)
Resource Allocation	Manages inputs like seeds, fertilizers	Farm equipment, purchase date	Decision support systems	Optimizes resource use	(Munz, Gindele, and Doluschitz 2020)

7.2.2 Data Collection and Storage

Data collection refers to the process of gathering data for analysis and visualization in FMIS. In the context of agriculture, data may be generated by field operations, sensors, remote sensing platforms, off-farm infrastructure, and other sources. Data may also be collected manually by the farmer or operator if the FMIS requests it. Therefore, data collection is one of the most critical parts of any FMIS. There are several means by which agricultural data streams may be collected, each offering different trade-offs between cost, accuracy, and timeliness. The cost of data collection may include the cost of hardware required, time taken by humans and machines, energy sources, connectivity for transfer, as well as the cost of maintaining the hardware in the long term. The benefits of each data collection method vary according to the type of data (Ammann, Walter, and El Benni 2022).

Finding efficient methods of storing agricultural and farm data is challenging. Data sources used in FMIS are usually different. They can be geospatial and time-series data from sensor technologies, images or remote sensing data, and agricultural workflows and lifecycle data. Data storage solutions should keep the data centralized, preprocessed, and accessible to ease usage in analytics and decision support for the end users. Data storage currently focuses on how to maintain the integrity of both the structure of the stored data object and the data entered by the user (Köksal and Tekinerdogan 2019). The selection of the database system used depends on various factors, including the default data collection system used, the user requirements related to system access, security, and platform compatibility, as well as the available budget. In data management, data persistence becomes the main concern and top priority to ensure that all entered data can be used at appropriate times. The persistence model is set according to the transmission mechanisms and the availability of backups for when the failure occurs. Data management and conservation system concepts are necessary conditions needed to facilitate the administration of the Farm Management Information System. Effective FMIS requires effective data management standards and mechanisms to support the accurate collection, storage, sharing, reuse, maintenance, and data discovery. Data and information management systems must follow the rules and protocols that enable system data sharing and exchange among different users. Data management standards and protocols that are used can be tailored to the scope of established data and applications. Architecture promotes data sharing to ensure data from a wide range of systems can be integrated (Husemann and Novkovic 2014). Data standards and protocols support data sharing and drive the integration of different systems.

7.2.3 Sensors and IoT Devices in Agriculture

Data analytics using field observations can help to optimize crop production by monitoring crop conditions and disease outbreak events, suggesting when to apply certain inputs, predicting the quantity before harvest time, as well as monitoring quality during the production season. Technology for data collection sees one of its major leaps with the Internet of Things, combining edge devices with software, connectivity, and application solutions. No longer is the data unlocking process limited to those with deep pockets who can fund a specialized project. Various sensors can be easily used in the field to measure the environmental variables for the crops, including temperature sensors to measure both air and soil temperature, humidity sensors for the air, light detection and ranging sensors or range-finding sensors for leaf area index, color sensors to refer to crop colors to help in pest detection, imaging sensors for both visible and near-infrared images for crop measurements, and specific sensors to detect or measure specific aspects of a plant's environment or crop growth (de Souza Filho et al. 2023).

In addition, robots have been developed that can carry multiple sensors to navigate through the field and collect data on soil and crop statuses, including manipulators for phenotyping purposes. Besides the contact sensing sensors such as thermometry and light interception, digital cameras on robotic platforms are also common in current use. AgriTech companies are also developing various smart IoT systems, with the capability of monitoring and controlling certain variables of the plants and their environment, leveraging analytics for better design of feedback loops and real-time adjustment of management parameters. Agencies are also developing national sensor networks to monitor soil, weather, and other parameters that are important for plants and animals (Nugawela and Sedera 2020). These sensor systems, when properly integrated with analytics together with data from other sources, enable faster and more accurate decision-making processes, which involve several processes and layers, as demonstrated in Fig. 7.3.

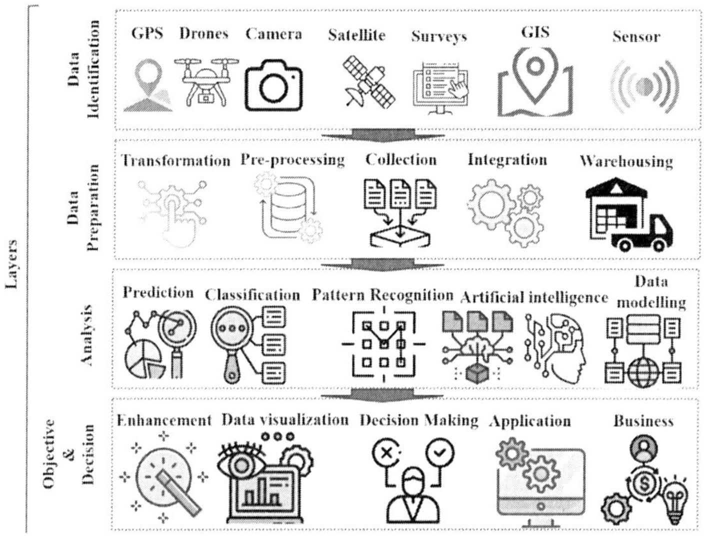

Fig. 7-3 Sustainable system data application and layers

7.2.4 Remote Sensing Technologies

The low spatial and temporal resolution of in-situ crop measurements makes it difficult to monitor and forecast crop production with high temporal resolution. Intra-seasonal information on crop development can provide early warnings for potential crop yield decline. Remote sensing has been increasingly used to monitor crop biophysical status by measuring electromagnetic radiation reflected from the crop canopy. The relationship between crop biophysical status and spectral features in different wavelengths has been well studied, and various vegetation indices have been developed to estimate crop parameters. The development of remote sensing technology in recent years has allowed agricultural data to be acquired in real-time with fine resolution, facilitating the implementation of more technically complicated and theoretically advanced algorithms (Vajjhala, Strang, and Bitrus 2020; Paraforos et al. 2017). Optical sensors, such as multispectral sensors mounted on satellites and unmanned aerial vehicles, are widely used in monitoring crop and soil spectral reflectance. SAR sensors can acquire real-time status of crop health and are beneficial for agricultural insurance. Polarimetric radar can not only detect crop canopy and soil moisture but also provide crop parameters for monitoring crop growth. Hyperspectral imaging, as an advanced remote sensing technology, has huge potential for monitoring crop physiology. It combines the advantages of both spectral and spatial resolutions, and the analysis of vegetation can correlate reflection in very narrow wave bands, so it has a greater capacity for detecting and mapping different vegetation and soil in detail (Köksal and Tekinerdogan 2019).

7.2.5 Data Pre-processing and Cleaning in Agriculture

A significant portion of the numerous errors that can find their way into agricultural data can typically be categorized as "dirty" data by default. Dirty data contain a wide variety of error types, and thus, these errors are especially ubiquitous in large data sets. Data errors can range from undetected tracking errors to variations in data collection conditions and routines. For example, malfunctions, contamination, and data entry errors can occur when sensors are damaged or broken. Farmers, service providers, researchers, or farm management information system developers may introduce different errors into the original data. Hence, data-level problems are likely to occur with the use of data at any scale, and inconsistency problems across reference tables can also be seen. Therefore, data-cleaning procedures for a wide range of outlier problems that can affect the final output of data analyses are required. Relevant machine learning and model-extracted data can be biased by unwanted outliers, primarily caused by device malfunctions, operator error, an evolving farm setup, or incorrect data entry. Different outlier techniques work well in various data collection environments. Outliers can cause various problems in prediction if one of the primary goals is to consider out-of-range values as noise or samples for prediction training (CATAL, KASSAHUN, and HOVİNG 2020). Predictive machine learning models can end up overfitting the model to training data.

Outlier analysis requires knowledge of the existing domain to comprehend whether an exception factors into a particular input. To decide the appropriate action when dealing with outliers, relevant changes in the data generation processes need to be recognized. Furthermore, it is expensive to change any data values, especially for cluster analysis. However, when handling outlier anomalies, the exclusion of only one specific observation may not sufficiently represent how to clean the larger dataset. With respect to the available data evidence, exclusion decisions depend on individual versus global anomalies and low versus high relative over/underrepresentation. Low fault rates, partially limited local anomaly clustering, and multivariate data complexity may also require special, more complex outlier identification techniques (Tummers, Kassahun, and Tekinerdogan 2021). For low-order farming systems, the amount and variation of available data often present a challenge that should be examined within input/output data to describe agricultural practices better. As a result, a very sizable portion of agricultural data can be classified as "dirty" and typically includes missing data, duplicates, and inconsistencies, among others. Furthermore, agricultural data can be generated at different stages of the agricultural production cycle, and different sensors, devices, platforms, and technologies are employed. Therefore, using operations and associated tokens of set intersection for a certain feature space allocation algorithm, the types of applied sensors can be contrasted at the data table level (Nikkilä, Seilonen, and Koskinen 2010).

7.3 Data Analytics Techniques in Agriculture

In contemporary agriculture, the provision of the most valuable information for farm management can be expanded by modern data analytics techniques. Data analytics offers several techniques. Based on the role played by them, these can be categorized as descriptive analytics, predictive analytics, and prescriptive analytics (Shafik 2024a). Descriptive analytics, as its name indicates, helps to describe what happened in the past. It helps to analyze historical data to gain insight for continuous and better performance. A farmer can utilize this method to generate and utilize farm-level management information based on historical data. For example, identifying the highest yielding or failure crop varieties or treatments based on average yield performance over the years (Ammann, Walter, and El Benni 2022). However, descriptive analytics provides valuable insights into the historical outcomes or events that happened on average. In contrast to descriptive analytics, predictive analytics is about the future. Predictive analytics helps to develop statistical models or employ machine learning algorithms using historical data to predict the future (Shafik et al. 2024). A farmer can utilize predictive analytics to forecast future events or outcomes in order to make farm management decisions proactively. For example, predicting the yield for the current year based on the historical relationship of yield with rainfall, temperature, and other grassroots indicators (Munz, Gindele, and Doluschitz 2020).

Prescriptive analytics provides prescriptions like actions from the huge amount of data analytics that has been done. It combines both descriptive and predictive techniques and describes actionable recommendations. As an example, it can recommend a set of actions from a variety of seeds based on soil type and temperature at the time of sowing. The techniques mentioned above offer great assistance in producing a model that can assist farmers in conducting their business based on previous data. Hence, it is crucial to select each technique based on the capital and capital financial impacts and objectives that need to be achieved accordingly. Descriptive analytics, as indicated by its name, has a descriptive role. It helps to summarize what has happened in the past on average. Crop area profile reports, cotton bulletins, economic surveys, agriculture censuses, and many government-published reports are examples. These reports help to inform stakeholders about the past trends of agriculture to make their investment decisions. Essentially, a farmer can use descriptive analytics to generate and use such information to manage the farm based on previous performance (Melzer, Bellingrath-Kimura, and Gandorfer 2023). Some of the practices include:

- Identifying the most and least-yielding crops.
- Best and worst crops for a season.
- Combination of different crops and the impact on total area yield.
- Pesticides or methods yielding maximum pests and diseases.

7.3.1 Descriptive Analytics

Reshaping agricultural data for analysis and insight generation necessitates descriptive, predictive, prescriptive, and cognitive analytics. Starting with descriptive analytics in the agricultural context, descriptive analytics aims to summarize, compress, or reorganize large and complex volumes of data into a form suitable for the extraction of more conclusive results. By using various statistical methods, descriptive analyses can provide insight into how agricultural producers have been performing in the past (Shafik 2025a). Descriptive analytics encompasses both annual and historical trend data, reporting on operational performances, including livestock turnover, efficiencies, gross margins, and crop yields, among others. The two main types of tools and technologies used to describe farm data are dashboards and reporting software. Both tools provide farmers with a highly visual platform to demonstrate key patterns and anomalies of data using graphs, charts, and mapping (Carrer et al. 2015).

Data visualization during descriptive analysis is essential when representing high-volume datasets. For positive decision-making, farmers need to understand the significance of the data displayed in a manner that can be explicitly relayed into actionable intelligence. A number of case studies have shown how descriptive analytics have led to further datasets being analyzed using predictive analytic methods. In one study, observational analyses revealed unique patterns in wilting behavior across 24 cultivars, while in another, plant biomass data were accumulated and showcased not only growth spurts but also interactive nutrient efficiencies when combined with drought stresses. In field trials, production information has been cleared of any errors on some days when energy production plummeted because no production was actually being undertaken. In another study, a descriptive overview of unique bovine reproduction EBVs based on 876 herd-rich genetic technologies clients in 2009 led to new expansion markets (Salami and Ahmadi 2010).

Descriptive analytics is further developed in order to confirm whether predictive and prescriptive findings draw the attention of experts who can use the findings to enhance practice. Others state that the development of descriptive investigations into predictive evaluations is a key turning point in the practical benefit of analytics in farming (Shafik 2024b). This type of analysis can be employed to formulate traction pneumatic tire inflation. Quite frequently, it is gleaned from farmers who have been able to utilize such findings best, knowing if treatment is having the desired results. Selected case studies make it particularly clear that the best time to use descriptive analytics is when exploring data that will assist with proving the capabilities of a technology (Paraforos et al. 2017).

7.3.2 Predictive Analytics

Predictive analytics enables farmers to make well-thought-out and precise decisions about the management of their farms. This decision support is based on predictions derived from historical data by using machine learning algorithms or statistical modeling techniques. Common applications in agriculture are predictions for crop yields or livestock. Predictive modeling can also be used to anticipate weather hazards or the occurrence of certain diseases and pests in order to take precautionary measures. Therefore, it might be a valuable instrument to improve overall sustainability in agriculture. A major prerequisite for the successful dissemination of such IT tools is the farmers' acceptance and trust in anticipation methods (Ammann, Walter, and El Benni 2022). It is known that high prediction accuracy can enhance acceptance and trust as it represents an important dimension for lack of trust in decision support systems in greenhouse automation, as illustrated in Fig. 7.4.

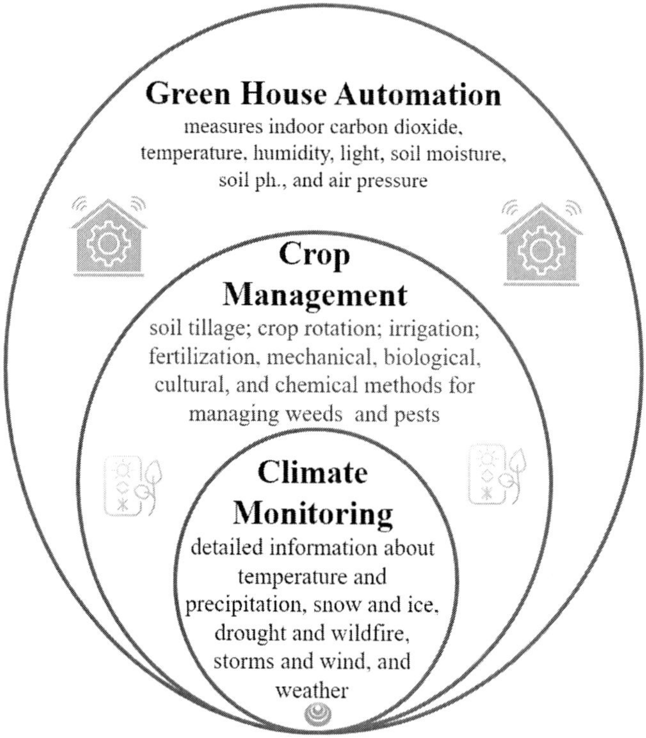

Fig. 7-4 Greenhouse automation

Predictive modeling always makes a statement about the future based on data from the past. The more extensive, comprehensive, and reliable the past data, which are used as a basis for the model, the more reliable and valid the derived predictions. Predictive modeling depends a lot on the quality and relevance of the input data. Tasks are mostly dealt with in the area of supervised learning, i.e., cases with already-known outcomes that are used to train the model. The main ingredients consist of past data. The collected historical data has to be cleansed, and a proper data set has to be picked. State-of-the-art machine learning algorithms define the data model based on machine learning techniques. Potential problems may arise at this stage, such as irrelevant or inaccurate data input, known as 'data bias,' which makes the results achieved with the model inadequate. A proper model has to be trained, evaluated and improved until the best accuracy is reached (Kassahun et al. 2022; Fountas et al. 2015). Once the output is trained, the same model is used on input data to predict possible outcomes in a real-time environment.

This is the high-level overall methodology behind a predictive model. The main challenge of modeling is that the results can only be as good as the data used to train the model. An additional limit is the necessity of advanced analytical skills, which are not widely used by most of the actors in agriculture, specifically farmers and practitioners. Predictive models must be widely understandable and user-friendly in order to reach a relevant public. Nonetheless, the currently available tools for agricultural predictive analysis pave a new pathway for future management and organization of agricultural production. The success of predictive analytics greatly depends on the application domain and the quality of the data used for prediction (Nugawela and Sedera 2020). A successful example is yield forecasting in agriculture, which has been significantly improved by using complex machine-learning models. Extensive results are now also published by various platforms and companies. With predictive analytics, future scenarios can be predicted, thus enabling farmers to optimize resource distribution and address future challenges, especially for feedback-based procedures (Shafik 2025b). Modern trends aim at:

a) integrating prescriptive and predictive methods to implement end-to-end decision-supporting management of an agricultural process in a fully unmanned IoT and Big Data environment; and
b) including IoT information acquisition sensors reachable through shares and Big Data processing for combining the high sensory dimension and the Big Data one.

7.3.3 Prescriptive Analytics

Prescriptive analytics is the final stage in the analytical approach, which not only performs predictions based on past data but also predicts what will happen in the future and recommends specific actions to take. This technique is built on the insights of both descriptive and predictive analytics. It suggests actions that can be taken to achieve the optimal states for decision-making (Vajjhala, Strang, and Bitrus 2020). This prescriptive decision framework consists of three main stages:

(1) Problem formulation (set private and objective constraints);
(2) Modelling and simulation (select a matching algorithm that can suggest the optimal amount of resources and operational adjustments); and

(3) Evaluation of policy suggestions and testing (acknowledge trade-offs and policy recommendations and involve the review of the set balance between the objective function).

Currently, several models and algorithms that recommend optimal resource allocation and operational adjustments are available, such as linear programming, integer programming, quadratic programming, and dynamic programming. Each algorithm has its strengths and weaknesses and also requires the necessary computational complexity. However, with the current condition of big data and data streams, future-state data analytics in real-time comes with additional challenges, such as the computation time needed to respond as fast as possible for quick decision-making and the ability to access and utilize a large number of sensor and actuator data. Furthermore, the role of data-driven management in farming is significant in supporting sensible agricultural practices, economically and environmentally, and also in identifying and eradicating the effects of bad farming practices, organic farm discrimination, and risk analysis (Munz, Gindele, and Doluschitz 2020).

On a worldwide basis, a number of platforms have been established to promote agricultural technology and best practices for soil and crop management, indicating that future farming is evolving towards data-driven management platforms. Three agribusiness case studies have successfully integrated a big-data farm management information system that provides

(1) one-click farming advice to produce the best crops, and the advice
(2) can lead to cost-effective and sustainable operations, and
(3) can empower the farmer to change a course of action that may provide a positive yield, use and replace natural resources, provide sustainable food, and build consumer trust.

They are FluroSat, Ranch Systems, and Primaris. The data value in this study is estimated using the company's invoice or free trial version, a demonstration of real-world application, and the latest scientific literature. This prescriptive analytics technology gives the farmer the opportunity to be proactive in managing data-driven decisions, therefore demonstrating how future agriculture will be predominantly rational (Carrer et al. 2015).

7.4 Applications of Data Analytics in Farm Management

Data analytics in farm management. Farm management information systems (FMIS) using data analytics are shifting from simply providing awareness about dynamic conditions in the field to demonstrating improved capabilities for management-level support to enable efficiency and informed decision-making. Furthermore, the emergence of farm data science as a discipline reflects the industry's growing understanding and expectation around the unique knowledge available by applying data analytics to farm data. This revolution in farm management, driven by data, is being applied to several farm management tasks, as highlighted (Nugawela and Sedera 2020).

Crop monitoring and yield prediction. Crop monitoring and yield prediction through algorithms such as machine learning are increasingly at the forefront of farm management. Several case studies show the business benefits on-farm when implementing the insights of such models. Livestock management. Gathering data on livestock performance allows for insights into health and weight gain over time, for example. Using this historical data, models can be built that predict future performance based on farm management variables. Resource optimization. Supporting variable rate application of fertilizers, herbicides, fungicides, and pesticides is a significant practical application of farm data science. For example, fertilizer recommendations can be based on accurate, spatially detailed field data, including potentially also remote sensing data (Ammann, Walter, and El Benni 2022). The same methods can be applied to irrigation scheduling in water use efficiency.

Resource optimization: Water management. Combining farm data with local forecasts and adopted processing models, a case study demonstrating the potential of analytic capabilities. Resource optimization: Fertilizer application. An optimization model was designed to estimate nitrogen leaching losses in potato farms for different irrigation systems based on detailed data analytics informed by soil moisture content and irrigation records. The model allows under and over-watered zones to be identified and can be used to guide farmers towards applying the right amount of irrigation in individual areas based on need and uptake. A significant body of case studies highlights the myriad benefits of data analytics in farm management systems, and, taken together; this evidence suggests that the implementation of such methods assists producers in feasible, day-to-day decision-making. The aforementioned demonstrates that data is starting to be utilized to support real activities on the farm rather than just awareness. Throughout these studies, most focus on one or more primary drivers of data generation: more robust planning, increased risk management, or enhancement of production (Bermeo-Andrade and González-Bañales 2022).

7.4.1 Crop Monitoring and Yield Prediction

Owing to the rapid change in global environmental conditions, scientific and technological advancements have been employed for crop surveillance systems. Advanced remote sensing-based soil and crop health assessment technologies offer real-time data on crop growth and environmental conditions. Information such as land use and cover metrics, vegetative indices, chlorophyll content, moisture indices, and temperature profiles are computed onboard satellites or aerial platforms from extensive satellite imagery and transmitted to ground stations. Clearly, there are vast amounts of data to be analyzed, and humans cannot handle this for instant decisions. Machine learning and deep learning models are there to analyze the historical data of various crops. Based on the land under each crop, their respective phenocycle measured by networks, and analyzed weather data, predictive analytical models are built using machine learning and deep learning models (Nikkilä, Seilonen, and Koskinen 2010). These models generate predictions on the output of various crops. By using these predictions, agricultural professionals can make relevant interventions in the growing season, such as planting, nutrition, pest management, and water application. The academia involved in the paths we have partnered with is also engaged in the deployment of machine learning and predictive analytics in yield formation. The crop insurance companies use big data analytics-built models to predict possible yield loss so that they can plan risk management. Several good studies and real-time case studies have been conducted worldwide to study the impacts of data analytics using satellite data in the field of crop prediction with good accuracy (Bermeo-Andrade and González-Bañales 2022). Overall, we can conclude that the accurate yield competition among the farming community can be increased by climate patterns that can be addressed by rapid warming and global demand from the population.

7.4.2 Livestock Management

Livestock management is one area where data analytics require a detailed specification of techniques used in the analysis of data for adequate livestock husbandry. Livestock management factors that can benefit from data analytics include health and nutrition monitoring, animal behavior tracking, management of livestock facilities, and more. There are obviously some ethical considerations regarding the amount of data that could be collected from each animal without violating animal rights. This consideration would have to be taken into account for data on collected animal behavior. A case study in livestock management provides an example of high-tech data gathering and advanced analysis used in pen-side animal management systems with a reduced number of animals in order to monitor autonomously an adequate quantity of animals required (Fountas et al. 2015). A careful data analytic experiment by the manufacturers or 'integrators' of the chip technology found this lower optimal value, which remains an adequate number of animals to monitor. However, if the use of smart visible tags for the difficult-to-detect estrous cows in pasture-based dairy farming is considered, the need for data analysis would be massive. The estimation of the best target group of animals to spend money on individually rather than on average for technology would again require data analytic input; this time, an economic analysis of improved efficiency and some research of sector structure gives farmers a perspective in terms of investment return for the electronic sensors and database system (Salami and Ahmadi 2010).

Similar state-of-the-art management and associated can be applied at the farm level. For example, farmers spend excruciating time monitoring and searching for lame cows that are in the early and late stages of being a welfare issue; thus, data can be used to identify the necessary measurements for resources to focus on the cows on a welfare trajectory. In the egg-laying sector, managing the aging layer hen brings the productivity challenge. A manager has to decide when to depopulate and repopulate young pullet birds consistently to satisfy the consumer's demand for eggs. Certain breeds mature at different rates, so deploying analytics to address stocking density for particular ages can give a better forecast of production timing. Mixing old and young layers has social implications, so knowing the number of old layers and social disturbances is an important factor to anticipate. Using real-time machine learning, a case study on smart sensors for environmental control, predictive nutrition, and more for an enterprise-wide PAM platform using a range of data sources that includes structured, semi-structured, and unstructured data for front-end diagnostic and prognostic computations (Köksal and Tekinerdogan 2019; Munz, Gindele, and Doluschitz 2020).

7.4.3 Resource Optimization

Society's demand for monitoring and managing the use of resources is increasing, especially in light of global challenges such as climate change. Efficiently deploying the resources is essential for the sustainability of agricultural activities during the current era of digital agriculture. Data analytics can support the utilization of scarce resources such as water, fertilizers, and energy to identify the optimal application amounts. Based on data-driven information collected from weather forecasts, soil moisture content, or plant health, usage patterns of certain resources can be detected, e.g., when energy is consumed and for which tasks. Associated with this, analytics tools are able to predict how much irrigation water plants need and when, and thereby advise farmers on the appropriate amount and exact time to rinse (Melzer, Bellingrath-Kimura, and Gandorfer 2023). Moreover, they also consider environmental information to verify if and when fields are suitably wet for planting in order to minimize soil compaction.

A particular focus when using resources in agribusiness should pertain to managing them in ways to protect the natural environment from potential harm. In the post-digital era, eco-friendly attitudes in consumers are expected to occupy more significance in purchasing decisions. In order to make informed decisions and provide support for conditions-based resource utilization, however, detailed monitoring of inputs needs to be carried out. A variety of case studies aim to optimize the use of resources in agriculture, and some of them have evolved into successful commercial services (de Souza Filho et al. 2023). Demand for efficient handling of resources is higher than ever, as disparities between expectations, financial outlays, and sustainability from the environmental angle are evident. Allocating funds for natural resource productivity rather than subsidies is also in sight. It is through sound resource management leading to improved productivity, which then links to sustainability, that the general public's food security will be solidified in the face of climate uncertainties. For data-informed and efficient decision-making to prevail in agriculture, however, understanding the variability in resources such as data quality, well capacities, and water supplies also needs to be managed alongside overall farm fertility and productivity. Additionally, digitization barriers such as isolation, policy conflicts, and even cyber threats come into play when handling resources in agribusiness (Ammann, Walter, and El Benni 2022).

7.5 Integration of Data Analytics with Precision Agriculture

The newest approach to agriculture suggests that inputs should be used in different amounts or compositions for the different parts of a field. This requires data. This is addressed by precision agriculture, which optimizes the application of seed, farm chemicals, water, and fertilizers via real-time data coming from the fields with the aim of increasing yields. Precision farming aids in keeping track of all local information by using technology that yields accurate and continuous location data. Additionally, in combination with in-field systems, such as computers, controllers, and software, it offers global connectivity to in-field equipment, further advancing the collection of important spatial data (Melzer, Bellingrath-Kimura, and Gandorfer 2023; Bermeo-Andrade and González-Bañales 2022).

In precision agriculture, all the position-specific data sent by sensors are gathered and stored in a data warehouse. Extremely sophisticated data analytics is then employed to determine the optimum amount and composition of inputs required at any point in the field. Modern farmers are required to reconcile a trade-off between conservation farming, which safeguards the environment, and sustainable intensification, which increases yield. They should have real-time information on soil health, weather forecasts, and market prices. Predictive analytics are useful here. Modeling can show the impact of different types and levels of input on soil and also identify any prolonged negative consequences (Carrer et al. 2015). There are encouraging indications of the utility of combining precision farming and data analytics gathered from case studies on precision weed management and precision nutrient management. However, the integration of data in general and standards and interoperability are still in the infancy. Such techniques should be broadened. Both predictions about farming and algorithms optimizing decision-making should be utilized. The approach should be prescriptive and should advise 'do this and not that' in real-time, based on the objective function settled by the farmers. This is the future of advanced farming (Nugawela and Sedera 2020).

7.6 Ethical and Privacy Considerations in Agricultural Data Analytics

The privacy and ethical considerations for using agricultural data every day have extensive dimensions and impact, critical for agricultural data processing to provide responsible services and maintain ecological management processes. Therefore, one must be aware of the issues from not only a technical perspective but also understand the perspective of agricultural stakeholders, how we can work together, or at least describe the status quo.

7.6.1 Data Misuse and Application

The potential for abuse of data by the collector and processor of the data is a major concern in agricultural systems. The issue of who owns the data is a challenging topic. The farmer/producer may own the 'raw' data that is generated from their farming activity, yet an independent company may have ownership of the methodology or algorithm used to analyze the data, and some may assert that they have partial ownership. The issue of consent to access data and the purposes for which that data may be used is not well addressed. Access to large amounts of data has the potential to give unprecedented insight into the operations of a company, and currently, there are no mechanisms that check, confirm, or verify the accuracy and ethics of the data that is collected. The issue of privacy with respect to the use of agricultural data is also a significant concern, as combinations of data can become sensitive. For example, soil nitrogen content and yield data can yield estimates of grain protein (Vajjhala, Strang, and Bitrus 2020). Not only is consent needed for its use, but also the trust of the farmer.

Many farmers and agribusinesses cannot easily navigate the data privacy laws and regulations for handling sensitive data. There is a challenge in the compliance aspect of privacy regulations for small to medium-sized businesses in trying to comply with privacy regulations. After the data is collected responsibly, there are many ethical considerations to be made about how the data is used to represent the situation being studied. The issue here is one of standardization, policy, procedures, and best practices. A challenge is how to represent and use the data in a non-exploitative way. Sometimes, some data may not be presented correctly when processed with a certain methodology. If data collection occurs without transparency and trust between data providers and data users, then all parties concerned can use the information to support their end goals, increasing the possibility of misuse of this data, and you can unknowingly be processing data that can potentially affect relationships within agricultural production (Melzer, Bellingrath-Kimura, and Gandorfer 2023). Therefore, it is important to develop trust and transparent policies to allow the farmer or agribusiness to ensure that backroom policies are in place.

For ethical data management, farmers, agricultural professionals, and the public need to observe their responsibility to take on the practices of being organically concerned for enhancing, auditing, verifying, and promoting a better and safer use of data. There needs to be an engagement of the agricultural community to consider the rights and privacy of data collected through agricultural production. Best practices should be guided to ensure that the data collected for algorithm development, management, and security provide safe, guaranteed accuracy and secure the ownership of the impact of the agricultural data. In most cases, best practices provide an overview and workflow options that carry significantly effective and practical aspects for each stakeholder's point of use. The extent to which each of these aspects is applied may fluctuate from case to case. Farmers need to be aware of their data assets and their rights. Agricultural professionals and farm advisers need to respect their professional impartiality towards their clients and maintain their credibility to be seen as professionals. There is a need for active engagement to ensure the standards are agreed upon and workable for all (Carrer et al. 2015; de Souza Filho et al. 2023). Farmers and agribusinesses need data privacy performance indicators to show that data protection is being maintained.

7.6.2 Regulatory Compliance for Agricultural Data

When applied to farm management information systems, comprehensive agricultural data analytics ensure regulatory compliance, promote transparency and accountability, encourage the adoption of improved baseline management practices, and provide access to data to solve unmet agricultural market needs more quickly. Several countries, states, and counties are starting to tease these principles out. The concepts that are currently being developed in some regions are global and can be used to guide the rest of the world. The resulting regulations quantifying agricultural activities will enable farm management units, resource management systems, as well as information technology systems engaged in agriculture to do a better, more accurate job. In particular, better and more accurate management, electronic monitoring, and data use are all certain to have a beneficial effect on the environmental impact of farming while enabling business optimization and market access (Tummers, Kassahun, and Tekinerdogan 2019; Nugawela and Sedera 2020). The current panoply of intensive and data-assisted agricultural activities is only part of a future view of the well-regulated intensification of farming worldwide. Optimally, these systems will include that compliance must be automated, linking resource monitoring and management systems to farm management information systems. The burden on farmers will be too high, which is the real pattern that will limit farm adoption of advanced environmental regulation. Yet this is also an area in desperate need of regulatory oversight and assistance. Insufficient agricultural activity oversight leads to negatively perceived effects, market interference, and distortions that again lead to unjust regulatory burdens on farmers. Regulations should be metrics-based (Köksal and Tekinerdogan 2019). Data fostering regulatory oversight should be in the public domain and available to all, as should the resulting benchmarks and regulatory procedures.

7.6.3 Data Privacy and Security in Agriculture

The rules that govern data privacy, data ownership, and data rights need to be thought out. In some jurisdictions, laws are still not designed to deal with the complexity of agricultural data. Agriculture is unique in the sense that it has weather, pests, and crop/field-specific knowledge captured in near-real time and at high resolution. General data privacy manipulation functions designed for conventional applications do not easily extend to the needs and demands of such farming practices. Governments need to lead the establishment of minimum guidelines and policies to encourage agricultural innovation at minimum expense to the farmer. However, growers and agribusinesses need a detailed understanding of where the correlations and synergy points exist (Melzer, Bellingrath-Kimura, and Gandorfer 2023). Take, for example, a farmer with crop/disease knowledge at a small resolution. This data would not have any synergy, as they describe different ecosystems. However, suppose the connectivity is established, and crop knowledge predictability has a high correlation across ecosystems. In that case, agribusinesses can potentially sell crop protection solutions that include this knowledge even when it was not collected in the ecosystem being considered, as long as a 'trust' mechanism can be established. There are many privacies versus utility trade-offs that have to be made. A clear and established land registry system is also key, as it is the core infrastructure that supports many environmental functions linked through agricultural data (Carrer, de Souza Filho, and Batalha 2017).

The absence of visual evidence relating to such long-term land changes has made it hard to position agricultural changes into justifiable evidence against competing environmental demands from society. Since large-scale agricultural monitoring and environmental liability are in their infancy, farmers may feel compelled or pressured in some form to enhance privacy. Additionally, not all businesses currently concerned with land data are citizens or residents of the region, and there will be concerns across geographic, regulatory, and governing laws. Merging land tenure with agricultural data while communicating exact location and land productivity has never been mandated at this scale, not just for small family-owned farms but for various divisions of multinational corporation landholdings. In contrast to conventional privacy scenarios, antagonist businesses can take liability across multiple areas and engage in coordinated attacks. Information technology farmers are also targets of choice for the exfiltration and monetization of intellectual property (de Souza Filho et al. 2023; Carrer et al. 2015).

7.7 Data Visualization Techniques for Agriculture

Recent years have shown a dramatic increase in the number of available data visualization techniques. Every day, researchers are proposing new visualization techniques to help analyze complex data. In agriculture, this is no exception. Given the advent of management information systems for farms and precision agriculture, vast amounts of data are being generated. While data are easy to collect, the real effort is in data analysis. Visualization plays a key role in data analysis, offering the possibility to grasp the underlying knowledge in data quickly. As such, a good visualization can greatly help decision-makers understand agricultural-related issues by providing visual clues that will help interpret the underlying data. This provides an overview of the state of the art in data visualization for agriculture and for management information systems in agriculture (Husemann and Novkovic 2014). The agriculture domain is experiencing a data revolution with the number of increasingly available data. With the current amount of data now available, it is possible to answer questions that were previously beyond our grasp. However, real effort is in the steps taken after data collection, where we organize data, interpret data, structure it, and try to find meaning in it. Visualizing the data using powerful and diverse visualization techniques can present the information in a way that the data analysis provides a useful and informative representation. Data visualization is a powerful tool to discover and interpret the information present in massive datasets, and as such, a wide variety of visualization techniques have emerged to tackle specific issues present in these datasets. The widespread availability of agricultural data has yielded an increasing number of proposals centered on agricultural concerns (Tsiropoulos et al. 2017).

7.7.1 Graphical Representations of Agricultural Data

This explores some useful methods for displaying agricultural data that may help consumers of information to understand the data and its interpretations better. With the increasing importance of technical and support facilities to the agricultural industry, agricultural businesses rely on computerized tools to help manage information resources. Farm businesses are complex systems that process a wide range of data required for productive farm management. These tools are referred to as FMIS. Various technologies like Geographical Information Systems and Global Positioning Systems help to monitor, control, supervise, document, and manage decision-making processes on a cattle farm. The complexity of agricultural information generation in livestock industries and crop-growing industries requires farm information systems that can organize, record, calculate, and display data to help assess and manage farming production efficiently and effectively. Thus, FMIS should present user-friendly methodologies and strategies to convert accumulated wasteful data into useful products easily. However, advances in database models, computer architecture, input device improvements, connectivity, GIS, and GPS make more dynamic and attractive data available for data visualization, criteria establishment, provision of efficient information, and statistical/diagnostic decision-making (Ammann, Walter, and El Benni 2022).

7.7.2 Interactive Dashboards for Farm Management

Agricultural Internet of Things (AIoT) based solutions are targeting data-driven insights and information delivery to increase farm productivity by implementing Farm Management Information Systems (FMIS). AIoT-based sensing solutions for smart farms generate large amounts of data. There is a need to adopt advanced analytics methods for enabling ontological representation and delivery of actionable farm insights. A data-driven IoT solution involving thing spaces, dataspace, and analytic data spaces derived from the farm big data enhances the management intelligence of smart farms. Semantic Big Data interaction frameworks enable deployable FMIS applications that encompass interactive dashboard view components for providing farm decision-makers with actionable farm insights. The malware detection and sanitization techniques for maintaining secure and predictable behavior in ontological farm descriptions are discussed (Carrer et al. 2015). Interactive dashboards support interactive real-time exploration, visualization, and actionable decisions to manage smart farms. They face multiple data and ontological farm descriptions extracted from the underlying data space or Sensor Big Data warehouse. A visual query interface is used to show multi-modal widgets and provide an interactive visualization application. Feature presentation shows the identification of patterns or abnormal visual observations. Decision recommendations provide guidelines and reasoning for actionable insights. Data and model details display and show describable aspects of data models used for the modeling process for ontological discovery, visualization, reasoning, and interactive intelligent dashboards (Ammann, Walter, and El Benni 2022). The proposed interactive intelligent dashboards rely on semantic representation, modeling, reasoning, data mining, and IoT-relevant datasets.

7.8 Big Data Technologies in Agriculture

Big data is characterized by volume, velocity, variety, and veracity, which require a novel approach to supporting agricultural business systems. Farm-level big data presents a unique challenge because data sources vary widely and can exist at greatly disparate scales, from highly spatial, temporal, and sensorial resolutions. With in-field sensors and globally connected tools, massive amounts of agricultural data are being generated, and it is growing. One of the tools that are widely used in precision agriculture is geographical information systems technology. From the data and information collected, geo-processing with spatial analytics, including remote sensing from satellite imagery, GPS, and GIS, is commonly used to derive spatial decision-making for precision farming (Munz, Gindele, and Doluschitz 2020).

There is little doubt about the ability of big data technology and analytics to improve farming information management and operational efficiency. The key technologies that make big data work together are natural language processing, statistical analysis, machine learning, cognitive computing, complex event processing, predictive and prescriptive analytics, facial recognition, speech recognition, vision and image recognition, and deep learning. In agricultural farming information systems, big data machine learning can address the farmer's actions by analyzing data from the tractor's location and sensors, as well as weather data. Field-specific and cloud-based software is used, capable of also processing data from hyperspectral sensors on drones (Nugawela and Sedera 2020). The results of this analysis are made into an easy-to-understand graphic format for the farmer's actions using analytic and visualization capabilities. With these capabilities, this research makes use of GIS-interfaced software operated on an Agri-IoT control center and in the cloud to control and monitor data collection from remotely managed IoT devices. With these tools, the study is able to characterize the level of soil moisture and to determine topography, using an assessment of surrounding land and moisture values, anomaly detection by processing sensor data collected. The ability of big data tools and their analytics to determine insights from vast, unstructured, unsecured farmer's decision-making data is able to address the various decision constraints that the farmer faces (Husemann and Novkovic 2014; Fountas et al. 2015).

7.8.1 Hadoop and MapReduce in Agriculture

Large agricultural enterprises in the modern era use the global positioning system and geographic information system to update information about farm activities over time. The massive datasets about agricultural problems have been known as big data in order to handle the heterogeneous data in large databases. Two revolutionary trends in big data analytics include distributed storage and processing, such as Hadoop for storing and MapReduce for processing. This chapter reports the findings of studies on Hadoop and MapReduce in the application for agricultural research or refers to them as agricultural data analytics (Köksal and Tekinerdogan 2019). The applications in Hadoop and MapReduce include:

(1) crop yield forecasting for classification problems,
(2) crop yield forecasting for regression problems,

(3) precision agriculture and crop management,
(4) disease management,
(5) livestock management,
(6) GIS-based agricultural systems, and
(7) other applications.

From the reviews, it can be concluded that Hadoop and MapReduce are suitable tools due to their ability to process massive datasets in crop yield forecasting. The global positioning system and geographic information system are remarkable technologies that can provide the ability in offline and online versions in order to frequently and instantly update information about farm activities. The consistently collected data from GPS and GIS then constitute massive datasets that refer to big data. A huge volume of data makes agricultural problems difficult to process. Two instrumental methods that assist with these problems are a distributed file system called Hadoop and a parallel programming model called MapReduce (Bermeo-Andrade and González-Bañales 2022). These have been developed and are maintained by large internet enterprises.

7.8.2 Spark and Flink for Real-time Data Processing

FMIS are becoming indispensable tools for modern farm operations. Collecting and processing large data volumes enables evidence-based policy and practices in food and agriculture systems and offers the potential to serve the global communities facing urgent problems, such as food security, the supply of bio-resources, climate change, and other sustainable development goals. This demands full deployment of ICTs, connected devices, and cloud computing technologies, and deep involvement in the new digital economy. In this context, research on precision agriculture and industrial crop protection applications is following methodological paths that have already been successfully adopted in many other smart farming sectors, where software services are replacing hardware devices. More specifically, for field-to-cloud automation of crop estimation, the on-demand processing of UAV-based sensing data needs to shift from batch processing to approaches that allow real-time operational modalities (Tsiropoulos et al. 2017). Currently, two dominant open-source big data analytics engines for real-time data processing are.

The two engines share the same basic data processing model that implicitly or explicitly organizes data into a sequence of records and is built on the ability to maintain an accurate size index for these data flows over time. Provides a better solution space for stateful data flows by extending the basic model with unnecessary sequence bounds that, for some data processing applications, can be significantly more serviceable than latency guarantees. Also allows users to control the deployment of the analytics pipeline stages within the system, which results in predictable, low-latency data flows that are not only more economical but, under certain conditions, also more reliable than what can be achieved. In both comparison points, unnecessary alignment and scheduling of periodic data extraction bursts maintain fixed latency and enable continuous traffic shapers even in processing settings (Carrer et al. 2015).

7.9 Statistical Analysis in Agriculture

The topic of statistics is not very popular among many students of agriculture and closely related disciplines, who often find it a necessary evil rather than a useful tool. The last few decades have seen many changes; however, these could have an interesting impact on this general perception. Many students of modern agriculture must now cope with sophisticated developments in molecular biology or the economics of agriculture, which inevitably rely heavily on statistics. In any case, after data collection, all scientists and producers need to analyze their data in order to discriminate between treatment effects and natural variability. All useful data can be processed and analyzed quantitatively, sorted, grouped, and summarized in numerical tables, represented graphically, or described by means of appropriate statistics (Tsiropoulos et al. 2017; Tummers, Kassahun, and Tekinerdogan 2019). For anyone who wants to make decisions in management, research, and extension, it is essential to know how to analyze the most common experimental situations in the context of agriculture.

Therefore, statistical methods are fundamental for a good education in agriculture. This is meant in two respects: to enable students, future researchers, and everyone who is supposed to manage information on modern farms to analyze existing data or to make new data available in an efficient and well-organized way. Control in agriculture has traditionally been implemented through field experimentation to minimize the effect of sources of natural variability that can hinder the comparison of different experimental treatments. However, examples from field trials in different countries reveal inadequate use of statistical techniques, leading to failed experiments, a waste of resources, and compromised results. Informal estimates seem to indicate that more than 50% of solid waste in toxicological research could be attributed to non-experimental mistakes, such as lack of design, non-random allocation, pseudo-replication, lack of control, and estimable bias (Paraforos et al. 2017).

7.9.1 Descriptive Statistics in Agriculture

Agricultural professionals understand the importance of descriptive statistics for providing insights into the properties of economic, agronomic, and financial variables. The variables that are generally studied by descriptive statistics can be easily derived by counting numerous qualitative measurements. Other variables are quantified through quantitative measurements. This includes information that provides characteristics and numerical associations that illustrate the central tendency, variability, distribution, proportions, and ratios on a per-unit basis. Descriptive statistics often provide the initial data examination and analyses that involve taking stock of the variable and also generating summary measures and tables about the variable properties (Ammann, Walter, and El Benni 2022). The data derived from descriptive statistical measures need to be fostered by follow-up investigations and analyses carried out by inferential statistics.

Descriptive statistics are the first set of analytics that provides information about the many variables that can be systematically collected across farms. Specifically, this type of statistical analysis is useful for the study of potential patterns, distributions, associations, and relationships of land use, yields, input usage, associated variables, and financial performance. These are key relationships that often help managers understand the complexities of creating on- and off-farm value-added activities. Insights derived from descriptive analytics help to identify trends and patterns that permit the development of more complex farm business management models. These models, in turn, enable an understanding of how one or more factors are related to each other, i.e., analyzing apparent relationships and differences between variables of interest (Nugawela and Sedera 2020). Educational outputs fostered by descriptive statistical analyses are valuable for decision-makers to understand better the farm and food system's relative economic and agronomic relationships.

7.9.2 Inferential Statistics in Agriculture

Inferential statistics determine the relationship between two groups and investigate the effect of an independent variable on a dependent variable. The relationships are first assumed or indicated by a simulation or regression modeling, and tests, correlation tests, or tests of means are then performed. Combinations of probability distributions with means and variances assigned to the independent and dependent variables are obtained from simulations. Models determine the correlation measure in a normal and beta distribution. In agriculture, future price movements or spatio-temporal relationships are determined. In GIS, statistics and spatial analysis are performed (Paraforos et al. 2017). Models and tools are developed that can determine field characteristics, and this provides not only information on soils and crop nutrition levels but also nutrient uptake, fixation, or applications as well. Balance decisions are determined by interval statistical models assigned to classification methods, decision trees, and settings, and data that link to decision-making will enable the system to suggest decision-support options and predict future scenarios. Decision trees are known for classifications and predictions of complex data sets when they are stored in tables (Köksal and Tekinerdogan 2019). Also, the statistical relationship in life sciences and chemistry relates to soil structure, color, morphology, and chemistry.

7.10 Future Trends and Innovations in Farm Management Information Systems

FMIS will experience a number of challenges regarding analytics in the future. The use of data-related analytical techniques like automated statistics, big data, business intelligence, data cleaning, data integration, data representation and visualization, descriptive statistics, embedding analytics, exploratory data analysis, geographic information systems, inferential statistics, learning analytics, machine learning, modeling and simulation, network analytics, open data, predictive analytics, prescriptive analytics, spatial data, text analytics are believed to be critical for FMIS functionality in the near future. In addition, a number of drifts, perturbations, innovations, and trends afford insight into the potential market dynamics for animal and crop production, where FMIS plays a crucial role (Fountas et al. 2015).

A small number of emerging technologies and trends will afford FMIS stakeholders considerable innovations and developments in data analytics and computational processes in the forthcoming years. Emerging data analytic trends directly based on web semantics and the future functionalities and needs of FMIS for all stakeholders can be noted as described in the following sections. Machine learning and, particularly, artificial neural networks, which can be considered an advanced sentinel for the predictive model, have been significantly used in agroecosystem or agronomic research. As they seem to be precise with a good capability of modeling nonlinear, complex agronomic systems that include multi-faceted agricultural and environmental variables, it is essential to use optimal developmental AI for future FMIS functionalities (Salami and Ahmadi 2010). Likewise, blockchain has the competence of constructing transparency and safe custody for datasets far beyond current practices.

7.10.1 Artificial Intelligence for Precision Agriculture

To develop the artificial intelligence necessary for precision agriculture, one must consider both the knowledge level and the manner in which knowledge is transferred. Artificial intelligence can function in a variety of ways at different levels of automation based on a knowledge base that it often uses at specific times. This base varies from a foundation based on a set of specific rules or training of an expert system under uncertainty using a basis for formulating concepts. When applied to precision agriculture, artificial intelligence assists producers in exercising their decision-making processes using existing scientific knowledge. Decision-making trees provide another reasoning facility for solving problems that a human expert can verify. Advanced levels of AI typically imply a greater knowledge level, enhanced reasoning facility, and increased cooperation among members of the work team (Köksal and Tekinerdogan 2019).

Intelligent databases can function as data warehouses, data marts, or data hubs connecting commercial DBMS to the information technology system that supports decision-making functions. They gain importance in the e-commerce era thanks to their interdisciplinary ability to interact with intelligent agents. Intelligent data mining searches for associations among data records. Its main uses include the discovery of data distribution, the acquisition of linguistic rules from segments of data and their substitution for numeric values in less general rules, aggregation to summarize associations detected in a complex phenomenon at a higher level, explanation to provide an understanding of the detected associations, and prediction to facilitate the use of intelligent agents along with expert systems. It concerns a guest expert system that can offer a quick and satisfactory response based on implicit and procedural knowledge, overcoming inherent difficulties in knowing all possible states and establishing general rules (Munz, Gindele, and Doluschitz 2020). More precisely, data mining can utilize these systems to aid in decisions made when certain conditions within structural environments are met.

7.11 Conclusion

AgriTech applications continue to disrupt the status quo and increase the demand for separable property rights and clear standardization systems. The link between farm and community becomes more important every time. One of the most important developments in agriculture is its suitability for traceability applications. This chapter introduces technology, discusses its potential applications to agriculture, compares them to existing practices, and concludes with losses. A distributed digital public ledger is used. After each spot, each transaction that is locked within the block is registered permanently and cannot be changed. It is an incorruptible digital record of economic transactions that can be customized to record not only financial transactions but also almost all-important values or dictations that could be named. In other words, it is a list of each block containing a specific number of transactions. The most revolutionary invention in architecture is that the technology uses so-called cementing and concepts of the difficulty chain to record the states of different devices and establish the credible order of transactions. Systems can now be programmed to record almost everything of value and provide not only documents but also money.

References

Ammann, Jeanine, Achim Walter, and Nadja El Benni. 2022. "Adoption and Perception of Farm Management Information Systems by Future Swiss Farm Managers – An Online Study." *Journal of Rural Studies* 89. https://doi.org/10.1016/j.jrurstud.2021.12.008.

Bermeo-Andrade, Helga, and Dora González-Bañales. 2022. "Appropriation Intention of a Farm Management Information System Through Usability Evaluation with PLS-SEM Analysis." In *Lecture Notes in Networks and Systems*. Vol. 236. https://doi.org/10.1007/978-981-16-2380-6_56.

Carrer, Marcelo José, Hildo Meirelles de Souza Filho, and Mário Otávio Batalha. 2017. "Factors Influencing the Adoption of Farm Management Information Systems (FMIS) by Brazilian Citrus Farmers." *Computers and Electronics in Agriculture* 138. https://doi.org/10.1016/j.compag.2017.04.004.

Carrer, Marcelo José, Hildo Meirelles de Souza Filho, Mário Otávio Batalha, and Fabiana Ribeiro Rossi. 2015. "Farm Management Information Systems (FMIS) and Technical Efficiency: An Analysis of Citrus Farms in Brazil." *Computers and Electronics in Agriculture* 119. https://doi.org/10.1016/j.compag.2015.10.013.

CATAL, Cagatay, Ayalew KASSAHUN, and Henk Jan HOVİNG. 2020. "Improving Farm Management Information Systems with Data Mining." *Balkan Journal of Electrical and Computer Engineering* 8 (1). https://doi.org/10.17694/bajece.555680.

Fountas, S., G. Carli, C. G. Sørensen, Z. Tsiropoulos, C. Cavalaris, A. Vatsanidou, B. Liakos, M. Canavari, J. Wiebensohn, and B. Tisserye. 2015. "Farm Management Information Systems: Current Situation and Future Perspectives." *Computers and Electronics in Agriculture* 115. https://doi.org/10.1016/j.compag.2015.05.011.

Husemann, Christoph, and Nebojsa Novkovic. 2014. "Farm Management Information Systems: A Case Study on a German Multifunctional Farm." *Ekonomika Poljoprivrede* 61 (2). https://doi.org/10.5937/ekopolj1402441h.

Kassahun, Ayalew, Robbin Bloo, Cagatay Catal, and Alok Mishra. 2022. "Dairy Farm Management Information Systems." *Electronics (Switzerland)* 11 (2). https://doi.org/10.3390/electronics11020239.

Köksal, and B. Tekinerdogan. 2019. "Architecture Design Approach for IoT-Based Farm Management Information Systems." *Precision Agriculture* 20 (5). https://doi.org/10.1007/s11119-018-09624-8.

Lewis, Tony. 1998. "Evolution of Farm Management Information Systems." *Computers and Electronics in Agriculture* 19 (3). https://doi.org/10.1016/S0168-1699(97)00040-9.

Melzer, Marvin, Sonoko Bellingrath-Kimura, and Markus Gandorfer. 2023. "Commercial Farm Management Information Systems - A Demand-Oriented Analysis of Functions in Practical Use." *Smart Agricultural Technology* 4. https://doi.org/10.1016/j.atech.2023.100203.

Munz, Jana, Nicola Gindele, and Reiner Doluschitz. 2020. "Exploring the Characteristics and Utilisation of Farm Management Information Systems (FMIS) in Germany." *Computers and Electronics in Agriculture* 170. https://doi.org/10.1016/j.compag.2020.105246.

Nikkilä, Raimo, Ilkka Seilonen, and Kari Koskinen. 2010. "Software Architecture for Farm Management Information Systems in Precision Agriculture." *Computers and Electronics in Agriculture* 70 (2). https://doi.org/10.1016/j.compag.2009.08.013.

Nugawela, Saliya, and Darshana Sedera. 2020. "Analysis of Farm Management Information Systems: Opportunities and Pathways for Future Value." In *Proceedings of the 24th Pacific Asia Conference on Information Systems: Information Systems (IS) for the Future, PACIS 2020.*

Paraforos, Dimitris S., Vangelis Vassiliadis, Dietrich Kortenbruck, Kostas Stamkopoulos, Vasileios Ziogas, Athanasios A. Sapounas, and Hans W. Griepentrog. 2017. "Multi-Level Automation of Farm Management Information Systems." *Computers and Electronics in Agriculture* 142. https://doi.org/10.1016/j.compag.2017.11.022.

Salami, Payman, and Hojat Ahmadi. 2010. "Review of Farm Management Information Systems (FMIS)." *New York Science Journal Review of FMISNew York Science Journal* 33 (55).

Schulze Schwering, Dorothee, Lukas Bergmann, and Winnie Isabel Sonntag. 2022. "How to Encourage Farmers to Digitize? A Study on User Typologies and Motivations of Farm Management Information Systems." *Computers and Electronics in Agriculture* 199. https://doi.org/10.1016/j.compag.2022.107133.

Shafik, Wasswa. 2024a. "Barriers to Implementing Computational Intelligence-Based Agriculture System." In *Computational Intelligence in Internet of Agricultural Things*, 193–219. Springer. https://doi.org/10.1007/978-3-031-67450-1_8.

Shafik, Wasswa. 2024b. "Biosensor-Based Drones Anomaly Detection Integration for Sustainable Agriculture Development." In *Achieving Food Security Through Sustainable Agriculture*, 283–316. IGI Global. https://doi.org/10.4018/979-8-3693-4240-4.ch012.

Shafik, Wasswa. 2025a. "Sustainable Agriculture and Diet for Urban Development." In *Sustainable Smart Cities and the Future of Urban Development*, 165–92. IGI Global Scientific Publishing.

Shafik, Wasswa. 2025b. "Sustainable Agriculture and Diet in the Metaverse Era: Sustainable Farming and Nutritious Eating in the Metaverse." In *Food in the Metaverse and Web 3.0 Era: Intersecting Food, Technology, and Culture*, 159–84. IGI Global Scientific Publishing.

Shafik, Wasswa, Ali Tufail, Rosyzie Anna Awg Haji Mohd Apong, and Liyanage Chandratilak De Silva. 2024. "Internet of Things for Smart Agricultural Practices." In *Internet of Things Applications and Technology*, 190–217. Boca Raton: Auerbach Publications. https://doi.org/10.1201/9781003458401-13.

Souza Filho, Hildo Meirelles de, Marcela de Mello Brandão Vinholis, Marcelo José Carrer, and Carlos Ivan Mozambani. 2023. "Adoption of Farm Management Information Systems (FMIS): The Case of Brazilian Sugarcane Farmers." *Information Development*. https://doi.org/10.1177/02666669231177864.

Tsiropoulos, Zisis, Giacomo Carli, Erika Pignatti, and Spyros Fountas. 2017. "Future Perspectives of Farm Management Information Systems." In . https://doi.org/10.1007/978-3-319-68715-5_9.

Tummers, J., A. Kassahun, and B. Tekinerdogan. 2019. "Obstacles and Features of Farm Management Information Systems: A Systematic Literature Review." *Computers and Electronics in Agriculture*. https://doi.org/10.1016/j.compag.2018.12.044.

Tummers, J., A. Kassahun, and B. Tekinerdogan. 2021. "Reference Architecture Design for Farm Management Information Systems: A Multi-Case Study Approach." *Precision Agriculture* 22 (1). https://doi.org/10.1007/s11119-020-09728-0.

Vajjhala, Narasimha Rao, Kenneth David Strang, and Nankyer Sarah Bitrus. 2020. "Contemporary Usage of Farm Management Information Systems in Nigeria." In *Recent Developments in Individual and Organizational Adoption of ICTs*. https://doi.org/10.4018/978-1-7998-3045-0.ch005.

CHAPTER EIGHT

ADVANCES IN BIOTECHNOLOGY AND CROP GENETICS

Abstract

Biotechnology and crop genetics advances have encouraged the elucidation of new knowledge platforms for sustainable agriculture. The combination of forces between technology and genetics concerning agricultural practices is demanding continued increases in crop production to meet the current global demand for 70% more food by 2050. The sustainable development of crops has received considerable interest from researchers in recent years, as evidenced by an increase in papers published on topics involving genetics, genomics, and plant technologies. The integration of genetics and molecular biology has also increasingly benefited cropping systems. The lesson here is that genetics alone, or genomics with various associated tools, cannot be separated from the technology platforms to develop new systems or subsystems for the propagation of plants. The transformation of plants by genetic modification or other molecular biology techniques has now been virtually global with regard to the major 'grain' crops. New traits added to crops can be for increased productivity or resistance through biotic or abiotic mechanisms or nutritional improvements for human health and well-being, as well as vegetative and reproductive breeding systems. Controversy has arisen over the use of enhanced biotechnology, and Europe has been particularly critical of modified crops, banning field trials and market approvals. However, even in Europe, this is likely to change given the media emphasis on the size and number of the global population and the requirements necessary for future food production. This essay offers, in qualitative terms, an accumulating insight into advances in biotechnology and crop genetics, reflecting a change over a single generation. The aims are to enhance knowledge capacity and, secondarily, career development primarily. These are global approaches with an emphasis on the 21st century. In sum, the future of the world is our collective emphasis because science is a worldwide profession. For future agricultural practice and sustainable crop production, it is in the hands of the next generation to determine the fate of global science.

Keywords: Agricultural Biotechnology, CRISPR Technology, Crop Improvement, Gene Editing Tools, Genetic Engineering, Genomic Selection, Molecular Breeding, Plant Genome Innovations, Transgenic Crops

8.1 Introduction

Biotechnology encompasses a variety of advanced tools and modern techniques, including genetic engineering, that are currently employed in crop improvement. Every technique in this field is unique, with its own set of advantages and disadvantages. These tools have helped plant scientists understand and utilize the genetic repertoire of various crops much faster than conventional breeding methods. Transgenic technologies have led to the development of a large number of genetically modified (GM) food and feed crops; so far, only major staple crops like rice, maize, cotton, and soybean (Moreto Guaitolini et al. 2024). A number of products and methods developed or used by plant scientists have come under intense public scrutiny, and the issues delineated are based more on socioeconomic rather than environmental concerns. Their initial focus may, therefore, have been misplaced; the technology development of tomorrow will have to start with the consumer first, meeting the end-use demands of a wary and skeptical market. Over the years, society has started accepting genetically modified foods. The substantial equivalence has led to the examination of a large number of such products based on their nutrition and any new protein produced as an allergen (Maxwell et al. 2024).

The issue mostly raised is whether the question of whether the machinery or tools of transformation cause the changes is addressed during the approval process. Also, the safety of the transformation method per se is not addressed. Many transgenic products are being used without these questions being answered. The farming community worldwide is showing interest in the first generation and other GM products based on creating resistance to pests, diseases, herbicides, abiotic stresses, or enriching the spike with nutrients. Transgenic germplasm of interest to them has therefore been developed with genes for resistance to pests or herbicides or to enrich the spike with nutrients (Peters 2024).

Recent advances in biotechnology and genetics have enabled the development of varieties of many crops with unique disease resistance or other attributes. Several of these crops have already been released, and many more will be released in the near future. The world's farmers have quickly adopted these new varieties, with significant acreages already planted. Countries that adopt regulations to encourage the development and adoption of new varieties will be among the first to enjoy the benefits of this powerful technology. Countries that are slower to adapt their regulatory systems may suffer reduced agricultural productivity and falling real farm commodity prices, both because they will not be able to produce as efficiently as the countries that have adopted this technology and because they will face real competition in agricultural markets from more efficient producers (Haley and Zawati 2024).

The emergence of modern biotechnology has been made possible by a series of scientific advances. These scientific advances in molecular genetics have built on discoveries going back to the time of Gregor Mendel and contemporaries who formulated the basic rules of genetics. Similarly, it was not until the 1950s that the structure of the DNA double helix was described. These early discovery scientists had no idea, of course, of where all this spectacular research would be in the relatively short period since these discoveries were first made. However, the application of this new knowledge by the next generation of scientists used this information to make major advances in recombinant DNA research methodologies, the creation of molecular breeding work in macro-organisms or tissue cultures, and somatic hybrids of two microbes (Moreto Guaitolini et al. 2024; Maxwell et al. 2024). These research breakthroughs and others like them represent the history of the advances in biotechnology.

8.2 Historical Developments in Biotechnology and Crop Genetics

The pedigree of modern biotechnology goes back to the time when man was making bread, brewing alcoholic drinks, and fermenting dairy products. The ancient Egyptians were baking bread using yeast, and the Babylonians preserved apples with the help of sulfur dioxide. The ability to modify and reproduce crops has taken place throughout history. Artificial selection of plants occurs when individuals save seeds from their best crop and sow them in the next season to improve them over time. In the late 1800s, the work of an Austrian monk provided the foundation for the inheritance and selective breeding of genetics (Haley and Zawati 2024). Soon after the discovery of the structure of DNA, biotechnology techniques emerged, increasing molecular biology. These biotechnologies influenced crop improvement; some of them include isolation of genes, cutting, and insertion of genes. By opening the pathway to understanding gene structures, this biotechnology has added new efficiencies and baselines in crop improvement programs. The containment of the human genome and the plant genome, including rice, makes it important to find new oilseeds for soybean development and oil palm to grow in the last two centuries. The science of biotechnology and commercial permits of seed in crop variety are intended for their release around the world (Mustapha Abdulsalam et al. 2024).

It cannot be overstated that while breakthroughs were made, several related factors pervaded the discourse and formed the basis for critical introspection that progressed with the development of science and technology as well as society. The major developments are presented from a historical perspective, which has shaped the growth and future of biotechnology in agriculture. For example, recombinant DNA technology led to the birth of biotechnology, which enabled farmers, for the first time in history, to improve crop production in a cost-effective and expedited manner (Zenin and Nekrasov 2024). Between 1996 and 2012, crop biotechnology that is led by herbicide-tolerant and insect-resistant traits has surfaced to occupy more than 1.3 billion hectares worldwide. The first bioengineered crop, herbicide-tolerant tobacco, obtained in the mid-1980s, could only demonstrate the possibility that new technologies can be applied to agriculture. Heat-tolerant, flood-tolerant, drought-tolerant, and bio-fortified crops are being entered into the pipeline as well. This was made possible through advancements in science-based regulatory policies (Maxwell et al. 2024).

Indeed, it was the regulations from governments across the world calling for the uniform federal triggering of recombinant DNA research that led the way for the early commercialization of a genetically engineered material. In Europe, there had been robust regulations to decide upon the safety of biotechnology-derived GM crop varieties. Biotechnology is basically a biomodification technology excluding the need for kinship. The implementation of biotech methods bypasses traditional breeding methods for the isolation of more easily detected genes. Given the historical development in plant genetic engineering, genetic modification has emerged as one of the amelioration strategies for crop productivity enhancement. In conclusion, scientists and society have had the wisdom to harness this technology in a direction that can bring benefits to farmers, consumers, and society (Tang, Liu, and Wu 2024). The time has come for a performance test of more complex transgenic traits using the most stringent scientific protocols, and the involvement of the scientific community in the process will make more products available in more countries based on the genetic transformation, as illustrated in Fig. 8.1.

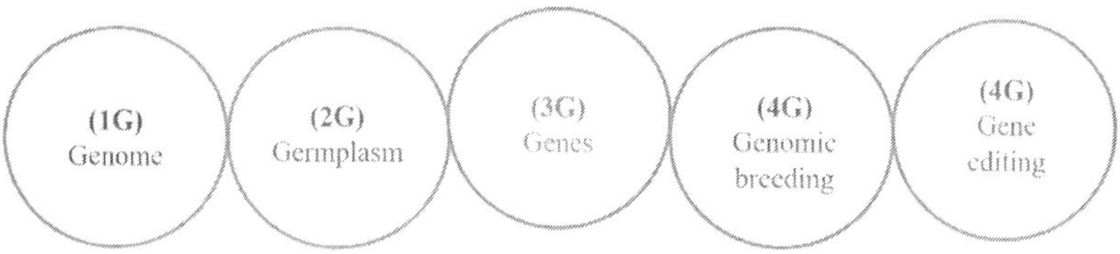

Fig. 8-1 Genetic transformation

8.3 Molecular Techniques in Crop Improvement

Progress in all areas of biotechnology has been staggering and impressive, and the revolution in the understanding of life has lifted the entire field of biology in the most dramatic way since the Renaissance. Nowhere has the conceptual understanding of biological processes been so dramatically transformed as in the area of genetics. The concept of a gene is no longer as clearly defined as it was eight years ago, and it is no longer possible to draw a clear boundary between the organism and viral and other genetic systems. Some genes are important in more than one way. In fact, some may not be important to generate any traits at all in the complete absence of others. Human genetics has also made fundamental contributions in identifying and understanding some relatively simple but important inherited disorders. In plants, gene mapping is now the most important development thus far (Deori et al. 2024; Gharatape et al. 2024). Only a few experimenters have been lucky enough to uncover the beneficial gene marker. This is also the case with other traits of interest to breeders.

The induction of mutations offers yet another important avenue for much-needed plant variation. We have all observed that nature has an enormous store of some potential genes in weeds, parasites, and other relatives. The problems in the introduction of these new genes into agricultural germplasm include the problem of recombination in proper linkage with worthwhile genes. The technology for this accomplishment in induction and transfer has become quite standard, with improved and still improving means of discovery and selection in the near future. The transformation of plant genetic systems is clearly the most dramatic of recent developments. It is only remarkable that we have succeeded (Bwalya, Olweny, and Chungu 2024). Now that the critical barriers in the transformation process have been identified, it is clear that we can improve micropropagation, selection in tissue culture, transformation, and regeneration to the point where every natural and created gene can be passed through this bottleneck. This changes not only what we can do but how we do it. Mutations can be moved from the breeding nucleus to choice commercial seed by allowing the natural crossing over during backcross and selfing. From a commercial point of view, engineered plants are quite different in the selectivity they imprint upon germplasm. We will no longer be content with vigorous, easy-to-harvest, durable, attractive, or digestible crop plants. These traits cannot compete effectively with the luxurious crop plants endowed with improved disease and herbicide resistance, improved insect resistance, complete chemical weed control, or directed assisted growth through genetic fertilization. We can now adapt the genetic capability of wild species and relatives to the improvement of our crops (Niederau et al. 2024). We can define the gene effect at the molecular level, and we can discern a safe and natural method of scavenging for the appropriate genes.

8.3.1 Genetic Engineering

The most publicly provocative and perhaps powerful tool of modern agricultural biotechnology is genetic engineering. Recombinant DNA technology, commonly known as genetic engineering, is a technique that allows the combination of DNA from disparate sources and allows for their incorporation into a host genome. Genetic engineering most likely offers important methodological advantages in the breeding of crop plants. Farmers plant crops for a desired combination of characteristics; they are not interested in the package of genes that impart this agronomic behavior. Traditional breeding methods are faced with the need to sort out the desired genetic characteristics among thousands of others brought along on the cross. They demand expertise in a wide array of plant sciences and consume years of inbreeding before a new variety can be utilized (Zheng et al. 2024). The potential for genetic improvement through genetic engineering has increased scientific and public interest. A major goal of current research in molecular genetics focuses on the isolation and characterization of specific genes from living systems. This interest has been stimulated by the ability to manipulate and transfer these genes by well-defined and reproducible procedures. Of particular importance is research directed toward the isolation and manipulation of genes that control specific plant processes. This research has reached a point where control of the expression of a large proportion of the genes required for plant growth and development is understood and predictable (Makulana et al. 2024). These developments have enabled us to consider the possibility of controlling the expression of any given gene in a target crop, such as sunflower, plant genetic sources, and genetic varieties results in new crop varieties, as illustrated in Fig 8.2.

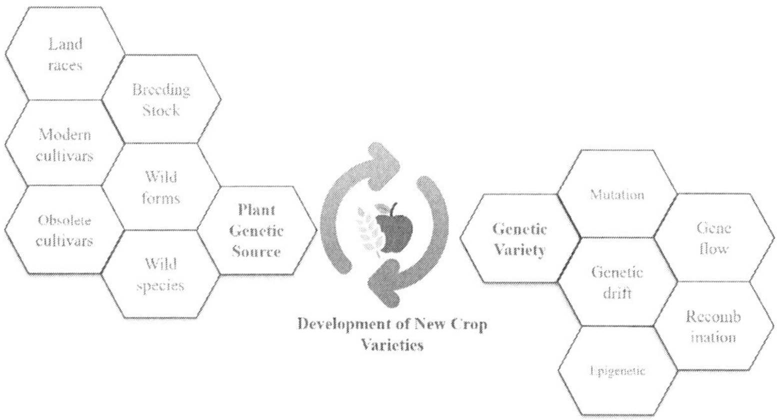

Fig. 8-2 Development of new crop varieties

8.3.2 Genomic Selection

The goals of crop improvement programs are to increase the genetic improvement of the traits of interest. The identification of individuals with the best breeding values is the first step in any indirect phenotypic selection. Genomic selection represents a recent advancement in plant breeding that allows for the prediction of the breeding value of lines. However, the ability of this strategy to promote the genetic gain of the population being selected depends on the complex intrinsic mechanism of the prediction, thus enabling the selection of line combinations with high breeding potential. Genomic selection quantifies genetic merit by modeling all genotypes across a particular set of conserved allele frequencies. It focuses on shrinkage estimation of the regression that allows whole-locus fit to aid the accuracy and selection of new, untested lines. This strategy can integrate the effects of many unpredicted effects and provide selection based on genomic estimated breeding value, combining the effects of many alleles, even undetected effects (Wang et al. 2024; Tang, Liu, and Wu 2024).

The genotyped data that are applied at specific marker intervals with known positions contain many chromosome segments with nucleotides of interest. The genetic map determines the genetic organization of the linkage groups and the distances between sites. Notably, the highly dense and informative map is commonly known as a genetic SNP map. Aspects related to genomic selection can be categorized into applied and theoretical considerations. However, the ratio of selection, the dense map, and the reference population involve an important balance (Deori et al. 2024). Moreover, issues regarding the genetic element that mainly contributes to population selection and the possibilities to control the references associated with a large study need to be more deeply investigated. Therefore, the aim of the theoretical part of the study is described as measuring the priorities of genomic selection to control linkage in reference to adaptive genetic regulation. Afterward, different lines will be assembled differently in a way that will allow estimation of the heritability of the interaction considering the given model (Xing et al. 2024).

8.4 Biotechnological Applications in Crop Breeding

The discovery of recombinant DNA methods for transferring genetic material from similar or disparate organisms has opened up new possibilities for the manipulation of crop species. The difficulty of genetic transformation varies widely between species. With most crop plants, transformation presents the great difficulty of moving the newly introduced genetic material through the tissue that determines key agricultural traits in order to produce the desired phenotype. These difficulties define the courses and context of the application of gene transfer in plant breeding. However, despite much evidence that single genes control agronomically important traits, there are no successful examples of engineered varieties that are improved by microprojectile delivery to agriculture (Wang et al. 2024). Although this technique was long promised as a method that could extend gene transfer to a larger number of critical species, the commercial application of gene transfer remains limited to those plant groups in which *non-agronomic* methods for gene transfer are based. Most commercial genetically engineered herbicide-resistant crops are canola, cotton, and soybean; these three crops account for about 99% of the global transgenic area (Deori et al. 2024). Most transgenic herbicide-resistant crops have been improved by the introduction of genes encoding a variant of the enzyme EPSP synthase that is resistant to the herbicide glyphosate, whereas crop tolerance to other herbicides has been conferred by *acetohydroxyacid synthase, dihydropteroate synthase, hydroxyphenylpyruvate dioxygenase, protoporphyrinogen oxidase, and 4-hydroxyphenylpyruvate dioxygenase genes*. Genes encoding two types of acetolactate synthase have been used for this purpose, one from Arabidopsis thaliana and the other from *Ochrobactrum anthropi.* Host plant tolerance to other herbicides, such as *bialaphos, imidazolinones,* and *sulfonylureas,* has been mediated through the introduction of bacterial, herbicide-specific genes (Makulana et al. 2024).

8.4.1 Marker-Assisted Selection

Classical breeding versus marker-assisted breeding: In classical plant breeding, hugely diverse genetic variations are generated in crop plants through hybridization and selection of progeny. However, in a breeding program that targets a specific gene modification, a major disadvantage is that not all progeny express the traits that the breeder targets. As a result, identifying and securing a progeny aligned with specific genes of interest takes longer. In contrast, in marker-assisted breeding, a breeder can discard all progeny that lacks the desired sequence without conducting phenotypic tests. Knowledge of DNA markers that are related to a targeted gene allows breeders to secure a genetically predictable progeny and thus reduce the number of breeding generations. The application of molecular tools that link a particular DNA fragment to a gene with known function in a breeding population provides enhanced options and precision in the development of new genotypes with added advantages (Mareguddikar et al. 2024).

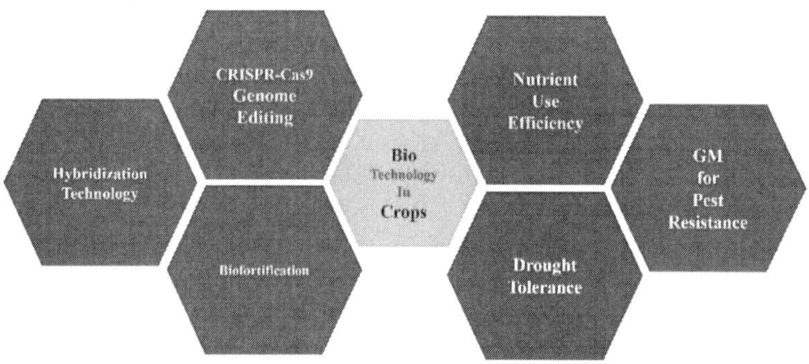

Fig. 8-3 The roles of biotechnologies in crop enhancement

8.4.2 Transgenic Crops

Genes can be moved between species within the same genus by crossing. Genes can also be moved between species by traditional breeding methods if the gene of interest is transferred by backcrossing and selection across generations. The introduction of foreign DNA into plants is a form of genetic engineering and is greatly facilitated by recent developments in recombinant DNA technology. In employing the gene gun, microprojectiles are bombarded into target plant tissue. The foreign gene of interest is also located on the microprojectile. The result of this procedure is that some cells in the target tissue take up the microprojectile and, in doing so, acquire the foreign gene. An alternative to using the gene gun is to transfer genes directly into plant cells in culture using a specific *bacterium* that causes crown gall tumors in trees and vines. The bacteria contain a tumor-inducing plasmid, which integrates a segment of cellular DNA in the chromosome of the plant species. The plasmid contains the tumor-inducing gene, which causes the plant cell to become tumorous, and genes necessary for the catabolism of plant hormones (Tiwari et al. 2024; Kumar et al. 2024). In the process, pieces of the plasmid are integrated into transfer cells surrounding the wound, and these foreign genes are expressed in the plant cell.

8.5 Tools and Techniques in Biotechnology Research

Conceptually, biotechnology research is tool and technique-oriented. Proponents are not really gene pushers or molecular surgeons, as critics would have it, but are trawling through various tools that they find at their disposal, tenuous and problematic as they may be. New tools, such as the gene editing technologies that have emerged from previous work with genetic engineering, are continually being added to or adapted to the research of biotechnology proponents. These are used to modify or generate genetic traits, not because they have somehow been pinpointed as what is needed or wanted by society (Mäkelin 2024).

Biotechnology research is providing useful tools to plant breeders, but the value of these genetic tools is not always self-evident. Genetic engineering techniques are capable of introducing new genes from unlike organisms. DNA from one or more donors can be inserted into the genome of an organism that finds the gene coding for the trait of interest situated at a particular location. Transgenic technology is a form of genetic engineering. The first transgenic plants made across scientific research laboratories during the 1980s were herbicide- or pest-resistant (Wippold et al. 2024; Lu et al. 2024). These plants were genetically engineered, showing enhanced characteristics as a result of the insertion of novel genes. Molecular markers are specific chromosomal locations near a gene or genes of interest whose patterns of inheritance can be easily followed. They can be rewarded in plant breeding by decreasing the time needed to develop new or improved crop varieties. They serve as the tools for genetic analysis to locate the gene or linked marker in the individual that carries the particular gene. In breeding, it is used for selecting individuals carrying the traits predicted by the marker. Molecular markers can also be helpful in plant breeding to identify different genes for the same trait that taxonomists ask for taxonomically different (Xing et al. 2024). Therefore, molecular markers are quite useful in the development of modern plant hybrids.

8.5.1 Gene Editing Technologies

Development of revolutionary gene-editing technologies, biotechnology has evolved greatly. These technologies are now being widely developed and have vast potential, especially in the crop area. Gene editing refers to a collection of methodologies having specific in-situ effects within the DNA of living organisms. In plants, the most used methods are *meganucleases* or homing *endonucleases*, zinc finger nucleases, transcription activator-like effector nucleases, or, more recently, the clustered regularly interspaced short palindromic repeats and associated protein. Gene editing, which is based on a mechanism that bacteria use to interfere with viruses, makes it possible to alter specific genes in organisms precisely (Kamga et al. 2024). It has been hailed as the game changer in gene editing because of its relative simplicity, continuum of techniques, and low cost. These attributes have led it to quickly overtake other techniques as the preferred method for gene editing in a wide range of organisms. Scientists have also succeeded in using this gene editing tool in crop plants to, for example, improve yield or disease resistance. More generally, the technical design of gene editing aims to develop any desired crop that has key traits for crop producers with efficiency and speed. These biotechnologies have ethical, public, and societal issues to consider when viewing risks and concerns (Mujyambere, Adomako, and Olympio 2024). Despite the promise of reducing off-target effects, whether in animals or plants, there are still potentials for off-target mutations that may pose downstream risks due to unwanted genotypes or low fitness of the resulting organisms; some innovative biotech solutions for crop improvements are illustrated in Fig 8.4.

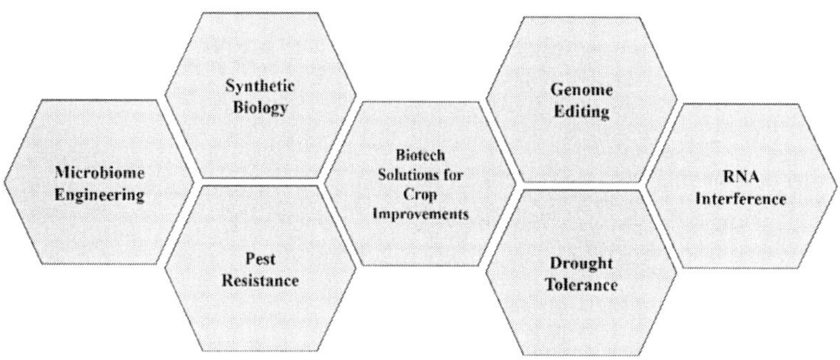

Fig. 8-4 Innovative biotech solutions for crop enhancement

Past research has speculated on the development of regulatory requirements for potential products from gene editing technologies. The issue of how to regulate and approve this new product is a challenge that policymakers have to address, especially if adopting very stringent regulations only has the effect of slowing down the use of these new gene editing technologies. However, as gene-editing technologies have continued to develop, the result has shown many positive results, even in food security. The gene modification of crops might greatly benefit the food safety sector, improving food yield, increasing resistance to changes in the environment, and decreasing pesticide usage. To be precise, there are already ongoing efforts to exploit gene editing to enhance trait development in a variety of crop species. There has been a spike in recent studies showing discoveries in developing improved crop disease resistance, increasing crop germination percentage, and enhancing crop yield using gene-editing technologies on cucumber, tomato, and rice (Windhagauer et al. 2024).

8.5.2 Genetic Engineering

Genetic engineering or recombinant deoxyribonucleic acid technology is a key component of crop biotechnology. Genetically engineered crops are those in which a new trait has been introduced by changing or adding a new gene to the crop. This is achieved by one of several methods or techniques. Successful genetic engineering can improve the quality or quantity of the harvested product, showing that the gene identified encodes the desired trait. Once identified, the functional gene can be inserted into a plant by a variety of gene transfer techniques, the most common of which are listed below. The herbicide-tolerant, pest-resistant, and high-oleic acid oil soybeans inserted with foreign proteins are just a few examples of this (Cong et al. 2024). Methods for inserting foreign genes into plants:

(1) Agrobacterium-mediated transformation
(2) Biolistic or particle bombardment
(3) Electroporation
(4) Liposome-mediated transformation
(5) Silicon carbide fiber-mediated transformation
(6) Ultrasonication

From what this expression "traditional biotechnology" refers to, it is appropriate to refer to selective breeding and hybridization, which provide farmers with genetically improved seeds that possess the beneficial characteristics of their parents. With the advent of genetic engineering in the early 1990s, plant breeders have begun accelerating traditional breeding by introducing genes from other, more distant varieties or related species. Genetically engineered or genetically modified organisms have become a polarizing and contentious issue, particularly in relation to food safety and environmental and ecological effects (Singh 2024). Government regulation considers this type of organism to be significantly different from crops and products derived from conventional breeding. As such, gene bioengineering crops are subjected to rigorous regulatory requirements and must demonstrate no harm to human health and the biological and physical environment. They are subjected to rigorous safety assessments before release. Key treatments undertaken by the regulated community include experimental work, applicant risk assessment and consultation, and regulatory approval and post-release supervision (Ricky Aprinaldi, Hery Bachrizal Tanjung, and Fuad Madarisa 2024).

In a broader context, ethical, socio-economic, and regulatory implications have been reviewed. Some of the benefits of adopting gene bioengineering in agricultural practice include increased productivity, increased nutritional content of crops, increased resistance of plants to pests and pathogens, increased crop resistance to viruses and bacteria, and increased tolerance of plants to drought and salinity. In 1993, in an independent review, the regulation of safety approval was regulated in government, information, and administrative practices relating to the regulation of biotechnology. More research and development in gene bioengineering is currently underway to improve the efficiency of multicellular crop varieties, including the fine-tuning of promoter systems that direct the expression of engineered genes in the desired tissues or at the desired stages of development. Other current genetic engineering approaches include morphological plasticity, trait genetic engineering, RNA interference, genome editing, synaptic biology, and metabolic engineering. The above approaches summarize that genetic engineering is a fundamental tool in plant biotechnology (Healey et al. 2024). Improved plant varieties with desirable features are made possible by choosing one or more traits from the several developed traits and through further breeding of the selected plant.

8.5.3 Transgenic Technology

Transgenic technology, the cornerstone of modern biotechnology, enables the creation of transgenic plants. "Transgenic" refers to the combining of genes from different organisms. The incorporated genes dictate a new trait or traits in the plant, usually for resistance to insects or tolerance to herbicides. A gene is incorporated into a plant through recombinant DNA techniques to create a genetically modified organism. Transgenic technology allows genes to be transferred from bacteria, fungi, viruses, insects, plants, and other organisms into plant genomes (Chaumier et al. 2024). Once incorporated into plant cells, the new genes are expressed or active. When a gene is expressed, it makes a product, such as a protein, that gives the plant its new trait. Successful transgenic crops include those with new traits for resistance to pests or herbicides or improved nutrient value. Some examples of insect-resistant transgenic crops include cotton, which is resistant to bollworm and budworm, and other crops with resistance to the European corn borer in corn, cotton bollworm in cotton, pink bollworm in cotton, and tobacco budworm in tomato (Nowak et al. 2024).

Transgenic technology has freed plant breeders from many of the "crossing barriers" that limited traditional crop breeding. Now, and into the future, any beneficial gene that imparts a desirable trait when expressed in the new host can potentially be moved into a plant, irrespective of its origin. Despite more than a decade and a half of research and field tests and the regulatory approval of hundreds of transgenic crops, there are concerns about the ecological and food safety of transgenic organisms. More than forty countries representing 195 million farmers are now growing transgenic crops. The international regulatory framework is still evolving, but all countries have safety regulations based on thorough risk assessment. Current research findings continue to address the early adoption of transgenic technology (Hernández et al. 2024). It is worth noting that research is underway to overcome, or at least lessen, consumer acceptance and market development issues for GM crops.

Transgenic plants, first sold in 1996, are an important part of the growing global system addressing food security. Offering such possibilities as resistance or tolerance to pests and herbicides, improved storage, processing, and nutritional quality, new modes of action for highly efficient crop protection, and less post-harvest spoilage, transgenic technologies will help ensure the future production of sufficient food with amazing visual benefits around the world. At the same time, we need to address issues that impede farmer choice and efficient access to these beneficial crops. This can be achieved through the development of organized deployment procedures to provide access to GM technology for all who would benefit from it (Valenzuela et al. 2024; Afordoanyi et al. 2024).

8.5.4 Molecular Markers

DNA markers, also known as molecular markers or simply "markers," are specific DNA sequences that are associated or linked to a phenotypic expression of interest, such as disease resistance, improved nutrition, or crop quality. Molecular markers have become an integral part of plant genetics and breeding to identify the location of genes or quantitative traits within a chromosome. With the advent of biotechnology, a number of techniques have been developed to characterize and identify these genes at the DNA level (Tremblay and Zhang 2024). They have found a number of applications in genetic research, both basic and applied, including:

(1) identifying and tagging genes,
(2) fingerprinting and genetic diversity,
(3) phylogenetic studies,
(4) QTL mapping,
(5) conservation of germplasm,
(6) seed purity testing and varietal release, and finally
(7) in marker-assisted selection.

Different methods exist to identify molecular markers, such as protein-based markers, DNA-based markers, and PCR-based markers. Since the advent of PCR, a number of molecular techniques and methods have been developed that use PCR technology, such as AFLPs, SAMs, S-SAPs, SRAPs, SNP, and microsatellite analysis. However, because of their high polymorphism, reproducibility, wide distribution in the genome, and suitability for automation, SSRs are becoming popular to study and identify diversity of QTL. The advent of marker-assisted selection is expected to speed up the development of improved varieties of crops. In general, marker-assisted selection makes the crossing process more "precise." Marker-assisted selection means selection based on molecular markers. Generally, the DNA markers used in marker-assisted selection are better or more reliable than using morphological markers or physiochemical markers (Bhat et al. 2024). Development of marker-assisted selection programs requires some general procedures such as:

(1) Association between phenotypic and genotypic data in large-scale populations,
(2) selection of the most superior markers,
(3) base population and parental clone evaluation and formation,
(4) laboratory testing and genotyping of pure lines and selection of F1 "hybrid" parents,
(5) large-scale field evaluation,
(6) "hybrid" clone selection, test development, and commercial release, and
(7) continuous evaluation and improvement.

Marker-assisted selection can shorten these procedures by creating a shortcut since it is based on DNA or genes that govern the specific trait or characteristic that breeders are interested in. The main or major genes and polygenes directly or indirectly control both genes and proteins. The selection of an improved or a good gene variant from one or both of each of the progenies has a great probability of carrying a variety of traits. Applications of marker-assisted selection also have significant impacts because they can change from one generation to the next, be specific to the gene and locus of interest, and, of course, maximize the probability of selecting improved or superior alleles and combinations of the genes. Not only efficient but marker-assisted selection is also cheaper because it only requires the creation of some specific primers and DNA amplification in the laboratory, as well as genotyping of necessary selection markers on a number of consistent, repetitive generations (Bahl et al. 2024; Baudrier et al. 2024). Turrets are widely disseminated in the populations to study population genetics and general selection for improvement in breeding.

8.6 Genome Editing Technologies in Crop Improvement

The marker-free binary vector pDE-CAS9 may be used for the introduction of the Cas protein into the plant protoplasts. To date, most of the work on the use of CRISPR/Cas9 has been devoted to the creation of monogenic mutations in plants. Various approaches, however, have been proposed for gene structure correction when the wild-type gene's DNA is homologous to a template that contains mutations compared to the target gene. The efficiency and characteristics of the use of various templates for the purpose of replacing plant gene fragments are still insufficiently studied, and different approaches are used in practice to introduce transgenic structures at the site of inactivated DNA regions (Ropero-Pérez et al. 2024). The development of efficient methods of non-homologous insertions of DNA fragments is of particular relevance.

The emergence of methods for the directed editing of plant genomes has created a good basis for the development of various applications for the introduction of changes to genomic regions via the CRISPR/Cas9 system. This system has been used for the introduction of changes into almost all genomic regions that have been studied up to now. In some cases, the tissue and the optimal developmental stage, the genetic and physiological status of somatic cells isolated from tissues, and the regeneration capacity of the transformed tissues should be taken into account (Shooshtari et al. 2024). Despite the limitations associated with the genetic and physiological status of organisms and the tissues and cells that can be used, a wide arsenal of methods has been developed and optimized. At the same time, more potential directions for the use of CRISPR/Cas9 editing systems as a whole and of targeted nuclease systems in crop breeding exist than have been so far investigated. Given the great importance of the expansion of genomic editing in plant breeding, both the further optimization of the existing methods and the development of new methods are particularly relevant (Yaschenko, Alonso, and Stepanova 2024).

8.6.1 CRISPR/Cas9

CRISPR/Cas9 technology is a technique that allows scientists to modify the genome of almost any organism. CRISPR/Cas9 consists of the enzyme Cas9 that can be programmed to cut a specific DNA sequence and guide RNA that guides Cas9 to a particular location. Usually, CRISPR/Cas9 induces a deletion or insertion that results in a frameshift and knockout of the targeted gene. Although large deletions can also be achieved, insertion of a tag, mutation of one amino acid, creating fusions, and base editing are all possible. By targeting multiple sites, genes can be deleted or excised. CRISPR/Cas9 is widely and rapidly adopted because it is easy to design, has a high targeting efficiency, does not leave behind a selection marker, and can manipulate almost any gene. In plants, this technology has been used to generate gene-knockout plants as well as gene-edited plants with altered, inserted, and deleted sequences. As would be expected, a variety of drought and salinity tolerance genes have been targeted in crops with generally successful results (Bhardwaj et al. 2024). Given the scalable nature of CRISPR/Cas9, plants edited with the same system can be used in further research for additional changes and combinations. The use of off-target sites is a major limitation of CRISPR/Cas9, and it is important to confirm the lack of off-target cleavage experimentally. This can be done by polymerase chain reaction amplification followed by deep sequencing. As mentioned, this technique is fast-paced, so there are many variants of Cas9 available and under development that aim to improve on factors such as targeting specificity, and this includes the use of other CRISPR system proteins in place of Cas9 (Bhat et al. 2024).

8.6.2 Transcription activator-like effector (TALE) nuclease (TALENs)

Transcription activator-like effector (TALE) nuclease (TALEN) is a fusion protein containing TALE repeats and a nonspecific nuclease domain.

TALENs bind the TALE repeat DNA target sites by the base number-targeting method, in which each TALE repeat recognizes one DNA base pair. TAL-effector nucleases are constructed by fusing the FokI nuclease domain to the C-termini of the TAL-effector repeat array. TALENs enter plant cells through recognition and cleavage of a specific targeted sequence, and the recombination of these fragments leads to the repair of double-strand breaks (DSBs) through the cellular DNA repair process via the DNA repair mechanism of nonhomologous end joining (NHEJ), thereby introducing a lesion or mutation into the targeted gene (Ropero-Pérez et al. 2024). The simplicity of the TALEN design opens the possibility of achieving single-character and multi-character gene modification of the target gene, achieving the desired trait in plant genomics. However, the use of TALENs in plant breeding is mainly associated with the transformation process using Agrobacterium-mediated transformation, biolistic particle bombardment transfer in plant protoplasts, and in vitro transcription of TALEN with protoplast vector transformation (Shooshtari et al. 2024). Furthermore, every time TALENs are used to target a specified locus, TALEN constructs need to be specifically synthesized.

8.7 Applications of Biotechnology in Crop Improvement

Biotechnology has various applications in crop improvement. This technological advancement is used to provide resistance against various diseases and pests caused by different biotic factors. Crop varieties are also developed that have the potential to perform under varied environmental conditions. Plants with resistance against abiotic stresses such as drought and salinity are also developed. Moreover, efforts are being constantly made to enhance the nutritional quality of different crops, which are considered staples in various parts of the world. Efforts are also made to enhance the shelf life of different horticultural crops through the use of biotechnological methods (Bhardwaj et al. 2024).

New and advanced methods in biotechnology have paved the way for producing transgenic plants that have dual resistance against more than one disease or pest. The technique of antisense technology has been used to develop a transgenic tomato that has dual resistance against early blight and late blight diseases. With the help of transgenic technology, the development of a single plant that is resistant to inoculation by 15 different viruses is also possible. These varieties are also called transgenic virus-resistant varieties. In this way, transgenic technology has enough potential to develop transgenic plants that simultaneously produce resistance against various diseases. For a long time, the use of chemical pesticides and fertilizers has led to deleterious effects on plants and the environment (Zenin and Nekrasov 2024). Pesticides also destroy beneficial insects in the process of plant disease or pest control. The consumption of chemical pesticides and fertilizers leads to many diseases like cancer, respiratory diseases, and bone diseases. Biotechnology is also used to develop plants that are resistant to various abiotic factors affecting crop yield. Biotechnological methods can be used to develop plants that are resistant to cold, heat, drought, and waterlogging. The gene involved in the process of carbohydrate formation helps in making the plant tolerant to waterlogging stress. Biotechnology can be used to develop plants for arid and semi-arid regions to protect them from drought stress (Evangel Chinyere Anyanwu et al. 2024). Biotechnology is also used to improve resistance against various pesticides.

8.7.1 Disease Resistance

This volume is focused on biotechnology strategies for improving crop production. Disease resistance is an important target for trait improvement. Many crop losses result from diseases and pests. The genetics of host-pathogen interactions have been intensively investigated in several model systems. The mechanisms underlying host resistance result from a plant's ability to recognize a pathogen and subsequently turn on defenses. The genetic dissection of these systems, together with genetic engineering, is providing the means to develop disease-resistant varieties for agriculture. Several methods can be used to insert foreign genes or to edit existing crop genes to provide resistance to specific threats (Liao et al. 2024). Disease resistance traits made through the use of genetic engineering and biotechnology have been commercialized and can significantly reduce yield loss due to biotic stresses.

Transgenic approaches can be designed to combat different types of plant pathogens. For bacteria, the macrolide antibiotic inhibits glutamine synthetase and blocks nitrogen metabolism, killing bacterial cells that infect plants. Several new wheat varieties have commercially useful levels of resistance to powdery mildew following the acquisition of new resistance genes. Trait improvements resulting in dramatic increases in value are an economic incentive for research and commercialization of transgenic, disease-resistant plant varieties. Molecular techniques developed to improve the translation from research to breeding for disease resistance in plants have beneficial global implications for primary production and food security. Creating plants resistant to pathogens and pests also reduces the dependency on agrochemicals, making agricultural systems more sustainable. However, there are some practical concerns. Most pest resistance genes are specific, and a new gene must be included in the crop for every pathogen species to be controlled. Even in the absence of this requirement, pathogens that attack a plant may evolve new strains that are not controlled by the current set of resistance genes (Wei et al. 2024; Bordel et al. 2024). Thus, it is important to continue researching new resistance traits. In the following paragraphs, a description on some genes that confer resistance to specific plant diseases.

8.7.2 Pest Resistance

The use of biotechnology in pest control and pest resistance has become widespread. Valuable crops can be severely damaged or reduced by pests, and a reduction of crop losses due to pests is an important factor for global food security. To protect crops from pests, a number of different biotechnological strategies have been developed, including the genetic engineering of plants to express a pest-resistant trait. Naturally, their viral pathogens help in the development of plants that can repel or attract pests in a number of different ways, or use several other natural pest control mechanisms. Strategies to achieve pest resistance are designed to reduce pesticide costs, increase profits, minimize environmental damage, and improve human health. There are many examples of successful pest-resistant crops (Gladkov and Gladkova 2024). The expression of toxic ribosomal inactivating proteins from transgenes has been shown to protect plants against beetles, weevils, caterpillars, and various other pests.

The widespread cultivation of insect-resistant crops has been shown to reduce damaging caterpillar populations and result in large savings for farmers. For example, in Arizona, planting cotton reduced the number of sprays to control damaging caterpillar populations, and similar results have been observed in Korea and Australian cotton-growing areas. The benefit of cotton to growers is not just due to the reduced yield loss and the lower cost of spraying; these transgenic plants are also good for the sustainable use of insecticides. It has also been shown that the cultivation of insect-resistant cotton has helped reduce the number of sprays needed for secondary pests in the neighboring food and fodder crops, so these "ecological services" mean that cotton can also benefit other species. Provide a region for planting crops (Mustapha Abdulsalam et al. 2024; Haley and Zawati 2024).

8.7.3 Abiotic Stress Tolerance

Abiotic stress tolerance aims to provide crops with the tools to thrive under less-than-ideal conditions. For example, drought and salinity alone are estimated to reduce average yield by 47 percent cumulatively; sustainable improvements could feed an additional 50 million people. The most common abiotic stresses are drought, salinity, extreme temperatures, and waterlogging. The ability to cope with one of these stresses is not a guarantee for another, but a number of genes and pathways provide a general increase in resilience. The best path forward to produce crops, particularly with climate change, is multi-stress solutions. Two main approaches can be supported to breed such a fit-for-future crop: genetic space exploration (usually by landraces or interesting wild relatives) or tolerance enhancement by precision breeding (Moreto Guaitolini et al. 2024).

Biotechnology contributes to and supports both. Gene editing has a clear, direct link to establishing controlled, precise abiotic stress tolerances. This can entail the formation of single mutations and alleles that, from the beginning, confer greatly improved tolerance. In genomics breeding, molecular markers can aid in the selection of the most well-suited varieties with a variation in stress-tolerance genes. Additionally, new traits can be introduced by other biotechnology techniques, such as allogamy breeding, mutagenesis, and genetic engineering. There are a few clear gene targets that improve the ability to cope with abiotic stress. One factor, the magnitude of the difference of the tested GM event when compared to the wild type, scored as drought tolerance by the Drought Resistance Integrated Assay System, shows clear, positive results. The event stack has, at five different time points, scored better or much better than all the other two drought-tolerant GM events tested by about 50 percentage units. Reducing exposure to abiotic stress now will increase future hunger. Sequencing these genes and using advanced breeding methods could further enhance these stress-tolerance traits to meet future yield needs (Zamora-González et al. 2024; Zenin and Nekrasov 2024). These genes can work alone or in concert.

This is supported by observing a cumulative, added effect of drought-resistance genes on yield protection in rice. The average increase of each stress avoidance transgene was 0.7 t per hectare. However, stacking all three genes increased the average yield advantage over the same wild type 2.1-fold; stacking genes could protect almost the entire yield. It is of utmost importance for future crop breeding to not only identify and deploy such multi-faceted traits and genes for stress reduction but also to test and ensure that there are no unforeseen negative effects of using either technique in the crop (Wei et al. 2024). It would also be prudent to invest in foresight programs for the potential long-term evolutionary impacts of altering abiotic stress production pathways or expression in dimension. A gene that has passed early selection is OsSAP17-1, running in sorghum. In the complete absence of water, the gene, which is triggered without the plant becoming overly stressed, saw a 32 percent improvement in grain yield. These positive results underscore the value that crop protection against drought offers (Peters 2024).

8.7.4 Nutritional Enhancement

The protein and fat content of food crops has not been studied as widely, as the variety in the literature already explains the physiology of the main processes involved in their metabolism, such as photosynthesis in the case of C3 and C4 plants. In contrast, the application of plant biotechnology to enhance the micronutrient content of staple foods holds significant promise for health and nutrition benefits, and this potential opportunity is well-marketed in the name of biofortification. Biofortification is defined as the use of a plant breeding strategy that increases the inherent nutrient content of a food crop. A major point of discussion in the theme of nutritional biofortification is the sources of essential vitamins and minerals. Fats, proteins, and carbohydrates must be eaten either alone or in combination to satisfy energy requirements (Zamora-González et al. 2024; Girin Hazarika et al. 2024). However, most modern diets are grossly deficient in essential vitamins and minerals, which the body cannot synthesize; they must be obtained only through diets.

To overcome such a serious malnutrition problem, still dominant in many developing countries, scientists in different organizations have taken up the project of identifying nutritional needs with which essential vitamins and minerals must be used. Then, as many transgenic events are to be concurrently created that significantly over-express a single trait, these traits can be pyramided into an elite germplasm background using classical breeding to combine multiple effects into a single line. Alternatively, three to five separate genes might be combined into a vector, and the recombinant genes can be introduced simultaneously (Evangel Chinyere Anyanwu et al. 2024). The first example of developing a nutritional crop is the development of Golden Rice with provitamin A, which is rapidly reviewed. It revealed that genetically enhanced crops for increased nutritional content could provide a subsistence farmer with both income and nutritional benefits. Consumer acceptance has been a powerful marketing tool for transgenic products. It will be important to determine what, if any, potential resistance there may be to such a tool in certain countries. Referring to the number of people who are marginally deficient in a given micronutrient is misleading because it fails to mention the number of people with full-blown clinical deficiencies and the number of people who have subclinical deficiencies of micronutrients (Gladkov and Gladkova 2024).

8.8 Omics Technologies in Crop Biology

The advent of high-throughput technologies has provided new tools for the comprehensive characterization of the genome, transcriptome, proteome, and metabolome in microorganisms, plants, and animals providing an overview of current omics technologies such as genomics, transcriptomics, proteomics, and metabolomics, which have revolutionized our understanding of gene function in crop biology. We describe recent advances and current limitations of these platforms and provide examples to show how the integration of various omics datasets may lead us to a better understanding of gene networks in plants. The rapid progress of high-throughput technologies such as high-density molecular markers, DNA arrays, and next-generation DNA sequencers has greatly revolutionized research in the field of crop biology (Bhat et al. 2024). With the overview of recent advances in omics technologies, including genomics, transcriptomics, proteomics, and metabolomics, as they apply to research in crop biology. We expect that readers can obtain an overall outline of these platforms.

8.8.1 Genomics

Genomics studies DNA sequences of genes and chromosomes and closely related details. It has potential applications on several levels: genes, gene products, single metabolic pathways, and even whole genomes. The last application can benefit from complete DNA sequences of crops. The more markers available, the more detailed view of the genetic map is obtained. This increases the probability of detecting a polymorphism between the crossing parents. The result is genetically founded estimates of the genetic effect of a closely linked marker and thus a powerful tool for marker-assisted selection for a wider range of traits. Because the number of alleles is limited for 'classic' genetic markers, only a fraction of the total number of DNA polymorphisms is made visible by 'classic' mapping. In contrast, because essentially the molecular bases of all genes are visible in a cDNA library, this library contains the molecular bases of practically all polymorphisms in RNA sequences, both allelic and non-allelic, that are expressed in tissues used for cDNA library assembly (Said, Isack, and Hassan 2024; Yaschenko, Alonso, and Stepanova 2024).

Other tools of genomics can be used to detect these polymorphisms within the cDNA sequences and to map them next to their loci. Genotypes that are identical to functional alleles of candidate genes can be used for MAS. QTL regions have different percentages of explained variance. Their percentage of explained variance was determined as follows: individual

QTL markers were each associated with the allele combinations of donor and recurrent parents observed in the populations of the validation work. These alleles were used as cofactors predicting the phenotypes of the corresponding lines. The percentage of variation explained by a QTL marker was defined as the percentage reduction of the residual variation induced by this cofactor for the trait in question. To add a confidence envelope for every size of the tort parameter, this analysis was repeated for different limits. For display reasons, the predicted variances were projected onto the proportion of explained variance (Lu et al. 2024; Mäkelin 2024). Their standard errors are indicated by error bars.

8.8.2 Transcriptomics

The study of global gene expression of an organism using high throughput techniques such as array analysis or sequencing has evolved into a new discipline known as transcriptomics. This promising field is increasing the understanding of many aspects of crops that have been little understood so far. There are various implements of this discipline, and intermediates between them have begun to be established. The diachronic shift through a dynamic elucidation of plant genes and genomes helped give birth to and strengthen this relatively new line of plant study. The advances in the methods of DNA microarray tools accelerated this process. First, at the beginning, it will develop continuous sequencing techniques of expressed genes that are communities of sequences corresponding to transcribed DNA of an organism, generally DNA that has had the introns removed (Zhao et al. 2024; Deori et al. 2024).

The development of DNA microarray elimination was the first method for the elucidation of gene expression due to their high sensitivity. Tiling arrays, micro-matrices, suppressive subtractive hybridization, and ordered subtracted libraries are other methods used for this. Some specialists accept that these tend to surpass as the most reliable technique. These methodologies allow the expression levels of virtually the entire genome to be examined at once, producing a comprehensive list of high-expressing genes, some of which have not been detected by previous methods. With these new methods, the understanding of traits, such as gene expression, interactions between gene products, the materials, and the processes involved with gene expression, and the stimulant agents of the expressed genes become more evident, as well as the differential expression of the genome and its regulation (Kimura et al. 2024).

8.8.3 Proteomics

The proteome is the entire complement of proteins expressed by a genome, cell, tissue, or organism at a certain time. These structures, which are the executors of genetic instructions, interact with each other and with the surroundings to create life. Proteomics research complements genomic research and is expected to lead to increased knowledge and understanding of the regulation of living systems. The main goal of proteomics is the comprehensive identification and characterization of proteins. The challenges for proteomics research are now to analyze intact proteins and their many functional and physical states in native cells, organs, and organisms, and generate the three-dimensional structures of proteins and protein complexes in their cellular environments (Lu et al. 2024). In the agricultural sciences, proteomics techniques are not only used for the identification of agricultural pests and the characterization of enzymes found in organs and tissues, but the main goals include the elucidation of protein expression profiles associated with agricultural quality traits and stress relief problems, functional genomics through protein analysis, determination of food allergens and allergenic degradation through food processing, identification of plant and animal diseases, food gene production and metabolite accumulation, safety assessment of food and evaluation of genetically modified foods, candidate protein studies on the pathogenicity of agricultural pests, and studies on improving the safety and quality of derived food, metabolomics (Windhagauer et al. 2024; Cong et al. 2024).

8.8.4 Metabolomics

Equivalent to genomics in the study of DNA and expressed sequence tags in the study of genes, metabolomics is a discipline that focuses on population-scale metabolic projects, where small molecule outputs are monitored and measured. Societal benefits of metabolomics applications, particularly with plants and microbial cells, include more efficient use of existing knowledge, better crops and agricultural goods, higher quality delivered food, and broader environmental protection by monitoring chemical contaminants at a lower cost for greater levels of compliance. Metabolomic advances offer an array of tools to assist plant breeders in the selection of new commercial plant varieties and countless opportunities to support landrace conservation activities. In microbial cells, metabolite modeling of genetic ideas provides the means to advance new metabolic engineering concepts for biofuel production, adversarial bioterrorism detection, or on-demand food and medical biomanufacturing applications (Healey et al. 2024). As with other disciplines, metabolomics shares ontological principles of organic systems under environmental influence, which include the study of small molecules in response to genetic alterations, revealed gene product regulation of the cellular machinery shaping the metabolic environment, the physiological impacts of exposure to toxicant molecules, the inheritance of certain metabolic services, and evidence of pathways that change small molecule pools under inputs directly linked to potential enabling steps that may lead over time to the evolution of novel metabolic structures. In contrast to other structural biology and systems biotechnology disciplines, metabolomics modeling has its significant challenges but is getting better at systems overviews and a building set of relevant modular parts, even though equivalent-to-gene functions and pathways generating maps are not completely characterized (Kumar et al. 2024; Tiwari et al. 2024).

Understanding the entire molecule network composition and connecting non-dynamic system parts is another big challenge for metabolomics, subsequent to the prediction of phenotypes through the network. At present, many of the activities of a metabolome, which are enriched by some molecular functional parts better represented in datasets, are dictating forward and reverse approaches for metabolite characterization. Data-processing bottlenecks, slow instrument data acquisition rates, and other high-throughput challenges continue to slow down experimental progress and promote metabolite-coverage imbalances. Some of these challenges have led, and continue to lead, to adaptive pathway landmark indications of structure, order, number, and amplitude, all to provide biennium-by-biennium estimates of metabolite classes or families that are experimentally accessible (Xing et al. 2024). Data planning and evaluation guidelines aimed at diminishing gaps and increasing metabolome coverage reliability are also in continuous progress and represent a natural transition adaptation for applying datasets in pathway-based or related life sciences problem-solving occurring at significant societal levels.

8.9 Regulatory and Ethical Considerations in Biotechnology

Robust regulatory frameworks are required to ensure that biotechnological products, especially foods, are (1) safe for human consumption and release into the environment and (2) effective for their intended purposes. In the USA, regulatory oversight of biotechnology in agriculture, and in particular genetically modified organisms, is conducted by three separate but interacting agencies, namely the FAO, the Environmental Protection Agency, and the Department of Agriculture. These organizations require testing of genetically modified organisms for human toxicity and allergenicity to establish their safety as food and for potential environmental effects before a product can be marketed. Because of specific concerns about genetically modified organisms being allergenic, international guidelines for the safety evaluation of foods derived from recombinant DNA plants have been established (Tang, Liu, and Wu 2024).

Ethical considerations in biotechnology relate to both the costs and the benefits of biotechnology. Generally, costs are marked in a number of ways. Environmental costs may be manifest in aspects such as the balance between increased productivity in monocultures and the concomitant decrease in diversity on the farm or in crop and product options for developing countries if they become technologically dependent on the products of northern research . Societal costs are marked in aspects such as issues of human rights, standards of living, and consumer rights in relation to the products of biotechnology. Claims about benefits usually relate to the possibilities of increasing the efficiency of food production, either through yield improvements or through other cost reductions in food production. These costs and benefits, called characteristics, help us to understand the overall health of our food system in its shift toward biotechnology (Zhao et al. 2024).

8.10 Biotechnological Approaches for Abiotic Stress Tolerance in Crops

Abiotic stresses, such as high salinity, drought, high temperature, submersion, and chilling, are major constraining factors for sustainable agriculture. These stresses cause distinct molecular and physiological changes that inhibit plant growth and productivity. Although conventional breeding and phenotypic selections have been partly successful in developing stress-tolerant crop varieties, their long generation time and phenotypic assessment hinder the pace of breeding programs. Recently, various genomic regions that are associated with abiotic stress tolerance have been identified in different plant species (Makulana et al. 2024). Moreover, rapidly advancing functional genomics has provided an invaluable resource for deriving crop plants with improved abiotic stress tolerance through genetic engineering approaches. Stress-responsive genes require highly regulated intracellular traffic for their positively or negatively regulated target gene expression or breakdown to achieve tolerance at the cellular level in different abiotic stress environments. This describes transgenic breeding strategies in developing crop plants that are resilient to abiotic stresses created with the novel biological insights provided by advances in the areas of xylem mobile protein identification (Mareguddikar et al. 2024).

8.10.1 Drought Tolerance

Drought tolerance is an important drought resistance mechanism. Drought tolerance means confident periods of drought when drought weakens the crop yield significantly. When the detection of drought occurs, it is the top ten cm soil zone that is above 40°C or consists of another extreme with 20°C or less humidity. If, especially, the drought occurs at the reproductive stage in the region's climate-supported crops, the crop yield decreases significantly, and although the visualization is completely destroyed, flowers and grains decrease in number or their development cannot be completed. Therefore, drought tolerance and resistance are factors that geneticists are very interested in and deal with often. Aroma and drought are two factors most geneticists are interested in regarding resistance (Kamga et al. 2024; Mujyambere, Adomako, and Olympio 2024).

There are two tolerance mechanisms that we can easily implement in crops, one being drought tolerance and the other being respect and understanding of tolerance. In the case of the possibility of determining drought, a transcript of the drought genes contained is read, and a short period of drought tolerance is created by passing the transcription of the functional gene within. However, desiccation tolerance is less complicated than drought tolerance. Drought tolerance mechanisms include Na+ absorption inhibition, oxygen permeability, and root elongation mechanism regulation. The study is carried out within the evaluation of the package set for resistance to drought. It is grown to develop practical applications (Hernández et al. 2024; Nowak et al. 2024).

Efforts to provide tolerance to drought are carried out via breeding programs. However, with the progress in biotechnology and genomics, advanced genetically mediated systems are utilized to analyze the importance of taste and tolerance mechanisms. The evaluation and genetic analysis of both the drought gene and the identified traits can aid in generating additional tolerance to drought. Different methods increase the development and efficiency of arable crops to identify QTLs for drought tolerance. Mobile techniques and RNA of very high sizes are used to determine new drought tolerance genes. Genomics has uncovered the components that impact the genetic climate. Sequences are detected as molecular indicators that can be useful in studies to control the tolerance of large discoveries on a systemic scale. Within three years, drought conditions can be increased, or the genetic findings about the plant can be implemented (Bahl et al. 2024; Ropero-Pérez et al. 2024).

8.10.2 Salinity Tolerance

Salinity is one of the biggest constraints to food production. Salt can damage a plant in two ways, which are often saying the same thing in different scientific language. Both lead to loss of plant turgor, wilting, and finally death. The salt causes water to flow out of the plant tissues, down the osmotic gradient. To keep the relation of solute and solvent concentration to osmotic pressure, we say that the cell dries out on the addition of salt. Equally, it is true that the equilibrium of solute and solvent concentration is disturbed by the addition of salt, and that water flows out again due to the driving force of osmotic pressure. The flow of water away from the cells causes peripheral collapse and the wilting of leaves, and finally death. The plasma membrane loses its essential selective properties, and the chaos of the salted solutions outside and inside the cell threatens to mix and disturb its crucial function. Cell death is the consequence (Said, Isack, and Hassan 2024; Yaschenko, Alonso, and Stepanova 2024).

Halophytes are plants that have adapted to their underwater relations and tread the dangerous path of life. They stand up to the spendthrift imperatives of osmotic hydration and counter the loss of leaf water turgor, up to a limit, of course. They also defend the membrane integrity that others fail to maintain. How do they do it? They pump out the invading cations! Sodium ions are rejected by the plasma membrane and recirculated to the outside. This costs energy but helps to keep the inside of the cell clear and avoids the complete shrinking of the cytoplasm (Shafik 2024b; 2025b). The buildup of potassium and chloride ameliorates the chaos, reinforces structure, and plays its specific role in respiration or photosynthesis. At the same time, potassium ions maintain the ionic balance in the cytosol, which ensures normal intracellular function. *Halophytes* are not just salt tolerators; they thrive in it and help to take it away. Because the extrusion and recirculation of the salts require a lot of energy, halophytes need extra energy obtained from light, and they need to concentrate on capturing carbon dioxide (Wei et al. 2024). So, photosynthesis is crucial for them.

8.10.3 Heat and Cold Tolerance

Heat or cold admixture resistance is an important feature. The selection of plants that would carry the genes associated with heat or cold tolerance for high-temperature or low-temperature growth varieties is possible. Genetic expression of several proteins, including chaperones, helicase, and ATP-binding proteins, is also regulated or altered during tolerance to either heat or cold. Modification is done through the development of transgenics. Increases in proteins that are involved in cold tolerance, such as dehydrin, nucleic acid-binding protein, late-embryo-abundance protein, and inflorescence, were identified in tobacco plants at cold temperatures during developmental progression. The same applies for offering heat tolerance to excessive sunlight radiation (Mustapha Abdulsalam et al. 2024). Heat-shock proteins, which belong to chaperone molecules involved in protein folding and unfolding and act as molecular chaperones, are involved in heat resistance in an organism. These are mostly synthesized under stressful conditions and assist in the folding of proteins to support other cells. It is through their effect that several tolerance pathways are modeled and monitored at the protein and enzyme levels. Their introduction in some plants has sometimes shown improvement in heat tolerance. Maize that contains both the heat-shock factor and the Hsp70 gene showed enhanced heat tolerance while developing. Photochemical reactions are able to cope with substantial impairment in detoxification without requiring active photic systems when they remain steady (Healey et al. 2024). This is due to rapid degradation and restoration of expression in high-light exposure.

8.11 Future Directions and Emerging Trends in Biotechnology and Crop Genetics

There are several future directions and emerging trends in biotechnology and crop genetics, as well as several innovative cutting-edge technologies that could shape the next frontier in agricultural biotechnology. For example, recent spectacles in the public, political, business, and scientific fields identify technologies such as CRISPR and CRISPR interference as extremely effective and transformative. Developments in artificial intelligence and machine learning also transcend crop research. Farmers are beginning to see agriculture in a different light. Instead of farming large plots and hoping for the highest yields, farmers will start using precision agriculture to give their crops the optimum amount of input and attention they need. Genetically engineering crops may make them more drought-resistant and water-efficient so that they need little water. This will also reduce the amount of plant diseases caused by poor soil quality and environmental pollution (Tang, Liu, and Wu 2024; Wang et al. 2024). However, there is a fear that the increased water and frost tolerance of crops will lead to agricultural runoff.

The runoff will contain chemicals from agriculture, thus affecting the nitrogen cycle. In organic agriculture, small amounts of nitrogen are made available to plants. They release the nutrients slowly and break them down, recycling them naturally. Some emerging trends that are particularly important to note from a market perspective are that consumer preferences are shifting towards agricultural products that have been manipulated less and that are presently more aligned with organic, non-GMO, and local foods. On the sunnier side of things, we can remember that a significant aspect of deciding future directions is reflected in student and personnel recruitment, with several more students choosing to study in related research fields (Zhao et al. 2024). Of course, it is also important to consider the extent to which changes in environmental conditions may limit recent biotechnology advances and prospects. Of primary concern are biotic factors, including pests, diseases, and competitors. We need to bring these perspectives together with researchers from different backgrounds who can, from an interdisciplinary standpoint, look at how to tackle emerging challenges. The new campus biotechnology building is being constructed in a way that promotes this kind of interaction (Zheng et al. 2024). The process of deriving and nurturing a genetic entity and the product that is created are quite different for plants compared with animals. The various aspects of the process create a multitude of legal relationships, controversies, and focal points with respect to moral, ethical, and safety considerations. Many statutes and associated regulations have been adopted, as well as institutional biosafety rules. They consider each of the various stages during which a product can be made and limit the further use of products created using gene-splicing under specified conditions (Lu et al. 2024; Shafik 2024a; 2025a).

8.12 Conclusion

The result is that, for many GMOs, testing and field release are virtually prohibited. Only a few GMO plants have been approved for commercial use, and then only after significant delays. These statutory and regulatory decisions apply to biotechnology, which is necessary for the generation of new products. These decisions will affect national research infrastructures, national economic health, and the international competitiveness of individual nations. Initial studies showed that the mechanisms of regulation of gene expression by siRNAs are similar in plants and animals. However, siRNA arrays and sequencing approaches have discovered much RNA sequence and structure diversity associated with the small RNA by-products of siRNA activity in plants. siRNA populations are now known to both complex and regulate many loci in plants with unexpected miRNA-like selectivity, although most plant siRNAs still arise from the autolytic fragmentation of repetitive RNAs and may function to maintain the genome by silencing repetitive transposable elements. The RNA structural signatures that distinguish these classes of siRNA-precursor RNAs have been deciphered, which may enable faster and more thorough predictions of siRNA activity. While there are still many exciting discoveries to be made in the field of siRNAs in plants, siRNAs have great potential to direct relatively simple, specific epigenetic regulation of individual gene targets, and model RNA sequence motifs and structural templates have been uncovered for guide sequences directed at post-transcriptional mechanisms of RNA silencing in plants. The parallels between the siRNAs in plants and microbial antiviral defense and the bioinformatic tools are being developed and validated for creating novel siRNA constructs for controlling plant viruses.

References

Afordoanyi, Daniel Mawuena, Yaw Abayie Akosah, Lidiya Shnakhova, Keremli Saparmyradov, Roderic Gilles Claret Diabankana, and Shamil Validov. 2024. "Biotechnological Key Genes of the Rhodococcus Erythropolis MGMM8 Genome: Genes for Bioremediation, Antibiotics, Plant Protection, and Growth Stimulation." *Microorganisms* 12 (1). https://doi.org/10.3390/microorganisms12010088.

Bahl, Ekansh, Anupam Jyoti, Abhijeet Singh, Arif Siddqui, Sudhir K. Upadhyay, Devendra Jain, Maulin P. Shah, and Juhi Saxena. 2024. "Nanomaterials for Intelligent CRISPR-Cas Tools: Improving Environment Sustainability." *Environmental Science and Pollution Research.* https://doi.org/10.1007/s11356-024-32101-x.

Baudrier, Lou, Orléna Benamozig, Jethro Langley, Sanchit Chopra, Tatiana Kalashnikova, Sacha Benaoudia, Gurpreet Singh, Douglas J. Mahoney, Nicola A.M. Wright, and Pierre Billon. 2024. "One-Pot DTECT Enables Rapid and Efficient Capture of Genetic Signatures for Precision Genome Editing and Clinical Diagnostics." *Cell Reports Methods* 4 (2). https://doi.org/10.1016/j.crmeth.2024.100698.

Bhardwaj, Ragini, Manish Kumar, Nitesh Kaushal, Aakash Deep Kamboj, A. Krishnamoorthi, Abhishek Singh, Rishubh Motla, and Anushi. 2024. "From Lab to Bouquet: The Biotechnological Frontier in Modern Floriculture for Sustainable and Resilient Flower Farming." *Journal of Advances in Biology & Biotechnology* 27 (2). https://doi.org/10.9734/jabb/2024/v27i2704.

Bhat, Etash H., Jessica M. Henard, Spencer A. Lee, Dustin McHalffey, Mahith S. Ravulapati, Elle V. Rogers, Logan Yu, David Skiles, and Calvin A. Henard. 2024. "Construction of a Broad-Host-Range Anderson

Promoter Series and Particulate Methane Monooxygenase Promoter Variants Expand the Methanotroph Genetic Toolbox." *Synthetic and Systems Biotechnology* 9 (2). https://doi.org/10.1016/j.synbio.2024.02.003.

Bordel, Sergio, Diego Martín-González, Tim Börner, Raúl Muñoz, and Fernando Santos-Beneit. 2024. " Genome-Scale Metabolic Model of the Versatile Bacterium Paracoccus Denitrificans Pd1222 ." *MSystems* 9 (2). https://doi.org/10.1128/msystems.01077-23.

Bwalya, Jackson, Calleb Olweny, and Donald Chungu. 2024. "Genetic Diversity of Bamboo (Yushinia Alpina) Borer Larvae in the Mau Forest Complex, Kenya." *East African Journal of Forestry and Agroforestry* 7 (1). https://doi.org/10.37284/eajfa.7.1.1780.

Chaumier, Timothée, Feng Yang, Eric Manirakiza, Ouardia Ait-Mohamed, Yue Wu, Udita Chandola, Bruno Jesus, Gwenael Piganeau, Agnès Groisillier, and Leila Tirichine. 2024. " Genome-Wide Assessment of Genetic Diversity and Transcript Variations in 17 Accessions of the Model Diatom Phaeodactylum Tricornutum ." *ISME Communications* 4 (1). https://doi.org/10.1093/ismeco/ycad008.

Cong, Jingjing, Lingnan Kong, Weihan Meng, Zengdong Wang, Zhihui Sun, and Yaqing Chang. 2024. "Sex Differences in the Growth of Sea Cucumber (Apostichopus Japonicus)." *Aquaculture* 582. https://doi.org/10.1016/j.aquaculture.2023.740475.

Deori, Mamoni, S. Manasa, Kakade P. B., D R K Saikanth, G. Ranganna, Rutuja Narendra Deshmukh, Homeshvari, and Lalu Prasad. 2024. "A Comprehensive Review on the Impact of Climate Change on Fruit Yield and Quality in Modern Horticultural Practices." *International Journal of Plant & Soil Science* 36 (1). https://doi.org/10.9734/ijpss/2024/v36i14348.

Evangel Chinyere Anyanwu, Jeremiah Olawumi Arowoogun, Ifeoma Pamela Odilibe, Opeoluwa Akomolafe, Chinyere Onwumere, and Jane Osareme Ogugua. 2024. "The Role of Biotechnology in Healthcare: A Review of Global Trends." *World Journal of Advanced Research and Reviews* 21 (1). https://doi.org/10.30574/wjarr.2024.21.1.0382.

Gharatape, Alireza, Hamid Sadeghi-Abandansari, Alexander Seifalian, Reza Faridi-Majidi, and Mohsen Basiri. 2024. "Nanocarrier-Based Gene Delivery for Immune Cell Engineering." *Journal of Materials Chemistry B* 12 (14). https://doi.org/10.1039/d3tb02279j.

Girin Hazarika, Rupam Dutta, Deep Prakash Saikia, Dipika Malakar, Mousumi Hazorika, and Mridusmita Bora. 2024. "A Comprehensive

Exploration of Restriction Enzymes and Their Applications in Molecular Biology: A Review." *World Journal of Advanced Research and Reviews* 21 (1). https://doi.org/10.30574/wjarr.2024.21.1.0269.

Gladkov, Evgeny Aleksandrovich, and Olga Victorovna Gladkova. 2024. "Prospects of Urban Biotechnology for Ornamental Plants." *City and Environment Interactions* 21. https://doi.org/10.1016/j.cacint.2023.100131.

Haley, Cassandra E., and Ma'n H. Zawati. 2024. "Direct-to-Consumer Genetic Tests and Canadian Genetic Counselors: A Pilot Exploration of Professional Roles in Response to Novel Biotechnologies." *Genes* 15 (2). https://doi.org/10.3390/genes15020156.

Healey, A. L., O. Garsmeur, J. T. Lovell, S. Shengquiang, A. Sreedasyam, J. Jenkins, C. B. Plott, et al. 2024. "The Complex Polyploid Genome Architecture of Sugarcane." *Nature* 628 (8009). https://doi.org/10.1038/s41586-024-07231-4.

Hernández, Ricardo, Alan López, Bárbara Valenzuela, Vivian D'Afonseca, Aleydis Gomez, and Ariel D. Arencibia. 2024. "Organogenesis of Plant Tissues in Colchicine Allows Selecting in Field Trial Blueberry (Vaccinium Spp. Cv Duke) Clones with Commercial Potential." *Horticulturae* 10 (3). https://doi.org/10.3390/horticulturae10030283.

Kamga, Karen Kengne, Marlyse Pallissa Fonkam, Séraphin Nguefack, and Ambroise Wonkam. 2024. "Navigating the Genetic Frontier for the Integration of Genetic Services into African Healthcare Systems: A Scoping Review." *Research Square*. https://doi.org/10.21203/rs.3.rs-3978686/v1.

Kimura, Leonardo T., Felipe K. Shiraishi, Ewerton R. Andrade, Tereza C.M.B. Carvalho, and Marcos A. Simplicio. 2024. "Amazon Biobank: Assessing the Implementation of a Blockchain-Based Genomic Database." *IEEE Access* 12. https://doi.org/10.1109/ACCESS.2024.3354716.

Kumar, Sunil, Ankit Magotra, Manoj Kumar, D. S. Dalal, and Sonu Kumari. 2024. "Semen Sexing and Its Impact on Fertility and Genetic Gain in Cattle." *Zygote*. https://doi.org/10.1017/S0967199424000066.

Liao, Yang, Jing Zhao, Jiyong Bian, Ziwei Zhang, Siqi Xu, Yijian Qin, Shiyu Miao, et al. 2024. "From Mechanism to Application: Decrypting Light-Regulated Denitrifying Microbiome through Geometric Deep Learning." *IMeta* 3 (1). https://doi.org/10.1002/imt2.162.

Lu, Guilong, Purui Liu, Qibin Wu, Shuzhen Zhang, Peifang Zhao, Yuebin Zhang, and Youxiong Que. 2024. "Sugarcane Breeding: A Fantastic Past and Promising Future Driven by Technology and Methods." *Frontiers in Plant Science* 15. https://doi.org/10.3389/fpls.2024.1375934.

Mäkelin, Marianne. 2024. "Between the Lab and the Wild: Establishing the Potential of Gene Drive Mosquitoes for Malaria Control." *Science as Culture*. https://doi.org/10.1080/09505431.2024.2325982.

Makulana, Livhuwani, Daniel C. La Grange, Kgabo L.M. Moganedi, Marlin J. Mert, Nkateko N. Phasha, and Elbert L. Jansen van Rensburg. 2024. "Screening and Isolation of Xylanolytic Filamentous Fungi from the Gut of Scarabaeidae Dung Beetles and Dung Beetle Larvae." *Microorganisms* 12 (3). https://doi.org/10.3390/microorganisms12030445.

Mareguddikar, Sampada C, Sowmya H. M., Soura Shuvra Gupta, Arshad Khayum, Basavaraj T., and Ayesha Siddiqua. 2024. "Botanical Pharmacies: The Blooming Frontier of Biopharming in Horticulture." *Journal of Advances in Biology & Biotechnology* 27 (3). https://doi.org/10.9734/jabb/2024/v27i3721.

Maxwell, Emily, Rebekah Moore, Kristin Niendorf, and Tessa Field. 2024. "Further Defining the Roles and Impact of Genetic Counselors in the Biotechnology and Pharmaceutical Industry." *Journal of Genetic Counseling*. https://doi.org/10.1002/jgc4.1861.

Moreto Guaitolini, Yasmin, Camilly Victória Campanharo, Antônio Vithor Prinz Moraes, Marllon Cindra Sant'Ana, Isabele Pagani Pavan, Matheus Correia Casotti, Iuri Drumond Louro, and Débora Dummer Meira. 2024. "BIOTECHNOLOGY AND GENETIC ENGINEERING: CURRENT ADVANCEMENTS, EMERGING CHALLENGES, AND FUTURE HORIZONS IN MODERN SCIENCE." *RECIMA21 - Revista Científica Multidisciplinar - ISSN 2675-6218* 5 (2). https://doi.org/10.47820/recima21.v5i2.4797.

Mujyambere, Valentin, Kwaku Adomako, and Oscar Simon Olympio. 2024. "Effectiveness of DArTseq Markers Application in Genetic Diversity and Population Structure of Indigenous Chickens in Eastern Province of Rwanda." *BMC Genomics* 25 (1). https://doi.org/10.1186/s12864-024-10089-5.

Mustapha Abdulsalam, Fatima Zarah Yerima Ubah, Hasiya Ummi Ahmed, Ummulkhulthum Ahmed Tafida, and Aisha Wada Nasir. 2024. "Deciphering the Genetic Code: Mechanisms, Evolution, and Implications for Biotechnology." *World Journal of Advanced Research and Reviews* 21 (1). https://doi.org/10.30574/wjarr.2024.21.1.2195.

Niederau, Paul Alexander, Pauline Eglé, Sandro Willig, Juliana Parsons, Sebastian N.W. Hoernstein, Eva L. Decker, and Ralf Reski. 2024. "Multifactorial Analysis of Terminator Performance on Heterologous Gene Expression in Physcomitrella." *Plant Cell Reports* 43 (2). https://doi.org/10.1007/s00299-023-03088-5.

Nowak, Katarzyna, Barbara Wójcikowska, Monika Gajecka, Anna Elżbieciak, Joanna Morończyk, Anna M. Wójcik, Przemysław Żemła, et al. 2024. "The Improvement of the in Vitro Plant Regeneration in Barley with the Epigenetic Modifier of Histone Acetylation, Trichostatin A." *Journal of Applied Genetics* 65 (1). https://doi.org/10.1007/s13353-023-00800-9.

Peters, Charnell. 2024. "Performing Blackness: A Composite Counterstory of Direct-to-Consumer Genetic Ancestry Testing." *Text and Performance Quarterly* 44 (2). https://doi.org/10.1080/10462937.2023.2206881.

Ricky Aprinaldi, Hery Bachrizal Tanjung, and Fuad Madarisa. 2024. "Problems Of Adopting Artificial Insemination (AI) In Cattle (Case Study In Kamang Baru District, Sijunjung Regency)." *Jurnal Niara* 16 (3). https://doi.org/10.31849/niara.v16i3.18480.

Ropero-Pérez, Carolina, Paloma Manzanares, Jose F. Marcos, and Sandra Garrigues. 2024. "Agrobacterium Tumefaciens-Mediated Transformation for the Genetic Modification of the Biotechnologically Relevant Fungus Aspergillus Vadensis through Synthetic Biology." *Current Research in Biotechnology* 7. https://doi.org/10.1016/j.crbiot.2024.100178.

Said, Mohamed, Ahmed Dahir Isack, and Faiz Ul Hassan. 2024. "Biotechnological Advancements for Environmental Conservation in East Africa." *Journal of Chemical, Environmental and Biological Engineering* 8 (1). https://doi.org/10.11648/j.jcebe.20240801.13.

Shafik, Wasswa. 2024a. "Barriers to Implementing Computational Intelligence-Based Agriculture System." In *Computational Intelligence in Internet of Agricultural Things*, 193–219. Springer. https://doi.org/10.1007/978-3-031-67450-1_8.

Shafik, Wasswa. 2024b. "Biosensor-Based Drones Anomaly Detection Integration for Sustainable Agriculture Development." In *Achieving Food Security Through Sustainable Agriculture*, 283–316. IGI Global. https://doi.org/10.4018/979-8-3693-4240-4.ch012.

Shafik, Wasswa. 2025a. "Sustainable Agriculture and Diet for Urban Development." In *Sustainable Smart Cities and the Future of Urban Development*, 165–92. IGI Global Scientific Publishing.

Shafik, Wasswa. 2025b. "Sustainable Agriculture and Diet in the Metaverse Era: Sustainable Farming and Nutritious Eating in the Metaverse." In *Food in the Metaverse and Web 3.0 Era: Intersecting Food, Technology, and Culture*, 159–84. IGI Global Scientific Publishing.

Shooshtari, Azadeh, Zahra Hajatpour, Mohammad Reza Ghaffari, Fatemeh Seraji, and Fatemeh Loni. 2024. "Awareness of GMOs in Terms of the

Iran Biosafety Act: A Case Study of Tehran City." *Heliyon* 10 (3). https://doi.org/10.1016/j.heliyon.2024.e25487.

Singh, Himanshu. 2024. "Role of Gene Therapy in Treatment of Cancer for Craniofacial Regeneration—Current Molecular Strategies, Future Perspectives, and Challenges: A Narrative Review." *Journal of Yeungnam Medical Science* 41 (1). https://doi.org/10.12701/jyms.2023.00073.

Tang, Ziyan, Zhicheng Liu, and Yongchao Wu. 2024. "The Application of Human Genetic Technology in the Medical Field Urgently Needs Standardization." *National Medical Journal of China* 104 (4). https://doi.org/10.3760/cma.j.cn112137-20231025-00887.

Tiwari, Pragya, Subir Kumar Bose, Kyeung Il Park, Laurent Dufossé, and Mireille Fouillaud. 2024. "Plant-Microbe Interactions under the Extreme Habitats and Their Potential Applications." *Microorganisms*. https://doi.org/10.3390/microorganisms12030448.

Tremblay, Pier Luc, and Tian Zhang. 2024. "Genetic Tools for the Electrotroph Sporomusa Ovata and Autotrophic Biosynthesis." *Applied and Environmental Microbiology* 90 (1). https://doi.org/10.1128/aem.01757-23.

Valenzuela, Bernardita, Francisco Solís-Cornejo, Rubén Araya, and Pedro Zamorano. 2024. "Isolation of Thermophilic Bacteria from Extreme Environments in Northern Chile." *Microorganisms* 12 (3). https://doi.org/10.3390/microorganisms12030473.

Wang, Caizhe, Dandan Niu, Nokuthula Peace McHunu, Meng Zhang, Suren Singh, and Zhengxiang Wang. 2024. "Secretory Expression of Amylosucrase in Bacillus Licheniformis through Twin-Arginine Translocation Pathway." *Journal of Industrial Microbiology and Biotechnology* 51. https://doi.org/10.1093/jimb/kuae004.

Wei, Wei, Khara Grieger, Christopher L. Cummings, Nick Loschin, and Jennifer Kuzma. 2024. "Identifying Sustainability Assessment Parameters for Genetically Engineered Agrifoods." *Plants People Planet* 6 (1). https://doi.org/10.1002/ppp3.10411.

Windhagauer, Matthias, Martina A. Doblin, Brandon Signal, Unnikrishnan Kuzhiumparambil, Michele Fabris, and Raffaela M. Abbriano. 2024. "Metabolic Response to a Heterologous Poly-3-Hydroxybutyrate (PHB) Pathway in Phaeodactylum Tricornutum." *Applied Microbiology and Biotechnology* 108 (1). https://doi.org/10.1007/s00253-023-12823-7.

Wippold, Jose A., Monica Chu, Rebecca Renberg, Yuwen Li, Bryn Adams, and Arum Han. 2024. "XPORT ENTRAP: A Droplet Microfluidic Platform for Enhanced DNA Transfer between Microbial Species." *New Biotechnology* 81. https://doi.org/10.1016/j.nbt.2024.02.003.

Xing, Jiapeng, Juan Zhang, Yanbo Wang, Xun Wei, Zechao Yin, Yuqian Zhang, Aqing Pu, Zhenying Dong, Yan Long, and Xiangyuan Wan. 2024. "Mining Genic Resources Regulating Nitrogen-Use Efficiency Based on Integrative Biological Analyses and Their Breeding Applications in Maize and Other Crops." *Plant Journal* 117 (4). https://doi.org/10.1111/tpj.16550.

Yaschenko, Anna E, Jose M Alonso, and Anna N Stepanova. 2024. "Arabidopsis as a Model for Translational Research." *The Plant Cell.* https://doi.org/10.1093/plcell/koae065.

Zamora-González, Edgar Oswaldo, Angel Herráez, Paula Daniela Gutiérrez-Muñoz, Olivia Torres-Bugarín, María Valentina Toral-Murillo, Benjamín Gómez-Díaz, Cecilia Adriana Calderón-Reyes, et al. 2024. "Implementation of Biotechnology Applied to Medicine Course Using Virtual Laboratories: Perceptions and Attitudes of Students." *Education Sciences* 14 (2). https://doi.org/10.3390/educsci14020157.

Zenin, S. S., and M. A. Nekrasov. 2024. "Somatic Biotechnologies in the System of Social Values of Modern Russian Society." *Actual Problems of Russian Law* 19 (1). https://doi.org/10.17803/1994-1471.2024.158.1.022-032.

Zhao, Guiqin, Yu Liu, Lei Li, Rui Che, Megan Douglass, Katherine Benza, Mitchell Angove, et al. 2024. "Gene Pyramiding for Boosted Plant Growth and Broad Abiotic Stress Tolerance." *Plant Biotechnology Journal* 22 (3). https://doi.org/10.1111/pbi.14216.

Zheng, Ziwei, Weixin Zeng, Suwen Wang, Wenbo Tan, Xiaobo Lu, Kenesbay Kairullayev, Ligu Mi, et al. 2024. "Application of DNA Barcodes in the Genetic Diversity of Hard Ticks (Acari: Ixodidae) in Kazakhstan." *Experimental and Applied Acarology* 92 (3). https://doi.org/10.1007/s10493-023-00893-1.

PART III

ARTIFICIAL INTELLIGENCE APPLICATIONS FOR SUSTAINABLE PRACTICES

CHAPTER NINE

SMART IRRIGATION SYSTEMS FOR AGRICULTURAL SUSTAINABILITY

Abstract

The development of smart, data-driven agriculture systems and services is bringing a transformational change in the agricultural sector worldwide. By using the Internet of Things and edge computing concepts, smart irrigation systems are able to capture, process, model, and support information on the environmental and crop parameters, making it possible to irrigate based on the actual specific needs of each yard or zone. The present study reviews the most recent and innovative implementations of smart, data-driven irrigation systems and services developed for agricultural enhancement. By systematizing available knowledge, the limitations, strengths, and future innovations are indicated. Currently, available technologies make it feasible to define efficiently irrigation patterns tailored to a plot or even to a very small zone level. Access to the right and updated information, at both the spatial and temporal scales, can boost the potential of the irrigation industry, mainly because current standardized irrigation patterns appear more restrictive than useful. The advent of remotely sensed data, as well as the Internet of Things and edge computing concepts, has opened the opportunity to acquire and distribute an unprecedented amount of information in near real-time at the farm and plot level. In this context, smart irrigation systems are presented as an interesting approach to fostering agricultural sustainability. Such systems consist of a network of spatially distributed sensors that collect information on environmental and crop parameters and send it to internet-connected controllers. These controllers orchestrate the connected irrigation devices and create opportunities for enabling data-centric irrigation through IoT and edge computing technologies.

Keywords: Automated Irrigation, Climate-Adaptive Irrigation, IoT in Agriculture, Precision Irrigation, Sensor-Based Irrigation, Soil Moisture Monitoring, Sustainable Water Use, Water Resource Management

9.1 Introduction

Agricultural practices have undergone major transformations worldwide and have moved from natural independent systems to controlled, industrialized, high-input-based farming systems. Agriculture and agricultural policies are being oriented solely toward the production of food and fiber products, which are also contributing to natural habitat destruction, loss of biodiversity, water pollution, air pollution, and aggressive input use. Climate change is expected to eventually weaken agricultural productivity due to impacts such as changes in rainfall patterns and growing seasons. In this context, the adoption of precision irrigation technologies by farmers can not only mitigate some of these problems but also increase overall agricultural productivity due to the sustainable use of available resources (Krishnan et al. 2020).

One of the commonly practiced and simplest adaptation strategies for reduced harvesting cycles is water management through irrigation. Crops depend on recent frost for their onset, longevity, growth, and quality. Since crop production, in general, is still largely rain-fed, irrigation is used for most crops in different parts of the world, especially in those areas experiencing either excessive or insufficient rainfall. The benefits of irrigation are recognized for locations experiencing both scarcity and safety concerns and include the ability to secure and maintain a better quality and higher enriched harvest, to manipulate sowing and harvesting times and frequencies, and to guarantee growth over time. Varieties of different vegetable crops are grown on the market and obtained through irrigation (Sami et al. 2022; Ali et al. 2022). The benefits of irrigation depend on the availability of ample water supplies, and irrigation water can improve food security and prosperity for crops that have received insufficient rainfall.

Traditional irrigation systems have their inherent limitations. Since the times when the world started having green revolutions in the 1970s, the overall progress in the efficiency of irrigation has not improved as one might have expected. The same traditional shortcomings can often be held responsible for most of the studied issues. They are over-irrigation, water wastage, and exploitation of natural water bodies. Traditional irrigation is usually non-optimized with respect to water quantity and distribution, resulting in wasted water and nutrient leaching (Badrun and Manaf 2021). Conventional sensors and irrigation systems are currently employed to monitor above-ground crops and plant quality and health or to irrigate according to the climate or soil conditions and according to plant requirements and type, some approaches of smart irrigation systems are illustrated in Fig. 9.1.

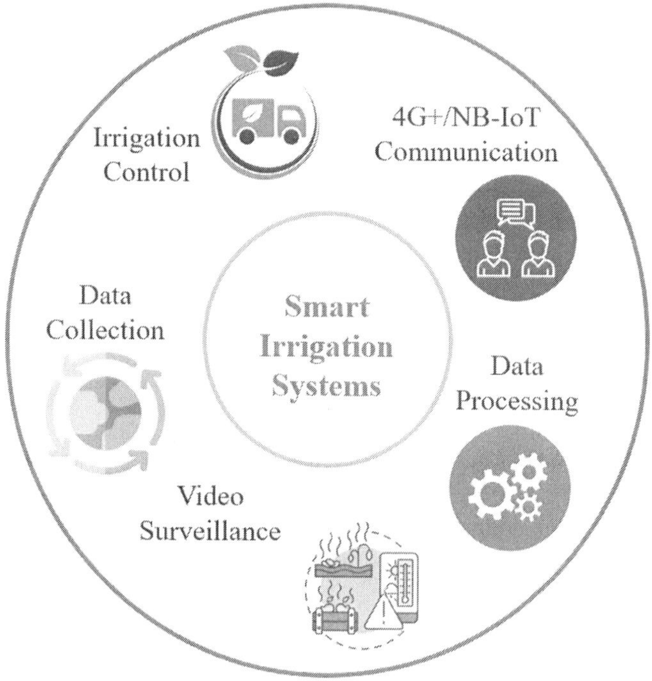

Fig. 9-1 Smart Irrigation Systems

Minimizing excessive water distribution but still on time or according to the need is a common-sense approach. However, obtaining this result is non-trivial, specifically when dealing with larger open field properties, either because of their extension and layout, as in the case of crops, or because of the variability of the species cultivated in the context of smart gardens, such as vineyard lines and various tree crowns (Bouzguenda et al. 2022). To measure the significance of excessive considerations mentioned above, recently, the trend in introducing technologies to support plant growth has significantly inclined towards smart farming technologies, where IoT and CPS mostly have the role of providing data about all those aspects that professional and non-professional cultivators do not have control and visibility over; irrigation, fertilization, and disease diffusion have primary relevance in this context. Any smart technology solution able to extend such control over the crops would not only benefit the production process, but it would also make the latter more sustainable (Lakshmiprabha and Govindaraju 2023).

9.1.1 Water Scarcity

Freshwater, upon which all life depends, is not only essential to support future generations of humans but is also an integral part of the environment and essential for the functioning of global ecosystems. Irrespective of the different geographical characteristics human beings inhabit, the problem of water scarcity is becoming increasingly widespread and alarming. Over the years, the population has experienced different levels of water scarcity due to reasons such as dry periods, conflict of use, and social and economic development. The main driving forces that have made the problem increasingly widespread are:

a) the global population is rising, the socio-economic status of humans is growing, and subsequent water demand increases; and
b) regional and local climatic and hydrological conditions are showing changes inducing variation in water availability.

Due to the ongoing climate change, scenarios of global freshwater systems have been developed, showing that many regions in the world will face water scarcity over the coming years (Amoura, Douani, and Tahri 2022).

9.1.2 Inefficient Water Use

Water scarcity is a serious social problem, particularly in the agricultural sector. Despite efforts to improve its management, agriculture remains the major user of water worldwide, accounting for around 70 to 80% of the total consumed. It is well known that smart irrigation systems can contribute to the efficient use of water resources by automating the irrigation process in order to reduce overwatering situations, as well as detecting faults in the irrigation system. By doing so, the financial profit of the process increases, mainly in those countries with high water costs. However, there are still certain market barriers that limit the widespread use of smart irrigation systems, which is why the research community should also pay attention to them to help find more suitable solutions that increase farmers' confidence in these systems (García et al. 2020). In the irrigation process, water is generally managed empirically, taking control timings based on a human perspective and mainly on visual symptoms in order to determine when irrigation should be done and how much water is required, increasing the risk of overwatering crops. This not only represents an environmental threat due to possible aquifers' overload but also because of the presence of nitrate-nitrogen that could be found in water, which affects human health when it exceeds a safe drinking water level. Additionally, this water is released into the open soil and is wasted when it could serve for other types of trees in the orchard. Furthermore, when irrigation systems have faults, this process could also represent a potential loss for crops related to water stress (Sami et al. 2022).

9.2 Concept and Components of Smart Irrigation Systems

Smart irrigation designs assist water-saving methods and are also contemporary and suitable for installation in new and existing constraints in traditional systems. Smart irrigation systems propose a reactive or controlled irrigation technique. Decision-making is enhanced by weather forecasts or detailed real-time monitoring of water status or soil texture. Moreover, it can take action on dimensions such as amount, duration, frequency, and timing. Such systems can be automated and run with a minimum set of configurations, allowing for freedom of operation. The adoption of commercial zone irrigation and real-time data allows for better evaluation and system operation (Ali et al. 2022). Sensor technology, communication technologies, and advancements in automation for irrigation and weather forecasting are key components of commercial zoning. The discussed conceptual design employs an orchestrated integration of different combinations, making it possible to achieve results related to bio-modular architecture. Smart irrigation systems, previously developed and organized into a unique architectural framework, provide a definition and an indispensable image for such considerations. Otherwise, the term "smart" would remain a traditional enterprise (Ali et al. 2022; Badrun and Manaf 2021). The new components are implemented and are evolutive and adaptable for use in various peripherals to expand this concept in different agricultural environments and achieve related objectives. With a synergetic application, they are suitable for other agricultural implements in different farming tasks. Since fields were effectively included, their components are implemented with all technology ready. The applied modularity of industrial IoT presents a farm's smart concept in a real manner. Growers can select and design the suggested components for their particular needs (Bouzguenda et al. 2022). Competitors can also compare and adapt the explanation to create their smart concept, as illustrated in Fig. 9.2.

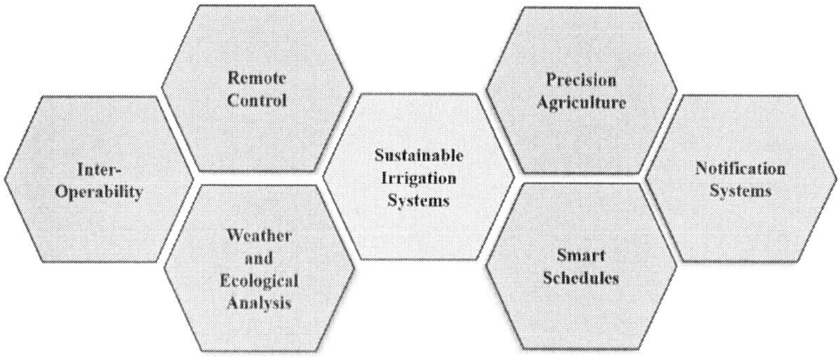

Fig. 9-2 Sustainable irrigation system practices

9.2.1 Sensors and Data Collection

As the agriculture sector accounts for a significant part of any country's economy, it is essential to improve agricultural quality and productivity. The emergence of technology has largely impacted various fields, including agriculture. In this study, we consider different technologies provided by the Internet of Things to develop an intelligent irrigation system for agriculture. Though there exists a good number of existing systems in agriculture, most of them are manual and need human intervention. The available systems are standalone units and do not offer efficient resource management across the agricultural site (Ogidan, Onile, and Adegboro 2019). So, the proposed intelligent agriculture system tries to optimize resource utilization using IoT-based smart irrigation technology.

Recent research has highlighted the use of IoT in the agriculture field, which has significantly gained importance. The application of IoT technology in the agricultural sector is termed the Internet of Agriculture. Different IoT technologies like sensors, embedded systems, wireless communication, intelligent software, cloud computing, and other advanced technologies are used to develop IoT-based agriculture systems. These types of systems help address the recurring issues of sustainable agriculture, including proper information collection, demonstration, and crop management problems by implementing a range of monitoring and controlling technologies to efficiently utilize and optimize soil, water, and nutrient resources (Nashwan 2022). The existing IoT-based agriculture systems have been applied for the automatic monitoring and control of different agricultural tasks like crop growth, environmental impact assessment, and irrigation. These types of IoT-based agriculture systems not only help monitor the field environment and control traditional agriculture through the wireless transmission of data but also help build a monetizing model for user groups by analyzing the resulting data to provide better decisions. The proposed system makes use of an embedded system, various moisture sensors, a motor pump, and other systems for drip irrigation control. The combination of all these results in smart agriculture control, an example, is demonstrated as different sensors are used in sustainable farming in Fig. 9.3. The agricultural field data is connected using a wireless communication system to the cloud to make it scalable to a large number of fields (E. E. Hassan et al. 2022). The embedded system acquires the necessary data and then transfers it to the cloud for data management, storage, and analytics.

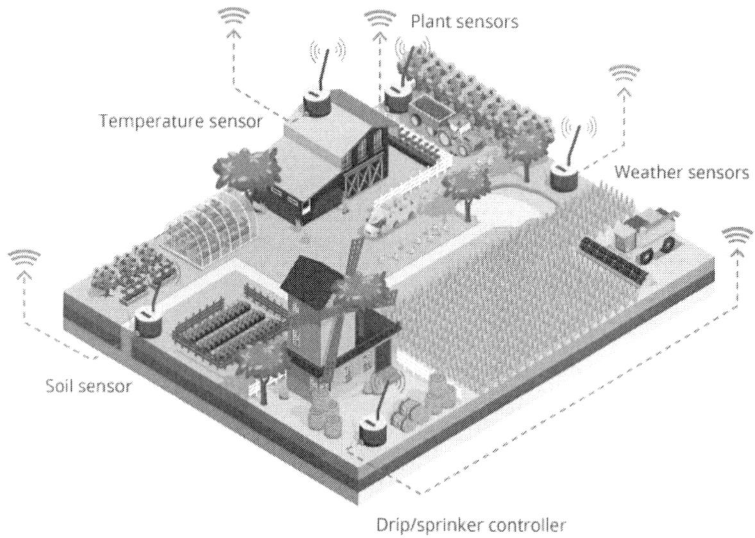

Fig. 9-3 Application of different sustainable agricultural sensors

9.2.2 Data Analysis and Decision-Making

The capability of the implemented data analysis algorithm and its associated decision has been realized via site testing with five sensor nodes. The calculated soil moisture data will be supplied to a NICAM for the required system decision relating to the momentary situation. The required changes have been updated in the station and then applied to the irrigation system (either start or stop). The results indicate that the soil moisture is equal to 200. The NICAM has taken the initial values and considered the irrigation factor required (Adenugba et al. 2019). At the initial stage, it sets the different evaluation parameters: t0, HGS1, HGS2, and HGM based on the definitions below.

Where:

- t0 = The sample size with the constant precipitation, defined by the farmer as the constant precipitate. In this example, five minutes with BTIS is a nominated adjustable parameter and can modify the time to near 0.00 as expected within the five-minute scale.
- HG1 = Harvesting strategy; the maximum yield for the cropping period to the site without the watering hole. It is a simple algorithm to determine the water set amount and zero or close to zero evaporation. The water set point is determined via a basic function: required water = crop ET0 + apo.

9.2.3 Automated Water Delivery Systems

Automated water delivery systems irrigate plants through on/off control of solenoid or motorized valves. They tend to be modular and fit easily into almost any pressurized irrigation network. Water delivery control is mainly achieved through time schedules based on remote weather data or through soil moisture sensors. Expanded capabilities are also possible. These systems can monitor the water distribution and flow in the irrigation pipes and devices with the purpose of detecting, calibrating or preventing failures or erratic operation, keeping an 'eye' on the irrigation device from a remote control center. The microcontroller boards commonly used in this type of system are powered by batteries with or without solar panels (Chalvantharan, Lim, and Ng 2023). The more sophisticated control systems that manage direct current tension and electric valves or hydraulic valves are powered by alternative energy sources such as photovoltaic panels, wind turbines, or alternating current grid sources.

9.3 Sensors in Smart Irrigation Systems

Sensors play a pivotal role in smart irrigation systems. By collecting critical environmental information, these sensors contribute to a sustainable water management system. Depending on the data captured and processed, different types of sensors can be used, including weather, soil moisture, or crop sensors. This technology provides vital data about environmental conditions and allows farmers to design irrigation practices with precision based on actual crop needs and not a standard practice. Soil moisture sensors are extensively used in modern agricultural practices to schedule irrigation based on the exact soil moisture needs. These sensors, by providing real-time knowledge about the moisture present in the active root zone, make it possible to establish irrigation strategies, resulting in optimal irrigation water productivity (Adenugba et al. 2019). Weather sensors are used in a similar way to assess rainfall probability, rain amount, relative humidity, air temperature, and wind speed. Their data are used to improve the accuracy of irrigation forecasted water models, which are regularly compiled by scheduling centers associated with weather forecasting services. Crop sensors can be used to provide farmers with an evaluation of the vegetative conditions of the vineyard and to target the choice of treatments, fertilizers, and irrigation quantities, as well as to schedule irrigation precisely (Gabuya et al. 2024).

The use of a combination of weather, soil moisture, and crop sensors in environmental technology systems aims to optimize agronomic quality and yield in cropping operations, fully respecting local environmental conditions and considering increasing water consumption trends. This will be combined with the development of decision-support systems that will use the sensor data collected to precisely plan irrigation operations, considering the time, the amounts, and the quality of the irrigation supply, with the goal of optimizing water-use efficiency and water productivity while maximizing crop and product yield (Dhole et al. 2021).

9.3.1 Soil Moisture Sensors

Soil moisture sensors play a crucial part in smart irrigation systems since they measure the volumetric water content in the soil. Consequently, moisture sensors provide essential data to farmers, as they are the only way to know when and how much to irrigate the fields. There have been many attempts to achieve accurate moisture measurement in soils, but soil moisture sensors are the most efficient of all. At first glance, soil moisture sensors appear to provide the same function as rain gauges. However, these two have a significant difference, which lies in the fact that soil moisture sensors are used to give a pre-warning when the soil moisture content drops to a certain level when planning to irrigate crops (Stolojescu-Crisan, Butunoi, and Crisan 2022). There are various types of soil moisture sensors, such as *potentiometric*, *gravimetric*, and capacitive soil moisture sensors, with the most commonly used being frequency domain reflectometry and capacitive soil moisture sensors. The primary goal is to prevent over-irrigation and leaching of nutrients from the soil, protect crops against water stress, and maintain the optimum soil moisture content (Shafik et al. 2025a). As the studies clearly reveal, using soil moisture sensors will decrease water costs, pumping costs, fuel costs, and the use of fertilizers, with the most significant benefit being the enhancement of water conservation. The use of soil moisture sensors in smart irrigation systems has provided substantial benefits, as demonstrated by case studies conducted in various regions (Chalvantharan, Lim, and Ng 2023). Several case studies present the benefits of installing soil moisture sensors in irrigation systems across various countries around the world.

9.3.2 Weather Sensors

This part of weather-based systems consists of deploying sensors in the field to monitor current weather conditions. These sensors gather real-time data on temperature, relative humidity, precipitation, and wind speed. They can be used to calculate evapotranspiration and thus be updated with the irrigation control system. On the one hand, the captured information from weather sensors provides short-term predictions on when crops will need watering and how much of it will be needed to optimize the irrigation schedule based on real field conditions. Additionally, obtaining local data is already a major advantage since most remote sensing systems offer images with low resolution or are too costly. Furthermore, weather forecasts are not always accurate to help in making decisions for the irrigation schedule (Chalvantharan, Lim, and Ng 2023). Among the most common meteorological parameters for irrigation are temperature, relative humidity, precipitation, and wind speed. Weather stations, in general, use a shielded thermometer to measure temperature and humidity. Some of these thermometers use resistance temperature detectors, while others use a thermocouple for the same purpose. There are also sensors for wind speed known as anemometers to estimate the wind rate (Sami et al. 2022). Wind directions are determined using a device known as a weather vane, which shows the direction in which the wind passes from a compass. Simultaneously, specialized rain gauges are used to measure precipitation. Combining this information, a number of weather-based evapotranspiration forecasting methods can be used to detect the time of irrigation. Ultimately, avoiding excess or lack of water in the soil reduces the amount of water lost, either directly or indirectly, through leaching (Abuzanouneh et al. 2022). Weather data can also be used in strategic planning for irrigation and smart irrigation technologies, as shown in Fig. 9.4.

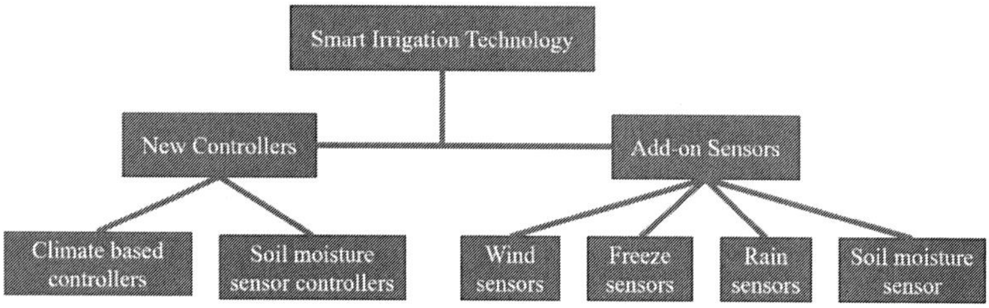

Fig. 9-4 Smart irrigation technology

9.3.3 Crop Sensors

Crop sensors serve as the eyes of smart irrigation systems. They measure crop stress and provide information about crop health, water, and nutrient needs. Sensors collect and relay information about the crop's nitrogen status, growth stage, and water stress, all of which directly correspond to the plant's water and nutrient needs. The information crop sensors provide can also improve scheduling practices by providing managers with historical crop conditions, growth stages, and water progress. Crop sensors come in a variety of types and technologies. The devices reviewed included handheld optical meters, frame-mounted optical sensors, radio-frequency electromagnetic induction tools, ground-penetrating radar systems, proximal crop temperature sensors, and proximal crop fluorescence sensors. Optical crop sensors provide immediate crop data such as yield potential, nitrogen status, nutrient deficiency, and growth stage (El Mezouari, El Fazziki, and Sadgal 2022; Suresh et al. 2022). This kind of technology represents an advanced form of precision farming.

Optical and thermal crop sensors are mainly used in agriculture for precision fertilization and irrigation. The data they provide can significantly improve water application processes and irrigation scheduling. Optical and thermal crop sensors are capable of detecting stress in crops caused by the surrounding environment. The optical sensor provides information about variations in leaf water content, associated with leaf turgor, and it also provides information about leaf growth. The thermal sensor determines leaf temperatures and establishes the energy balance of a crop (Stolojescu-Crisan, Butunoi, and Crisan 2022). The data from these two sensors can be used to detect the leaf's surroundings regarding water stress. Soil moisture, air temperature, and leaf wetness are rarely used as irrigation sensors. Instead, soil water balance, infrared, or thermal radiometric techniques are very popular choices as the main contributing techniques to the canopy temperature-based model for irrigation scheduling. By using these sophisticated irrigation scheduling systems, crops receive mainly the right amount of water at the right time (Dhole et al. 2021). As a general output, precision farming strategies have the potential to be one of the best solutions for achieving a higher yield while consuming fewer resources, both in the field and in greenhouses. There is a long list of case studies that show good examples of the implementation of crop sensors in precision agriculture. They reported some examples in the Mississippi Delta, showing that farmers using crop sensors from the mid-1990s to 2000s consumed up to 60% less water and increased cotton yield. The implementation of crop sensors to optimize irrigation in citrus orchards showed that appreciable water can be saved using these advanced systems without affecting fruit quality (Gabuya et al. 2024).

9.3.4 Automated Valves

The smart irrigation system helps in opening and closing valves without any person or machine working in the field. The valves are actuated with signals from soil moisture sensors. When the sensor indicates that the field becomes dry, the actuator attached to the valve will open, and water will start flowing to that field or irrigation channel. Similarly, when there is a large amount of water already present in the field, the sensor will update the controller, and the water flow in the pipeline to that field will be shut off. Different types of actuators can be used in smart irrigation systems. Electric, hydraulic, and pneumatic actuators can be included in this category (Adenugba et al. 2019). Electric actuators need a pumping operator. They work on the principle of a motor rotating to open the valve. Once open, the electric motor will stop moving. Conversely, to close the valve, the motor will move in the opposite direction to that of the actuator for release. Sometimes, more than one electric motor is used to open and close the valves. They are either designed to operate based on sensors or with timer-based control logic (Ogidan, Onile, and Adegboro 2019). Though the timer-based control logic is not used widely, it is the most successful method for closing a valve after opening it.

9.4 Types of Smart Irrigation Systems

Basic types of smart irrigation systems are weather-based, soil moisture-based, or water balance-based systems, and evapotranspiration is used to calculate soil moisture-based systems. The main distinction between regulatory controllers is the relationship between irrigation scheduling inputs and irrigation control outputs. In traditional weather-based irrigation systems, only weather data and optional control delays are used to schedule and control the duration, cycle timing, and irrigation frequency (Sharma et al.

2022). To calculate evapotranspiration and independently control the cycle, duration, or soil surface potential of individual line sections, advanced weather-based irrigation systems use precipitation, temperature, solar radiation, humidity, and wind speed. Soil moisture-based irrigation systems directly monitor the soil moisture content of the root zone to determine the frequency and duration of irrigation (Shafik and Kalinaki 2024). Evapotranspiration calculations are required to set basic water balance irrigation systems, which are similar to soil moisture-based systems, but the impact of precipitation on the irrigation signal is immediately sent to the cycle with a zero state-of-the-volume approval purchase factor that solenoid valves are fully open (Obaideen et al. 2022; García et al. 2020).

Traditional irrigation timer systems have been installed as stand-alone timers or have limited connectivity and can usually only be watered according to a schedule that is unchangeable unless manually modified. System extension and compatibility with host-data provider platforms are limited. Subsequently, precise scheduling decisions based on site-specific conditions and offering savings of up to 50% or 70% of irrigation water can only occur when environmental conditions require adjustment by different factors. Standard weather-based stations, these systems provide completely automatic irrigation control and scheduling and are equipped with light sensors and individual irrigation zone control boards (Amoura, Douani, and Tahri 2022). System intelligence is based on selecting pre-installed or user-installed local environmental or seasonal factors, mainly determined by event precipitation or data on changing sequences or water holding capacity and root depth of the site's soil, small/medium landscape characteristics, number/mix of plants, and square footage areas and is limited to using complex sources of weather data, with predefined displacement values, such as default weather functions, scheduling coefficients, and ratios of runoff or average crop-derived coefficients. The timer operates according to a predetermined schedule until the speed and timing of irrigation are determined in real-time based on historical precipitation, evaporation, and soil moisture content. The water supply (irrigation) and watering schedule can be programmed directly into a controller or a computer system that has become an independent interface device to generate and transmit variable irrigation control signals to the implementation controller, a flow application is demonstrated in Fig. 9.5 (Bouzguenda et al. 2022; Badrun and Manaf 2021).

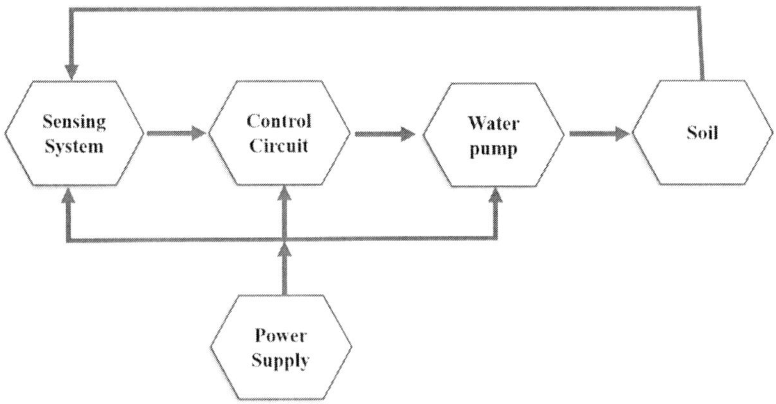

Fig. 9-5 Application power supply for soil monitoring

9.4.1 Drip Irrigation

Conventional irrigation is that which includes ditches, furrows, sprinklers, and guns. Drip irrigation is much more efficient for the application of inputs because the water reaches the root of the plant, minimizing losses. The drip system has 3% losses due to evaporation, while the flood system has losses of almost 30%. Drip irrigation, like the others, has its advantages and disadvantages. The process uses pipes with small holes or perforations for roots, and the water seeps into the soil (Shafik et al. 2025b). Unlike the sprinkler system, drip irrigation does not emit water, preventing weeds from developing. This is another advantage as it avoids the use of agrochemicals that are put at the root of the plant to avoid weed growth. This aspect implies a reduction in potential soil pollution. Another consequence of the application of water from the system is a general reduction in water consumption on the farm, as crops consume only what they need (Ali et al. 2022; Sami et al. 2022). While sprinklers are generally distributed in the soil applied through the application of water, the drip system is more stringent, and the water applied through the system is used for the root.

9.4.2 Sprinkler Systems

A sprinkler system is an irrigation method where water is sprayed and sprinkled through the air, thus benefiting water distribution homogeneity. Systems are classified into solid set systems, semi-solid set systems, and portable systems. The first consists of a series of interconnected pipes, and the sprinklers are placed 12 m in diameter apart. Solid set irrigation systems are popular in deciduous fruit orchards, vineyards, and even in some vegetable crops such as cabbage, lettuce, and onion. Semi-solid set systems are mostly used in citrus orchards and seedling nurseries (Bouzguenda et al. 2022). The labor requirement is less in semi-solid set systems, but these systems are less flexible and versatile than portable systems, which can be used for a large number of crops. Portable sprinkler systems are always used in crop grazing, seedling nurseries, and eucalyptus for charcoal production. A critical factor for an irrigation system is the correction of the defects. The main steps to take during the installation phase to take advantage of the benefits that these systems can provide are checking the connections and conditions of the couplings and pipes, verifying the diameters of the pipes that can lead to pressure loss during their monitoring, checking the choice of the head based on the local climatic conditions, ensuring that the points are buried or have sprinklers that minimize interference, and making sure that these are properly adjusted, as well as the hit breaker being properly adjusted (E. E. Hassan et al. 2022; Ogidan, Onile, and Adegboro 2019).

9.5 Benefits of Smart Irrigation Systems

Smart agriculture in irrigation offers benefits for growers, the community, the environment, and the economy. When implemented correctly, agricultural irrigation systems using smart irrigation technology can have the potential for significant water savings, reduction in nutrient runoff and evaporation, improved crop and fruit quality, and reduced cost of irrigation. This can benefit the community, the environment, and the economy. A well-managed agricultural farm can consequently also improve the well-being of the community and the economy. This can be done by ensuring that nutrients are applied and retained by the plant for yield. This discusses the benefits derived from utilizing the new smart irrigation technologies system. Currently, many growers are turning to precision irrigation and advanced technologies (García et al. 2020). Using the smart system in conjunction with established irrigation best practices could help growers as well. Smart precision irrigation means not watering based on industry norms but on the needs of each unique crop. Smarter systems allow growers to enhance their crop yield and quality with less water by increasing control over water supply and efficient water use. Some benefits of smart agricultural systems include reducing water and energy use, improving irrigation efficiency, managing diseases, and reducing runoff of sediments. Reducing water and energy use through smart irrigation systems is important for growers (Nashwan 2022; Ogidan, Onile, and Adegboro 2019). There are multiple potential positive gains when the agricultural industry adopts and uses advanced smart irrigation systems.

9.5.1 Water Conservation

Water is a precious and finite resource without which life is unsustainable. All living organisms depend on water, which is essential for the regulation of body temperature, digestion, transport of nutrients, and for many other activities. Also, can we produce abundant quality crops without the proper management of irrigation? The answer is no. Yes, we can use this invaluable resource for irrigation very effectively and properly by applying the smart irrigation system concept. The main objective of this text was to discuss the water needs for irrigation and also to facilitate the selection of suitable irrigation technologies for commercial farmers and smallholders (Obaideen et al. 2022). To address the water supply and food security challenges associated with water scarcity, a new smart irrigation system adequate for all scales of farmers needs to be designed and implemented. The design, development, and large-scale implementation of the smart irrigation system should be facilitated by systems engineering principles and concepts. The smart irrigation system presents significant opportunities for the advancement of the economic development objectives of the world's population and its knowledge of peak potentials. The results of the study will help individual farmers make an informed decision on how to apply commercial or smart irrigation technology without the need to change crop type or food quality. In general, the study will play a role in planning and applying water distribution plans for successful food security. The smart irrigation system technology uses accurate sensor measurements to provide information about climate, soil, and irrigation(Obaideen et al. 2022). This smart irrigation system aims to optimize farmers and the planet's livelihood.

9.5.2 Improved Crop Yields

Although the ultimate objective of irrigation scheduling may not be to improve crop yields, a well-designed soil water sensing network, combined with irrigation control systems, is capable of mitigating seasonal damage from over- or under-irrigation. In the case of over-irrigation, nutrient leaching can be avoided. When the plants actually require water, the irrigation system will provide for the necessary applications at the right time so that the detrimental effect of water stress on flowers, fruit setting, and ultimately yield resulting from the limited reservoir of plant-available water is reduced. Limiting stress-induced stomatal closure caused by achieving mild water stress, irrigation management allows for more CO2 to be taken up by leaves, which in turn helps in maintaining the photosynthetic carbon metabolism and maintaining a balance between energy production and consumption in plant cells, improving the growth of plants and delaying senescence (Amoura, Douani, and Tahri 2022). It is becoming increasingly clear that reports of yield loss due to mild water stress may have been exaggerated, and a rather more balanced criticism is needed. If growing conditions and relative humidity, air temperature, and irradiance during the period of mild stress are optimal, potential yield loss is indeed quite modest. Under some unfavorable conditions, however, moderate stress levels cause a significant reduction in both vegetative and reproductive growth. Nonetheless, the main advantage of the continuous monitoring of the field is the minimization of yield loss resulting from periodic water stress during critical growth periods. The effect of salinity can also be detected at an early stage, and irrigation strategies can be adjusted to ensure economical water use to alleviate substantial yield loss. Farmers who are committed to producing high-value horticultural and floricultural crops would not only be interested in improved yield but also in the production quality, which could be optimized using information from soil water sensors (Bouzguenda et al. 2022).

9.5.3 Economic Savings

Various estimates attribute the loss of water for inefficient irrigation management to a significant portion of the total water used for agriculture, meaning substantial savings could be realized by implementing the smart irrigation system. By directing water only where it is required and when it is required, such a system could optimize the use of this known limited resource. Watering the plants according to need or through controlled water levels also affects their agricultural yield. The economic advantages of using these management techniques result from an increase in crop yield and can be achieved at significantly reduced operating costs (Ali et al. 2022; Krishnan et al. 2020). When irrigation is unnecessary, it does not occur, and the utility costs are not present; over-watering is discontinued, and its outcomes, along with environmental pollution, plant disease risk, and plant productivity, are reduced. Additionally, resources associated with crop irrigation are presumably reduced when using available technology. However, the price of the necessary hardware, initial setup, and maintenance should not only be affordable, it should also be manageable. There are a variety of systems that, when combined and connected, can contain any or all of the features mentioned to make them effective (Arul Anitha, Stephen, and Arockiam 2019).

9.5.4 Energy Efficiency

It is important to make the most of precious energy resources. Nevertheless, the study of the energy efficiency of irrigation systems is usually excluded from the general advisories. This is even more relevant at the time we are sending these lines because there has been a very clear position in favor of new irrigation installations that reduce energy consumption. The rise in oil prices, high electricity costs, and the increasing demands of the population for a larger supply of water affect the cost and availability of energy. The optimization of irrigation systems is becoming increasingly important with rising fuel and labor costs, and government policies that promote the conservation and improvement of energy efficiency. Trade-offs are often possible between the investment in energy-efficient irrigation equipment and changes to existing water delivery systems that result in greater operational efficiencies. The energy balance of the traditional pressurized irrigation system is generally unfavorable (Tace et al. 2022). This is because the high-performance losses occur in the heads investigated in the system. The energy balance of pressure systems could be improved through the application of modern technologies. New equipment such as pumps, motors, variable capacity vehicles, alternatives to high-performance overhead irrigation systems, and innovative control systems all applied to the irrigation system can contribute to a significant reduction in the amount of electricity consumed per unit of work done. Similarly, the reform of delivery systems such as pipelines and increased network pressure contributes directly to energy savings because network losses increase when network pressure increases for both production and distribution systems (Anbarasi et al. 2019). Recent advances in technology can and must be used to provide smooth and efficient design, installation, and operation of modern irrigation systems; selected irrigation techniques are demonstrated in Fig 9.6.

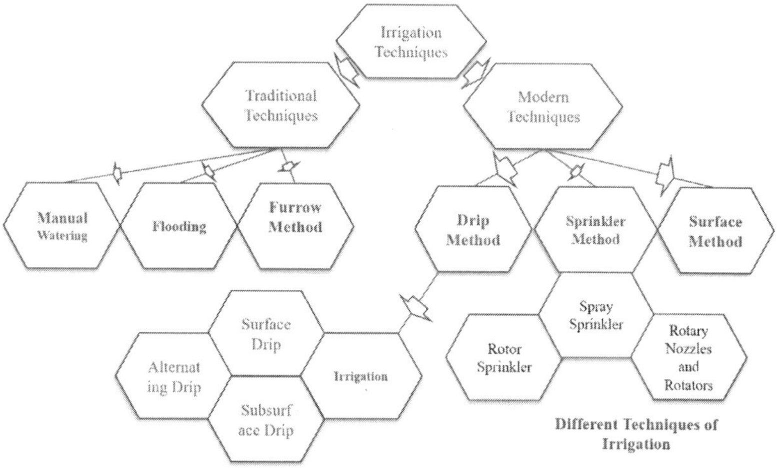

Fig. 9-6 Irrigation techniques

9.6 Case Studies and Examples of Successful Implementation

The use of drip irrigation systems has led to the sustainability of the horticultural industry in arid regions. The study also sought to benchmark the project regarding sustainable irrigation practices. Sustainable irrigation practices included the utilization of residual soil moisture at a 22% depletion level, the calibration of irrigation volumes per drop in seven different climate stages, and the irrigation schedule closely aligned to weather data. The project highly ranked the use of drip irrigation systems in managing water resources above the use of programs designed to reduce runoff losses. The affordability of drip irrigation systems for horticultural small-scale farmers for sustainable crop production was made possible through a program (Edodi, Ogidan, and Amusan 2022). The low-cost drip system covered crop production of less than a hectare. Vegetable production is one of the oldest irrigation practices in the world. In arid and semi-arid regions, the most common irrigation systems are surface and sprinkler irrigation systems. Adoption of SDI compared to the conventional methods of macro-irrigation is on the rise among smallholder farmer communities in Africa due to concerns about high water wastage through surface runoff and deep percolation. Contrary to this claim, there are now emerging concerns about the improper management of SDI systems leading to water wastage through deep percolation and runoff (Bhandari 2021). The drip irrigation system is efficient compared to macro-irrigation systems due to high water use efficiency leading to increased yields. Disadvantages of standard drip irrigation systems are the high cost of implementation and foul subsurface water due to poor water filtration systems and the clogging of emitters. Therefore, a system that promotes water and energy-efficient measures is necessary for irrigation practices. Smart Irrigation Systems implement real-time data to actuate irrigation events based on soil conditions and atmospheric data, reducing water wastage (Abdulaziz and Croock 2023). The system enables data transfer between farms.

9.6.1 Precision Agriculture Techniques

The practice of precision agriculture involves monitoring and actively modifying agricultural variables at small-scale spatial resolutions. These techniques are well-suited for use in smart solutions to address global food production challenges and remove unique constraints on traditional agricultural practices. The use of precision agriculture has been shown to increase yields in multiple settings. Techniques used in precision agriculture can be both non-automated, where the farmer makes decisions based on the supplied information, and process automation, where sensors and actuators initiate actions. The former category, "non-automated precision agriculture," typically involves large farms with inhomogeneous fields and combines satellite and drone-based environmental monitoring, which is particularly useful for large operations where a human operator can address plant-specific needs (Elgaali et al. 2023). Based on the analysis of the assembly of images or remote sensing data, many insights can be provided to the farmer so that effective decisions on water, fertilizer, pesticides, and growth stimulator usage can be made, resulting in resource optimization and increased yields. Data mining, machine learning, and pattern recognition techniques can analyze these remote observations and make predictions on crop growth or declare crop type, disease, and health. The former falls into the more general category of supervised learning and centers around the problem of remote sensing (Sinugo and Longe 2021). Water is a critical issue for precision agriculture, so specialized techniques have been developed to optimize its distribution. Modified sprinkler and drip irrigation control systems have been successfully used for precision-based agriculture. Specialization beyond circular coverage areas can be achieved by using multi-agent irrigation strategies. In such schemes, multiple robots equipped with GPS and distributed sensing work together to irrigate large areas efficiently (Scarlatache et al. 2021). Other multi-robot systems can be utilized to inspect vast areas, analyze crop health, and make precise field predictions, enabling precision irrigation monitoring soil, plant and ecology, as traits are demonstrated in Fig. 9.7.

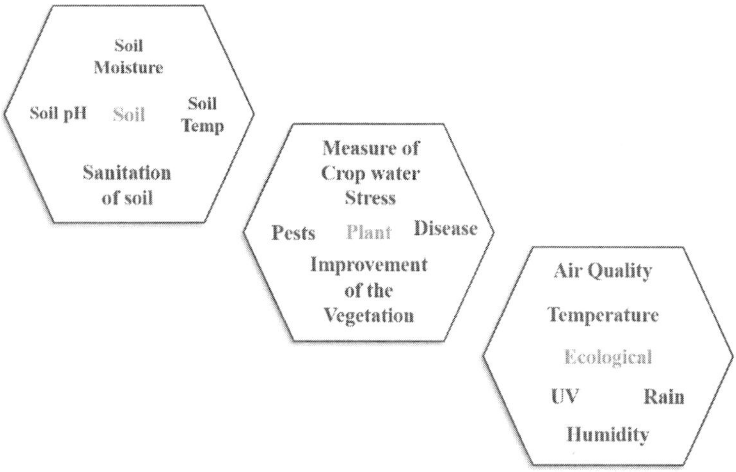

Fig. 9-6 Traits of Soil, Plant, and Ecology

9.6.2 Israel's Drip Irrigation Revolution

Israel is a global leader in irrigation technologies and practices, and this leadership is traditionally traced back to the invention of the modern drip irrigation method by Israelis in the 1950s and 1960s. But from the arid Jordan Valley in the south to the snow-covered summits of the Golan Heights in the north, Israelis had been developing techniques for irrigating crops in arid to semi-arid conditions long before the initial breakthroughs after the establishment of the state of Israel in 1948. Archaeological evidence tells us that as far back as three millennia ago, Jewish farmers sowed barley and other crops harvested for animal feed and production on expanses of markedly barren landscape that they transformed into productive fields irrigated with flowing or diverted rainwater (Sunehra 2019). Some distant descendants of these enterprising settlers are still farming the same or similar lands today. Over the centuries and millennia between the ancient Jewish agriculturists and contemporary kibbutz members, multiple generations with ties to what is now the modern-day state of Israel developed a wide range of agricultural techniques that proved to be adaptable and suitable both for the relative scarcity of usable natural resources and in some cases also for other dry to semi-dry climates worldwide (Hari, Rahmarestya, and Harsono 2021). It is no surprise, therefore, that in the years following the introduction of the modern drip irrigation method, Israelis would create a plethora of innovative cultivation and precision agriculture techniques. Many of these techniques were made possible even during later decades when there were scarcities of fresh water in the country or they were developed with innovative use of less water thanks to the marriage of new and traditional technologies. Today, in the age of climate emergency, the descendants of those ancient Jewish farmers are working together as innovators of agricultural and livestock farming techniques to continuously refine pioneering smart irrigation systems that inherently contribute to a circular and increasingly energy-efficient national economy in the water, food, energy, and possibly also the nutrients and biomaterials cycles (K. A 2020).

9.6.3 California's Sustainable Groundwater Management Act

Designing water management systems is a highly complex task, and it is influenced by policies that dictate the rules and behavior of the underlying human and natural systems. In the sphere of water, measures that lead to efficient water utilization need to be sought by empowering local communities. However, one wishes to describe governance arrangements that promote both efficiency and equity in water supply, a serious problem arises: what should be done when governance arrangements fail to produce these results? The solution to problems of over-exploitation of water or environmental degradation has often been top-down regulatory intervention through large utilities or state organizations (Froiz-Míguez et al. 2020). However, an increasing number of countries are shifting power to the local level in the management of resources, and interest is growing in the management of water supplies at the local level. The Sustainable Groundwater Act is the most significant alteration in water policy in California over the last 100 years and is intended to promote agricultural sustainability by making local agencies responsible for the management of individual groundwater units. However, the statutes in the act do not specify or standardize the actions that local agencies predominantly perform. Our purpose is to develop a principled approach that empirical data can be used to distinguish success from failure for the crucial job of maintaining safe operating space for groundwater levels. We also recommend that the state of California issue guidelines to local Groundwater Sustainability Agencies, guidelines that provide clarity regarding what actions the local agencies must mostly undertake to comply with the statute (Sasi Kumar et al. 2023).

9.7 Economic and Environmental Impacts

In 2008, the total value of agriculture from farms in the U.S. exceeded 300 billion dollars. Irrigation directly contributes to the success of many economically important crops in the U.S. Economically, the increased use of less costly energy sources for irrigation pumping, the reduction or elimination of labor by scheduling irrigation events automatically, and higher yields created by improved irrigation schedules allow the increase in farm size without access to large capital. Efficient use of water resources, increased crop yield, labor savings for irrigation events, and electrification capital are some of the benefits associated with the use of smart irrigation systems. There could be additional savings in maintenance and associated benefits, including reduced long-term environmental impact, reduced greenhouse gas emissions, and perhaps participation in trading markets of carbon (Abba et al. 2019; Zaman 2022; Masseroni, Arbat, and de Lima 2020). The projected annual water savings can be realized from a reduction in the state's demand for irrigation water, equal to half of total losses that occur when water is distributed in the fields, and assuming that once practical systems improve the capture and utilization of precipitation. The potential for water savings projected has many implications, such as the potential to increase agricultural demand for water from cities, impose restrictions, and reallocate water previously used for agriculture to municipal and environmental needs. Such a large contracted state water supply will cascade down through the local districts to the region's irrigation water supply entities, creating challenges and opportunities as they seek to address the impacts caused by changes in demand from their users and the potential for disruption of the reliability of the water delivery system (Jain 2023; Krishna, et al. 2020).

9.7.1 Cost Savings

One significant advantage of a fully functional precision irrigation system is saving costs from limited resource usage such as water and fertilizers. Water savings can be about 30 percent. Lamps are a significant consumer of energy in a commercial greenhouse. Knowledge of crop characteristics could be used to inform changes in growth conditions, facilitating a reduction in energy consumption. Saving water, considering its contamination or scarcity in certain areas, is an important issue. However, investment in smart irrigation technology is currently high, and its widespread reuse is currently limited (Quimbita, Toapaxi, and Llanos 2022). We claim that intelligent greenhouse technology needs to be further advanced so that costs can be reduced and new implementation models promoted not only in the industrial greenhouse sector but also in the amateur greenhouse sector. Moreover, the parameters concerning crop health, such as the presence or absence of infected crops, could be monitored. In greenhouses, a smart system based on the position of the lamps and the position of the crops could be realized for cost savings by reducing the time of operation of such systems and repositioning the lamps to optimize them. Growers use a lot of irrigation water every day, and periods of unusually high consumption of water trigger growers to think about potentially changing strategies for irrigation (A. Hassan et al. 2021). Such behavior change is particularly evident, as growers think they will not have to pay high prices since they use both energy and water in extensive quantities. In particular, such decisions should be made daily. However, not taking such a decision might be dangerous because plants can be affected by water stress if their values are too low, and the most important parameters for plant yield control are not balanced (Zaina, Fadli, and Hosseini 2022).

9.7.2 Reduction of Water Pollution

Besides contributing to the reduction of eutrophication, the reduction in the volume of irrigated water by precision irrigation systems may be particularly relevant in order to minimize the impact related to the presence of certain pesticides and emerging pollutants found in water. In fact, the concentrations of these pollutants in the water from irrigation percolate, locally called "concentrated water bodies," are highly variable, including those that contain sources of organic matter, like the compost basin. Particular flow and temperature ranges can lead not only to a decrease in the effectiveness of degradation but also to the gradual accumulation of these substances or their metabolites and the increase in their concentrations in drainage flows (Farooq et al. 2021). The remaining question is not only whether sustainable agricultural activities based on precision irrigation may contribute to lower concentrations of pollutants in irrigation water but also whether these systems can be more environmentally benign. Families of pesticides such as herbicides, insecticides, fungicides, and others can be controlled or mitigated as with nutrients, mainly using low water volumes and fertilizing substances and the improvement of the uniformity and distribution of irrigation coverage efficiencies. The most suitable intelligent measurement, monitoring, and warning technologies for flow, volume, and head loss, including soil moisture, wind, canopy, and atmospheric conditions, were applied. The selection of both pre-irrigation and intermediate nutrition levels allowed the ability of the irrigation system to meet the requirements of the crop and of environmental flows (Ndunagu et al. 2022).

9.8 Barriers to Adoption

Technology adoption is a complex process, and even technology that is recognized to be efficient and required can take years to become widely adopted. Smart irrigation systems in agriculture come with several challenges and barriers. Adoption by the farmer of sophisticated irrigation systems, especially if those systems are automated, is a slow process, and its success will partly depend on the education level and the previous experience of the farmer, as well as company-oriented actions of farm advisors, suppliers, and irrigation dealers. Smart irrigation systems should assure farmers' control through real-time monitoring of applied irrigation water, local weather forecasts, the ability to remotely act on an irrigation system, and the freedom to modify irrigation decisions (Krishna, et al. 2020). Only through an increased understanding and correlation between data from smart irrigation systems, crop and weather, and mainly their economic impact, will advisors and farmers increase trust and confidence in the systems and optimize the adoption. There are several reasons why farmers are reluctant to adopt smart irrigation technology. First, these systems are expensive, including frequent adjustments of sensors and maintenance. Second, options should be provided to tailor the system to the farmer's needs and must therefore provide these options in a modular, scalable, and upgradeable architecture. Third, differences in neurocognitive ability between farmers could decrease the readiness to use smart systems, and the complexity of the feedback necessary to convince the farm manager to accept the advice requires a sizable effort in analysis. Fourth, further development is required to define and optimize platforms to connect the smart irrigation systems and GIS to provide the services of a cloud-based irrigation water management platform (Jain 2023; Masseroni, Arbat, and de Lima 2020; Zaman 2022). Obviously, technical problems with incorrect irrigation management and unsatisfactory results may hamper farmers' trust in these systems.

9.8.1 Initial Investment Costs

In technology-based irrigation systems, the initial investment includes the cost of the technology and, potentially, their installation, as well as the installation of a control system that will integrate the technological systems in the most efficient way possible. The cost of the sensors is increasing as new sensors are developed to measure new parameters and to increase the spatial variability of their readings. Moreover, in most cases, several sensors are required to properly monitor the behavior of the crop and its environment, and each system may require a few hundred euros for purchase and installation. Additionally, we must consider the periodic calibration or replacement of these sensors (Abdulaziz and Croock 2022). These sensors or actuators usually require regular maintenance, and the accessories are usually renewed after a certain period of time. Depending on the lifespan of different sensors and the location of the irrigation system, the operational cost with sensors is highly divergent. Many studies are available to predict the break-even point of the savings in water, labor, energy, and chemical costs when using the most common technology. Most examples are based on proposed solutions and are not based on real market prices. Although each user has a specific way of managing their systems, in the majority of studies, shared irrigation strategies and areas of usage of sensor systems, automation, and irrigation management devices are depicted (Bazaluk et al. 2022).

These reviews provide financial budgets to future or current users. Therefore, the present needs are to clarify, in real market prices, the range of acceptable water-saving technologies by different segments of users. Additionally, specific guidelines must inform potential buyers of the real chances of irrigation success. The tool, which was developed to release new smart technologies into the market, establishes a predesigned architecture for implementing the technologies and measures the benefit of decision-making in the pilot irrigation action. The methodology includes the definition of measurement references, sensor accuracy, and efficacy at field water and permits an assessment through a set of indicators whose association can lead to specific measures to improve irrigation equipment or design (Arul Anitha, Stephen, and Arockiam 2019). Actually, the most advanced technology is technically complicated and, for this reason, important technical support is essential. Guidance and support, or even the existence of an external agent, are needed to overcome the barriers regarding user support. For proper motivation, farmers need very optimized business models in which a reasonable payback time for the investment—usually between one and five years—has been clearly demonstrated. Users cannot, by themselves, undertake the necessary investment. On the other hand, because we mainly notice the early innovators and early adopters of cutting-edge irrigation systems, financial support may change the attitude of these potential users, converting them into leads, trials, buyers, and adopters (El Mezouari, El Fazziki, and Sadgal 2022; Dhole et al. 2021).

9.8.2 Lack of Technical Expertise

Another challenge associated with adopting precision agricultural technology in smart irrigation systems is the lack of technical expertise. Scaling advanced technology in large domains that require such domains, such as precision irrigation technology, relies heavily on technology adoption at the hands of people with limited expertise on the subject. Water availability and pollution are global challenges, and in the near future, experts predict that groundwater management will be extremely relevant for economies around the world. A sustainable solution for water management will require a commitment from the agricultural ecosystem. To include this ecosystem, the technology must be easy to use, affordable, and of high quality. This outlines the challenges associated with using advanced technology systems, such as smart irrigation systems, that require a fair degree of technical knowledge to operate. It then details different technical support options that can be offered to ensure better adoption of these technologies (Adenugba et al. 2019). The challenge that smart irrigation systems face is that they are complex and lack user-friendly interfaces. Therefore, it becomes necessary to support the implementation of such technologies.

9.9 Future Trends and Innovations in Smart Irrigation

The future of the smart irrigation system lies in its ability to enable evaporation-based feedback while relying on very little power and remote weather-based operations. This involves the use of a simple supercapacitor and charge controller that use no power to keep accurate logs of weather conditions. This technology should greatly increase access and expand life-sustaining crops in the Saharan and similar climates, which have traditionally proved challenging to feed. All systems created for more widespread use and commercialization are targeted. The transpiration levels of the irrigation system have historically been optimized (E. E. Hassan et al. 2022). Work to optimize water absorption would be eminently useful given the percentage of the world using smart irrigation systems that rely on subsurface irrigation. An energy harvest strategy built into the recharge module would allow the solar panel to be packaged into a sleek, lightweight design, making it almost solar panel free of cost. Battery life cycles and replacement costs would immediately decrease. Such developments will extend the shelf life of these systems. Intelligent irrigation technology demonstrates great promise for the future: Smart irrigation is fundamental to the efforts to provide all people with safe and healthy food while focusing on horticulture, preserving the world, and uplifting communities around the globe. By promoting eco-friendly practices for irrigation, these initiatives aim to raise money for businesses while also offering value for the environment (García et al. 2020). Optimizing water usage at both operational and energy levels results in cost savings and improvements in food quality. With precision irrigation, a software program combats climate change by conserving water and reducing physical labor. In combination, these tools ensure the sustainable growth of crops and also the reduction in environmental impact related to water usage. The main focus of the latest research strategic plan is on the creation of the next generation of irrigation technology. In drought-sensitive environments, these platforms will bring the variables and visualizations to optimize the position and the timing of the irrigation when the crops are already water-stressed near the wilting point (Ali et al. 2022).

9.9.1 Integration with Internet of Things (IoT)

The applications of precision agriculture require all devices to communicate with each other to collect and analyze data. IoT is expected to be an essential tool for smart irrigation, operating automatic irrigation systems that use directly networked sensors or actuators. For smallholder farmers, IoT is used to inform irrigation time or tools to determine crop water requirements, while for large companies, IoT can support data mining and the use of artificial intelligence techniques. Smart irrigation systems based on IoT can also offer opportunities for supply chain upgrading in agriculture, for example, through certification projects. However, there are still technological challenges associated with the adaptation of smart irrigation systems to various types of crops and soils, in the face of weather-exposed conditions, as well as the scalability required to be effectively adopted at large farms (Chalvantharan, Lim, and Ng 2023; Adenugba et al. 2019). Overall, new IoT-based architectures can contribute to smart irrigation systems for more sustainable agriculture, able to support every link in the production chain, from smallholders to large companies.

9.9.2 Artificial Intelligence in Irrigation Management

In general terms, thinking about precision irrigation without imagining intelligent systems is quite complex, since this scenario will always be marked by a high degree of artificial intelligence. Therefore, technologies derived from the aforementioned paradigms that include reasoning, learning, and supervision techniques should always be incorporated into new concepts of smart agriculture. Nevertheless, there is currently an absence of intellectual techniques related to new paradigms, specifically applied to long-standing paradigms in addition to traditional views like expert systems. On the contrary, all the proposed approaches agree on the requirements of these management paradigms and can establish real-time systems for efficient irrigation management. An important factor of investigation for the growth of intelligent systems in agriculture is the coverage of international scientific literature, which has been growing since along this trend (Jain 2023). However, a growing percentage of these techniques are related to generic machine learning and not specifically to agriculture; therefore, there are many opportunities for specialized techniques applied to specific scenarios addressing food security challenges, sustainability, climate issues, and others. These techniques can be applied at different computational levels, from embedded systems used in sensor acquisition, transmission, and actuation systems to highly efficient cloud systems used in stored data mining techniques trained with specialized knowledge. All the proposals mentioned include a large number of collected data, in addition to different typologies of irrigation, and mainly incorporate the role of several sensors, usually used to acquire information from the production process and make real-time decisions (Wanyama et al. 2023; Farooq et al. 2021).

9.10 Challenges and Limitations of Smart Irrigation Systems

Smart irrigation systems offer a variety of benefits, making it possible to cultivate plants even under water stress. However, such systems are not without limitations and challenges. For example, the technologies used in smart irrigation systems often face problems when installed and operated far from the grid, due to different reasons, yet mainly due to the high maintenance needs. Some technologies, like solar panels installed to charge batteries, are able to support water pumping systems even in remote areas, yet the type of soil should be suitable for installation away from civilization. The communication protocols and the sensors also have worse performance if far from civilization. Another important challenge is the cost: the technologies normally used for smart grid irrigation are relatively expensive (Abdulaziz and Croock 2022; Pallav 2022). Thus, the advantages are undermined by the initial investment, especially for small-scale or poor farmers. Other important drawbacks and challenges are: the accuracy, which still lacks despite the variety of sensors available on the market; incomplete knowledge about the best algorithms for irrigation systems, as well as the best algorithms in models that predict the dynamics of the water in the soil; difficulty of installation, configuration, and fine-tuning of the system; and energy consumption. The latter is particularly important, as the majority of irrigation systems rely on photovoltaic panels or diesel generators. Making those systems efficient in terms of energy consumption is crucial, especially in a world where water resources are becoming rarer and more valuable.

Finally, it is important to highlight that the heterogeneity of the water resources that are used by sectors other than agriculture, in contrast to the homogeneity of drinking water, often relegates agriculture to a second position in the management of this resource (Quimbita, Toapaxi, and Llanos 2022; El Mezouari, El Fazziki, and Sadgal 2022). These challenges have presented attractive opportunities for research and development, aiming at implementing high-performance, low-cost solutions to smart irrigation systems.

9.10.1 Initial Investment Costs

The main cost associated with any irrigation system is the initial investment cost incurred to purchase, install, and prepare the system, which includes the mechanical, hydraulic, and electronic systems. Initial investment costs for each specific automated irrigation system have to be balanced against the costs of labor, energy, water, and risk of crop loss and compared with the potential increase in yields or the monetary value of the products in the specific local farming conditions. Initial investments in an automated pressurized irrigation system typical of greenhouses are around 25,000 € to 50,000 €/ha. The high costs of installing advanced smart irrigation systems are a major disadvantage to their adoption. However, variations in automation levels can also introduce variations in initial investment costs (Sharma et al. 2022). Detailed information was collected and is summarized about the initial investment costs of center pivots, rotating booms, linear systems, and localized/encompassing irrigators. Calculations must be made to identify the most efficient and convenient irrigation solution for increasingly different soil types, irrigation purposes, and accompanying water management practices. After purchase, operation costs for pumping water with electric, diesel, or natural gas engines must also be considered. Irregular and steep incline fields need a higher number of expensive pumps, generating increased investments and also maintaining fueling costs. After installing a fully automated fertigation system, higher yields in any given crop can result in significant economic increases, while also solving environmental pollution caused by leaching within nutrient-rich areas (Amoura, Douani, and Tahri 2022; García et al. 2020).

9.10.2 Technical Expertise Requirements

To survive fierce competition, a smart farm needs to employ diverse human resources. As various technologies are used in agriculture, technical expertise is required not only in farming but also in managing the cloud system and secure data. Specifically, knowing how to manage cultivation and grow light conditions is necessary. Managing climate conditions to maximize growth efficiency depends on a myriad of complex algorithms. In addition, big data is frequently applied for analysis. The cloud layer acts as a decision support center that processes data and applies algorithms that give specific operating instructions to the farm layer. Being the foundation for the best growth in plants, horticulture and its understanding are essential to define the correct light conditions, proper watering, and nutrient supplementation (Krishnan et al. 2020; Sami et al. 2022). In the decision support center, data must be continuously updated based on the changed growth conditions and must be analyzed in real-time. Constant analysis of the data patterns and live plant status allows the grow center to remind the cloud control of specific plants that require attention, supplement certain nutrients, apply specific lighting conditions, or alter the spraying. Most importantly, a fully optimized plan for smart farms that can consolidate profits depends on comprehensive models, requiring statistical data analysis. Therefore, data from a variety of sources also need to be collected and analyzed in the cloud. In light of this, security standards should be considered (Adenugba et al. 2019). Large amounts of data must be stored with support for a distributed and parallel network on multiple layers that has high reliability and stability, thwarting various network attacks.

9.10.3 Data Security Concerns

In agriculture and environmental applications, "smart" solutions involve heavy reliance on internet-attached sensors and remote data processing. Especially for privacy or critical infrastructure applications, public trust in such systems is extremely important. All data transmission should be encrypted (Shafik et al. 2023; Shafik 2025). Many sensor networks are in remote places and can be load-bearing if traveled to. This means that data should be cryptographically protected during transmission and then verified upon receipt. For systems managed by a third party, this means that two-stage integrity is needed to ensure that both the transaction history and application behave according to expectations (Obaideen et al. 2022; Sharma et al. 2022). Privacy laws enhance the need for this sort of technology. Others increase the requirements for covered services. In some cases, holding the sensor or the processing in a more controlled location can reduce the attack surface. In environments such as house yards, fencing around the sensors may be enough. For a user holding the sensor, transmitting the sensor data via encrypted SMS may provide a measure of additional respite. Providing secure key exchange methods to permit the pairing of multiple smart irrigation sensors can further reduce attack surfaces. Ensuring that access to the data is done with the least privilege helps to protect data collected in these systems. Use SSL and credentials to eliminate attacks. Data stored in the cloud is effectively both backed up and unprotected if not encrypted. Server authentication is critical in these systems (Ali et al. 2022).

9.10.4 Policy and Regulatory Frameworks

Irrigation policy has three main functions. First, it aims to develop and manage water resources in order to secure the agricultural sector and other sectors reliant on water for sustained productivity levels. Secondly, the policy addresses spatial or temporal imbalances in water resource availability and demand. Policy strives to allocate resource use against different institutional claims and rights for competing uses of water. Advancements in technology have been mainly focused on the supply of water, such as conveyance, control, and application processes (García et al. 2020). However, policy approaches to enable and sustain technological advancements that support efficient water demand management, particularly in agricultural user segments, have not progressed at the same pace. Assessing digital agricultural technologies, especially those involving applications with water savings, includes several sectors such as agriculture, water, telecommunications, and meteorological services, among others(Jain 2023; Masseroni, Arbat, and de Lima 2020). Policy impediments are likely to indirectly and directly affect several sectors within the wider economy. Indeed, successful innovation results in the creation of new market opportunities, which is likely to positively affect growth, productivity, job creation, and economic output measurements (Abdulaziz and Croock 2022). The most effective policy interventions encourage demand-driven technology innovation or diffusion that addresses specific digital agricultural technology needs of the agricultural sector. Experimenting with regulatory and policymaking apparatus that simultaneously address anticipated technology capabilities and needs of a complex, cross-discipline sector is likely to realize competition and long-term sustainability benefits as a result (Abba et al. 2019).

9.11 Conclusion

Rapid population growth and changing consumption patterns pose a big challenge in the provision of food, water, and energy, with agriculture being one of the most demanding human activities. A future with diminishing resources due to climate change, high contamination, and possibly a decrease of arable land implies solutions that contribute to the sustainable development of society. Information and communications technology are starting to play a vital role in an area where the need for inspiration is endless. We are smart individuals, and we want smart industries and smart societies, which implies smart agriculture. Smart irrigation systems are a reality with a bright future. Currently, we are in the phase of developing the connection technologies and suitable sensors. Even though the process has already started, much work is still required for the design of smart irrigation solutions, with a layout adapted to each specific area. Those solutions must integrate in a single platform, different but complementary knowledge such as irrigation, climate models, plant-atmosphere, and environmental interaction. All that knowledge needs an intelligent approach to decision-making, maximizing the use of the collected data and favoring the control process Simula cannily, maximizing the use of resources optimally. The adoption of smart irrigation should be pushed by governments, associating technologies, and financial plans. A future is so bright that we need to invest smartly, plan fines, and transit irrevocably to smarter agricultural practices.

References

Abba, Sani, Jonah Wadumi Namkusong, Jeong A. Lee, and Maria Liz Crespo. 2019. "Design and Performance Evaluation of a Low-Cost Autonomous Sensor Interface for a Smart Iot-Based Irrigation Monitoring and Control System." *Sensors (Switzerland)* 19 (17). https://doi.org/10.3390/s19173643.

Abdulaziz, Wid Badee, and Muayad Sadik Croock. 2022. "Optimized Power and Water Allocation in Smart Irrigation Systems." *Periodicals of Engineering and Natural Sciences* 10 (3). https://doi.org/10.21533/pen.v10i3.2998.

Abdulaziz, Wid Badee, and Muayad Sadik Croock. 2023. "Review on Cyber-Security for Optimized and Smart Irrigation Systems." *Telkomnika (Telecommunication Computing Electronics and Control)* 21 (1). https://doi.org/10.12928/TELKOMNIKA.v21i1.24234.

Abuzanouneh, Khalil Ibrahim Mohammad, Fahd N. Al-Wesabi, Amani Abdulrahman Albraikan, Mesfer Al Duhayyim, M. Al-Shabi, Anwer Mustafa Hilal, Manar Ahmed Hamza, Abu Sarwar Zamani, and K. Muthulakshmi. 2022. "Design of Machine Learning Based Smart Irrigation System for Precision Agriculture." *Computers, Materials and Continua* 72 (1). https://doi.org/10.32604/cmc.2022.022648.

Adenugba, Favour, Sanjay Misra, Rytis Maskeliūnas, Robertas Damaševičius, and Egidijus Kazanavičius. 2019. "Smart Irrigation System for Environmental Sustainability in Africa: An Internet of Everything (IoE) Approach." *Mathematical Biosciences and Engineering* 16 (5). https://doi.org/10.3934/mbe.2019273.

Ali, Salamat, Rabia Nawaz, Sharoon Azad, Muhammad Sadiq Orakzai, Saqib Amin, Zohaib A. Khan, Faraz Akram, and Usman Masud. 2022. "Solar Powered Smart Irrigation System." *Pakistan Journal of Engineering and Technology* 5 (1). https://doi.org/10.51846/vol5iss1pp49-55.

Amoura, Mohammed, Mustapha Douani, and Toufik Tahri. 2022. "Study of a Smart Irrigation System for Water Managing for Potatoes in Chlef Region, Algeria." *Journal of Water and Land Development* 54. https://doi.org/10.24425/jwld.2022.141560.

Anbarasi, M., T. Karthikeyan, L. Ramanathan, S. Ramani, and N. Nalini. 2019. "Smart Multi-Crop Irrigation System Using IOT." *International Journal of Innovative Technology and Exploring Engineering* 8 (7).

Arul Anitha, A., A. Stephen, and L. Arockiam. 2019. "A Hybrid Method for Smart Irrigation System." *International Journal of Recent Technology and Engineering* 8 (3). https://doi.org/10.35940/ijrte.C4826.098319.

Badrun, Burhanuddin, and Murshal Manaf. 2021. "The Development of Smart Irrigation System with IoT, Cloud, and Big Data." In *IOP Conference Series: Earth and Environmental Science*. Vol. 830. https://doi.org/10.1088/1755-1315/830/1/012009.

Bazaluk, Oleg, Valerii Havrysh, Vitalii Nitsenko, Yuliia Mazur, and Sergiy Lavrenko. 2022. "Low-Cost Smart Farm Irrigation Systems in Kherson Province: Feasibility Study." *Agronomy* 12 (5). https://doi.org/10.3390/agronomy12051013.

Bhandari, Aakash. 2021. "Research Article on Smart Irrigation System Using IOT." *International Journal for Research in Applied Science and Engineering Technology* 9 (12). https://doi.org/10.22214/ijraset.2021.38830.

Bouzguenda, Mounir, S. Rajamohamed, M. H. Shwehdi, and Adel Aldalbahi. 2022. "Solar Powered Smart Irrigation System Based on Low Cost Wireless Network: A Senior Design Project Experience." *International Journal of Electrical Engineering and Education* 59 (4). https://doi.org/10.1177/0020720919860414.

Chalvantharan, Arivintharan, Chun Hsion Lim, and Denny K.S. Ng. 2023. "Economic Feasibility and Water Footprint Analysis for Smart Irrigation Systems in Palm Oil Industry." *Sustainability (Switzerland)* 15 (10). https://doi.org/10.3390/su15108069.

Dhole, Bhavna, Pratiksha Patle, Onkar Patole, and Suprriya Lohar. 2021. "Solar Powered Smart Irrigation System." *International Journal of Advanced Research in Science, Communication and Technology*. https://doi.org/10.48175/ijarsct-1506.

Edodi, Michael U., Olugbenga Kayode Ogidan, and Akinwumi Amusan. 2022. "Smart Irrigation System: A Water and Power Management Approach." In *Proceedings of the 2022 IEEE Nigeria 4th International Conference on Disruptive Technologies for Sustainable Development, NIGERCON 2022*. https://doi.org/10.1109/NIGERCON54645.2022.9803183.

Elgaali, Elgaali, Jamil Al Titi, Ahmed Ismail, and Omer Alhajri. 2023. "Smart Irrigation System Using Arduino." In *2023 Advances in Science and Engineering Technology International Conferences, ASET 2023*. https://doi.org/10.1109/ASET56582.2023.10180638.

Farooq, Qazi U., Muhammad T. Naqash, Abdelkader T. Ahmed, and Bilal A. Khawaja. 2021. "Optimization of Subsurface Smart Irrigation System for Sandy Soils of Arid Climate." *Modelling and Simulation in Engineering* 2021. https://doi.org/10.1155/2021/9012496.

Froiz-Míguez, Iván, Peio Lopez-Iturri, Paula Fraga-Lamas, Mikel Celaya-Echarri, Óscar Blanco-Novoa, Leyre Azpilicueta, Francisco Falcone, and Tiago M. Fernández-Caramés. 2020. "Design, Implementation, and Empirical Validation of an IoT Smart Irrigation System for Fog Computing Applications Based on Lora and Lorawan Sensor Nodes." *Sensors (Switzerland)* 20 (23). https://doi.org/10.3390/s20236865.

Gabuya, Alden Q., Fernando N. Mangubat, Victorino H. Patindol, Johnrel M. Paglinawan, and Kent Marcial L. Catubis. 2024. "Improved Growth of Coffee Seedlings (Coffea Canephora) under SMART Irrigation System." *Journal of the Saudi Society of Agricultural Sciences* 23 (2). https://doi.org/10.1016/j.jssas.2023.09.007.

García, Laura, Lorena Parra, Jose M. Jimenez, Jaime Lloret, and Pascal Lorenz. 2020. "IoT-Based Smart Irrigation Systems: An Overview on the Recent Trends on Sensors and Iot Systems for Irrigation in Precision Agriculture." *Sensors (Switzerland)*. https://doi.org/10.3390/s20041042.

Hari, Ivony, Elita Rahmarestya, and Harsono Harsono. 2021. "Development of IoT Based Smart Irrigation System with Programmable Logic Controller." *International Journal of Agriculture System* 9 (1).

Hassan, Ahmed, Hafiz Muhammad Abdullah, Umar Farooq, Adil Shahzad, Rao Muhammad Asif, Faisal Haider, and Ateeq Ur Rehman. 2021. "A Wirelessly Controlled Robot-Based Smart Irrigation System by Exploiting Arduino." *Journal of Robotics and Control (JRC)* 2 (1). https://doi.org/10.18196/jrc.2148.

Hassan, Elia Erwani, Leong Lek Chung, Mohamad Fani Sulaima, Nazrulazhar Bahaman, and Aida Fazliana Abdul Kadir. 2022. "Smart Irrigation System with Photovoltaic Supply." *Bulletin of Electrical Engineering and Informatics* 11 (1). https://doi.org/10.11591/eei.v11i1.3338.

Jain, Ravi Kant. 2023. "Experimental Performance of Smart IoT-Enabled Drip Irrigation System Using and Controlled through Web-Based Applications." *Smart Agricultural Technology* 4. https://doi.org/10.1016/j.atech.2023.100215.

K. A, Olatunji. 2020. "A Mobile Phone Controllable Smart Irrigation System." *International Journal of Advanced Trends in Computer Science and Engineering* 9 (1). https://doi.org/10.30534/ijatcse/2020/42912020.

Krishna, M., A. Vinitha, S. Ravinder, and G. Akhilesh. 2020. "SMART IRRIGATION SYSTEM USING RASPBERRY PI." *Turkish Journal of Computer and Mathematics Education (TURCOMAT)* 11 (3). https://doi.org/10.61841/turcomat.v11i3.14485.

Krishnan, R. Santhana, E. Golden Julie, Y. Harold Robinson, S. Raja, Raghvendra Kumar, Pham Huy Thong, and Le Hoang Son. 2020. "Fuzzy Logic Based Smart Irrigation System Using Internet of Things." *Journal of Cleaner Production* 252. https://doi.org/10.1016/j.jclepro.2019.119902.

Lakshmiprabha, Kakkanallur Ethirajan, and Chinnathambi Govindaraju. 2023. "Hydroponic-Based Smart Irrigation System Using Internet of Things." *International Journal of Communication Systems* 36 (12). https://doi.org/10.1002/dac.4071.

Masseroni, Daniele, Gerard Arbat, and Isabel Pedroso de Lima. 2020. "Editorial-Managing and Planning Water Resources for Irrigation: Smart-Irrigation Systems for Providing Sustainable Agriculture and Maintaining Ecosystem Services." *Water (Switzerland)*. https://doi.org/10.3390/w12010263.

Mezouari, Asmae El, Aziz El Fazziki, and Mohammed Sadgal. 2022. "Smart Irrigation System." In *IFAC-PapersOnLine*. Vol. 55. https://doi.org/10.1016/j.ifacol.2022.10.125.

Nashwan, Shadi. 2022. "Secure Authentication Scheme Using Diffie–Hellman Key Agreement for Smart IoT Irrigation Systems." *Electronics (Switzerland)* 11 (2). https://doi.org/10.3390/electronics11020188.

Ndunagu, Juliana Ngozi, Kingsley Eghonghon Ukhurebor, Moses Akaaza, and Robert Birundu Onyancha. 2022. "Development of a Wireless Sensor Network and IoT-Based Smart Irrigation System." *Applied and Environmental Soil Science* 2022. https://doi.org/10.1155/2022/7678570.

Obaideen, Khaled, Bashria A.A. Yousef, Maryam Nooman AlMallahi, Yong Chai Tan, Montaser Mahmoud, Hadi Jaber, and Mohamad Ramadan. 2022. "An Overview of Smart Irrigation Systems Using IoT." *Energy Nexus* 7. https://doi.org/10.1016/j.nexus.2022.100124.

Ogidan, Olugbenga Kayode, Abiodun Emmanuel Onile, and Oluwabukola Grace Adegboro. 2019. "Smart Irrigation System: A Water Management Procedure." *Agricultural Sciences* 10 (01). https://doi.org/10.4236/as.2019.101003.

Pallav, Priyanshu. 2022. "SMART IRRIGATION SYSTEM FOR VERTICAL FARMING." *INTERANTIONAL JOURNAL OF SCIENTIFIC RESEARCH IN ENGINEERING AND MANAGEMENT* 06 (05). https://doi.org/10.55041/ijsrem13575.

Quimbita, Wilmer, Edison Toapaxi, and Jacqueline Llanos. 2022. "Smart Irrigation System Considering Optimal Energy Management Based on Model Predictive Control (MPC)." *Applied Sciences (Switzerland)* 12 (9). https://doi.org/10.3390/app12094235.

Sami, Maira, Saad Qasim Khan, Muhammad Khurram, Muhammad Umar Farooq, Rukhshanda Anjum, Saddam Aziz, Rizwan Qureshi, and Ferhat Sadak. 2022. "A Deep Learning-Based Sensor Modeling for Smart Irrigation System." *Agronomy* 12 (1). https://doi.org/10.3390/agronomy12010212.

Sasi Kumar, G., G. Nagaraju, D. Rohith, and A. Vasudevarao. 2023. "Design and Development of Smart Irrigation System Using Internet of Things (IoT) - A Case Study." *Nature Environment and Pollution Technology* 22 (1). https://doi.org/10.46488/NEPT.2023.v22i01.052.

Scarlatache, Florina, Gheorghe Grigoras, Vlad Andrei Scarlatache, Bogdan Constantin Neagu, and Ovidiu Ivanov. 2021. "A Hybrid Methodology Based on Smart Management Energy Consumption in Irrigation Systems." *Electronics (Switzerland)* 10 (22). https://doi.org/10.3390/electronics10222864.

Shafik, Wasswa. 2025. "Sustainable Agriculture and Diet in the Metaverse Era: Sustainable Farming and Nutritious Eating in the Metaverse." In *Food in the Metaverse and Web 3.0 Era: Intersecting Food, Technology, and Culture*, 159–84. IGI Global Scientific Publishing.

Shafik, Wasswa, and Kassim Kalinaki. 2024. "Societal and Ethical Implications of Technology-Enhanced Agriculture and Healthcare: An African Context." In *2024 IST-Africa Conference (IST-Africa)*, 1–11. IEEE. https://doi.org/10.23919/IST-Africa63983.2024.10569306.

Shafik, Wasswa, Ali Tufail, Chandratilak De Silva Liyanage, and Rosyzie Anna Awg Haji Mohd Apong. 2023. "Using a Novel Convolutional Neural Network for Plant Pests Detection and Disease Classification." *Journal of the Science of Food and Agriculture*. https://doi.org/10.1002/jsfa.12700.

Shafik, Wasswa, Ali Tufail, Liyanage Chandratilak De Silva, and Rosyzie Anna Awg Haji Mohd Apong. 2025a. "A Lightweight Deep Learning Model for Multi-Plant Biotic Stress Classification and Detection for Sustainable Agriculture." *Scientific Reports* 15 (1): 12195.

Shafik, Wasswa, Ali Tufail, Liyanage Chandratilak De Silva, and Rosyzie Anna Haji Mohd Apong. 2025b. "An Enhanced Deep Convolutional Neural Network for Plant Disease Detection and Classification: Elevating Sustainable Agriculture." In *Artificial Intelligence and Data Science for Sustainability: Applications and Methods*, 297–322. IGI Global Scientific Publishing.

Sharma, Nitin Kumar, SachinKumar, Pradeep Kumar Yadav, and Ekata. 2022. "Standalone Smart Irrigation System: An Analysis of Reliability Incorporating BPNN Technique under Various Failures." *Universal Journal of Agricultural Research* 10 (4). https://doi.org/10.13189/ujar.2022.100411.

Sinugo, Rinae, and Omowunmi Mary Longe. 2021. "Solar-Powered Smart Irrigation System." In *6th International Forum on Research and Technology for Society and Industry, RTSI 2021 - Proceedings*. https://doi.org/10.1109/RTSI50628.2021.9597219.

Stolojescu-Crisan, Cristina, Bogdan Petru Butunoi, and Calin Crisan. 2022. "An IoT Based Smart Irrigation System." *IEEE Consumer Electronics Magazine* 11 (3). https://doi.org/10.1109/MCE.2021.3084123.

Sunehra, Dhiraj. 2019. "Web Based Smart Irrigation System Using Raspberry Pi." *International Journal of Advanced Research in Engineering and Technology* 10 (2). https://doi.org/10.34218/IJARET.10.2.2019.006.

Suresh, P., R. H. Aswathy, Sridevi Arumugam, Amani Abdulrahman Albraikan, Fahd N. Al-Wesabi, Anwer Mustafa Hilal, and Mohammad Alamgeer. 2022. "Iot with Evolutionary Algorithm Based Deep Learning for Smart Irrigation System." *Computers, Materials and Continua* 71 (1). https://doi.org/10.32604/cmc.2022.021789.

Tace, Youness, Mohamed Tabaa, Sanaa Elfilali, Cherkaoui Leghris, Hassna Bensag, and Eric Renault. 2022. "Smart Irrigation System Based on IoT and Machine Learning." *Energy Reports* 8. https://doi.org/10.1016/j.egyr.2022.07.088.

Wanyama, Joshua, Paul Soddo, Prossie Nakawuka, Peter Tumutegyereize, Erion Bwambale, Isaac Oluk, William Mutumba, and Allan John Komakech. 2023. "Development of a Solar Powered Smart Irrigation Control System Kit." *Smart Agricultural Technology* 5. https://doi.org/10.1016/j.atech.2023.100273.

Zaina, Sara M., Fodil Fadli, and Seyed Morteza Hosseini. 2022. "Evaluation of Smart Irrigation Systems in Hot-Arid Climates for Green Roofs and Walls: Case of Doha, Qatar." *Smart and Sustainable Built Environment* 11 (2). https://doi.org/10.1108/SASBE-11-2021-0201.

Zaman, Rokon Uz. 2022. "Design and Development of Smart Irrigation System for Watering Rooftop Garden." *Review of Plant Studies* 9 (1). https://doi.org/10.18488/rps.v9i1.2948.

CHAPTER TEN

SUSTAINABLE CROP MONITORING AND DISEASE DETECTION

Abstract

The agriculture industry must monitor the growth and health of crops to enable timely action in high-yielding and sustainable production. Accurate growth monitoring further helps in estimating the expected yield before the actual harvesting. The possible loss in production due to various factors such as climate change, growth, and environmental conditions can be predicted through the incorporation of these futuristic inventories for improved yields. A delay in monitoring the production process and the health status of the crop will result in yield deficits or a drop in crop quality, with a direct impact on the shortage of production. At the same time, disease detection is also crucial for eliminating the entry of pests or pollutants into agricultural land for modified gene spread; if proven positive, it may affect the export of agricultural products overseas. Furthermore, ascertaining the stage of development and related environmental conditions during biological activity is a crucial input not covered so far in studies. Research on extensive automated crop monitoring from satellites requires an interpretable synergetic solution with technological advancement. Current interest in the monitoring of agriculture has triggered research in the domain of providing vital information about the health status of crops. It is compelling for researchers to come up with solutions for existing challenges. The likely future use of crop health condition systems in precision agriculture, climate change, and also in the healthy living index, food quality, and facility management are future tools and outputs that can be used for reshaping the future of agriculture. The need for the future, as per the research, is to make the tool smarter with the combined use of genetics and diagnostics agency data. This can be used as an end-to-end solution covering all the grids for the benefits.

Keywords: Disease Prediction Models, Drone Surveillance, Early Pest Detection, IoT-Enabled Sensors, Machine Learning Algorithms, Precision Agriculture, Real-Time Data Analytics, Remote Sensing Technologies, Spectral Imaging Techniques, Sustainable Farm Management

10.1 Introduction

Agricultural experts have long advocated the need to have effective monitoring of crops and detection of diseases. The economic aspects of farming cannot be disregarded. It is therefore required to check the health-wise performance of the crop. Timely identification of any problem usually leads to a full-grown solution. This eventually saves excessive losses for the farmers. Monitoring of the cropland ultimately plays a vital role in ensuring food security (Deelip and Govinda 2023). A healthy crop grown by a natural process can produce abundant grains and other products. Healthy land maintains a balance in nature and ensures proper ecological cycles. Diseases in human beings can also be avoided by consuming organic food gained from plants grown in a pesticide-free environment. Crop disease detection is a kind of eye operation or X-ray analysis for plant diseases that cause changes in their leaves or stems' structures. Technology can help monitor disease patterns efficiently and precisely (Patil et al. 2023).

Integration of computer vision and information system technologies can improve the quality of agri-food products; at the same time, this integration may be able to provide environmental benefits, like reducing the volume and threat of agrochemicals. In earlier times, ways to monitor crop and tree diseases included the placement of pesticides and fertilizers. By routinely monitoring them, diseases of any limit can be diagnosed and further treated to increase plant production (Shafik 2024). This technique has opened up new dimensions in planting experimentation. It is thus feasible to maintain valuable plots provisionally. It cannot be accepted as a substitute for human intelligence and judgment, but it can help to improve the quality and yield of the product by taking corrective actions through timely information. The increasing population has brought a greater demand for better-quality crops. Integral advancements that are made on a daily basis are being given to the field of crop monitoring and disease detection (Deelip and Govinda 2023; Prasad and Thyagaraju 2024).

10.2 Traditional Methods of Crop Monitoring and Disease Detection

Surveillance of any form begins with assets. In crop management, the assets are the crops. Hence, farmers have tools to monitor the expenses, and these tools, over time, have and will continue to improve. The application of these technologies adapts to the needs of the current time. By definition, the word "disease" can be synonymous with "symptom." Hence, many of the traditional monitoring methods mentioned can also function as disease detection methods. Visual inspection is one of the earliest and most obvious methods for farmers to monitor plant health. In this method, individuals assess the plant based on what can be seen with the naked eye. This is generally symptom-based. Diseases reveal themselves in a variety of ways. In some cases, the leaves show a more visible blotched dying pattern. In others, the entire plant may die. As such, a variety of visual forms may be a part of visual inspection (Ojo and Zahid 2023). Visual forms may also be vibrant earlier than molecular changes, allowing for earlier detection. There is no requirement for additional equipment. This method can be low-tech. At the same time, it is also prone to subjective observation. There is no standard for what looks healthy and what looks infected. The research has not been able to identify how much of the crop is affected by symptoms. Visual forms may also be similar regardless of the disease. In some cases, the disease may look the same across different plant species. It is important to limit inaccuracies that occur with humans measure what they see in a crop. These can all deny the need for the use of additional equipment for better accuracy. Trends in farming include interest and dedication to crop rotation and organic farming. With that in combination, there may also be reduced pesticide use, which may require extra attention to crop health (Jamadar, Sharma, and Wagh 2023). More than ever, history is telling us that one failure can have strong economic undulations.

10.2.1 Visual Inspection

Visual inspection is a cornerstone of traditional crop monitoring. It is the process of examining plants for signs of disease or nutrient deficiencies performed by farmers or crop advisors in growing seasons. Characteristics of interest are compared with those of healthy plants, and informed decisions are made about whether a certain percentage of plants needs spraying or if a field needs further assessment (Shafik 2025). Although this process can be immediate and is available to anyone, it may fail to detect subtle symptoms that are too small or covered by tillers. In this case, complementary technologies are needed to improve the accuracy of inspection. Regional differences in light, soil, climate, and the subjective nature of the assessment of a percentage of infection or class of disease are different for each plant disease, crop, and observer (Ojo and Zahid 2023). In larger farms, where daily scouting of crops is conducted, visual inspection can be performed internally, increasing the accuracy of inspection. Indeed, the experience, accuracy, and skill of a grower may be of most use in effectively identifying and quantifying disease within a crop. Conversely, visual inspection can be time-consuming, and it may not always be efficient in the early detection of crop disease. In such instances, it is advisable to gather data of other types, such as soil samples, damaging agents, and initial crop conditions. Environmental conditions such as light, humidity, and temperature and the ease of image acquisition are some of the critical aspects that must be considered when setting up imaging acquisition for plant diseases, there are sampled plant disease datasets that are publicly available for AI model testing, as illustrated in Fig. 10.1 (Prasad and Thyagaraju 2024).

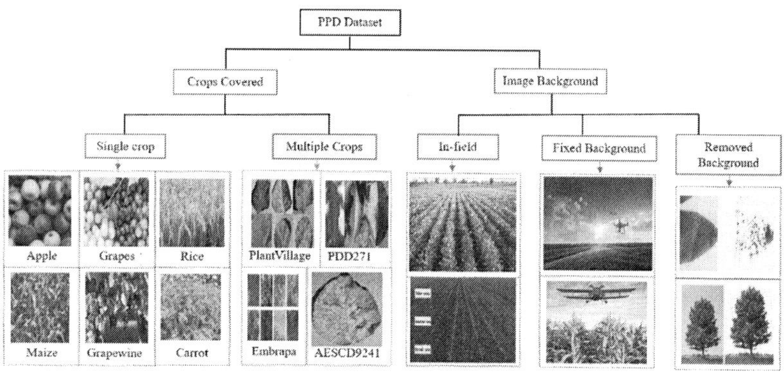

Fig. 10-1 Plant disease datasets and their categories

10.2.2 Symptom-Based Identification

Symptom-based identification is a further extension of the traditional monitoring methods in which the typical signs or patterns of a particular disease or pest are recognized. The main advantage of this technique is that it does not require high-cost technology but relies on the extensive practical knowledge about local agriculture that exists in the form of technical leaflets, user manuals, or handbooks. It can be used to train new farmers. This is the method that many apps also employ. However, there are several disadvantages (Sarah et al. 2024). First of all, as it is required to make a diagnosis based on a symptom, there may be similar symptoms for different diseases or pests. Restricted symptom sets evolved to have to pivot in the user interface or warn the user that they might consider needing a secondary set of symptoms to help differentiate. This technique can misdiagnose a real problem and be far too slow and uncertain for a true expert. In the case of having differently varied and erratic environments, the symptom-based method may not describe the whole complex details of the problem. Many shade netting cases have been documented to show the remobilised 'old tuning cv' of the L1 and L2 magnitudes finally ruined by magnificent new or fresh pot experiments. In many such cases, the street lighting can be dressed easily by an additional tension of less than 1% on the above of the cooler light blue light blocks (Samson Adekunle et al. 2024). The integrated, sophisticated and modern DNA and machine learning-based techniques are best used in conjunction with the low-cost symptom-based methods to confirm the precise answer. Understanding the technology behind diagnostics will clear the way for understanding what they can be used for and how they might be used smartly. Also, in this section, it is important to lay down, as a finished synopsis of current research, the dynamics of natural sowing parameters and outcomes of disease. This is by no means a topic of merely practical interest. It is also a genetic topic and should be handled in a genetic and phylogenetic context (Hairani and Widiyaningtyas 2024). It is also important as a major factor governing the development of resistance of the varieties.

10.3 Remote Sensing Technologies

Remote sensing technologies are an innovative advancement that has begun to be used for crop monitoring and disease detection activities. The term remote sensing describes the gathering of data about agricultural landscapes from artificial flying objects. These include satellites far from Earth, planes flying at commercial flight altitudes, as well as small, low- and slow-flying airplanes and rockets. Remote sensing technologies are particularly well-suited for viewing and monitoring croplands from above, where surveillance can be conducted more quickly than by foot or ground transportation, and observation can be made at an optimal vantage point - directly overhead (Sofuoglu and Birant 2024). With remote sensing, monitoring of large areas with multiple crops and plant species can be accomplished efficiently, and regions of interest can be identified more quickly than possible by employing traditional monitoring techniques. Diseases and other micro- and micro-biological incursions that may otherwise not be seen by the naked eye may also be detected, albeit indirectly, using these technologies. Croplands and the states of the featured crops may also be measured and modeled more quickly and accurately than when done manually (Shafik et al. 2024).

Currently, several remote sensing technologies are finding applications in agriculture, including satellite imagery, aircraft and low-altitude UAVs and contemporary hyper- and multispectral imaging spectroscopy. With improved technology, the modern field of remote sensing is now able to provide a new multifaceted and detailed perspective on the state of the crop, turf, or other plant system of interest at the time of data collection, as well as inferences about local field topography, drainage, and soil state, and information, often at the landscape level, portraying some aspect of the environmental stress and insult evident to the crops (Weber et al. 2024; Shafik, Tufail, Namoun, et al. 2023a). The most recent iterations of remote sensing technologies incorporate some mixture of passive reflectance and fluorescence, atmospheric gases, LIDAR, and movable data collection systems to plot a large region and thus reduce the variability that might degrade the overall information content. As remote sensing technologies are used to map spatial and temporal gradients in any of these entities, the hope is that this new technology will augment, or even replace, traditional genomics and phenomics technologies to identify new genes and genetic effects on stress tolerance. For agriculture, the appeal of these modern systems is the possibility of quickly identifying those locations within a field that are struggling under the stresses imposed by the environment and other insults. Such a tool would enable precision agriculture technologies the ability to manage the problem ahead of time, rather than reacting to losses after they occur (Gahlowt et al. 2024; Shafik, Tufail, Liyanage, et al. 2023).

10.3.1 Satellite Imagery

A significant advancement in agriculture is the monitoring of vast fields and aerial surveillance from space using satellites. Remote sensing through satellites is data-rich and provides a panoramic perspective. The satellites are classified on the basis of orbits and data type, such as geostationary or polar orbiting – capturing instantaneous images or swaths, respectively, and viewing them over time to gain insight. Triangulating information obtained from multiple satellites and ground-based sensors provides better analysis. Signals sensitive to greenness, photosynthesis, and, therefore, edible biomass production are termed vegetation indices (Hairani and Widiyaningtyas 2024). The data involved includes:

- Multispectral: blue, green, red, and near-infrared wavelength reflective spectrum into crop reflectance.
- Synthetic Aperture Radar (SAR): microwave signals penetrate soil cover, casting a different light on crop topographies by enabling all-weather, all-day imaging.

Satellite imagery depicts exceptional coverage and can be utilized to monitor crop health or irrigation requirements across large swaths of vast cropping territories. Nevertheless, due to reduced resolution, interpretations are subject to errors. To be interpreted, ground-scale applications call for sophisticated data analytics (Shafik et al. 2025). Crop health investigations on a larger scale can greatly benefit from satellite imagery synchronization with traditional ground-based visual observations. Evolving satellite technology, competition, and affordability delineate its advancements (Marco-Detchart et al. 2024). Integrating multiple connected data sources can terminate conventional crop scouting techniques, fostering more fruitful and resilient crop management strategies.

10.3.2 Unmanned Aerial Vehicles (UAVs)

UAVs, also known as drones, play a vital role in sensing and monitoring agricultural fields and detecting biotic and abiotic stresses in crops. By equipping them with several sensors, a platform is created to gather field information in real-time and directly aid precision agriculture. Currently, UAV platforms equipped with different types of cameras and sensors are used to provide and segregate data on organisms and conditions of crops. Drones are advantageous in providing easily accessible and high-resolution images at short notice (Sai Sharvesh, Suresh Kumar, and Raman 2024). They have been tested and have benefited pest detection, fertilizer detection for weeds, crop monitoring, mineral nutrition assessment, and real-time stress detection in crops.

There are normally two types of sensors mounted on a UAV: non-imaging and imaging sensors to acquire high-quality and comprehensive images of a given field. For UAVs, there are multispectral and commercial sensors usually attached to a UAV, such as a camera. The multispectral cameras are typically attached to the UAV for exploring the assemblage and biomass estimation in crops and for detecting crop diseases. However, thermal and infrared sensors are initially attached to detect and monitor any fungal diseases and infections, especially among fungi in the leaves of wheat. The use of UAVs automatically adds several advantages, including avoiding the need to walk in vast and waterlogged fields, ease of access, and the ability to capture images quickly with much higher resolution (Shafik, Tufail, Namoun, et al. 2023b). In addition, observations of crop health and disease analysis can be made through the frequency of UAV flights in specific frequency bands and times. In general, the costs of purchasing a premium UAV, battery lifespan, and regulatory restrictions for operating a UAV are costly and highly variable. Deadline and real-time issues may arise when handling large datasets. Hence, an accurate and quick algorithm is necessary to compute large aircraft datasets efficiently. Integrating UAVs with satellites or any other remote sensing systems boosts the efficiency of monitoring large crop fields (Hari, Singh, and Singh 2024).

10.3.3 Hyperspectral Imaging

Many types of imaging systems are used for monitoring crops, but one of the most sophisticated is hyperspectral imaging. Hyperspectral imagers, or sensors, are able to capture a broad spectrum of wavelengths of electromagnetic radiation. The images that result from these sensors contain data that can be used for a wide variety of applications (Shafik and Kalinaki 2024). When the imaging systems are applied to crops, their specialized design – which records far more wavelengths than ordinary cameras – can provide a significant amount of detail regarding the health of the plant. Importantly, as crop disease can sometimes exist at early stages in which no symptoms are visible to the human eye, a technology that can detect and diagnose these diseases is quite valuable. Hyperspectral imagers can reveal several characteristics about plants, the implications of which can describe or interpret the changes that have occurred. Hyperspectral imaging systems typically contain several parts (Katti et al. 2024). Commonly, they include a light source to illuminate the plant, some optics for the conditioning and manipulation of light, and a spectrometer, which resolves light into spectral components. A smaller portion of the spectrum, the traditional visible portion of the spectrum, is what we typically perceive as light. As light is absorbed, transmitted, or reflected by the leaf and moves through the plant, it carries information that can tell us more about the plant, with useful information for diagnostics, such as the chlorophyll content or the presence of certain pigments. All this raw 'picture' information must be analyzed to derive information that can help growers (Shende et al. 2024). A critical challenge to the widespread use of hyperspectral imagers in agriculture is that the systems generate massive quantities of data that must be interpreted in order to be of use. A suite of algorithms is needed to do this, which can range from straightforward averages and ratios of different wavelengths to more complex math and machine learning. Because of this processing demand, the algorithms that accompany hyperspectral imaging can be difficult to use without experience. To date, the technology has seen applications in precision agriculture as a means of providing information such as detecting pests, providing estimates of crop yields, uncovering nutrient imbalances or deficiencies, and assessing the effect of different planting depths of seeds (Kumar and Vijayakumar 2024; Kothari et al. 2024).

10.4 Machine Learning and Artificial Intelligence in Disease Detection

AI is enabling decision-making at scales and speeds previously unattainable. Machine learning is a fundamental tool within AI designed to analyze complex datasets. By using historical data to identify patterns and relationships, models can make predictions based on real-time inputs. The advantages of this for disease detection are obvious: by using knowledge of existing outbreaks, a model could predict when and where outbreaks will occur in the future. A wide range of machine learning algorithms are already being utilized within disease detection. Image recognition algorithms are particularly popular, being able to classify crop health based on visual input. Furthermore, deep learning networks have evolved in recent years, significantly enhancing the accuracy of disease diagnosis using image data (Bouacida et al. 2024; Shafik et al. 2024). While deep learning models offer promise for disease detection, the combination of AI and computer vision also offers scope for efficient automation of routine monitoring tasks. As a result, farmers can gain rapid insights into their farms, leading to potentially faster or real-time decision-making. Nevertheless, integrating AI into agriculture is not without its challenges. Despite the technical possibilities, the protracted ethical, societal, and economic implications of AI implementations remain uncertain. In the field of plant disease diagnosis, fairness and transparency are increasing concerns, particularly concerning bias in the models, as well as the potential for the technology to be weaponized. Regardless of these concerns, AI and machine learning continue to grow in popularity across agriculture as a proactive solution to pest, disease, and crop monitoring (Bhagat et al. 2024; Derisma, Rokhman, and Usuman 2022). Indeed, the application of computer vision and AI tools has already enabled monitoring capacity to improve, providing timely data that directly informs decisions to respond, mitigating yield loss potential, different types of expert systems are illustrated in Fig. 10.2.

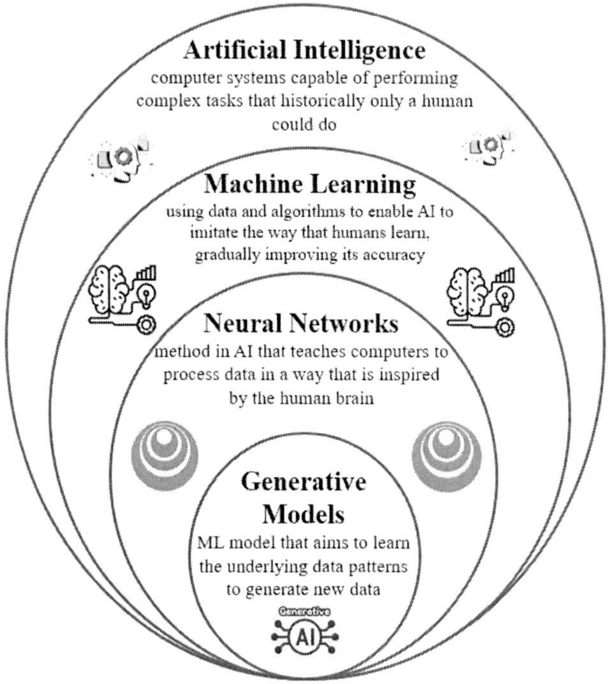

Fig. 10-2 Expert systems

10.4.1 Image Recognition Algorithms

Image recognition algorithms have opened a new frontier in crop disease detection, where computer vision-based techniques are utilized for the automated analysis of plant images. These techniques leverage algorithms that have been trained using large datasets and are capable of achieving state-of-the-art performance in terms of plant disease detection and localization. The output of a disease recognition system is the classification of an image into one of three states: healthy, with a pest, or with a disease. A system equipped with these recognition algorithms goes through a number of consecutive steps that include image preprocessing, pre-trained feature extraction, disease feature identification, statistical measurements, and a discriminant classification that distinguishes between multiple classes, such as healthy or the disease type (Kothari et al. 2024). There are multiple algorithms in image preprocessing that are used to improve the results of plant disease detection systems, such as color transform algorithms that transform the image color space, denoising filters that remove image noise, segmentation algorithms that are used to differentiate between the background and the plant, and texture analysis that highlights and enhances the disease texture features. Image recognition technologies can help promote the worldwide application of agricultural technology, reduce losses, and promote the intelligent upgrading of agriculture, but at the same time, this technology puts forward higher requirements for data acquisition and system integration. Deep learning techniques in recent years have significantly improved the robustness and generalizability of image recognition systems and are still rapidly changing and evolving (Derisma, Rokhman, and Usuman 2022; Korra et al. 2024). Each change brings people closer to the ideal algorithm: efficient, accurate, and able to adapt to the particularities of each situation; selected types of plant diseases are illustrated in Fig. 10.3.

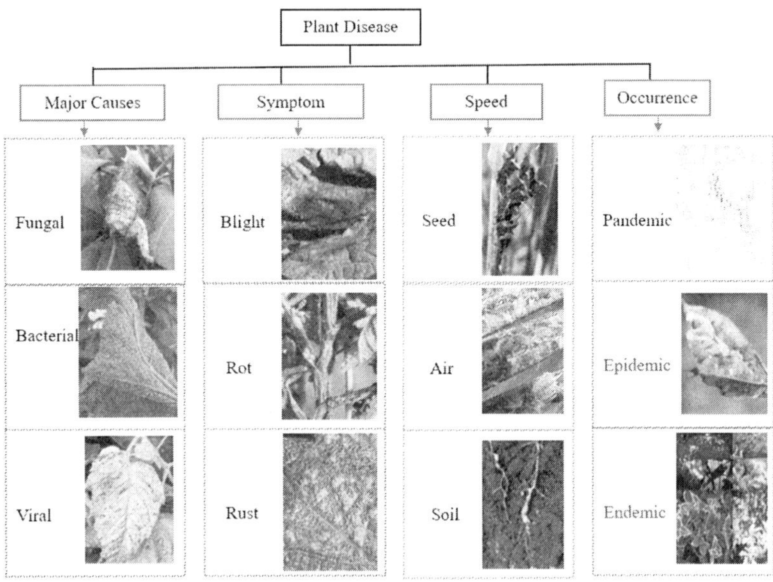

Fig. 10-3 Plant diseases

10.4.2 Deep Learning Models

In disease detection literature, deep learning has significantly improved detection rates. Deep learning (DL) networks possess the potential to improve crop disease detection accuracy. One popular deep learning architecture is the Convolutional Neural Network (CNN), which has shown significant improvement in normal image analysis and computer vision tasks and has the adaptability for detecting diseases and their severity based on crop images. Deep learning models are capable of learning complex and abstract features from a vast amount of data as raw input without performing the manual extraction of features. Deep learning, especially CNN, has been used to identify diseases in several crops. Overall, deep learning has a better accuracy rate compared to traditional machine learning algorithms. Deep learning methods require substantial computational resources (Jafar et al. 2024). When insufficient data is available, they have a greater chance of overfitting. Training quality and diversity play a significant role in the effectiveness and generalization of the resulting CNN's performance. To ensure the performance of the disease detection model, the training set should include a more diverse collection of healthy and diseased samples of a range of ages. Deep learning takes advantage of CNN architecture's ability to learn abstract features from data and has the potential to be a valuable tool for automating plant disease diagnostics, which is in the early stages of progression. The major objective of agricultural community experts is to enhance quality, increase productivity, and ensure sustainability. By using techniques of early detection, researchers and farmers can boost agricultural production by minimizing the impacts of diseases (Chin, Ng, and Palanichamy 2024a; Goyal, Verma, and Kumar 2024). In many viewpoints of agricultural productivity, deep learning models are recognized as a transforming tool, and this is traced back to the early 1950s, as demonstrated in Fig. 10.4.

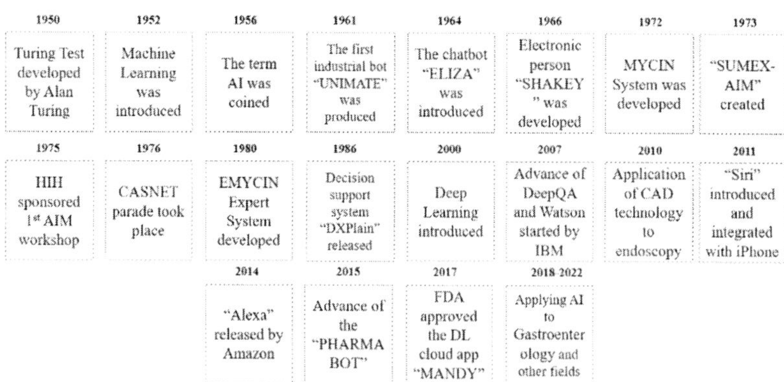

Fig. 10-4 The revolution of artificial intelligence

10.5 Internet of Things (IoT) in Agriculture

The application of the Internet of Things (IoT) technologies in agriculture has gained considerable attention in the last few years. Indeed, IoT has touched several aspects of agriculture, from precision farming to aspects like animal feeding and alert systems. In particular, IoT has provided a framework to connect many devices and sensors such that the continuously collected data is utilized efficiently using the methods of cloud computing. This aspect is particularly helpful in crop disease identification. In practical applications, there are a number of sensors that can be fitted in the field, and when connected, they make up a network (Demilie 2024). This sensor network makes it possible to know in real-time the condition of crops and the state of the field effectively. The information obtained from the sensor data can be connected to remote terminals, and after the aggregation of data, it can be used for decision-making. Many applications in real life have demonstrated that the IoT has provided a strong base for agricultural monitoring systems, including crop monitoring and disease detection. The images of the crops taken using UAVs together with sensors are an added advantage in obtaining the plant and the crop's information (Ananthi et al. 2024).

Other technologies like cloud computing and machine learning also aid in the identification of diseased crops and, subsequently, various management technologies related to precision agriculture. In a plant disease monitoring system, diseased plant leaves are recognized, and diagnosis is subsequently carried out using machine learning techniques. IoT-based disease monitoring, in combination with field imaging, makes it a compelling case for the farmer or the scientist to make earlier exploitation or suppression of the plants in the field by positioning UGVs in infected regions, and thus, the use of IoT can be extremely useful in agriculture and farming. Although the use of IoT is gaining importance with trained engineers to conduct the monitoring of crops and the management of data, there is a necessity to develop a data management system that aligns with ethical guidelines. There are some challenges in adopting IoT in agriculture (Thivya Lakshmi, Katiravan, and Visu 2024; Verma et al. 2024). The free flow of data can be invaded, and this poses a serious threat to the farmer. The policy measures of the right to data and privacy policy should be adopted to protect the farmers. Other challenges include the cost of infrastructure to install technologies, lack of awareness, the existence of gaps in the technologies and the smart policies, as well as the running costs for technology maintenance. In addition to the stresses, a major challenge is adapting the technologies to smallholder farmers. The various recent technologies are all based on developing precision farming. The usage of precision farming means that a large amount of data is being collected from the field and that this data can be used for decision-making. Technologies in agriculture, including IoT, can contribute to resource savings and food safety. In terms of environmental monitoring, IoT can aid the optimal usage of water, soils, and fertilizers by monitoring the dynamics of temperature, humidity, wind speed, and much more in agricultural fields. In combination with imaging, it is possible to optimize the usage of the resources while keeping a balance with the optimal quality of the crop. In the near future, IoT and sensors that collect the data can be connected to many devices. The agriculture monitoring system is an IoT-based system that is an ongoing trend in the current world, and the use of the Internet of Things in agriculture would be beneficial and effective (Pakutharivu et al. 2024; Maurya, Mahapatra, and Rajput 2024). Interconnecting technologies in the IoT era can enhance agricultural practices.

10.5.1 Sensor Networks

One of the key elements of IoT in agriculture systems is the use of a sensor network. There are different types of sensors, from physical sensors to perception sensors, dedicated to measuring environmental parameters associated with crop conditions. Weather sensors measure sunlight, temperature, relative humidity, and precipitation. Satellite sensors reveal the vigor of crops and differences within fields. Ground-based sensors measure indicators like moisture and temperature, and components of ground-based monitoring networks play several roles, including providing crop data for real-time monitoring to support farmers and precision agricultural methods better, bringing the chance to both avoid crop conditions that favor infection and more efficiently apply therapeutic fungicide sprays, reducing the spread of fungicide resistance (Morchid et al. 2024). When applied to vineyards, there could be savings of up to 100 million dollars. While the use of sensor networks can bring many benefits to farming, there are also challenges. In order for the environmental data measured by a sensor to be the most accurate, the sensor must be taken out of the field and sent to a lab for calibration, which is a time-consuming process. A second challenge is providing the equipment needed for sensor networks with long battery lives owing to their remote locations. More research is necessary to reveal such systems' reliability and ease of use for different growers, to demonstrate that this approach would be generalizable to other regions and farming practices. There is also a need to illustrate such systems' potential to be scaled up to regional or national crop-disease monitoring efforts and to illustrate how they could be integrated with existing farm practices and decision models (Prasad and

Thyagaraju 2024). New sensing technology provides opportunities to improve data collection efficiency. Overall, the use of sensor networks enables smarter farming using machine learning, as shown in Fig 10.5.

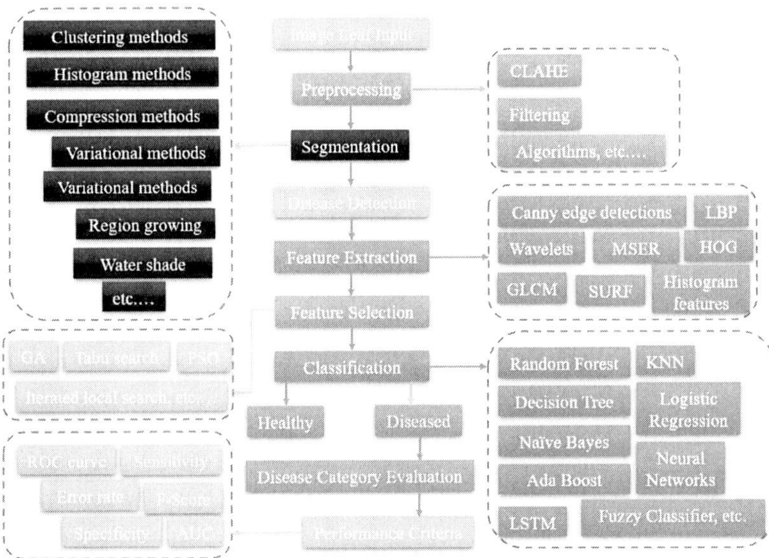

Fig. 10-5 Process of image classification

10.5.2 Data Analytics

In agriculture, large datasets from farm monitoring, consumer preferences, or soil and weather information are being collected. To handle, process, and understand these vast amounts of data, sophisticated methods have to be applied in data analytics. Data mining is part of data analytics and is defined as the process of discovering patterns in large data sets. Statistical data analysis includes the generation of descriptive and informative summaries of the data to be analyzed. By using statistical means, relationships among certain crop stress phenomena can be proved or disproved. In addition to these descriptive methods, there is a growing focus on predictive modeling. This part of data analytics looks ahead and thereby provides a helpful tool for decision-makers at the management level (Patil et al. 2023; Deelip and Govinda 2023).

Real-time data analytics are important because questions raised during crop observations cannot be answered at a later time. The fast data transfer between smartphones, tablets, or other end-user devices is important to ensure successful and immediate feedback to the user about the health of plants or soil status. In site-specific farming and precision agriculture, the tools and methods of data and information analysis are many and varied. With rapidly gaining success in enhancing business, these operations also present a great opportunity for new business and breakthroughs in yield, efficiency, or the ability to intelligently manage the application of nutrients and control plant and crop pests. Sustainability, as one of the primary principles underlying the future vision of agriculture and farming systems, will derive significant benefits from the gains achieved as a result of advanced data and information management (Pakutharivu et al. 2024). There are a number of challenges and issues surrounding data and the management of information important to agriculture and horticulture with regard to impending revolutions in the field. In parallel with the development of data analytics, the digital cloud has also been developing. Cloud-based solutions are offered extensively by large companies and infrastructures. These cloud-based platforms hold most or all of the raw agricultural field data and weather data, for example, and supply processed data to a farmer's device (Ojo and Zahid 2023).

Many farmers are using these systems, and many are unaware that they are. In accessing this data, there is often a cost. Data studies mention the cost of and supply of data on a regional scale or to large organizations rather than at a national level. Many factors can make it difficult to have a trustworthy and seamless flow of data back to a farmer's device. This comes back to the challenge of maintaining data quality in the collection and how it is processed by either the cloud platform provider or other third-party software providers. Furthermore, data analytics can also suffer due to issues such as the accuracy of the measurements and how well the data being used for analysis represents the true field conditions (Patil et al. 2023). To further investigate how successful and profitable the use of data analytics by agriculture is could be an area of research for the future. The next stage of advanced modeling and software is artificial intelligence and machine learning. These areas have a high volume of data in big cloud systems, collected over some time, and will continue to grow. For on-farm sensors, the ability of an AI or machine learning system to spot patterns in data and anomalies or provide support in decision-making is beneficial and could become more so in the future (Jamadar, Sharma, and Wagh 2023). The approach and methods under development are outlined in the next section.

10.5.3 Decision Support Systems

The integration of data analytics with user-friendly interfaces for novices in agriculture and across a range of devices is essential for the technology to allow farmers and farm advisors to manipulate digital data and develop insights into what the future might look like after a range of decisions have been made. Real-time visualization and analytics have the potential to become crucial and timely rain simulation mapping and possibly real-time/short-term forecasting with sensor data, not just for crop yield. The proposed system can act as a decision support system advising on general farming practices that aid in reducing crop losses, crop diseases, and pest invasions. The decision support system can forecast what agriculture will look like in different seasons based on simulations (Ojo and Zahid 2023). Collaborative platforms and clusters were also indicated as instruments that are required for easy access and shareability of data to multiple stakeholders in the sector. Data platforms can integrate with other systems for monitoring the data for real-time disease tracking and analytics to assist with treatment plan predictions. Although these systems offer significant gains, some hurdles have to be overcome, not least of which is the skill in interpreting the data and applying it to the farmers' particular fields. Systems that can be calibrated, based on large datasets, to a new farmer's field data and modeling in order to link and integrate with existing digitalization tools. Data within farm management systems is often available but does not provide a good segregation of data for the prospects being examined in this project (Patil et al. 2023; Deelip and Govinda 2023). The system will provide more data, which is particularly beneficial in smallholder farming because data poses a number of challenges, in particular, the growing demand for large amounts of data and lack of internet connectivity. Together, these can compromise the insights possible and the interpretation of data. The computational capabilities of the system would significantly improve the ability for longitudinal data to be stored and analyzed while the system has a range of applications that can support different parts of the comprehensive agriculture improvement system as defined. To secure the data, a secure access control mechanism can be implemented (Chin, Ng, and Palanichamy 2024a). Clearly, data analytics are anticipated as the key ingredient in shaping the future of agricultural practices depending on the plant anomalies, as illustrated in Fig 10.6.

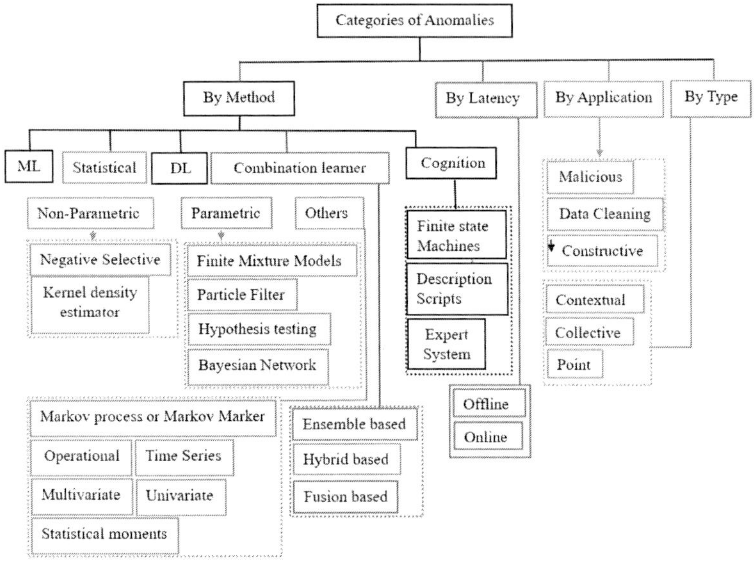

Fig. 10-6 Categories of anomalies

10.6 Challenges and Limitations

A major concern in the implementation of present-day technologies is the apprehension around data security and privacy. While the amount of data and information is increasing at a dramatic pace, farmers are hesitant to lose proprietary rights, especially as cyber threats become increasingly pervasive in modern society. Faced with an assortment of data coming in from different sources, getting these diverse data platforms to work together may present another potential limitation of current disease detection and crop monitoring. Infectious diseases, weather predictions, soil conditions, and many other dependent variables may need to interact (Thivya Lakshmi, Katiravan, and Visu 2024). Interoperability issues may be difficult to surmount and challenge the validity of some machine learning-based statistical studies if external conditions are not reported upfront. Ultimately, the utilization of multiple, incompatible devices among diverse geographical regions with a wide array of support teams, software, and hardware all seem to present challenges for the present era of smart farming with the use of data and predictions. Indeed, digital divides based on technological literacy and availability of resources, as well as shared trust and knowledge, should be addressed and better understood. Furthermore, the cost of obtaining and maintaining such advanced forms of technology does not come without a price. At first, the use of data to monitor crop disease may be cost-prohibitive. While the cost of certain handheld technologies tends to be significantly less expensive, the cost of infrastructure, computational expertise, and maintenance often required in post-processing does not come without financial implications (Maurya, Mahapatra, and Rajput 2024). The commodification of smart technologies for farmers is therefore to be done with caution and, indeed, during times of over-expectation, an awareness of limitations and increased calls for critical actualization is needed.

10.6.1 Data Privacy and Security Concerns

As IoT is crucial for agricultural development, the collection of large amounts of data goes hand in hand – which can end up being sensitive regarding data privacy and user behavior. Indeed, possible concerns regard crop monitoring data, the potential negative uses of this information in case it falls into potentially malicious hands, and also all derived data regarding early warning systems for diseases. In that regard, some works consider the collection of data as being influenced by the so-called "data privacy paradox," i.e., if potential users have concerns regarding data privacy, they are not willing to deliver the data. This shall be a primary concern: protection mechanisms could be time-consuming to implement, and some stakeholders could prefer to rely on other sources of information (Patil et al. 2023). Regarding the first question, certainly landmark frameworks and regulations can be an inspiration in ensuring data safety also in IoT for agriculture. However, these regulations do not specifically comprehend some critical aspects of IoT for agriculture, such as "zero-rated" data. This addresses this limitation by providing an integrated cross-sectoral approach with exploratory data analysis, regulatory framework reflections, and validation activities, aiming to help protect privacy, confidentiality, and data in general (Shafik, Tufail, Liyanage, et al. 2023). In particular, this report can provide evidence of the interconnected dimensions that regard privacy, IoT in agriculture, and data confidentiality and security in agriculture IoT. While addressing the appropriateness of the measures mentioned above to protect against potential physical or logical attacks on security, we acknowledge that single measures can be bypassed; thus, in the optimal scenario, they should be designed to work in synergy alongside a multi-stakeholder approach involving and respecting the rights of concerned stakeholders (Maurya, Mahapatra, and Rajput 2024).

10.6.2 Integration and Interoperability Issues

These issues are some of the many obstacles that need to be considered when a digital transformation is planned in crop monitoring. Within a certain surveillance level, different technologies can and should be used, each relying on a specific type of information and procedure. Although they represent a guarantee of flexibility and freedom in their usage, the lack of integrated protocols and formats for data exchange can lead to wasteful situations and impair the reliability of monitoring results. Producers and stakeholders should cooperate to harmonize these devices and services at different levels, all leading to the same information platform for annotations and prescriptions. Actually, platform-as-a-service models are strictly connected to a certain agricultural brand, and they do not communicate with each other through a suitable unique service and cooperative and open service-oriented computing can be seen as the appropriate ICT solution in terms of technology for system integrability (Jamadar, Sharma, and Wagh 2023). The continuous reading of information coming from territories and the continuous operations based on remote platforms will involve sharing large amounts of data among different platforms, software, and companies, and sometimes with specific authorities. In this sense, even the use of new technology could be enough to provide naturally open system communication if adopted in the LPIS. The main challenge is in sharing information among systems claimed to be functioning properly and cooperating in the same environment and predefined levels of service (Ojo and Zahid 2023).

Nevertheless, it must be highlighted that operability can also be inhibited by proprietary behavior, even for effective computer networking systems. A didactic example is given by commercial precision farming tools, which often require payment for the production of copy-protected cards, meaning the devices cannot even be used without agreeing to the producer's conditions. Practitioners and stakeholders are in need of long-term support for system evolution and collaboration in technological development among the main companies operating in the field. Encouraging access to information, competitions, and an extended view of the common interests related to the standard supply chains can lead to an increasing level of trust in open integration solutions and to a wider degree of readiness for the adoption of ICT many-to-many services. Further developments should involve new or evolved regulations on system collaboration for innovation (Patil et al. 2023). The final aim is to consider the agri-system as a unique entity, where data flow between and among existing systems and platforms for monitoring purposes is mandatory in order to achieve an integrated and complete description of the agri-phenomenon.

10.7 Future Directions and Emerging Trends

In-field data collection is advancing with connected mobile devices like mobile phones. It is conceivable that with larger mobile device screens and data glove interfaces, symptoms displayed on the crops could be taken by smallholder farmers, researchers, and customers, and then after analysis, information could be returned to them directly with informative diagnostics and potential interventions. Novel systems include terahertz lining scan cameras, which use this type of light to determine where water is in tissues, and cameras loaded with different lighting to view the underlying veins of leaves not visible to the naked eye, which can help determine the rapid plant stress. Research continues to develop more extremophile-derived farm-monitoring bacteria with field tests occurring in their efficacy. Related technologies have been listed as a key area to watch out for under emerging agri-technologies. These are a set of incentives and rules that manage a system and are used to drive collaboration toward an outcome or yield (Deelip and Govinda 2023; Patil et al. 2023). In agriculture, for example, it might be possible to see tokens issued for environmental monitoring on the farmland between livestock and crop traders and a financial institution.

10.7.1 Blockchain in Agriculture

Blockchain is considered a potential technological innovation that can improve the agricultural industry by delivering transparent operations, traceability management, and supply chain efficiency. Blockchain technology can securely store and share data on an interconnected digital ledger that various supply chain stakeholders associated with the agricultural industry can access and use. Blockchain can help trace the source of an agricultural product, its processing facilities, and the intermediaries handling the product before it even reaches the consumer. The public availability of such information on the blockchain system would prevent fraud, mislabeling, and contamination and, therefore, ensure the quality of the product and possible trace back in case of disease outbreaks. In sum, by embracing blockchain technology, the agricultural supply chain can become more transparent, where the information on crop and animal processing protocols can be shared among various players, leading to food safety. Consequently, it will benefit customers with healthier and safer food (Jafar et al. 2024; Chin, Ng, and Palanichamy 2024b). In the context of end users, blockchain enables significant benefits for tracking crops in real time while products move from the farm through the supply chain to the end consumer. Blockchain, combined with IoT sensors, can ensure the product's conditions—transparency in shipment, including temperature, on-site storage, and handling compliance, as well as blocking unauthorized environmental changes by enabling consumer-assisted conflict resolution. It can also facilitate automating transactions among farmers, warehousing, and transportation companies using smart contracts. So far, a couple of challenges are associated with the adoption of blockchain in agriculture, including technological readiness, expertise in handling, and data quality (Demilie 2024). Similarly, like all digital technologies, privacy concerns are part of adopting the blockchain system in farming, food, and feed products. Imported solutions from the developed world may not always be suitable for the existing rural and farming situations in developing countries. In the same line of thinking, concepts of construction and application of technology-focused solutions can be deemed a way to increase entry barriers on imports of agricultural products from developing countries to developed countries. Based on some empirical support, the adoption of blockchain technology has improved the areas of economic improvement, ecosystem, social, and production in small and medium agricultural enterprises and improved collaboration among supply chain actors. Some studies concluded that the application of blockchain technology is improving engagement, trust, and performance across various components of the supply chain (Kumar and Vijayakumar 2024).

10.7.2 Precision Agriculture

Would it be more apt to regard agriculture as a foreground rather than a background activity in development? Those who advocate precision agriculture certainly think so. Precision agriculture involves the application of data-driven techniques to ensure that crops receive exactly what they need when they need it. Site-specific management creates management zones within a field that are managed together, even though they may be relatively small. Most often, the interventions in these management zones are made on the basis of differences in apparent soil electrical conductivity, but other techniques, such as remote sensing and record-keeping systems, are also used. Remote sensing can be used in many aspects of crop monitoring and management, from locational disease detection to estimation of aboveground biomass. The area under crop that any individual ground-based sensor covers is limited (Bouacida et al. 2024). Ground-based sampling should thus focus on highly variable areas, which can be identified using remote sensing. Such information can be used in conjunction with ground-based demographic data, the output of remote sensing, historic yield data, and meteorological data to produce a viable model with which to predict yield. This type of remote data is used to create an intervention in the field on the basis of variability. GPS technology is a key underpinning of precision agriculture because it provides a method for tracking individual machines and the precise locations of what they are doing in a field (Shafik, Tufail, Namoun, et al. 2023b). Position can be determined to within one centimeter. GPS and remote sensing are used to provide spatial soil and other environmental data that can be transformed into decision-making information. The aim has been to evolve an ability to manage a field on a sub-field basis and hence manage a field according to the needs of the site rather than the requirements of a composite. This principle is encapsulated by the notion of all fields being zoned at any point in time (Samson Adekunle et al. 2024).

10.7.3 Technological Supply Chain Advancements

Advancements in crop monitoring and disease detection are evidently still evolving with rapid technological innovations. Some of the future breakthroughs may be directed towards soil management and improvements, making crops more resilient to diversities in climate change as well as elements produced by global warming. Relying on smart agriculture technologies, there could be an increased scale of remote process control and monitoring facilities, in turn, enabling the ease of accessing electronic data and information. AI for mechanical harvesters is likely to be a trend in the coming years (Hari, Singh, and Singh 2024; Katti et al. 2024). Even smarter precision planting will be based on variables such as soil composition, landscape of agricultural lots, weather, and moisture of growing conditions by addressing constantly growing big data. The application of blockchain technology can ensure security and transparency in the data shared within agricultural supply chains. In the age of technology and digital data, the main implication is data security and sharing based on trust among participants, the main characteristic of a blockchain. As blockchain promises to enhance transparency, producers and manufacturers could employ such technologies to make exclusive products with verified quality (Shende et al. 2024). Tailor-made solutions will be more widespread.

10.8 Conclusion

As evidenced by the discussions in this chapter, systems enabling sustainable agriculture require bottom-up input from the people most connected to food production in a given area: the farmers. Indigenous knowledge is a key part of that view of reality, compounded with the fact that most of the world's small-scale farmers are still practicing many of the integral strategies that Indigenous peoples have been utilizing for millennia. Inclusive agricultural monitoring systems that seek to quantify farmer-based observations of disease and other crop health pathology, including current technological approaches, can serve as a bellwether for understanding potential solutions and a prelude to how future technologies might be integrated. Emerging sensor technologies are showing promising potential in identifying occurrences of a multitude of crop ailments affecting commercial crops, with farmers having an enormous appetite for assimilating technological advancements into their practices. Research is only beginning to investigate the integrity of a multitude of promising technologies and methods for implementation, but several are certainly off to a promising start. While it is true that the world is constantly changing, and new diseases occur on a near-constant basis, it is this opinion that for many of the symbiotic bacterial and fungal diseases seen in agriculture, we are on the correct trajectory for ushering agriculture into an era of biotic disease resilience across an increasingly variable and ever-changing global landscape. As we continue to refine and develop the agriculture of that future, it is the expectation and hope of many that the associated revolutionary technology remains open-access, decentralized, and hyper-specific to the communities they serve. As technology and data collection practices become increasingly accessible and beneficial to producers, it is only the spread of these technologies and collaboration between researchers, policy-makers, and local interest groups that can lead to the unlocking of their full potential. Moving into the future, research on the development and integration of inexpensive and new sensors and production practices can serve as an invaluable resource for assessing technology's contributions to greater agricultural sustainability.

References

Ananthi, N., V. Balaji, M. Mohana, and S. Gnanapriya. 2024. "Smart Plant Disease Net: Adaptive Dense Hybrid Convolution Network with Attention Mechanism for IoT-Based Plant Disease Detection by Improved Optimization Approach." *Network: Computation in Neural Systems.* https://doi.org/10.1080/0954898X.2024.2316080.

Bhagat, Sandesh, Manesh Kokare, Vineet Haswani, Praful Hambarde, Trupti Taori, P. H. Ghante, and D. K. Patil. 2024. "Advancing Real-Time Plant Disease Detection: A Lightweight Deep Learning Approach and Novel Dataset for Pigeon Pea Crop." *Smart Agricultural Technology* 7. https://doi.org/10.1016/j.atech.2024.100408.

Bouacida, Imane, Brahim Farou, Lynda Djakhdjakha, Hamid Seridi, and Muhammet Kurulay. 2024. "Innovative Deep Learning Approach for

Cross-Crop Plant Disease Detection: A Generalized Method for Identifying Unhealthy Leaves." *Information Processing in Agriculture*. https://doi.org/10.1016/j.inpa.2024.03.002.

Chin, Pei-Wern, Kok-Why Ng, and Naveen Palanichamy. 2024a. "Plant Disease Detection and Classification Using Deep Learning Methods: A Comparison Study." *Journal of Informatics and Web Engineering* 3 (1). https://doi.org/10.33093/jiwe.2024.3.1.10.

Chin, Pei-Wern, Kok-Why Ng, and Naveen Palanichamy. 2024b. "Plant Disease Detection and Classification Using Deep Learning Methods: A Comparison Study." *Journal of Informatics and Web Engineering* 3 (1). https://doi.org/10.33093/2024.3.1.10.

Deelip, Mali Shrikant, and K. Govinda. 2023. "ExpSFROA-Based DRN: Exponential Sunflower Rider Optimization Algorithm-Driven Deep Residual Network for the Intrusion Detection in IOT-Based Plant Disease Monitoring." *International Journal of Semantic Computing* 17 (1). https://doi.org/10.1142/S1793351X22400165.

Demilie, Wubetu Barud. 2024. "Plant Disease Detection and Classification Techniques: A Comparative Study of the Performances." *Journal of Big Data* 11 (1). https://doi.org/10.1186/s40537-023-00863-9.

Derisma, Nur Rokhman, and Ilona Usuman. 2022. "Systematic Review of the Early Detection and Classification of Plant Diseases Using Deep Learning." In *IOP Conference Series: Earth and Environmental Science*. Vol. 1097. https://doi.org/10.1088/1755-1315/1097/1/012042.

Gahlowt, Priya, Durgesh Kumar Tripathi, Surendra Pratap Singh, Ravi Gupta, and Vijay Pratap Singh. 2024. "GABA in Plants: Developmental and Stress Resilience Perspective." *Physiologia Plantarum*. https://doi.org/10.1111/ppl.14116.

Goyal, Praveen, Dinesh Kumar Verma, and Shishir Kumar. 2024. "Plant Leaf Disease Detection Using an Optimized Evolutionary Gravitational Neocognitron Neural Network." *National Academy Science Letters*. https://doi.org/10.1007/s40009-023-01370-4.

Hairani, Hairani, and Triyanna Widiyaningtyas. 2024. "Augmented Rice Plant Disease Detection with Convolutional Neural Networks." *INTENSIF: Jurnal Ilmiah Penelitian Dan Penerapan Teknologi Sistem Informasi* 8 (1). https://doi.org/10.29407/intensif.v8i1.21168.

Hari, Pragya, Maheshwari Prasad Singh, and Amit Kumar Singh. 2024. "An Improved Federated Deep Learning for Plant Leaf Disease Detection." *Multimedia Tools and Applications*. https://doi.org/10.1007/s11042-024-18867-9.

Jafar, Abbas, Nabila Bibi, Rizwan Ali Naqvi, Abolghasem Sadeghi-Niaraki, and Daesik Jeong. 2024. "Revolutionizing Agriculture with Artificial Intelligence: Plant Disease Detection Methods, Applications, and Their Limitations." *Frontiers in Plant Science*. https://doi.org/10.3389/fpls.2024.1356260.

Jamadar, Riyazahemed A., Anoop Sharma, and Kishor Wagh. 2023. "Combining Generative Adversarial Networks with Transfer Learning for Deep Learning-Based Pomegranate Plant Leaf Disease Detection." In *Lecture Notes in Networks and Systems*. Vol. 650 LNNS. https://doi.org/10.1007/978-981-99-0838-7_42.

Katti, Jayashree, Gulbakshee Dharmale, Swapnaja Amol Ubale, Radha Deoghare, and Shiv Havgirao Sutar. 2024. "IoT Innovations in Cotton Plant Disease Detection for Sustainable Agriculture." *International Journal of Intelligent Systems and Applications in Engineering* 12 (15s).

Korra, Sampath, T. Bhaskar, N. Ramana, Sreedhar Bhukya, and Nagunuri Rajender. 2024. "An Efficient Guided Backpropagation Approach for Detection of Plant Diseases Using Deep Learning Models." *International Journal of Intelligent Systems and Applications in Engineering* 12 (10s).

Kothari, Sonali, Pooja Bagane, Manasvi Mishra, Saloni Kulshrestha, Yashika Asrani, and Vrinda Maheswari. 2024. "CropGuard : Empowering Agriculture with AI Driven Plant Disease Detection Chatbot." *International Journal of Intelligent Systems and Applications in Engineering* 12 (12s).

Kumar, G. Nandha, and V. Vijayakumar. 2024. "Texture Features-Based Detection of Plant Leaf Diseases Using RM-SVM." *SSRG International Journal of Electronics and Communication Engineering* 11 (1). https://doi.org/10.14445/23488549/IJECE-V11I1P110.

Marco-Detchart, Cedric, Jaime Andrés Rincon, Carlos Carrascosa, and Vicente Julian. 2024. "Evaluation of Deep Learning Techniques for Plant Disease Detection." *Computer Science and Information Systems* 21 (1). https://doi.org/10.2298/CSIS221222073M.

Maurya, Ritesh, Satyajit Mahapatra, and Lucky Rajput. 2024. "A Lightweight Meta-Ensemble Approach for Plant Disease Detection Suitable for IoT-Based Environments." *IEEE Access* 12. https://doi.org/10.1109/ACCESS.2024.3367443.

Morchid, Abdennabi, Marouane Marhoun, Rachid El Alami, and Bensalem Boukili. 2024. "Intelligent Detection for Sustainable Agriculture: A Review of IoT-Based Embedded Systems, Cloud Platforms, DL, and ML for Plant Disease Detection." *Multimedia Tools and Applications*. https://doi.org/10.1007/s11042-024-18392-9.

Ojo, Mike O., and Azlan Zahid. 2023. "Improving Deep Learning Classifiers Performance via Preprocessing and Class Imbalance Approaches in a Plant Disease Detection Pipeline." *Agronomy* 13 (3). https://doi.org/10.3390/agronomy13030887.

Pakutharivu, P., D. Sasirekha, V. Devaraj, and Rahul Sanmugam Gopi. 2024. "Improving Plant Disease Detection Using Super-Resolution Generative Adversarial Networks and Enhanced Dataset Diversity." *Journal of Advanced Research in Applied Sciences and Engineering Technology* 35 (2). https://doi.org/10.37934/araset.35.2.144157.

Patil, Rutuja Rajendra, Sumit Kumar, Ruchi Rani, Poorva Agrawal, and Sanjeev Kumar Pippal. 2023. "A Bibliometric and Word Cloud Analysis on the Role of the Internet of Things in Agricultural Plant Disease Detection." *Applied System Innovation* 6 (1). https://doi.org/10.3390/asi6010027.

Prasad, S. R., and G. S. Thyagaraju. 2024. "Leaf Analysis Based Early Plant Disease Detection Using Internet of Things, Machine Learning and Deep Learning: A Comprehensive Review." *Journal of Integrated Science and Technology* 12 (2).

Sai Sharvesh, R., K. Suresh Kumar, and C. J. Raman. 2024. "An Accurate Plant Disease Detection Technique Using Machine Learning." *EAI Endorsed Transactions on Internet of Things* 10. https://doi.org/10.4108/eetiot.4963.

Samson Adekunle, Temitope, Morolake Oladayo Lawrence, Oluwaseyi Omotayo Alabi, Adenrele A. Afolorunso, Godwin Nse Ebong, and Matthew Abiola Oladipupo. 2024. "Deep Learning Technique for Plant Disease Detection." *Computer Science and Information Technologies* 5 (1). https://doi.org/10.11591/csit.v5i1.p49-56.

Sarah, Mazari, Meddoudi Abdlemadjid, Benameur Sarah, Hadjadj Yacine, and Chikr El Mezouar Miloud. 2024. "Evaluating the Effect of Super-Resolution for Automatic Plant Disease Detection: Application to Potato Late Blight Detection." *Multimedia Tools and Applications*. https://doi.org/10.1007/s11042-024-18574-5.

Shafik, Wasswa. 2024. "Biosensor-Based Drones Anomaly Detection Integration for Sustainable Agriculture Development." In *Achieving Food Security Through Sustainable Agriculture*, 283–316. IGI Global. https://doi.org/10.4018/979-8-3693-4240-4.ch012.

Shafik, Wasswa. 2025. "Sustainable Agriculture and Diet in the Metaverse Era: Sustainable Farming and Nutritious Eating in the Metaverse." In *Food in the Metaverse and Web 3.0 Era: Intersecting Food, Technology, and Culture*, 159–84. IGI Global Scientific Publishing.

Shafik, Wasswa, and Kassim Kalinaki. 2024. “Societal and Ethical Implications of Technology-Enhanced Agriculture and Healthcare: An African Context.” In *2024 IST-Africa Conference (IST-Africa)*, 1–11. IEEE. https://doi.org/10.23919/IST-Africa63983.2024.10569306.

Shafik, Wasswa, Ali Tufail, Chandratilak De Silva Liyanage, and Rosyzie Anna Awg Haji Mohd Apong. 2023. “Using a Novel Convolutional Neural Network for Plant Pests Detection and Disease Classification.” *Journal of the Science of Food and Agriculture*. https://doi.org/10.1002/jsfa.12700.

Shafik, Wasswa, Ali Tufail, Abdallah Namoun, Liyanage Chandratilak De Silva, and Rosyzie Anna Awg Haji Mohd Apong. 2023a. “A Systematic Literature Review on Plant Disease Detection: Motivations, Classification Techniques, Datasets, Challenges, and Future Trends.” *IEEE Access* 11:59174–203. https://doi.org/10.1109/ACCESS.2023.3284760.

Shafik, Wasswa. 2023b. “A Systematic Literature Review on Plant Disease Detection: Techniques, Dataset Availability, Challenges, Future Trends, and Motivations.” *IEEE Access* 11 (June).

Shafik, Wasswa, Ali Tufail, Chandratilak De Silva Liyanage, and Rosyzie Anna Awg Haji Mohd Apong. 2024. “Using Transfer Learning-Based Plant Disease Classification and Detection for Sustainable Agriculture.” *BMC Plant Biology* 24 (1): 136. https://doi.org/10.1186/s12870-024-04825-y.

Shafik, Wasswa, Ali Tufail, Liyanage Chandratilak De Silva, and Rosyzie Anna Haji Mohd Apong. 2025. “An Enhanced Deep Convolutional Neural Network for Plant Disease Detection and Classification: Elevating Sustainable Agriculture.” In *Artificial Intelligence and Data Science for Sustainability: Applications and Methods*, 297–322. IGI Global Scientific Publishing.

Shende, Devyani, Laxman Thakare, Rahul Agrawal, and Nikhil Wyawahare. 2024. “Automatic Disease Detection for Various Plants Leaf Using Image Processing Techniques and TensorFlow Algorithm.” In *Lecture Notes in Electrical Engineering*. Vol. 1115. https://doi.org/10.1007/978-981-99-8661-3_36.

Sofuoglu, Cemal Ihsan, and Derya Birant. 2024. “Potato Plant Leaf Disease Detection Using Deep Learning Method.” *Tarim Bilimleri Dergisi* 30 (1). https://doi.org/10.15832/ankutbd.1276722.

Thivya Lakshmi, R. T., Jeevaa Katiravan, and P. Visu. 2024. “CoDet: A Novel Deep Learning Pipeline for Cotton Plant Detection and Disease Identification.” *Automatika* 65 (2). https://doi.org/10.1080/00051144.2024.2317093.

Verma, Shradha, Anuradha Chug, Amit Prakash Singh, and Dinesh Singh. 2024. "Plant Disease Detection and Severity Assessment Using Image Processing and Deep Learning Techniques." *SN Computer Science* 5 (1). https://doi.org/10.1007/s42979-023-02417-5.

Weber, Sven, Manuel Stothut, Lisa Mahla, Alanah Kripp, Lena Hirschler, Nina Lenz, Anneke Junker, Sven Künzel, and Henrik Krehenwinkel. 2024. "Plant-Derived Environmental DNA Complements Diversity Estimates from Traditional Arthropod Monitoring Methods but Outperforms Them Detecting Plant–Arthropod Interactions." *Molecular Ecology Resources* 24 (2). https://doi.org/10.1111/1755-0998.13900.

CHAPTER ELEVEN

SUPPLY CHAIN OPTIMIZATION THROUGH ENABLING TECHNOLOGIES FOR SUSTAINABLE AGRICULTURE

Abstract

Globalization and organizations are facing increased competition, which compels businesses to restructure their supply networks. This reformed supply chain infrastructure requires a comprehensive and advanced framework. An optimum owned or controlled value chain cannot be used as a competitive edge for a long period. In multitiered supply chains, this impediment can only be overcome through multi-entity collaboration. Hence, the main focus of businesses is to improve supply chains by moving toward an integrated approach rather than isolated ones. The objective of this chapter is to introduce some of the traditional and emerging challenges facing supply chains and to discuss a variety of enabling technologies. Then, we represent the information for further improvement of supply chain performance by vertical integration opportunities. Modern supply chains face a plethora of traditional challenges, including, but not limited to, demands for cost reduction and lead time reduction coupled with demands for levels of product and service quality comparable or superior to those delivered in the past and continually shifting global competitive conditions including the ever-increasing emphasis on supplier consolidation and globalization. Supply chains also face a myriad of new and novel challenges, including engineered variability of products, increasing product proliferation with reduced life cycles and more extensive regulatory conditions, and network modeling and optimization requirements, including protection from surprise developments. One of the biggest global challenges is to integrate existing supply chain management software solutions to provide a horizontal communication interface that provides seamless data exchanges. The challenges identified above indicate, in part, the complexity of our global and competitive environment and the need for technologies that are specifically designed and evaluated with an emphasis on enabling supply chain optimization, network integration, improved efficiencies, and enhanced supply chains overall.

Keywords: Agricultural Sustainability, Blockchain in Agriculture, Data-Driven Decision Making, Enabling Technologies, Green Supply Chains, Precision Agriculture, Smart Logistics, Supply Chain Optimization, Sustainable Agriculture

11.1 Introduction

Supply chain management (SCM) is the management of the flow of goods, information, and money as a product makes its way from the raw material stage to the end consumer. SCM aims to streamline the entire process of producing, storing, and delivering products to retailers and customers. One of the most important ways in which this can be achieved is through the better coordination of the activities that must take place among a complex network of companies that make up a typical supply chain entity. In a basic framework, the SCM process includes suppliers, manufacturers, and distributors, who move goods between multiple companies before they eventually find the end customer in a retail store (Sofi and Zahoor 2023). Coordination between a number of the members in the supply chain is often necessary to exponentially improve the performance of the chain and for it to consistently produce better value at lower cost. This need for supply chain coordination and management has driven enormous research and technological development from the 1970s to today, turning SCM and logistics into a vast, complex, and incredibly valuable segment of the global economy. Over the last few decades, supply chain management (SCM) has been getting more attention (Di Martino et al. 2023).

The need for SCM is for the purpose of meeting service levels and costs and, more importantly, to gain a competitive advantage. As we know, supply chain management has been defined as the network of different entities to deliver products and necessary services from the supplier to the consumer. It also defines and manages the procedures, the quantity, and the mix of the materials, how the materials need to be stored, how they are delivered, and when they are delivered. SCM also provides strategy-based coordination and integration of these functions within a company and across companies into the supply chain as a holistic approach (Sasikala and Bharatula 2024).

Supply chain management, in broad terms, is the strategic planning, execution, and monitoring of specialized logistic services within a supply chain network. These services include, among others, the acquisition of raw materials, customer demand forecasting, production planning, product/component/inventory management, and distribution, warehousing, transportation, and customer order management. The objective of optimized supply chain management is to support competitive advantages and an organization's overall strategies with service-level goals, operational performance efficiency, and financial effectiveness. Critical to achieving this objective and organizational benchmarks is the management of three key supply chain flow categories: raw materials/components, in-process, and finished products (Zhao, Wang, and Pham 2023). Furthermore, the perception of an organization's supply chain should not be perceived solely as a linear entity composed of chronological functions but instead as a collaborative mesh of interconnected and supportive organizations and processes for all downstream and upstream participants and activities. Supply chain management's purpose has sub-processes, including planning, acquiring, making, moving, selling, and customer order service. In addition, an effective SCM practice should establish clearly defined roles within a management consensus and ensure that supply chains are managed to and from finished goods sources. Regional SC network policies may be internal (within an organization between business units or operating companies) or external (extending among trading partners) (Ramasamy et al. 2024).

Although supply chains comprise a wide array of activities, the following four components are considered vital for the success of an integrated supply chain. The principal upstream player in a supply chain is the supplier, also known as the first-tier supplier. Suppliers feed their products or pitch raw materials to manufacturing companies, i.e., the manufacturers or original equipment suppliers. They, in turn, distribute the semi-finished products or finished goods to another tier, known as distributors or wholesalers. Finally, distributors pass on the products to various retail outlets where customers may eventually purchase them (Lakshmiprabha and Govindaraju 2023). Logistics involve the management or coordination of the flow of information, material, and money between different companies. Inventory management is, of course, a key activity in supply chain management and is viewed as an interface between the actual physical movement of goods and the flow of information related to it. The flow of information is a vital link that ties together the supply chain elements and is viewed as a fifth component in the chain. Information about customer requirements must efficiently flow upstream through the supply chain network so that appropriate production and inventory adjustments can be made. In the same manner, actual inventory and delivery status should be communicated in order to enable proper scheduling. All these chains and connections together make the supply chain demands grow in geometric progression (Dharashive and

Sawale 2024). This dynamic interrelationship involving various companies, from raw material suppliers to end customers and the activities it involves, forms the concept of supply chain management. The efficiency of interconnected partners is the preference that leads to the simple word "optimization." The explanation of the "cost effect" has become the driving force for constructing the strategy that performs the correct quantile backfill. The performance enhancement offered by a single player gives the efficacy of the whole system. This is the limit on the way extended optimization can produce the assuring result (Lakshmiprabha and Govindaraju 2023). Thus, the purpose is to pinpoint various key factors or processes where technological construction may provide the best recycling effect so that the entire network becomes more recurrent, as illustrated in Fig 11.1.

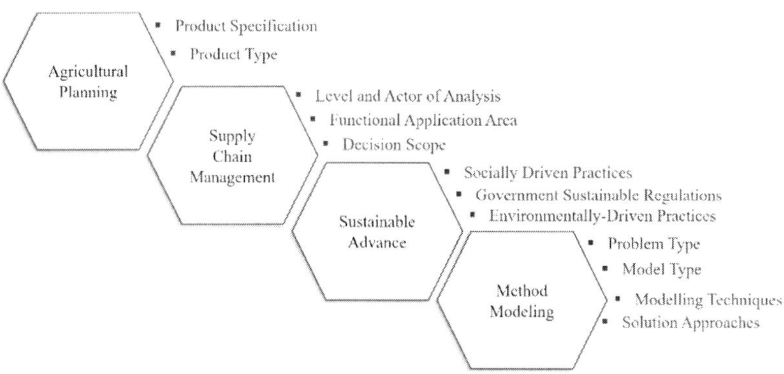

Fig. 11-1 Supply chain optimization and its traits in Agriculture

Supply chain optimization is vital for the agriculture sector, as it can contribute to addressing the various challenges related to food security, environmental protection, and climate change. In addition to environmental sustainability and economic viability, the supply chain should be inclusive among all actors from farm to consumer. Technological development is imperative for society to achieve efficient supply chains. There is a need for enabling technologies that perform smart agriculture processes, reduce the use of natural resources, increase environmental quality, and reduce the environmental impact of the agri-food supply chain. Efficient supply chains multiply the benefits of optimal inputs, reduce wastage, increase quality and safety, optimize land utilization, encourage the production of multiple crops according to soil and climate, source larger markets, and increase yields and farmer incomes. Producers get the best price, while consumers receive quality produce without any agent-deducted charges (Zaina, Fadli, and Hosseini

2022). Technology development may lead to trade impacts in local agri-food supply chains. Therefore, fast exchanges of technology might not be a better approach, as this can lead to large-scale production that may result in environmental degradation. Moreover, trade may lead to a loss of diversity in local ecosystems, further undermining the operations of agri-food systems. The available ICT could be used to harness information that can be utilized to design a sustainable agri-supply chain. The leading question is centered around "What are viable ways to optimize agri-food supply chains using existing technological and institutional tools?" or "How may existing technologies help in reducing trade liberalization impacts in agricultural supply chains?" The main objective of the research is to identify possible approaches and methodologies that can be used in optimizing current agri-food supply chains and suggesting ways and procedures to take advantage of liberalizing trade offers rooted within a normative framework of sustainable agriculture issues (Chalvantharan, Lim, and Ng 2023). Expertise from a range of different disciplines is applied, providing a wide-ranging portfolio of options for developing an approach to this essential optimization aspect of agricultural trade. Semi-structured interviews, participatory research, ICT matching, and case studies have thus been used as research strategies. Data has been collected from both the developing and developed worlds (Zhao, Wang, and Pham 2023).

11.2 Theoretical Framework

Supply Chain Management is an integrative management philosophy that emphasizes integration and management in relationships between and among companies, suppliers, and customers. Its main purpose is to deliver the right amount of products at the right time and in the right place in response to customer demands based on the fast synchronization of supply chain flows. The principles of supply chain management raise a number of research topics. In the topic of optimization of agricultural supply chains, theoretical models have been widely used to solve complex questions based on different perspectives adopted to describe agricultural supply chains because researchers have different research goals. Our literature survey indicated that many models in the present literature do not fully consider some important aspects, such as enabling technologies, which are important parts in the optimization of agricultural supply chains. This research sets a theoretical framework (Masseroni, Arbat, and de Lima 2020). The center of this framework is the theoretically built principles of supply chain management that emphasize the integrated flow of products, information, and finance, as well as evaluate supply chain performance maximization by reducing the gap between customer value and supply chain cost. Notably, in this interest, this framework also reveals that enabling technologies are the technologies that are integral parts of the principles of supply chain management given their important role in the effectiveness of supply management principles. It can be stated that the role of enabling technologies is to increase the effectiveness of supply management principles; when different enabling technologies are implemented, their potential benefits can be increased. This framework also indicates the interaction between technology and flow to increase efficiency (Gamal et al. 2023). From the definition of interest and perspective, this gives a subtle principal review of two viewpoints of supply chain management principles and highlights the integral role of enabling technology in increasing the effectiveness of supply chain management using data analytics, as demonstrated in Fig 11.2.

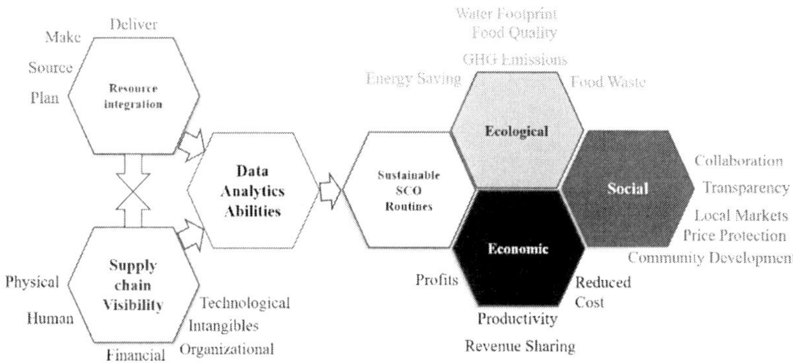

Fig. 11-2 Application of Data analytics abilities for sustainable supply chain management

11.2.1 Supply Chain Management

A supply chain is a network of suppliers, manufacturers, distributors, retailers, and consumers working together to meet the needs of a target group as efficiently as possible. The agricultural supply chain is a network of interconnected and interdependent components that process and transform inputs into outputs that can be traded in the market through respective value chain stakeholders from farmer to consumer. The management of this network is called supply chain management. Pioneer investments in supply chain management have been substantial in the area of reducing supply costs, but now supply chain management has substantial implications for sustainable agriculture.

Supply chain functions include, among others, sourcing (for example selecting suppliers), procurement, storage, transport, transformation, distribution interchange with consumers, and information sharing (García et al. 2020). The performance of agricultural supply chains is jointly determined by the performance level of each stakeholder in the process from the sourcing area up to consumption. To enhance the efficiency of supply chains, the stakeholders have begun to pay more attention to how they can collaborate rather than pursue individual gains independently. Several problems may occur in the supply chain arrangements. These include rapid fluctuations in demand due to price slashes, such as buying in fractional quantities, and dislocated consumption time and place. The existence of these problems has affected the performance of the supply chain. Supply chain management has been viewed as a means to enhance stakeholders' capabilities for increasing their responsiveness and flexibility toward uncertainty and risk (Suresh et al. 2024).

11.2.2 Disruptive Technologies in Agriculture

The rapid digital evolution in agriculture that was initiated years ago is mainly driven by technological discontinuities in affordable information and communication technologies, rural internet coverage, social media, and big data. These technologies are grouped under the term enabling technologies that drive innovations in agriculture and have potential and scalability for direct efforts towards efficiency, economic growth, food quantity, and food quality. The adoption of enabling technologies in agribusiness greatly contributes to a balanced supply chain through cost reduction, a lower price of goods to final customers, and better quality of the products and services offered. The way they apply and combine enabling technologies for specific use cases can minimize farmers' risks, protect scarce resources and the environment, including water and climate, and increase farmers' profits. Therefore, applying advanced enabling technologies in agriculture can give birth to a new digital agriculture paradigm that nurtures the health of the soil, plants, farmers, food quality, consumers, rural development, and the earth (Younes et al. 2024). The potential of enabling technologies is so great that farmers' traditional customs and knowledge would finally be allied with scientific progress in digital agriculture. This can provide recommendations to farmers about accurate crop monitoring, precision real-time crop management, optimization of the bio-production process for accurate and efficient resource management, efficient and effective pest control, inputs and outputs quality, agronomical services, and capturing economic and environmental aspects at the farm and landscape levels. Although digital agriculture and enabling technologies have prosperous impacts on agriculture, many farmers and agribusiness companies have still not been able to apply these enabling technologies directly (Vallejo-Gómez, Osorio, and Hincapié 2023). They rely mostly on capability development and local agricultural conditions.

11.3 Challenges in Supply Chain Management

Many organizations implement supply chain management techniques to optimize the flow of materials and information, reduce operational costs, and improve performance. However, contemporary supply chain management faces numerous simultaneous challenges. These challenges can be characterized as traditional and emerging. Traditional challenges arise due to three focal areas: transportation delays, inventory inaccuracies, and demand variability. Challenges in these focal areas have proliferated into a multifaceted hierarchy of interrelated issues that require top-notch management to navigate successfully. Emerging challenges may be segregated into the following primary areas: external concerns related to globalization, technological concerns related to change and the shift of enabling technologies, internal concerns linked to business-to-business e-commerce business strategy that targets developing technological superiority, and associated social concerns of changing consumer expectations (Abdulaziz and Croock 2022; Vallejo-Gómez, Osorio, and Hincapié 2023).

Operations and supply chain management face a myriad of challenges in the modern business environment. These multifaceted challenges have converged over time into complex operational systems that are dispersed over a variety of geographical regions. Stemming from this situation is the realization that supply chains are increasingly vulnerable to a large number of unexpected disturbances. Navigating effectively through these challenges often means the difference between venturing and performing tolerably or poorly in any given year. These multifaceted challenges represent unique yet interrelated variables. In-class discussions have shown that a variety of challenges related to supply chain management can be identified (Sadiq, Wanke, and Azad 2024). By categorizing these challenges into the traditional and emerging varieties, we can develop a deeper level of comprehension of the interrelated operational hurdles that need to be managed more effectively in businesses today, as shown in Fig 11.3.

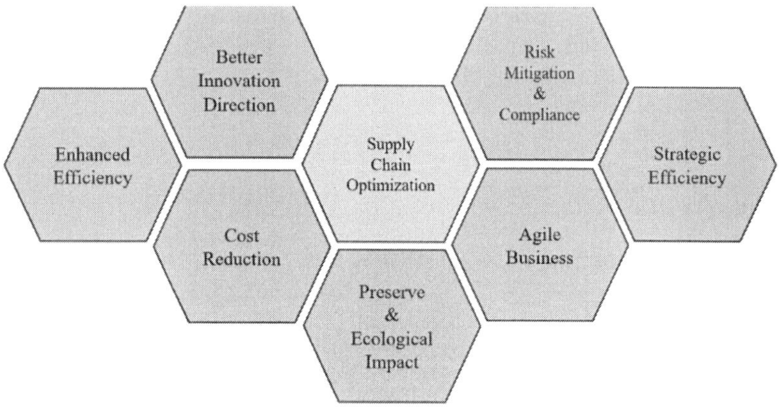

Fig. 11-3 Supply Chain Optimization

11.3.1 Traditional Challenges

Many different factors can affect the success of a supply chain. Optimal selection of the strategies and tactics for addressing the impact of these factors is the focus of supply chain management. Inefficiencies in any part of the supply chain, whether in management, logistics, or manufacturing, have a leverage effect on the supply chain as a whole, so supply chain performance has become an important revenue enabler. In the past, companies have struggled with some traditional challenges, which have resulted in operating costs reaching up to 40% of any firm's revenue. These challenges include the presence of excess inventory in the supply chain that did not contribute towards the value of the final product (Nashwan 2022). Variations in customer demand had an impact on the stock inventory, leading to forecasting inaccuracies of up to 10% of the January sales volume, causing the company also to hold unused stock. Other issues that companies have been experiencing were related to supplier reliability. Inefficient, slow, or costly services would significantly decrease customer satisfaction levels. To solve these issues and to get ahead of the competition, the executive team adopted this powerful, state-of-the-art supply chain model software tool. When a simple linear regression model was done, the results indicated that an increase in the company's reliability could lead to a doubling of the profit, reaching up to R10 million. Traditional challenges in supply chains show inefficiencies within logistics, manufacturers, and management processes, resulting in high operational costs (Scarlatache et al. 2021; Lunstad and Sowby 2024).

11.3.2 Emerging Challenges

Driven by an evolving business landscape, companies are facing increasing challenges related to their supply chains. The effects of globalization, the ongoing trend of digital transformation, the rapid adoption of new technologies, and the development of capabilities for generating and analyzing supply chain data have led to more complex, interconnected supply chains that rely on information, digital assets, and trusting relationships between partners. Supply chains have also become more vulnerable due to a number of external disturbances, where both primary and complementary data are subjected to extensive threats, thus increasing the number of potential joint effects. Information in supply chains represents a vulnerable point susceptible to cybersecurity threats and subsequent reverberations through the network. Another challenge relates to actions taken by governments, who are committed to increasing both the volume and intensity of rules and regulations designed to remedy flaws in the existing compliance-incentive systems (Zaina, Fadli, and Hosseini 2022).

Regulatory compliance is expected to have major effects on supply chains, in many cases falling under the master responsibility of suppliers or supply chain partners. The logistics industry and its use of road, air, and sea account for at least 1,000 million tons of CO_2 emissions annually. In an increasingly global world, both experts and the general public show increasing concern related to climate change and its negative effects on logistics and supply chains. The linkages between these challenges are shown by the dispersion of air through several high-echelon points, presumably through close partners, demonstrating the risk transfer to a partner higher in the analytical ecosystem. These potential issues and threats can cause significant disturbances among organizations and their customers. Thus, companies must be agile and develop informed strategies to accommodate the dynamism of the business environment. Companies must move to relational models where they design supply chain systems capable of performing regardless of and taking advantage of risky or uncertain future systemic events (Lakshmiprabha and Govindaraju 2023; Ramasamy et al. 2024). Not only is it important to have good risk assessments to help understand the nature of the risks an organization is facing, but organizations also need to understand the readiness of their supply chain system to take advantage of newly identified vulnerabilities resulting from new ways of doing business, legislation, and regulation. Specific industry standards, practices, and prescriptions, especially in engineering frameworks, underline how companies should review their sourcing strategies and supply chain design in light of changing business environment constraints (Sofi and Zahoor 2023). This environment, specifically the emerging trends described earlier, is discussed in further detail in the following section.

11.3.3 Seasonality and Variability

Seasonal demand is probably one of the most remarkable characteristics of agrifood supply chains. For most crops, planting and harvesting decisions have to be made yearly based on demand and climate expectations, which results in very frequent demand forecasts. To make matters worse, these forecasts have to be done without the actual product information at the beginning of the sales season. Slow-moving products, perishable goods, and long manufacturing times increase the vulnerability of the supply chain to these errors. The perishability in the agrifood supply chain has been tackled as a challenge and also as a competitive advantage through enabling technologies for optimization. One of the consequences of seasonality is that incomes and costs for farmers are concentrated in a single point of the year, which has implications for the willingness of farmers to withhold production until next year if they perceive that prices are too low (Masseroni, Arbat, and de Lima 2020). A lower bound for the sales price is determined by this willingness that is, in turn, influenced by the production but also by the farm's financial situation. The character of agricultural production, with a concentration of most commercialization at one time of the year, imposes difficulties in planning and organizing the chain, as well as placing a significant risk on the chain. This seasonality is dictated by the cycle of the plant or animal, which is influenced by genetics and agrotechnology, and the weather, which is unpredictable and varies from year to year. Also, the different logistical competencies needed by the various chain partners to plant, grow, harvest, process, transport, and market the product over time cause internal supply chain inefficiency (Younes et al. 2024). All supply chain members that are forced to deal with the seasonality and variability are actually forced to supply, bring, and operate stock, with appropriate inventory management needed.

11.3.4 Transportation and Logistics

An essential component of any supply chain is the logistics framework necessary to transport goods from the farm gate to the consumer at a reasonable price with adequate quality information. The cost of transportation in an agricultural supply chain, at the start of the chain, impacts the overall profitability of the industry. Logistics is also characterized by a high level of regulation and uncertainty associated with interstate, local, and on-farm biosecurity, granting credentials for food transport, fuel pricing, oil supply, and the movement of vehicles. The ongoing problem of a lack of critical infrastructure, particularly in regional areas, coupled with high transport costs, constantly rates as the most significant cost of business on the farm. Reducing these costs reduces farm gate prices. A significant portion of the population in developing regions lives more than 90 minutes away from the nearest city because of inadequate transport and other infrastructure (Obaideen et al. 2022). The importance of fresh food reaching the final consumer with a reasonable shelf life, in turn, affects what a producer would invest in capital to manage speed to market, as illustrated in Fig 11.4.

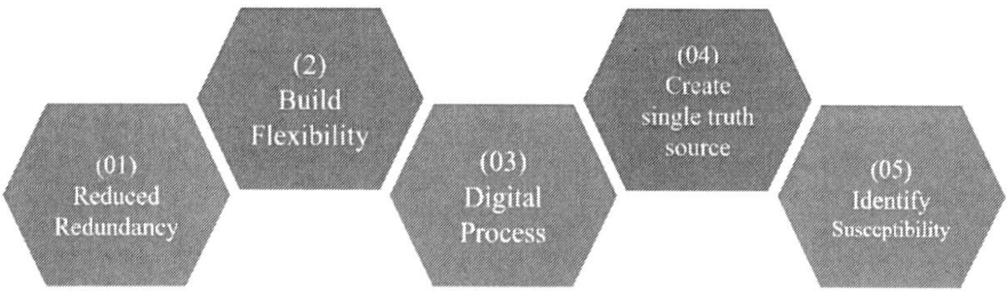

Fig. 11-4 Five supply chain management processes

Sustainable transportation systems aim to minimize greenhouse gases and the use of fossil fuels. Given the challenges associated with overproduction, transportation-based waste management is imperative. One of the significant problems for producers lies in the high perishability rates of fruits, vegetables, and other temperature-sensitive products. Fresh products lose from 10% to 45% of their mass, most of it being water content, during transportation. Ineffective transportation strategies also lead to substantial waste since the demand for products is difficult, if not impossible, to predict. Slow-moving goods are usually sold at a steep discount before they lose their market value entirely. Given the importance of efficient logistics, shippers often prefer to keep the products under their watch or contract with a logistics organization for further transportation and handling. The use of collaboration and synchronization among producers in order to create full loads has been suggested to be one of the likely strategies that could be used to improve efficiency in transport between producers (Abdulaziz and Croock 2022). This is particularly relevant for the delivery of inputs such as feed and chemicals and the return delivery of grain.

11.4 Enabling Technologies in Supply Chain Optimization

The supply chain can generate about 75% of the shareholder value in a firm, and logistics represents the major part of this chain. This suggests that intelligent logistics can eventually improve shareholder value. Enabling technologies plays an important role in logistics and creates visibility, awareness, and intelligence in supply chain operations. Advanced technologies are now available that can support innovative ways to work with distribution operations and trigger the transformation toward a more strategic supply chain function. From a technical perspective, enabling technologies provide the possibility to collect real-time data, which can be used to create insight and support logistics operations (Abdulhamid et al. 2024). Several enabling technologies can be identified, including sensor technology, networking technologies, identification, tracking and tracing systems, standardization and regulation of these systems, automation, and optimization systems. Two advanced technologies create new possibilities for the supply chain. First, developments in the field of the Internet are creating connectivity possibilities that we have never experienced before. Second, improvements in data analytics provide a foundation on which we can build solutions that help make smarter decisions faster. Combined, these two elements bring a significant new and extended capability to optimize business processes, leading to a step change in performance. Supply chain optimization is able to improve connectivity across all internal and external partners of the supply chain, as well as insight generation to guide decision-making throughout the supply chain. An organization must have an innovative mindset, a holistic planning process, and strong logistics to excel in the competitive environment (Abdulaziz and Croock 2023). Logistics must also be integrated with the innovation value chain process to generate value for the firm.

11.4.1 Internet of Things (IoT)

In recent years, the IoT has been positioned as a potential game changer in logistics, inspiring changes in the way supply chain processes are managed. The IoT is the concept in which day-to-day objects are equipped with the capability to connect to the Internet, allowing them to send and receive data through digital networks in order to enable new opportunities for information, analytics, and automation. By equipping assets with identification and data processing capacity, 'smart' products can be tracked and digitally monitored through the supply chain. Such digital tracking gives rise to new possibilities for further automating supply chains, inventory levels, and even reorder points (Bazaluk et al. 2022). The IoT can gather and process vast amounts of data to enable a range of applications at every stage of the supply chain. This connectivity will encompass both goods and people and is influencing the design of supply chain management, where the main goal is to enable accurate, precise, reliable, and timely physical flow with data-driven containment of uncertainty. IoT data can enable operations monitoring which triggers real-time corrective actions, providing efficient and effective input parameters for replenishment algorithms, and reducing food wastage through dynamic expiration calculations. 'Smart' items with IoT devices and sensors may, for example, include temperature, humidity, position, and impact sensors, RFID, NFC, or BLE chips for identification and any form of data storage, smart tags with audible information, LED displays, or even an LCD screen for relevant information (Lunstad and Sowby 2024).

11.4.2 Big Data and Analytics

Big data stored in databases contains a plethora of knowledge that can help in decision-making. Big data can significantly enhance the decision-making process at various levels of the supply chain. This subsection studies the development of characteristics of big data. Predictive analytics is extensively used in this area to study customer behavior, inventory, and demand forecasting. In simple terms, big data is data that cannot be stored or processed traditionally. It has three distinct Vs of size, i.e., volume, variety, and veracity, and by some, it is also described with an additional V, i.e., velocity. In retail, various sources of big data include customer verbal interactions, chat sessions, emails, social media interactions, web blogs, and articles. All of these interactions and their analyses can be used to analyze customer feedback; for example, customer dissatisfaction can be extracted from data where only one or two are dissatisfied customers (Zaina, Fadli, and Hosseini 2022). Apart from customer interaction data, various other operational metrics and measurements are stored via big data storage. From this huge amount of data, businesses derive the best possible combination of goods and services that are more attractive to their customers. Access to multiple online sources generates a large amount of data. Based on this data, businesses can track the buying behavior of consumers, which can be compared with the number of times a product is bought. Primarily based on this data, supply chains can control inventory levels. If the time taken to ship the products and goods decreases, the inventory level will rise in the future, sampled trends are illustrated in Fig 11.5. Apart from making supply decisions, the use of big data is helpful in pricing decisions as well (Ramasamy et al. 2024).

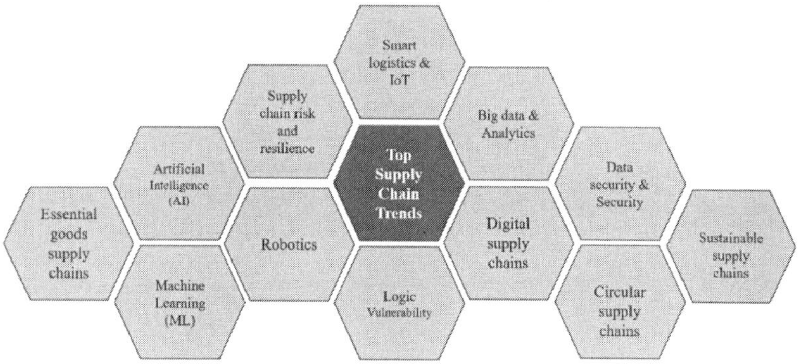

Fig. 11-5 Top supply chain trends

The scope of analytics is broadly classified into three categories, i.e., descriptive analytics, predictive analytics, and prescriptive analytics. Descriptive analytics is a set of techniques that analyze historical data to identify trends, patterns, and relationships. Descriptive analytics quantifies observations to summarize or represent them in a manageable form. Descriptive analytics helps supply chain businesses understand what has happened, and it also uses past performance data gathered from customer feedback, sales, and other sources to set the foundation for business activities. Predictive analytics is the use of statistical algorithms and machine-learning techniques to identify the likelihood of future outcomes based on historical data. Predictive analytics attempts to forecast future events by examining the historical data stored in databases (Sasikala and Bharatula 2024). It mainly examines the relationships between different data elements to forecast future values of a dependent variable based on changes in the values of independent variables. Prescriptive analytics is a set of techniques that seeks to determine the best alternative from all feasible solutions. Prescriptive analytics extends beyond prediction to not only forecast what will happen in the future but also why it will happen, and offers specific actionable recommendations for taking advantage of the forecast or dealing with an adverse outcome. Prescriptive analytics can continually take in new and external data to re-prescribe (Lakshmiprabha and Govindaraju 2023).

11.4.3 Precision Agriculture

The signals from satellites helped to develop autonomous agricultural vehicles and geo-referenced soil sampling. Remote sensing provides nutrient status, hydration in plants, and much more. Along with GPS and remote sensors, data analytics systems to process and visualize data are essential. In principle, this technology makes it possible for a farmer to manage a field as a collection of small plots with different conditions rather than an aggregation where the variability is hidden. Site-specific activities can then be optimized. Practitioners of precision agriculture often describe the four Rs: right rate (of seed, nutrients, chemicals, and even water), right place, right time, and right variety. Usually, input-intensive agriculture does not maximize efficiency or top-line financial benefits. The anticipated benefits are the reduction of input as a result of efficient management. This is both economically and ecologically profitable because there is an increase in yield as well (Sadiq, Wanke, and Azad 2024; Abdulaziz and Croock 2023). There is also an increased adoption of organic farming and the certification of sustainable practices mainly because of such data-leveraging technologies. Precision agriculture is of great importance as it allows us to minimize the environmental impacts and, in general, to manage the value of ecosystems and the services they provide. Clearly, the potential and current applications of digital technologies in agriculture are extensive. There are considerable challenges, including data management and protection, better integration with other digital and enabling technologies, regulatory and policy-making, enabling infrastructures, and the provision of digital skills, training, and user support. Indeed, it seems highly probable that digital technologies that focus on precision agriculture will have a disproportionately high uptake from farmers who are strongly oriented toward the practice of sustainable agriculture (Gamal et al. 2023; García et al. 2020).

It has been demonstrated that the uptake of such precision tools transfers financially. One review confirmed the direct financial benefits as greater than the costs associated with implementing the technology. One key barrier to the uptake of precision agriculture is that the skills and aptitudes required are diverse and non-trivial. High levels of numeracy combined with non-technical skills such as an independent ability to make decisions and interact with others effectively. An interesting sign of the importance of digitalization to the environment relates to the growing interest in agriculture, as the big companies make efforts to develop a strong foothold in agribusiness. A growing issue in agriculture is the huge density of chemical and nutrient inputs, with associated high risks and inputs of fuel and transport costs. We can now extract data from the soil that used to require invasive methods and apply this to detection and yield prediction. It seems very probable that the digital toolbox will be used to develop a precision agriculture where all farm operations are managed in a more optimized and eco-friendly way (Abdulaziz and Croock 2022; Abdulhamid et al. 2024). It is a high research priority to support the development of such practices.

11.5 Integration of Enabling Technologies in Supply Chain Management

The integration of technologies within the supply chain management practices of business organizations, known as enabling technologies, can help to optimize and automate the supply chain, as well as result in better-informed decision-making. IoT, big data, and robotics process automation are among the emerging technologies that can facilitate the reshaping, re-evaluation, and transactional re-execution of the supply chain processes to make them optimized, self-driven, and agile. The successful integration of technological enablers is a strategic decision. For example, the executive-level management of a multimodal logistics global integrator decided that process automation was a necessity for the organization to create a customer-led supply chain. The functional approach highlights various operations wherein the integration of enabling technologies is performed. One function brings a big picture view, wherein the approach to integration is from the strategic and managerial decision-making (Nashwan 2022). A big pharma company has adopted product lifecycle management software, integrating R&D with the production, supply chain, and market intelligence units. By choosing that software, the company is able to use the advice from a validation project. Several cases point to the relevance of enabling technologies in the organization being studied. There are, however, some issues that need to be kept in mind when integrating enabling technologies. Change management and resistance to change are important barriers to IT adoption. Thus, the mindset of managers should be one of change management, who are able to involve a wide range of supply chain members in the implementation of the change. The findings show that the capabilities of the supply chain can be enhanced through the technologies adopted (Chalvantharan, Lim, and Ng 2023).

The successful adoption of such approaches by a wide variety of organizations shows that there can be no 'correct' system for integration. Consequently, to illustrate that while it will be difficult for one organization to directly learn from the experiences of another, because such systems must be case-specific, it is still possible to benefit from these studies. This section shows how different enabling technologies can be selected and used in integration strategy. The case studies are applicable to organizations and industries in different parts of the supply chain, interested in a variety of technologies to improve supply chain efficiency (Lakshmiprabha and Govindaraju 2023; Dharashive and Sawale 2024).

By studying a variety of different organizations, from the resilient supply chain in automotive manufacture to carrier and retailer-integrated services, to the supply point tracking of oil changes, this offers a completely different perspective providing practicable guidance to organizations when attempting to plan and construct enabling technology initiatives. Each of the case studies in this research has also come to appreciate the complexity associated with the transformation of organizational buying and selling experiences through the intelligent and strategically customized application of innovative tools. At a 'high level', our case studies will reflect the advances that organizations can hope to make if they are willing to make available the resources necessary to deliver upon a vision that is supported by four strategic LVs (Abdulaziz and Croock 2022). One organization presents a supply chain that is in large part suited to the organization’s philosophy by developing refined resilience. This robust supply chain, with its formal and informal lubrication, drives a constant process of supply chain improvement. Another organization is similarly resilient through the strength of its marketing engines qualified to integrate with offering services from carriers and, more recently, retailers. The platform delivers a fully functional and robust toolkit to a large user base designed to sell oil changes through thousands of skilled and semi-nomadic installers within the framework of a very stubborn dual franchise network (Chalvantharan, Lim, and Ng 2023).

11.6 Impact of Enabling Technologies on Supply Chain Performance

A large share of studies has been devoted to investigating efficiency and cost reduction. From the results of the preceding, it appears that a more efficient supply chain allows quick response to market demands, lowers inventory levels and material handling costs, improves the speed of material flows, and decreases the number of lost sales. In addition to the reduction of operating costs and efficiency improvements, the capability to share real-time data directly from one participant to another enhances asset utilization. It has been observed in this study that the participants believe in and have evidence of real performance improvements from exploiting these technologies. Concerning the methods of evaluating the performance of supply chain technology, the scholars suggest various sets of KPIs depending on the factors that mostly affect performance. The costs of logistics, response times, market demand reactions, service to the customer, extension of technology into the organization, competitive performance, cost of supply chain operation, asset utilization, and risk management at the top-management level can be used as KPIs of the supply chain. Furthermore, increased visibility and transparency are enabled by information system advancements that allow the further utilization of valuable resources, which are important for the successful operation of companies in these competitive markets.

Simulation allows companies to 'see' the real impacts of changes to their operations on costs and service levels and provides valuable information to decision-makers about trade-offs. Currently, a significant percentage of companies use supply chain modeling and simulation technologies.

11.6.1 Efficiency and Cost Reduction

Traditionally, one of the main drivers of investment in enabling technologies has been to provide the potential to increase the efficiency gains achievable within a supply chain. Many technologies are available that are focused on the efficient streamlining of business operations. Warehouse automation systems, such as goods-to-person, are relatively recent examples of this type of technology receiving considerable attention in the industry. At a management level, real-time data and business intelligence technologies have long been established to allow the efficient interpretation and utilization of key performance indicators and other operational statistics to manage across multiple sites. These efficient technologies are used to gain the highest possible advantage from existing supply chain structures, reducing loss and waste and increasing the output of the supply chain. Any given goods or information may be seen as requiring a certain amount of processing and movement. Any time that this requirement is reduced represents an inefficiency saving. Typically, there is a cost associated with each unit of processing and movement of goods and information (Lakshmiprabha and Govindaraju 2023).

As a result, reductions in this time required directly translate into cost savings. Streamlining technologies are, therefore, directly translated into cost savings in many instances. Clearly, reducing the cost base of the supply chain will automatically increase the profits of the company using the chain. Processes and technologies that promote the efficient movement of products will also help companies to become more responsive to demand. As previously discussed, a responsive supply chain is more able to grow than a purely efficient one. While there may be some capability for an efficient supply chain to compete on cost, this is typically short-lived in a global market where the lowest possible cost is often the arbiter of supply. The cost advantages of an efficient supply chain are rarely long-lived. Companies may also struggle, in the long term, to respond to both retailers' and consumers' demands without a responsive aspect to their supply chain. The advantages of the marriage of efficiency and responsiveness give some insight into why many of the published business case studies for supply chain technologies and processes promote the view that faster, more efficient supply chains are the answer to profitability, growth, and market share (Sasikala and Bharatula 2024; Zhao, Wang, and Pham 2023). There is, however, no one-size-fits-all set of solutions; different strategies are appropriate for different types of supply chains.

11.6.2 Visibility and Transparency

One of the keys to the optimization of a supply chain system has always been information. Visibility in the supply chain rises through the use of enabling technologies, including sensors and devices, collaboration software, yield sourcing technologies, and blockchain. Greater transparency and a better view of the overall supply chain can make it easier for organizations to increase trust and collaboration among partners, as well as provide an extra layer of visibility and control out to the final point of sale. This includes advanced tracking that enables all parties to visually see and confirm the authenticity of their purchased products, which is very important to brand owners and their supply chains. Blockchain's distributed ledger and automated smart contracts will be instrumental in helping businesses automatically manage the measures in a transparent, traceable, and secure way. Enhanced operational transparency will also drive greater customer satisfaction with the timely and relevant information that can be shared transparently, including, among other things, expanded viewing and selection options for tracking purchase shipments in a B2C world or machine assets in the case of machines. Supply chain visibility can directly and indirectly impact a company's overall performance as well as its competitive position (Di Martino et al. 2023). By having real-time visibility, a company has better control over its supply chain as it can detect problems at an early stage and thereby mitigate them. Real-time visibility provides the organization with up-to-date information on the shipment status and identifies the impact on the supply chain. By being able to assess the impact at an early stage, the organization can react immediately and either put in place a solution or, equally, bring another line if necessary.

There is risk all around us, and that includes supply chain risk. Whether it's cyberattacks, natural disasters, stockouts, or other catastrophes such as fraud, organizational entities must balance the need to protect their supply chain investments with a laser focus on cost control. Regulatory requirements mean increased demands on businesses to have more information and to produce it more quickly and report on it. Implementing IoT capabilities in supply chains can dramatically help in many different types of risk management, whether it's hazardous materials management, asset management, or perishable risk management. Supply chain and distribution transparency provide more than just increased visibility to prevent spoilage; it also enhances the usability and trustworthiness of producers (Sofi and Zahoor 2023). It increases the collaboration capabilities of all partners in the supply chain. Ensuring the success of traceability programs helps keep more collaborative supply chains working together to deliver what the final customer wants. The level of visibility and transparency differentiates from organization to organization. Their adoption can result in better operational performance as well as offering competitive differentiation. In effect, this will lead organizations to optimize their operations (Abdulaziz and Croock 2023).

11.7 Future Trends and Innovations in Supply Chain Optimization

The future of supply chains is closely intertwined with the latest developments in technology. Innovations in artificial intelligence and machine learning have been advancing at an alarming rate in many aspects of corporate business life. Much of this is still in the early stages of exploration. A particularly hopeful aspect of their application to supply chain management emerged with intelligent predictive engines of accumulated data over a long period. Companies are now attempting to use AI and machine learning in predictive analytics for supply chain management to help with trade-offs and optimize flow, output, expenditures, and 3D printing demand for better decision-making. AI and machine learning in supply chain management can optimize downtime, predict breakage, and reduce repair costs in machinery. As blockchain technology moves beyond cryptocurrencies, it is set to revolutionize the supply chain and transform transparency and trust in all elements of the supply chain (Nashwan 2022). Companies are looking at decentralized public ledger technology to improve visibility, constraints, and transferability of information and enhance skill, safety, and active devices. Software and consulting companies are starting to develop blockchain supply chain processors. Many companies are beginning to market blockchain system services involving consensus. This growing popularity of blockchain as a vehicle for transparency is of increasing interest within supply chains, not just in manufacturing but also in agri-food. These early adopters become smarter and stay a step ahead of the competition. It is essential to know future trends for supply chain optimization, as they are dynamic and shift in accordance with evolutionary market forces. Providing a company with this new knowledge positions it to optimize the supply chain in their respective industries (Ramasamy et al. 2024).

11.7.1 Expert Systems

Artificial intelligence (AI) and machine learning (ML) have the transformative power to drive supply chain optimization profoundly. AI can include complex decision-making processes such as those involved with supply chain decision-making. AI is enabled when all this data, processes, and systems are put together to serve a specific business function. AI systems can peruse the massive amount of structured and unstructured data to reveal insights, offer suggestions, and ultimately take action. AI-powered techniques automate the interaction between data and systems to deliver insights. AI drives highly sophisticated data-related processing, enabling the handling of big data. Data scientists and trained AI models draw on the technology of machine learning (ML), which finds patterns in data. AI and ML-based supply chains help automate standard processes. In the supply chain, these techniques and the underlying data can be leveraged to optimize many areas, such as forecasting and inventory management, transportation management, procurement, logistics, warehousing, and demand management (Sofi and Zahoor 2023; Shafik 2024e).

At the same time, AI could reduce oversupply margin errors through better demand forecasting. The broad set of data available for analysis will enhance these forecasts. Excluding manual tasks in the supply chain that AI can manage will save time and, therefore, money. Labor costs can be reduced by automating menial tasks that AI can handle effectively, freeing up human resources for more mission-critical activities. When AI automates routine decision-making processes in a supply chain, the higher the rate of return on humans remains focused on targeted, actionable results stemming from analyzing AI findings or executing improved supply chain decision-making based on AI output. Such advanced capabilities of AI can lead to optimized talent management and workforce planning. AI will statistically predict the optimum times to execute certain tasks or move products from inventory. As the ML algorithms that power AI become more sophisticated and more data are available, the high demand for emerging supply chain capabilities can be met. There are some business and ethical considerations with AI to consider. AI-powered supply chains are the future (Shafik 2024a; 2024c). These advances in technology can be a game-changer, and organizations that adopt these technologies will have a competitive advantage.

11.7.2 Blockchain Technology

At its core, a blockchain is a distributed ledger composed of records called blocks, which are linked together using cryptography. Usual blockchains gain their high-security properties by taking full advantage of main conceptual theories like decentralization, immutability, and consensus. Decentralization, where no single entity operates, is a guarantee against corruption or manipulation of facts. Immutability ensures that once a record is created in a blockchain, it can neither be altered nor erased. Consensus on a blockchain ensures that transactions posted on a blockchain are legitimate. By combining public key infrastructure, hash functions, consensus systems, and peer-to-peer networks, any organization with access to the internet can build a permissioned blockchain system to accommodate its record-keeping needs. Applications have become known in areas such as supply chains, identity verification, and voting systems (Shafik 2024d; Das et al. 2023).

However, various challenges exist. One challenge is the scalability of blockchain networks. Another challenge is the interoperability of different blockchains. It is expensive to transition to blockchain technology from existing systems, and small organizations often do not have enough resources to accomplish the conversion. With a verifiable chain of custody for products, a blockchain can help make the supply chain transparent. However, not all supply chains can be accommodated by a blockchain. Blockchain technology allows the establishment of comprehensive networks of suppliers with techniques to ensure their trustworthiness. Since the blockchain technique has appeared in the supply chain domain where no similar efforts have been examined earlier, it is believed that blockchain is one of the technologies that may matter for the redesign of future supply chains (Sadiq, Wanke, and Azad 2024). The future of supply chains needs trust, resilience, curiosity, and an inclination to innovate.

11.8 Summary and Recommendation

It is a useful division of work, as the technologists can focus mainly on the end-user actors in the value chain, while rural sociologists have the time and knowledge and can devote more energy to working with farmers in the vertical supply chain of primary production where they can work with stakeholders of operator networks. They provide a useful platform for early engagement and for drawing out new spaces of innovation in developing a shared vision of the project goals, drafting guiding principles and governance, and fostering an active community of practice and bilateral links among developers and early adopters, who can work ahead on value creation and business model design for the new horizontal supply chain and get the data to drive the social life cycle assessment of the pilot chain in relation to its social objectives.

11.8.1 Summary of Findings

The role of enabling technologies in making agriculture's supply chain more sustainable and agile, as well as some of the framework conditions that will make this possible, have been identified. With the development of enabling technologies, the benefits to stakeholders all along the supply chain of a 4.0-enabled agricultural supply chain are expected to start multiplying. Enabling technologies will not only allow traditional and new players to move from complicated, time-consuming, manual chain activity aggregation and data transactional processes but also provide instant access to extensive data collection and analysis possibilities, real-time tracking, and quality controls based on this data for cost-cutting and efficiency improvements. Improvements across the chain depend on a number of gateway and framework conditions that we need to ensure are in place in order to truly unlock and extend these benefits. No matter how good the technology is, the business case to invest in is critical. Enablers for and barriers to the implementation of these technologies have also been reported. Issues around the cost-friendliness and simplicity of filming and technology accessibility were crucial (Lunstad and Sowby 2024). Thus focus on stakeholders' perceptions of weak links in the value chain, as well as their confidence in data security and willingness to share data, was highlighted. Voluntary sharing of data across the network is linked to trust in the security of the infrastructure.

First, it has provided insights into the benefits of enabling technologies, with the use of complete traceability and targeted data-led operational decision-making being seen as the most beneficial. Cost-cutting and greater return on investment were cited, as well as the possibility to decrease the environmental effects of agriculture (Shafik 2025b). These are really important factors because they align with the objectives of the agricultural supply chain, innovation, and enhancement of the operating model. The linkage of these results will assist our stakeholders, as well as those who manage goods, services, and business competitors in their portfolio, in a more sophisticated manner to differentiate the true focus and selling points that must be emphasized when promoting this research in combination with others. There are a number of critical challenges that need to be addressed by our stakeholders if the goals and potential of technologies are to be achieved (Chalvantharan, Lim, and Ng 2023). Firstly, the cost of building the infrastructure is essential since it will affect the overall viability of the supply chain and its deployment. Due to this, affordability can be linked to and is still evident now. This applies not only to the technology infrastructure but also to the wages and minimum capabilities required to cover the workers. Given the increasing competition, this constitutes a significant problem for some businesses.

Thirdly, self-imposed resistance and reliance are also problems with the ability to adopt enabling technology in SMEs. All of this is linked to system lock-in, which is a related issue. In comparison to the evidence in the market, the efficiency benefits of 4.0 are perceived to be driven by the proposals from some organizations and networks to provide agile and open access, irrespective of whether the data portfolio and folklore are a unique sales concern or an underused asset. Networking and cooperation with stakeholders allow for a better understanding of the efforts required to modernize and digitize the small-to-medium-sized environment (Nashwan 2022).

11.8.2 Practical Implications

Our findings provide insights into how farmers and agribusinesses can integrate enabling technologies into the decision-making process to increase the efficiency and sustainability of their operations along the agricultural supply chain. In the previous sections, we highlighted barriers to and facilitators of the integration of enabling technologies in agriculture and food production. These are essential when considering the next steps involved in making the supply chain more sustainable. In this subsection, we provide some practical recommendations to overcome those challenges (García et al. 2020).

The integrated enabling technologies are able to provide an opportunity for every stakeholder involved in the chain to optimize their processes. However, this will require collaboration among stakeholders within a supply chain to exchange information and have processes tailored for integration (Shafik 2024b). This reiterates the importance of collaboration and cooperative practices, internally and externally, within an agribusiness. We suggest that to address the challenge of data silos and create a digitalized agri-food supply chain, a process approach should be taken, and reasonable expectations should be set about the time and resources needed to complete the transformation (Vallejo-Gómez, Osorio, and Hincapié 2023). Sharing non-sensitive data like market information, pest outbreaks, and weather conditions is an example of incentive alignment with network users. Policies and incentives to motivate the sharing of personal information and reward investments in these technologies are also recommended. Political support is needed to accelerate these changes. We also recommend that NGOs, agribusinesses, and policyholder stakeholders work together to make recommendations that suit all the parties involved as well as the environment (Zhao, Wang, and Pham 2023).

11.8.3 Lessons Learned

From the reviewed literature, a number of lessons have been learned related to the implementation of enabling technologies in the supply chain to promote sustainable agriculture. These lessons are as follows:

1. Enabling technology experiments is essential for initialization purposes.
2. Experimentation with enabling technologies with different operators, for example, an alternative broker, is beneficial.
3. The use of enabling technologies for sustainable agriculture in interactive settings shows interesting results.
4. It is desirable to conduct the experiment when the forest products supply chain is in operation.
5. Flexibility in the use of these three enabling technologies is required for practical reasons.
6. If limiting the use of three enabling technologies for social validation is required, the system should be allowed. Seventy-four percent of participants agreed with this option.
7. There were three main challenges for assessment, namely the business environment within the supply chain settings, resistance to change by operators, and the sustainability of interventions, including the impacts outlined in the evaluation reports that are freely accessible.

Lessons have been learned during the two stages of solicitation from callers' perspectives, either engaging directly or receiving feedback on alternative technologies that were offered. These are as follows. It is important, as well as interesting, to gather views about the use of enabling technologies in supply chain experiments and settings and to determine whether they can be implemented in real-life scenarios (Shafik and Kalinaki 2024). The views expressed were based on the limited exposure to and understanding of the technologies. The system has to be flexible since there needs to be some learning and adaptation of how it functions. Willingness to use the system for industrial forestry management would be limited if it was designed to be stand-alone and not interrelated to a business or commercial software system. Participants will only engage if there are some substantial benefits to the adoption of the subject technology or successful, proven pilots for similar real-life conditions that provide good outputs and inputs to the application user and further ways forward in the actual commercial decision-making process. Benefits were interpreted from an operator's perspective and are thus related to its utility and functionality (Shafik 2025a). Participants in forest product and farming supply chains stressed that it would be useful to work with stakeholder groups to assess the benefits and trade-offs of the applications, particularly in relation to applying cumulative effects rather than just facility-level assessments. This echoed informal feedback indicating the need for some validation of the improved effectiveness of the approach to support better decision-making.

11.9 Conclusion

This chapter has taken a closer look at enabling technologies and how they can be used to optimize supply chains. In summary, there are many examples of traditional challenges that supply chain managers have always struggled with, such as bullwhip effects and increasingly more emerging challenges, such as customer brand expectations or climate change and natural catastrophes. Enabling technologies can help businesses understand, prevent, or minimize these challenges. Therefore, the strategic position of an organization should be chosen to match the intended use of technology; focus strategies need a reasonable level of technological intervention, which also needs to match the changing role of people in supply chains. A proactive approach is recommended to make use of technology as an entry barrier, ensuring long-term success. As a first attempt in this area, the following concluding recommendations are presented for insurgents:

1. Integration - the enabling technology should be integrated with internal and external business partners;
2. Forward integration can be the prologue to a co-creation platform;
3. Create a business model that provides value to the business itself and can be attractive to potential customers and
4. Appropriation of resources and competencies: when appropriate, investment can help to ensure differentiation, making the organization's marketplace offer more difficult to imitate.

Further research is required to corroborate these recommendations for practitioners and to test their relevance. The whole area of the strategic role of technology represents an enticing domain for the completion of studies in supply chain management.

References

Abdulaziz, Wid Badee, and Muayad Sadik Croock. 2022. "Optimized Power and Water Allocation in Smart Irrigation Systems." *Periodicals of Engineering and Natural Sciences* 10 (3). https://doi.org/10.21533/pen.v10i3.2998.

Abdulaziz, Wid Badee, and Muayad Sadik Croock. 2023. "Review on Cyber-Security for Optimized and Smart Irrigation Systems." *Telkomnika (Telecommunication Computing Electronics and Control)* 21 (1). https://doi.org/10.12928/TELKOMNIKA.v21i1.24234.

Abdulhamid, Alhassan, Md Mokhlesur Rahman, Sohag Kabir, and Ibrahim Ghafir. 2024. "Enhancing Safety in IoT Systems: A Model-Based Assessment of a Smart Irrigation System Using Fault Tree Analysis." *Electronics (Switzerland)* 13 (6). https://doi.org/10.3390/electronics13061156.

Bazaluk, Oleg, Valerii Havrysh, Vitalii Nitsenko, Yuliia Mazur, and Sergiy Lavrenko. 2022. "Low-Cost Smart Farm Irrigation Systems in Kherson Province: Feasibility Study." *Agronomy* 12 (5). https://doi.org/10.3390/agronomy12051013.

Chalvantharan, Arivintharan, Chun Hsion Lim, and Denny K.S. Ng. 2023. "Economic Feasibility and Water Footprint Analysis for Smart Irrigation Systems in Palm Oil Industry." *Sustainability (Switzerland)* 15 (10). https://doi.org/10.3390/su15108069.

Das, Puja, Moutushi Singh, Dimitrios A. Karras, and Deepsubhra Guha Roy. 2023. "Block-A-City: An Agricultural Application Framework Using Blockchain for Next-Generation Smart Cities." *IETE Journal of Research*. https://doi.org/10.1080/03772063.2022.2162982.

Dharashive, Prashant, and Manish Sawale. 2024. "Smart Irrigation System Techniques Using Artificial Intelligence and IoT." *International Journal of Intelligent Systems and Applications in Engineering* 12 (11s).

Gamal, Yomna, Ahmed Soltan, Lobna A. Said, Ahmed H. Madian, and Ahmed G. Radwan. 2023. "Smart Irrigation Systems: Overview." *IEEE Access*. https://doi.org/10.1109/ACCESS.2023.3251655.

García, Laura, Lorena Parra, Jose M. Jimenez, Jaime Lloret, and Pascal Lorenz. 2020. "IoT-Based Smart Irrigation Systems: An Overview on the Recent Trends on Sensors and Iot Systems for Irrigation in Precision Agriculture." *Sensors (Switzerland)*. https://doi.org/10.3390/s20041042.

Lakshmiprabha, Kakkanallur Ethirajan, and Chinnathambi Govindaraju. 2023. "Hydroponic-Based Smart Irrigation System Using Internet of

Things." *International Journal of Communication Systems* 36 (12). https://doi.org/10.1002/dac.4071.

Lunstad, Nathan T., and Robert B. Sowby. 2024. "Smart Irrigation Controllers in Residential Applications and the Potential of Integrated Water Distribution Systems." *Journal of Water Resources Planning and Management* 150 (1). https://doi.org/10.1061/jwrmd5.wreng-5871.

Martino, Beniamino Di, Luigi Colucci Cante, Antonio Esposito, and Mariangela Graziano. 2023. "Towards a Methodology for the Semantic Representation of Iot Sensors and BPMNs to Discover Business Process Patterns: A Smart Irrigation Case Study." In *Lecture Notes in Networks and Systems*. Vol. 570 LNNS. https://doi.org/10.1007/978-3-031-20029-8_24.

Masseroni, Daniele, Gerard Arbat, and Isabel Pedroso de Lima. 2020. "Editorial-Managing and Planning Water Resources for Irrigation: Smart-Irrigation Systems for Providing Sustainable Agriculture and Maintaining Ecosystem Services." *Water (Switzerland)*. https://doi.org/10.3390/w12010263.

Nashwan, Shadi. 2022. "Secure Authentication Scheme Using Diffie–Hellman Key Agreement for Smart IoT Irrigation Systems." *Electronics (Switzerland)* 11 (2). https://doi.org/10.3390/electronics11020188.

Obaideen, Khaled, Bashria A.A. Yousef, Maryam Nooman AlMallahi, Yong Chai Tan, Montaser Mahmoud, Hadi Jaber, and Mohamad Ramadan. 2022. "An Overview of Smart Irrigation Systems Using IoT." *Energy Nexus* 7. https://doi.org/10.1016/j.nexus.2022.100124.

Ramasamy, Palanisamy, Nagarajan Pandian, Krishnamurthy Mayathevar, Ramkumar Ravindran, Srinivasa Rao Kandula, Selvabharathi Devadoss, and Selvakumar Kuppusamy. 2024. "Design of Arduino UNO Based Smart Irrigation System for Real Time Applications." *International Journal of Reconfigurable and Embedded Systems* 13 (1). https://doi.org/10.11591/ijres.v13.i1.pp105-110.

Sadiq, Md Ijaj, Peter Wanke, and Md Abul Kalam Azad. 2024. "Smart Irrigation Systems Enabled with Internet of Things: A Bibliometric Review." *International Journal of Process Management and Benchmarking* 16 (1). https://doi.org/10.1504/IJPMB.2024.135751.

Sasikala, S, and S D Bharatula. 2024. "IoT-Based Smart Irrigation System Based Adaptive Radial Deep Neural Network (ARDNN) Algorithm Applicable for Various Agricultural Production." *International Journal of Intelligent Systems and ...*.

Scarlatache, Florina, Gheorghe Grigoras, Vlad Andrei Scarlatache, Bogdan Constantin Neagu, and Ovidiu Ivanov. 2021. "A Hybrid Methodology

Based on Smart Management Energy Consumption in Irrigation Systems." *Electronics (Switzerland)* 10 (22). https://doi.org/10.3390/electronics10222864.

Shafik, Wasswa. 2024a. "Artificial Intelligence and Machine Learning with Cyber Ethics for the Future World." In *Future Communication Systems Using Artificial Intelligence, Internet of Things and Data Science*, 110–30. Boca Raton: CRC Press. https://doi.org/10.1201/9781032648309-9.

Shafik, Wasswa. 2024b. "Barriers to Implementing Computational Intelligence-Based Agriculture System." In *Computational Intelligence in Internet of Agricultural Things*, 193–219. Springer. https://doi.org/10.1007/978-3-031-67450-1_8.

Shafik, Wasswa. 2024c. "Biosensor-Based Drones Anomaly Detection Integration for Sustainable Agriculture Development." In *Achieving Food Security Through Sustainable Agriculture*, 283–316. IGI Global. https://doi.org/10.4018/979-8-3693-4240-4.ch012.

Shafik, Wasswa. 2024d. "Blockchain-Based Internet of Things (B-IoT): Challenges, Solutions, Opportunities, Open Research Questions, and Future Trends." In *Blockchain-Based Internet of Things*, 35–58. Boca Raton: Chapman and Hall/CRC. https://doi.org/10.1201/9781003407096-3.

Shafik, Wasswa. 2024e. "Deep Learning Impacts in the Field of Artificial Intelligence." In *Deep Learning Concepts in Operations Research*, 9–26. New York: Auerbach Publications. https://doi.org/10.1201/9781003433309-2.

Shafik, Wasswa. 2025a. "Sustainable Agriculture and Diet for Urban Development." In *Sustainable Smart Cities and the Future of Urban Development*, 165–92. IGI Global Scientific Publishing.

Shafik, Wasswa. 2025b. "Sustainable Agriculture and Diet in the Metaverse Era: Sustainable Farming and Nutritious Eating in the Metaverse." In *Food in the Metaverse and Web 3.0 Era: Intersecting Food, Technology, and Culture*, 159–84. IGI Global Scientific Publishing.

Shafik, Wasswa, and Kassim Kalinaki. 2024. "Societal and Ethical Implications of Technology-Enhanced Agriculture and Healthcare: An African Context." In *2024 IST-Africa Conference (IST-Africa)*, 1–11. IEEE. https://doi.org/10.23919/IST-Africa63983.2024.10569306.

Sofi, Shabir A., and Saniya Zahoor. 2023. "Internet of Things-Enabled Irrigation System in Precision Agriculture." In *Smart Innovation, Systems and Technologies*. Vol. 327. https://doi.org/10.1007/978-981-19-7524-0_18.

Suresh, P., G. Jenifa, S. Srithar, G. Johncy, and R. H. Aswathy. 2024. "ASI (Agriculture Smart Irrigation) Multiparameter Optimization System for

Precision Agriculture." *International Journal of Intelligent Systems and Applications in Engineering* 12 (8s).

Vallejo-Gómez, David, Marisol Osorio, and Carlos A. Hincapié. 2023. "Smart Irrigation Systems in Agriculture: A Systematic Review." *Agronomy*. https://doi.org/10.3390/agronomy13020342.

Younes, Abiadi, Zouhair Elamrani Abou Elassad, Othmane El Meslouhi, Dauha Elamrani Abou Elassad, and Ed dahbi Abdel Majid. 2024. "The Application of Machine Learning Techniques for Smart Irrigation Systems: A Systematic Literature Review." *Smart Agricultural Technology* 7. https://doi.org/10.1016/j.atech.2024.100425.

Zaina, Sara M., Fodil Fadli, and Seyed Morteza Hosseini. 2022. "Evaluation of Smart Irrigation Systems in Hot-Arid Climates for Green Roofs and Walls: Case of Doha, Qatar." *Smart and Sustainable Built Environment* 11 (2). https://doi.org/10.1108/SASBE-11-2021-0201.

Zhao, Wei, Meini Wang, and V. T. Pham. 2023. "Unmanned Aerial Vehicle and Geospatial Analysis in Smart Irrigation and Crop Monitoring on IoT Platform." *Mobile Information Systems* 2023. https://doi.org/10.1155/2023/4213645.

CHAPTER TWELVE

SUSTAINABLE AND DIGITAL AGRICULTURE: A BROADER ECONOMIC, POLICY AND TECHNOLOGY PERSPECTIVE

Abstract

Agriculture is a crucial part of our existence, and in present times, it is pertinent to discuss farming from various perspectives, ensuring that it remains sustainable. The link between technology and sustainability is key to investigate. We present an in-depth analysis of this critical relationship, namely how today's agriculture can benefit from the introduction of classic and digital green practices covering both broader societal values, such as food safety, environmental protection, citizens' well-being, and private stakeholder benefits, such as economic advantages for farmers and agro-industry. The rationale of this background paper is, therefore, to integrate these different dimensions, spanning from economics to public policy and technology, to provide a critical discussion in a policy-oriented framework that can become a valuable reference for experts and stakeholders working in the agricultural and rural development sectors. The document investigates a comprehensive body of research that integrates these different perspectives that are briefly presented here. Globally, food production is expected to increase significantly to meet new demands. Furthermore, there is increasing pressure from climate change, which is creating new geographical patterns of food production. Agriculture requires innovative solutions to solve these challenges, and today's technology provides new solutions. It is, therefore, particularly important to focus on agriculture in the debate on sustainability.

Keywords: Agricultural Policy, Agricultural Sustainability Policy, Agri-Tech Innovations, Climate-Resilient Farming, Data-Driven Agriculture, Digital Agriculture, Economic Development, Precision Farming, Smart Farming Solutions, Sustainable Agriculture

12.1 Introduction

Agriculture has continuously evolved over centuries and, despite major productivity increases and methodological shifts, has ultimately increased food per capita and, thus, economic progress and development. The agricultural sector is of utmost importance in both low- and high-income countries, and despite increased aggregation and vertical divestiture in the agricultural value chain, it still represents the main source of income, as well as a major business area. The agricultural sector in European countries is vital for food security and the stability of rural economies, avoiding irreversible urbanization and the transition of young people. In light of this, agricultural economics has evolved to focus on financial, economic, and environmental performance at the farm level, emphasizing sustainable and efficient management that accounts for policy and risk management as well (Kovács, Botos, and Felföldi 2024). In addition, food safety and quality to address consumer attitudes towards high-quality agri-food products have also been investigated.

In addition to these, several related topics have grabbed the attention of politicians, political scientists, lobbies, and decision-makers. In these respects, environmental sustainability has been a topic of particular relevance since the 1980s, culminating in global, regional, and local commitments. In this complex environment, achieving the competitive and sustainable intensification of current and future agrifood systems is difficult. While technological innovation has historically played a crucial role in food enhancement and proposes possible ways to reduce environmental impact, traditional thought has been increasingly raising concerns about the ability of traditional practices and systems and their management by modern standards rather than addressing the problems of contemporary agriculture (Li, Xie, and Su 2024). Today, on the one hand, policymakers are promoting the sustainability of agriculture and food management within the Circular Bioeconomy Framework as a concept that joins economic growth, social inclusion, and environmental sustainability. This framework assigns particular significance to the bioeconomic sector when attempting to reconcile diverse socio-economic and environmental stages in different areas of interest.

The scope of this essay is a broader discussion involving sustainable and relative digital agriculture coupled with various policy and technological aspects as practiced across the world (Langer and Kühl 2024). Accordingly, the objectives of the chapter is a part, are set as follows:

(i) To analyze the main recent positive technological, economic, and policy advancements for increasing agricultural sustainability; and

(ii) identify the main remaining challenges against agricultural sustainability.
(iii) To reach the first main objective, the paper explores the main economic drivers and political support underpinning the on-field development of sustainable agricultural practices.
(iv) To reach the second main objective, the paper discusses the relevance of technological transfer in the context of agriculture, considering the large heterogeneity of the actors involved, which include a high percentage of micro and small-sized family farmers in Asian and African countries.

A detailed discussion of the several dimensions of agricultural sustainability is presented to meet the two main objectives outlined above. It shows the need to go beyond traditional microeconomic and agro-ecological studies for discussing policy measures to promote increased sustainability in agriculture. Increased use of cross-sector and cross-disciplinary studies in the field would enlighten the complementariness of policy and technological actions through the supply chain and the potential for welfare enhancement at large. In light of the above, best practices adopted to comply with minimum sustainability standards on the part of several supply chain actors are analyzed (Thotho and Macheso 2024). The importance of the above objectives yields a multifaceted analysis of the many issues concerning links between agriculture and sustainable development in various countries discussed within the chapter.

12.2 Economic Perspectives

In the global economy, agriculture is much smaller than the manufacturing and services sectors, but it has a disproportionate bearing on economic development. The smaller the size of the agriculture sector, the higher the overall economic growth. The prime focus of modern agricultural policies should be the linkage between agricultural practices and economic development, not just in terms of producing cheaper food. In most developing countries, many people still derive their income mainly from agriculture, and consequently, it contributes significantly to the gross domestic product. The last few decades have witnessed farmers taking a stronger interest in profits earned from value addition. Rural employment opportunities and farm export revenues are directly proportional to the growth of the food processing segment (Hackfort, Marquis, and Bronson 2024).

To realize these achievements and further universal local market integration, there are some critical areas to be taken care of. Most of the time, farmers face difficulties due to inconsistencies in market prices, are unaware of market access and relationships, and hope to be saved by an effective government support system. It is crucial that the functioning of the market is transparent and systematic, allowing farmers to derive benefits from it. The importance of the spread in income and its contribution among the developing and leading crops and farmers mainly plays a vital role in agricultural productivity (Prasetianto, Maharddhika, and Trimus 2024). Many economists see agricultural development through price incentives. Valid arguments can be made on both sides of this issue. Agricultural growth indeed needs to be translated into viable markets through policy measures and active investment in necessary infrastructure. Secondly, agricultural growth can have a major negative effect on farm and land yield, at least for a long time. Sustainable farming talks more broadly about making agriculture viable rather than just economically sustainable. Fostering this type of development offers great promise for the growth of crops and income, in addition to the success of the agricultural sector (Boonlue et al. 2024).

12.2.1 The Economic Importance of Agriculture

Agriculture plays a crucial role in the lives of millions of people around the world. First, the agricultural sector is the primary source of healthy and diverse food products. It serves as the base of food supplies that can satisfy the internal demand within countries and also the cross-border demands in international trade. Second, the global agricultural sector provides a significant percentage of the employment generated in developing countries, which represent the majority of the world's population. Many poor and unemployed citizens rely on the agricultural sector to access minimum necessities and to create job opportunities. Agriculture can provide raw materials as inputs for many industries (Prasetianto, Maharddhika, and Trimus 2024). The development of the agricultural sector will increase the surplus of employment and income, which can be transferred to other economic sectors, such as manufacturing and services of the agricultural ecosystem, as demonstrated in Fig 12.1.

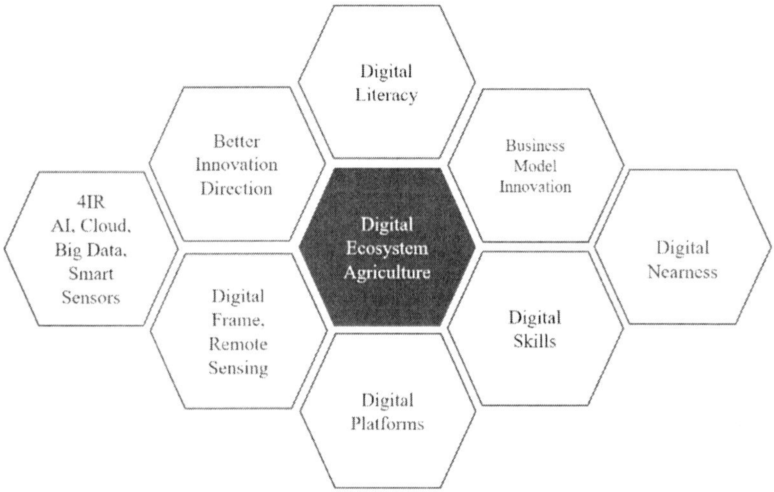

Fig. 12-1 Digital ecosystem agriculture

The economic importance of the agricultural sector can be observed from a number of perspectives. From the perspective of the country, agriculture is the primary sector that drives national economic growth and ensures the stability and sovereignty of the country. This triggers the activity of producing basic industrial goods and serves as a supporting factor for the export-import process. From the perspective of the community, agriculture is an important sector that is a source of life and livelihood for the people, providing food, raw materials, and leather. Owing to the interconnected agricultural development on both a local and international level, countries or communities must develop clean agriculture as a basis for food stability and sufficiency (Y. Zhang 2024). Sustainable agriculture can ensure the long-term food production capacity of a country. On the contrary, the agricultural production process that is not in harmony and performed continuously will gradually decrease the productivity of the land, air, and farming. Governments are, therefore, obliged to provide precise and efficient policies and infrastructure so that the agricultural sector continues to grow in accordance with the carrying capacity and the needs of the people. Policy makes a large contribution to the success and failure of the effectiveness of the agricultural sector (Tirrell, Millar, and Basso 2024).

12.2.2 Challenges and Opportunities in Sustainable Agriculture

Over the past 25 years, increased awareness of regional and global challenges has generated huge interest in agriculture to address sustainability on a broad scale. While many experts from the natural and environmental sciences have underlined the challenges, socio-economic aspects such as the status of resource-poor farmers in Asia and Africa have seldom been addressed. Sustainable practices also suffer from scale inefficiencies and competition between marketing chains. Even well-designed projects may not succeed unless a particular strategy is in place to create a balanced demand for an increase in agriculture's contribution to economic, environmental, and social wealth. Investment in environmental friendliness leading to carbon sequestration, water recharge, and assured ecosystem benefits needs to be estimated as part of any science and technology policy, assuming that technical efficiency has almost reached its zenith (J. Zhang et al. 2024). It provides an opportunity for agriculture to be connective, making traditional systems, bioinformatics, and biotechnology for traceability and consumers' access to delivering food as a service integrated with science and technology.

The need for eco-regional planning, taking into account the regional environmental endowments and addressing the core environmental issues specific to the region, informed several multi-state consultations to develop regional perspective plans, which consider the eco-regional endowments. The challenges of the shift of technology from green to gene, clinching the TRIPS agreement implementation to climatic compulsions, the creation of conducive policy towards functional regulatory systems, and harmonized trade regimes bring new understanding in technology ex-ante roles (Lodhi and Shah 2024). It, thus, delinks the advancement of biotechnology from the larger context of new trends in rural-urban relations, multinational and transnational investment, and people movements. Thus, addressing biological, cultural, and ethical questions would not be a disaster-centric choice. This way alone can establish new scales of economizing that are not divorced from scientific advances. This is the challenge of going beyond nature, wealth, sex, and technological centeredness. Current knowledge has strengthened economic evidence of outpacing the productivity of agricultural externalities by increasing risks in agricultural entrepreneurship, as shown in Fig. 12.2 (Ryskeldi, Shelomentseva, and Mirkovic 2024).

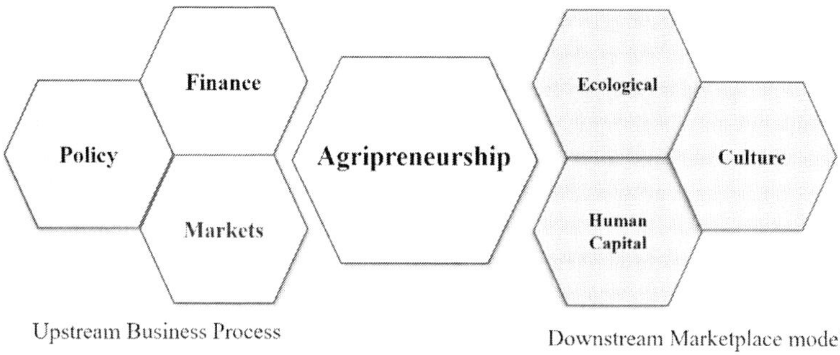

Fig. 12-2 Components of agricultural entrepreneurship

12.3 Policy Frameworks

Sustainable agriculture, digital agriculture, and a bio-economy all rely on and operate according to existing policy frameworks. Clearly, there are critical dimensions to policy in the areas of appropriate land or transition governance. The policies need to be linked down from public owners and the land managers delivering change, reaching agreements with those policy-makers and holders of the land. This is both a national and an international issue. This dual elongated period of transition relating to agricultural policies and trade agreements is not in the place deemed to be critical, with accelerating climate change and increasing societal demands under a cloud of continuing global conflicts. Whatever the international effort will be, policies are also needed at a national level to support and guide the sustainable development of the agricultural sector. There are a number of initiatives around the world to develop some form of policy or strategic plan for ensuring the sustainable development of agriculture (OKOH, Patricia. A. and OLADOKUN, A. 2024). As agriculture utilizes both the public goods of land and natural resources and also needs to function properly within social partnerships, which are constantly changing, there are a number of challenges in policy, particularly when attempting to govern sustainability. Of course, the type of policy needed is itself contingent on the particular model of sustainability that is being implemented. Moreover, policy may need to be multi-scalar and a combination across national, regional, and bioregional/local (Gong et al. 2024). Given that the development of a policy based on sustainability necessitates a change in operating culture and behaviors, this section considers the sum of the parts from economic and market disincentives and regulatory policies.

12.3.1 National and International Policy Initiatives

Underpinning all processes of action and strategies are national and international policy initiatives that are both explicit and implicit in the development of sustainable agriculture. They provide the context within which these processes operate. These policies take a number of forms, including legislation and regulatory changes, trade measures, incentives and taxes, advisory and technology development programs, and national strategic plans and frameworks. Each country has its variations in policy positions in agriculture, land use, and the environment due to specific histories, political systems, economies, and socio-cultural values. It is beyond the scope of this chapter to delve deeply into the formulations and agendas of any one country or region, or any specific combination of countries (Colaço et al. 2024). However, we note that 'success stories' can be characterized by three policy elements. The first is the embracing of the triple bottom line of economic, environmental, and social parameters for measuring agricultural profitability and performance. Second, intervention in markets to recognize that markets do not always transmit the full costs of production to consumers. Third, programs that recognize the importance of new technologies, knowledge, and people power, particularly on-farm and in production (Giannakopoulos et al. 2024).

There are also a number of multilateral organizations involved in policy formulation and advice around agriculture and the environment. For example, numerous agreements influence agricultural trade and border measures. One of the leading sources of unique, unbiased, relevant, actionable data and analysis provides policy recommendations for economic, social, and development issues across the globe. It is an intergovernmental portal for policies attempting to resolve issues such as feed-in tariffs and related fiscal policy, public spending on power stations, renewable programming policy, research and development, and skills and capabilities across low and high technological agriculture, net carbon investment and carbon tax, improving water and energy harvests and reuse, technical collaboration and partnerships between neighbors (Martin and Schnebelin 2023). In summary, it is a dense policy area, with numerous institutions and international structures attempting to shape common or harmonized policy as well as national-level action. There is always a balance between treating agriculture and land as just another economic activity and recognizing a central aspect of national uniqueness and integrity. Government policy can be separated into 'public' policy (developed by policymakers and bureaucrats) and 'private' government policies developed jointly by industry and public sector policymakers. The adoption of policy innovations is contextual and complex with no one straightforward way of successful adoption (Hong, Chen, and Wang 2024).

In addition, understanding adoption contexts enables government policy-makers to develop policies that maximize the likelihood of farmers' technology adoption in new farming segments, as illustrated in Fig. 12.3.

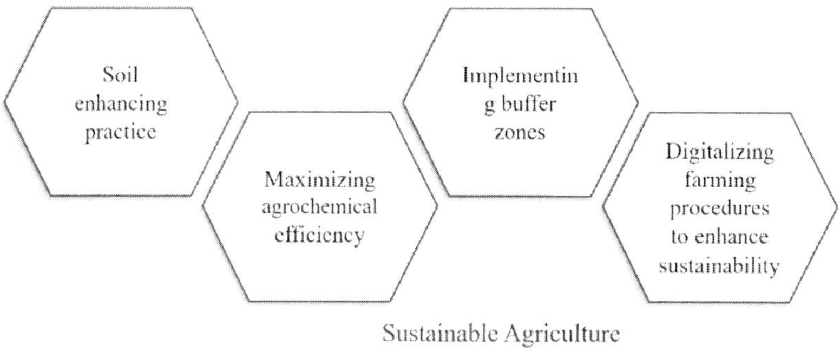

Fig. 12-3 Sustainable agriculture and its merits

12.3.2 Regulatory Challenges and Solutions

One of the main barriers to the adoption of sustainable agriculture, which improves efficiency, resilience to climate change, and protects the environment and human health, is the regulatory framework for agriculture, the environment, and public procurement. The regulatory hurdles often come in the form of bureaucracy, lack of transparency and coherence, and the presence of conflicting regulations. As becomes evident from stakeholder workshops, the industry often cites the regulatory framework as the major limitation to adopting innovative solutions, as the lengthy, expensive, and unpredictable process can deter innovations, turning investment elsewhere (Mhlanga 2024). Besides this, compliance with existing regulations is itself costly, limits opportunities for product diversification, and necessitates significant monitoring for enforcement. Moreover, evidence shows that even existing legislation is poorly enforced, failing to reflect the findings of inspections, monitoring, and reporting at the national and community level. Modifying or removing these obstacles to innovation, market-entry, and market uptake of sustainable agricultural technologies is a prerequisite if changes in economic and environmental performance are to be realized. The one-size-fits-all approach to setting regulations may have been more economically and administratively feasible in the past but cannot accommodate the levels of innovation and differentiation necessary for sustainable agriculture in the future. There are many examples of forward-thinking, proactive regulatory measures in the pipeline across the globe. The next section goes into more detail on a number of these, which are intended to use the law to shape a different science and technology pathway for our future. The principal activities where proactive policy is required are outlined, alongside examples showing that a different approach is indeed possible: invest in the science needed to help shape regulation, work with all stakeholders, including regulators, to create mutually agreeable scenarios for change, and lobby for regulatory change (Kim and Heo 2024).

12.4 Technological Innovations

Over time, the increasing use of digital technologies in every process of agriculture is transforming the agricultural landscape. The main focus of these technologies is efficiency and sustainability, and digital technologies have become the backbone for upscaling and operationalizing the concept of sustainable intensification. We describe the potential of digitalization in agriculture, in particular digital agriculture, and discuss the possible pathways of how and why digitalization is key to further enabling sustainable intensification in agriculture. We illustrate our arguments by presenting future digital agriculture technologies such as precision farming, autonomous tractors, cattle breeding, data analytics, and agro-meteorology. It is important to note that integrating such technologies into existing agricultural landscapes poses challenges for predominantly small-scale agriculture during the technological transition phase, and these need to be adequately addressed (Rasyid and Mumpuni Ningsih 2024). Despite the need for technological solutions, we also caution against the sole reliance on technological innovations and highlight the role of sustainable and proactive agricultural practices in mitigating the sectoral challenges associated with digital agriculture. We illustrate strategies on how to use digital agriculture to foster a balanced approach to farmer-driven adaptive learning. We conclude by engaging in a discussion on balancing sustainability with technological progress. We propose smallholder-friendly digital agriculture methods in order to present not only grand solutions for scaling up food production but also to show pathways for farming practices that can deliver a sustainable and ethical 'digital' future for all stakeholders. We use a set of grand solutions, tools, and case studies to present the potential future technological farming advancements that are possible to scale across the world. How to design and implement these pathways should differ based on local context, keeping farm-level modules as the fundamental building blocks of the complex overall system (Bolfe et al. 2024). A systems approach can provide a useful framework to build and deploy pathways in organic farming, as shown in Fig 12.4.

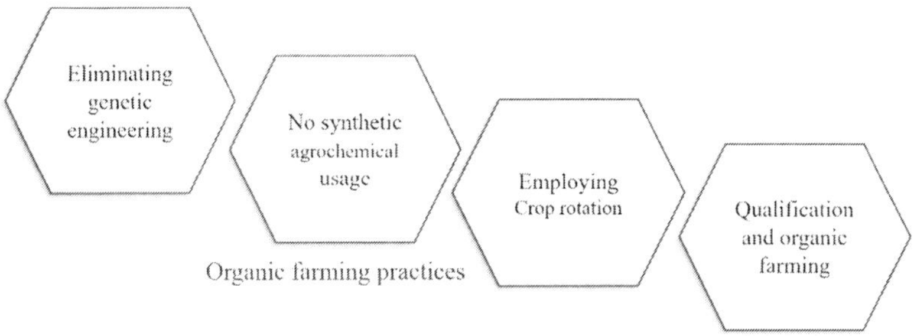

Fig. 12-4 Organic farming practices

12.4.1 Digital Technologies in Agriculture

Digital technologies, including IoT, sensing, automation, robotics, precision and smart farming, big data, machine learning, artificial intelligence, blockchain, and distributed ledger, are enabling a revolution in agricultural practices. Farm management software, sensors, and numerous Internet of Things devices are now available. They offer numerous benefits, including enhanced monitoring and decision-making using AI. The diverse data-centric applications of digital and precision agricultural practices hold the potential to improve farming through the optimization of resource use and production efficiencies. These digital applications extend beyond the farm to support the digitization of the supply chain, which helps in integrating all the stakeholders in the agri-food chains and in running the business efficiently and profitably (Ahmad et al. 2024).

Particularly worth consideration is the concern about exposing farmers to the risk of private data ownership and data usage. On the other hand, with improved adoption, the digitization of the farm or precision agricultural practices can widen the digital divide between farmers who can access and use technology and those who cannot, creating a particularly problematic situation for already marginalized smallholders. With all farmers accessing appropriate digital technologies and receiving proper training and capabilities, the benefits can be shared by all. Carefully coordinated policies are needed to ensure equitable access, appropriate training, capacity building, and extension services to ensure sound applications. Use of advanced technologies: A number of examples and case studies show a holistic approach in the application of digital technology within broader sustainable farming systems, one that has the potential for significant economic and environmental benefits beyond just its energy benefits (W. Wang and Mei 2024).

12.4.2 Emerging Technologies for Sustainable Agriculture

The emerging new technologies that are fundamentally changing the agricultural business ecosystem include Vertical farming: Growing crops in vertically stacked layers in controlled indoor environments, eliminating the need for fertile farm soils, allowing increased accessibility to fresh fruits and vegetables, and reducing food miles. This method uses significantly fewer pesticides and reduces nutrient run-off and agricultural water consumption to almost zero, allowing for year-round production of crops. Genetic engineering enables the incorporation of beneficial traits into plants and animals through laboratory techniques to splice pieces of DNA from other organisms or species. This facilitates individuals and communities from developing countries to enjoy a nutritious diet from foods derived from bioengineered plants that are hardy enough to grow where they live, sustainable enough to protect the environment, and safe enough to feed their families, looking forward to improved food security and protection of the world's natural resources (J. Wang, Lin, and Zhang 2024).

Industrial and fly larvae farming is an exciting innovation in which, just like earthworms, fly larvae will be used to convert bio-waste into proteins, fats, and other valuable ingredients. Research and development in emerging technologies for agriculture is often commercially driven, which implies that only those products that are economically viable and acceptable to society are pursued. It is here that scientists and policymakers come into action, evaluating what is acceptable while also creating awareness of what the industry is doing. The risk should not always be tolerated; alternatively, cautious optimism needs to be adopted. Worldwide funding to support sustainable and emerging technology depends on government policy, priorities, and resources, in addition to the number of private investors and collaborations among industries (Symeonaki, Maraveas, and Arvanitis 2024). The commercial utility guarantees its staying power in the marketplace, and regarding cost-effectiveness, it is definitely going to become more useful and marketable. A small case study to quantify the demand is a South African company that will start operating its first fly factory in South Africa to study the energy of flies occurring at certain densities and how efficient they are at waste management and mass protein production. In conclusion, hybrid systems with a significant biotechnological component can effectively lead to the development of genuine, safe, and commercially viable advances in biological control with smart applications, as illustrated in Fig 12.5 (Singh et al. 2024).

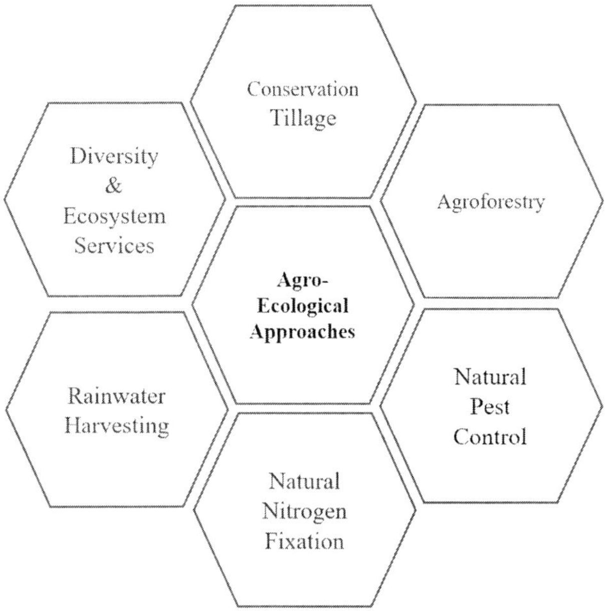

Fig. 12-5 Agro-ecological approaches

12.5 A Broader Economic, Policy, and Technology Perspective

Sustainable agriculture requires investors and policymakers to go much further than the narrow domain of efficiency and yield, central to the Green Revolution and so-called "productivism." Are there any alternative practices related to agriculture that could directly contribute to a certain kind of environmental stewardship? The answer is, of course, yes. Practices such as agroecology, permaculture, or other approaches directly promote soil health and biodiversity above and below the soil, and we refer to these as "sustainable" approaches. But we also need to move away from the concept of "practices." "Practices" can contribute to sustainability if they respect the principles we have mentioned above and if they are also responsible for maintaining a wide range of ecosystem services while at the same time maximizing production and efficiency. Practices are good, but they are not enough on their own (Beksultanova and Dzhankhotova 2024). Technologies can be very useful in some contexts, but they can also be harmful and must be evaluated on a case-by-case basis.

In this part, we describe and discuss the various agricultural practices that propose or claim to be more sustainable than modern conventional agriculture, and we also discuss to what extent they are economically viable or not. These practices are developed specifically to avoid using particular products or minimize their usage in a given context or for a given objective. There have not been many practices developed explicitly to increase soil carbon sequestration, a practice that is sometimes weaponized by those who oppose sustainable farming practices: if the practice does not involve soil carbon sequestration, it is not really sustainable. More often than not, these alternatives have come from local knowledge, tradition, or socio-historical practices that evolved, and often, new practices have come from rejected "modern" practices, generating a new set of "alternative farmers" and the demand from a new range of "alternative consumer" (Metta, Dessein, and Brunori 2024; J. Wang, Lin, and Zhang 2024).

12.5.1 Agroecology and Permaculture

Agroecology is a holistic approach to sustainable agriculture that encompasses many of the same principles described in the section on regenerative agriculture. The primary focus of agroecology is the integration of ecological principles into farming practices. Permaculture, a broader environmental philosophy, focuses on designing agricultural systems that emulate natural ecosystems, relying on perennial plants and diverse species as the building blocks of sustainable agricultural systems. Agroecology attempts to support whole, complex farming systems and supply chains, whereas regenerative agriculture promotes the benefits of at least a percentage of production occurring within a purely on-farm context. Many approaches to 'sustainable' agriculture, such as agroecology, organic and biodynamic farming, and permaculture, share several similar principles, and in many cases, there are more similarities than differences between the approaches (Singh et al. 2024). Agroecology is community-integrated and bolstered by the social sciences, governance, NGOs, and policy, whereas regenerative agriculture has largely grown through on-farm systems and siloed industries.

Agroecology offers the potential for increased economic gain from the farm as biodiversity and the ecological services from agroecosystems increase and as supply chain premiums meet the growing consumer demand for diversity, having a well-outlined sustainability ethic. The incorporation of greater promotion of agroecology and permaculture principles into sustainable agriculture methodologies and practices is wholeheartedly encouraged, as these systems can increase direct and indirect biodiversity protection on the farm. Agroecology is one of the most effective tools enabling multiple, lasting changes to the food supply for the world, as well as enhancing biodiversity outcomes. Agroecology rejects the application of expensive and superfluous industrial inputs like fossil fuel-sourced fertilizers and chemicals. Agroecology also promotes more resilient, diverse farms with less-bedded supply chains, which can increase national food security. Agroecology can benefit profitable reform from the edges, as this can allow more farmers to participate if there is no pressure to release the same output as has been enjoyed or pressured by governing bodies (Beksultanova and Dzhankhotova 2024; Metta, Dessein, and Brunori 2024). Agroecology can support farmers to remain on and enjoy farming for the multiple benefits that central features of living in semi-remote areas provide.

Agroecology requires the following consolidation of activities to stay current on the latest strategies and methodologies: attending agroecology conferences and workshops, joining community or provincial agroecology organizations, reading books and publications, participating in learning walks at agroecological farms, and phoning to receive agronomy help from the team, including experienced teachers, farmers, and food production judges. Agroecology breeds intense loyalty, as the concept has the potential to support millions of farmers worldwide, regardless of where in the globe they are located. Agroecology has been misrepresented, and instead, people refer to the instant and radical change from stoichiometric to ecological farming (Y. J. Wang et al. 2024). The shift can occur quickly; however, it takes time to build an ecological function for maximum nutrition from the soil.

12.5.2 Precision Agriculture

Precision agriculture is a strategy to manage variability in the fields with information and communication technologies. Precision agriculture has the expertise to provide food and support to preserve environmental quality. Despite the emphasis on digital solutions, mainly embedded in sensors, guidance systems, geographic information systems, and automation approaches, the combination and interactivity between different agricultural management elements have risen. Precision agriculture seeks to develop collected performance figures in electronically transmittable data such as Global Positioning Systems, computer systems, and odometer values. Precision agriculture focuses on gathering data that can generate better profitability, system efficiency, and, at the same time, environmental sustainability (Kovács, Botos, and Felföldi 2024). Precision farming increases farmers' abilities to input crop or livestock inputs in the appropriate location at the required amount, thereby reducing waste of fertilizer, water, and the environmental impact these inputs have. Precision agriculture provides important physical, economic, and environmental advantages when implemented in the right ways. While developed countries have examined the benefits of digital agriculture, some rural areas in developing countries have reported benefits from using digital agriculture. Using precision agriculture, it is possible to manage inputs such as water, fertilizers, and chemical pesticide use in relation to particular site qualities to lessen costs, minimize such use, increase yields, and defend the environment in particular situations (Langer and Kühl 2024). Millions of small-scale entrepreneurial farmers from developed to developing countries could benefit from digital precision and knowledge agriculture.

12.5.3 Farm Management Software

Farm activity planning, inventory management, financial bookkeeping, budgeting, and investment planning are just some of the most regularly found functions in agricultural management software. However, some of the software extends the functionality to include, for example, pest and disease records, machinery maintenance or use records, mapping functionality, data integration, gross margin reporting, balance sheet reporting, bank reconciliation, weather recording, and data analytics, to name just a few. Generally, simple software programs may have only the capacity to plan activities, store records against the plan, and may be able to do some basic reporting against the stored data. However, as the activity plan is just a plan and relies on data input by the user, the reporting is not normally live from a database (Prasetianto, Maharddhika, and Trimus 2024). Additional functionalities may offer automated reporting live from a linked database to pull through real-time data for decision-making or compliance. Integrating data from other standalone databases or programs is also a relatively new area emerging. Data can be imported/exported in various forms, including Excel and CSV formats. Software with the capacity to pull data in from other sources, such as weather data, financial data, or machine telematics, may aid in workflow, compliance, decision-making, and reporting. The software can link to an accounting or payroll package or have a mobile recording system (Y. Zhang 2024). They can provide a full link to the farm as it is also part of a retail branch within the agricultural industry. One digital platform allows users to access the functionality for free or can buy pro tools. They have something that is relatively low cost with apps on the marketplace and companion apps for field information such as soil sampling and precision apps. The possibility exists then to develop companion apps within our team that can make use of the information flow, including seed placement apps. With sync capabilities between mobiles and web-based features, it would be possible for a team leader to monitor information collected in the field instantly and make the necessary decisions (Lodhi and Shah 2024).

12.5.4 Drone Technology

Drones, also known as unmanned aerial vehicles (UAVs), are rapidly gaining recognition due to their potential benefits for not only sustainable agriculture but also many areas such as search and rescue, mapping, meteorology, atmospheric studies, and environmental studies. For agriculture, drones offer a cost-efficient, flexible, and timely technology to help with operational field decisions and activities (Shafik 2024b). They can be used for crop monitoring in terms of water, nutrients, and pest management, field scouting, detecting microclimates, genetic studies, infrastructural and environmental water monitoring, mapping, yield prediction, emergency response, early-stage crop insurance, planting or spraying of seeds, and precision spraying of agrochemicals or water for irrigation. Drones with the capability to plant trees as seed capsules being dropped from the sky could transform the agricultural sector once scaled up (Gong et al. 2024). Overall, as unmanned aerial vehicles (UAVs), drones have significant potential to revolutionize modern farming practices. Provided that it is carried out legally and under the existing guidelines, using drones in agriculture incurs minimal risk. However, the rules related to drone flights are set by each country, and they may differ depending on whether the drone will be used for commercial or private purposes. In order to operate a drone for commercial purposes in most countries, pilots must have a Remote Pilot Certification, which can be obtained after completing security awareness training and passing an Applicable Knowledge Test (Li, Xie, and Su 2024; Kovács, Botos, and Felföldi 2024). In agriculture, the use of a drone is beneficial in terms of reduced costs and time management.

12.5.5 Organic Farming Techniques

Organic farming is a production system based on specific principles designed to promote sustainable agriculture while ensuring that products are guaranteed for consumers free from substances that can harm human health and the environment. Organic farming is, above all, a production method based on the idea of soil health, understood as motivating the development of new breeds of resilient plants besides developing a healthy microbiota to decompose organic matter, mineralize nutrients, de-repress plant immune systems, and sequester carbon. Soil health means the preservation and promotion of soil biodiversity, that is to say, the attempt to maintain the heterogeneity of many forms of living organisms and to understand their interactions with respect to plants in order to promote and preserve beneficial functions (Thotho and Macheso 2024). The strategies are deferring plowing, diversification through crop rotation and the integration of a proper livestock sector, closing the nutrient cycle, i.e., preferably using organic fertilizers and making maximum use of compost, and protecting and promoting natural enemies by diversifying the cultivated plants. Any conservation strategy in agriculture has multiple positive outcomes. To begin with, the adoption of sustainable techniques reduces the number of inorganic fertilizers and pesticides, a very expensive budget item for farmers, and at the same time increases the value of the final product, which can be sold at a different price on the market, as happens in the case of organic food (Prasetianto, Maharddhika, and Trimus 2024). Consumer studies have shown that the main driver behind buying organic food is the association of organic products with a healthier diet. On the producer side, there are many constraints, such as an exponential increase in the cost of producing healthy plants due to bureaucracy related to the heavy and complicated process of obtaining organic certification for themselves and their land, the organizational aspect of marketing goods like organic products, and the weak point of entering highly competitive markets. Despite the constraints, some successful stories already show that economic competitiveness can start from selling small quantities in the local market and farm shops directly. Given the profitability of organic products, the production of these putative "functional" plants could be an added value. Public policies could support this pathway (J. Zhang et al. 2024). Unfortunately, this is not yet the case in most countries, especially in Europe.

12.6 Case Studies

We can use such peer-reviewed case studies as examples of agricultural strategies in different socio-economic settings that have achieved some success and highlight the mix of strategies and 'actor ecologies' that contributed to that success. Such cases can be used as exemplars because the contextual technologies and engagement strategies may well be applicable, but also to highlight how differences in the distribution and quality of technologies in each location inevitably led to different policy and economic outcomes. Examples of successful digital agriculture to date: cases that demonstrate a significant contribution to efforts to promote digital and sustainable approaches to global food security. This issue describes digital farming and food production in Norway, and this issue provides evidence of sustainable digital agriculture in different African countries (Gong et al. 2024).

Report on experiences in digital extension in Brazil and the potential benefits of integrating traditional and digital models in one rural extension service. Both sets of authors identify key factors for success and the use of ICT-based approaches that have improved access for farmers to agricultural extension support. Discuss an approach to the identification of barriers, incentives, and solutions for increasing the adoption and scaling of digital technology in agriculture. Three case studies identify farmers who have adopted an innovative ICT-based approach for reducing fertilizers and increasing straw utilization. These cases describe different paradigms in their approach to innovation and identify lessons from experience. Report on the development and performance of the system and summarize the relevance of the prototype for improving nutrient recycling. They highlight lessons for improving technology acceptance and acknowledge the appetite for more benchmarking data to help assess the reliability and financial gain from nutrient recycling processes (Symeonaki, Maraveas, and Arvanitis 2024). Uncertainty over the move from climate neutrality ambition to actual effect reveals the important role of subsidies. This issue identifies subsidies and taxes as necessary tools to increase the sustainability of the food chain. The experiences of ancestors in what is now low-lying Netherlands clearly show where we come from and explain why produce is too cheap nowadays. The bottom line seems to be that 'negative shocks' (degradation, climate change, drought, floods, trade disruption) can always happen and that agriculture and land management, as well as naturalist responses, must be based on the acceptance of uncertainties (Puntel, Thompson, and Mieno 2024).

12.6.1 Successful Sustainable Agricultural Projects

This section critically reflects upon successful projects that employ strategies and methodologies, providing a useful illustration of how the results described earlier may materialize. The cases presented here showcase that successful practices are those that work interactively within a locality or through linking producers with buyers that have a commitment to sustainable agriculture or simply providing better returns to the community (Shafik and Kalinaki 2024). They provide contextual information, the sustainable agricultural practice, technology and method employed, the desired outcomes, how involved stakeholders are in project management, the impacts in terms of income generation, in-kind benefits, and job creation, the constraints faced in implementation, and the lessons learned. These case studies mainly focus on local or regional interventions and are diverse in geographic area and crop type, as well as the level of stakeholder involvement. Several describe innovative contributors to the final products or outlets for which the producers are able to sell their output as a result of their use of sustainable practices (Y. J. Wang et al. 2024). Some lessons are emerging from selected projects that can be applied to others working in the same or similar areas of sustainable agriculture; examples include paddy rice producers in Korea and Fiji; vegetable producers in Vietnam, Fiji, and India; coffee producers in Vietnam; mushroom producers in Fiji; fruit producers in Australia and New Zealand; market gardeners in New Zealand; and bush tucker and other food producers in Australia. The selected cases show that the successful practice varies by stakeholder and project. Furthermore, they demonstrate that continuing engagement with the community or stakeholders in project implementation is critical to success. Based on these and other observations, it is possible to make generalizations about some commonalities of those projects that do well. In all of these cases, producers increase in some way the yield or quality of the product, including ecolabel derivatives, through the use of particular methods of production (Beksultanova and Dzhankhotova 2024; Singh et al. 2024). These, in turn, command a premium from the final customer. The buyers, who are sometimes part or in their entirety the end consumers, in return, positively market the product. All projects have the involvement of the community or participating stakeholders. All projects have management in place in terms of, at the very least, a good communicator and an individual making logistical arrangements. One of the overriding lessons is that the less complex, the better (Giannakopoulos et al. 2024).

12.6.2 Innovative Digital Agriculture Initiatives

The FarmOS project collaborates to develop open-source, web-based platforms for farm management tailored for different designs of farming systems in the US, Europe, Africa, and Asia. The main goal is to aid the data-driven farming innovation known as 'precision agriculture' to serve as a pathway for the development of innovative digital tools and practices with the potential to enable more productive farming activities . Following the increasing demand worldwide for organic and sustainable farming products by consumers due to excessive use and environmental impact of chemical fertilizers and other synthetic inputs, researchers and policymakers raise concerns about the most appropriate digital technologies that can potentially help improve the performance of farmers according to organic farming standards, rules, and regulations (Boonlue et al. 2024; Y. Zhang 2024). The

Landscapes for People, Food, and Nature Initiative Encourages practitioners and scientists to develop digital farming solutions. This initiative offers projects that apply digital technology to strengthen the control of farming decision-making on intensive farming properties and gather big data analytics for the selection of digital devices and platforms that could assist in this process. This work draws upon suggestions of digital solutions both on and off the farm to create a comprehensive network that could provide support, natural infrastructure, and a broader reporting system of ecosystem service monitoring. The decisions of the project team and community partners based on these case study results will be instrumental in the further development of on-the-ground digital technologies and applications in the near future (Ryskeldi, Shelomentseva, and Mirkovic 2024). Barriers encountered on the ground in trial testing of IoT in these case studies include the high cost of developing interfaces for data transmission, start-up expenses, technology training, and the inclusion of smaller farmers without access to this technology. Additionally, the project focused on accessibility issues in creating platforms that will not only provide benefits to larger or more technologically advanced farming systems but also to the smaller-scale farmer who needs this technology the most. These important devices can be used for other sites, applications, or industries. As such, the 'FarmOS' design provides an opportunity for integration. Unlike other approaches, the platform is flexible and can be configured by those who trial it for a wide range of purposes. Given this potential for adapting the platform, the connections we have made during the Landscapes for People, Food, and Nature project will continue to guide and inform our future endeavors, including the further refinement and pilot testing of 'FarmOS' with on-the-ground partners (Mhlanga 2024).

12.7 Future Directions

This part outlines future directions. First, we discuss emerging trends and innovations. Then, we highlight potential challenges and require proactive strategies. Finally, we provide a set of considerations.

- Emerging Trends and Innovations Environmental and social pressures are forcing global shifts in agriculture towards more sustainable practices. This trend is likely to continue in the future. These changes are closely related to shifts in technology. Emerging frontier technologies are quickly reshaping agriculture. The global policy, economy, and technology landscape are all interacting and will drive emerging trends in sustainable and digital agriculture into the future (Meng et al. 2024).

- Research priorities in sustainable agriculture must balance productivity to meet future food demands with environmental protection. Agro-ecological intensification shows early promise in achieving this balance, but it is not the only option. For instance, some consider circular agriculture as the pathway forward, whereby agriculture interfaces more effectively with other systems (Y. J. Wang et al. 2024).
- Tough Challenges Ahead Anticipating and responding to future challenges is essential in international science and policy. The world can be transformed in the next few decades. Common strategies and guidelines for policymakers, industry practitioners, civil society, and academics are therefore required to navigate the future successfully (Bolfe et al. 2024).
- What are potential pitfalls and dead ends? Below we propose a set of considerations to avoid dead ends and to promote effective transitions to more sustainable systems. Our goal is to drive innovation in sustainable agricultural development.
- We must remain agile and open to emerging innovations in the production, protection, and policy of and for agriculture. New policy frameworks are needed at national, regional, and international levels so that human ingenuity and advances in knowledge can drive global agricultural development while not diminishing the life-support system of planet Earth. From policymakers through industry practitioners to civil society, we need a common framework for better interpretation of global trends, their implications, and strategies to respond effectively (Puntel, Thompson, and Mieno 2024).
- What are the biophysical and social scientific research priorities? These need to be as exciting as new technologies. The upcoming generation of technologies will work with ecology to shape high-performance agri-food methods to combat increasing food loss and waste and eliminate food poverty in well-to-do countries. Recognizing value addition and ecosystem service contributions are critical for primary producers in the new era of agriculture. We need a vision of research capacity and capability in sustainable agriculture that looks one to five generations ahead (Meng et al. 2024).

12.7.1 Trends and Forecasts in Sustainable and Digital Agriculture

Current trends in agriculture are predominantly concerned with sustainability, an area that has been steadily gaining momentum in the past two decades. One major reason for this is that agriculture is a significant driver of deforestation and degradation of ecosystems, loss of biodiversity, and a major source of greenhouse gas emissions. In response to these global challenges, societies are calling for a financial and business model for farming that prioritizes environmental stewardship and the supply of diverse, nutrient-rich foods with transparent, ethical supply chains. Moreover, they have a growing appetite for functioning ecosystems that can offer them fresh water, sequestration, pollination services, flood control, and beautiful landscapes (Ryskeldi, Shelomentseva, and Mirkovic 2024). Scientific literature that reviewed regulatory trends plays down the significance of sustainable issues and places the technological question at the core of future trends. Remarkably, smart agriculture, precision agriculture, and regenerative agriculture—a global trend that has recently gained traction—do not only encompass soil-centric dimensions, but also include broader ecosystems with the aim of producing food from healthy, living soils. Technologies are undeniably one way to attain this, in addition to facilitating the exchange of information across agricultural systems and impacted industries. However, there is always the possibility of tools that currently connect broadband and gigabytes to fail in helping make sound environmental, economic, and social decisions at scale (Colaço et al. 2024).

The role that digital technology plays in agriculture is paramount for Swiss stakeholders. It aims to serve a population that is growing up in one of the most digitalized places on earth. The digital economy and society create the conditions that facilitate innovation, already imposing unwavering demands for honing knowledge, adopting new tools and methods, expansion of network services, and various other areas of expertise (Shafik 2024a). Digitalization also constitutes the core value of agriculture since to adapt and develop sustainable farming systems without losing productivity and find answers that work best under Swiss constraints, the diversification of food and agriculture could be a means to create computing capacity and help culture organisms and support lifecycle. Crucially, such diversification should be—or, better yet, should consist of—biological and circular farming production. This is why the digital economy and society, agriculture, societal values, and the biosphere are inextricably intertwined (Kim and Heo 2024; Mhlanga 2024). In the Swiss context, fostering discussions around the transformations that jointly occur in these four drivers is crucial to gleaning a robust understanding of the ontologically and epistemologically complex challenges and opportunities that agriculture faces.

It is crucial to integrate digital technological opportunities with sustainable farming practices, cultures, and economics. Evidence confirms the need for policies that support the development and implementation of effective and resilient innovative and knowledge-intensive solutions among farms, forest landowners, and other relevant actors to promote new and evolving attitudes and values. As a corollary, the adoption of new methods and tools can also improve the efficiency, productivity, and profitability of sustainable agriculture. Finally, any sustainable practice, however small or large, requires a commitment to continuing research and collaborations among relevant and interested stakeholders. Practicality must walk hand in hand with flexibility and imagination, all the while improving production standards, social goods, and planetary health (Bolfe et al. 2024; Ahmad et al. 2024).

With these points in mind, we recommend the following strategies and approaches. Strategically, environmentally and socio-economically, support the development of digital and technology solutions that promote environmental and social sustainability, increased productivity, and profitability on farms and improve the quality of life for farm owners and rural communities. Shift to a more research-focused (though highly practical) emphasis. Encourage university, industry, government, and non-profit research that expands interdisciplinary and translational research with a strong emphasis on collaboration among all stakeholders (Shafik 2025). A morphing of successful outcomes based on internal and external evaluation is essential. Advancements and public and trade acceptance require both the methodical and quick adaptability of grower-to-community perceptions and attitudes. Facilitate the general development of public-private initiatives and structures that make policy-based public revenues for externalities available to farmers and other investors via public-private partnerships that use environmental service markets to leverage private-sector funds (Ahmad et al. 2024; Rasyid and Mumpuni Ningsih 2024). The basic design of the system must recognize both the transaction costs and benefits, and an evaluation and auditing process must be built into the cost of the environmental service.

12.8 Conclusion

The chapter has discussed the multi-directional connections among the economic factors behind the development of agricultural practices, policy choices required for aligning the behavior of different stakeholders with the public interest, and the existing and potential role of technologies in supporting such a transition. This points to a high degree of interconnectedness among economic, policy, and technological issues in the move towards sustainable agriculture. Analyzing both the drivers of unsustainable practices in the first part and the factors behind the suboptimal choices in the transition to sustainability in the second part, the essay provides a number of options that can turn the main challenges into opportunities, underlining the enormous potential that a return to sustainable agriculture entails. The essay also offers inclusive definitions of what the concept of sustainable agriculture embodies, and where technological innovations can be of help, better synthesizing energy and food-related aspects. It has explained how and to what extent unsustainable practices are relevant for future sustainable agricultural growth, highlighting pressure points and policy shifts that hold the promise of a substantial move towards sustainability in multiple agricultural regimes. Furthermore, the essay has formulated a number of recommendations for possible future developments, including the importance of a commitment at various levels within public and private institutions in order to drive the selected options, the design of optimal policies providing appropriate incentives to all relevant stakeholders; the selection of profit-maximizing alliances and their interdependencies in the process towards sustainable options; and a closer look at crucial questions that still require some additional intense research, in order to devise a controlled transition towards sustainability. Hence, the conclusion stresses the role of continuous adaptation, embedded into an active portfolio, in dealing with an evolving realm of agricultural challenges.

References

Ahmad, Babor, Md Abdur Rouf Sarkar, Fahima Khanom, Rozina Yeasmin Lucky, Mou Rani Sarker, Md Golam Rabbani, Shraboni Rani Rani Ray, Md Naimur Rahman, and Md Nazirul Islam Sarker. 2024. "Experience of Farmers Using Mobile Phone for Farming Information Flow in Boro Rice Production: A Case of Eastern Gangetic Plain." *Social Sciences and Humanities Open* 9. https://doi.org/10.1016/j.ssaho.2024.100811.

Beksultanova, Aybika I., and Petmat M. Dzhankhotova. 2024. "Personnel of Digital Transformation in Agriculture." *Sociologia y Tecnociencia* 14 (1). https://doi.org/10.24197/st.1.2024.106-114.

Bolfe, Édson Luis, Daniel de Castro Victoria, Edson Eyji Sano, Gustavo Bayma, Silvia Maria Fonseca Silveira Massruhá, and Aryeverton Fortes de Oliveira. 2024. "Potential for Agricultural Expansion in Degraded Pasture Lands in Brazil Based on Geospatial Databases." *Land* 13 (2). https://doi.org/10.3390/land13020200.

Boonlue, Surapon, Maneerat Manyuen, Jariya Neanchaleay, Atipat Boonmoh, and Vitsanu Nittayathammakul. 2024. "The Inquiry-Driven VCoP Model to Promote Digital Agriculturalists' Learning Competencies in the Agriculture 4.0 Era." *International Journal of Information and Education Technology* 14 (1). https://doi.org/10.18178/ijiet.2024.14.1.2019.

Colaço, A. F., B. M. Whelan, R. G.V. Bramley, J. Richetti, M. Fajardo, A. C. McCarthy, E. M. Perry, et al. 2024. "Digital Strategies for Nitrogen Management in Grain Production Systems: Lessons from Multi-Method Assessment Using on-Farm Experimentation." *Precision Agriculture* 25 (2). https://doi.org/10.1007/s11119-023-10102-z.

Giannakopoulos, Nikolaos T., Marina C. Terzi, Damianos P. Sakas, Nikos Kanellos, Kanellos S. Toudas, and Stavros P. Migkos. 2024. "Agroeconomic Indexes and Big Data: Digital Marketing Analytics Implications for Enhanced Decision Making with Artificial Intelligence-Based Modeling." *Information (Switzerland)* 15 (2). https://doi.org/10.3390/info15020067.

Gong, Siyu, Ziye Sun, Bo Wang, and Zhigang Yu. 2024. "Could Digital Literacy Contribute to the Improvement of Green Production Efficiency in Agriculture?" *SAGE Open* 14 (1). https://doi.org/10.1177/21582440241232789.

Hackfort, Sarah, Sarah Marquis, and Kelly Bronson. 2024. "Harvesting Value: Corporate Strategies of Data Assetization in Agriculture and Their Socio-Ecological Implications." *Big Data and Society* 11 (1). https://doi.org/10.1177/20539517241234279.

Hong, Xianna, Qiuhua Chen, and Nan Wang. 2024. "The Impact of Digital Inclusive Finance on the Agricultural Factor Mismatch of Agriculture-Related Enterprises." *Finance Research Letters* 59. https://doi.org/10.1016/j.frl.2023.104774.

Kim, Steven, and Seong Heo. 2024. "An Agricultural Digital Twin for Mandarins Demonstrates the Potential for Individualized Agriculture." *Nature Communications* 15 (1). https://doi.org/10.1038/s41467-024-45725-x.

Kovács, Tamás, Szilvia Botos, and János Felföldi. 2024. "What Does Digital Transformation Do in Agriculture? A Systematic Literature Review of Agri-Food Sector's Digitalisation." *Journal of Agricultural Informatics* 14 (2). https://doi.org/10.17700/jai.2023.14.2.707.

Langer, Greta, and Sarah Kühl. 2024. "Perception and Acceptance of Robots in Dairy Farming—a Cluster Analysis of German Citizens." *Agriculture and Human Values* 41 (1). https://doi.org/10.1007/s10460-023-10483-x.

Li, Hanbing, Shouhong Xie, and Mingwei Su. 2024. "Does Digital Technology Innovation Promote Low-Carbon Development in Agriculture?: A Spatial Econometric Analysis Based on 31 Provinces in China." *Environmental Science and Pollution Research International* 31 (3). https://doi.org/10.1007/s11356-023-31369-9.

Lodhi, Yagya Kumari, and Dr. Amisha Shah. 2024. "Review on Agriculture 4.0: Exploring the Digital Transformation of the Agriculture Supply Chain." *International Journal of Agriculture Extension and Social Development* 7 (1). https://doi.org/10.33545/26180723.2024.v7.i1e.224.

Martin, Théo, and Éléonore Schnebelin. 2023. "Agriculture Numérique : Une Promesse Au Service d'un Nouvel Esprit Du Productivisme." *Natures Sciences Sociétés* 31 (3). https://doi.org/10.1051/nss/2023046.

Meng, Jiajia, Baoyu Zhao, Yuxiao Song, and Xiaomei Lin. 2024. "Research on the Spatial Dynamic Evolution of Digital Agriculture—Evidence from China." *Sustainability (Switzerland)* 16 (2). https://doi.org/10.3390/su16020735.

Metta, Matteo, Joost Dessein, and Gianluca Brunori. 2024. "Between On-Site and the Clouds: Socio-Cyber-Physical Assemblages in on-Farm Diversification." *Journal of Rural Studies* 105. https://doi.org/10.1016/j.jrurstud.2023.103193.

Mhlanga, David. 2024. "Digital Revolution in African Agriculture." *SSRN Electronic Journal*. https://doi.org/10.2139/ssrn.4697324.

OKOH, Patricia. A., and OLADOKUN, A. 2024. "USING DIGITAL RESOURCES IN AGRICULTURE AS A STRATEGY FOR CURBING HUNGER CRISIS IN NIGERIA." *Nigeria Journal of Home Economics (ISSN: 2782-8131)* 12 (10). https://doi.org/10.61868/njhe.v12i10.320.

Prasetianto, Mushoffan, Rizkiana Maharddhika, and Syakilla Eka Putri Lestari Trimus. 2024. "The Digital-Mediated Extensive Reading on English Language Learning of Agriculture Students." *Journal of Education and Learning* 18 (1). https://doi.org/10.11591/edulearn.v18i1.21176.

Puntel, Laila A., Laura J. Thompson, and Taro Mieno. 2024. "Leveraging Digital Agriculture for On-Farm Testing of Technologies." *Frontiers in Agronomy* 6. https://doi.org/10.3389/fagro.2024.1234232.

Rasyid, Harun, and Gumoyo Mumpuni Ningsih. 2024. "The Role of Digital Technology in the Transformation of Agriculture Toward Smart Farming." *Journal of World Science* 3 (1). https://doi.org/10.58344/jws.v3i1.523.

Ryskeldi, Olzhas, Valentina Shelomentseva, and Milika Mirkovic. 2024. "The Economics of Digital Tools in Kazakh Agriculture." *International Journal of Innovative Research and Scientific Studies* 7 (2).

https://doi.org/10.53894/ijirss.v7i2.2629.

Shafik, Wasswa. 2024a. "Barriers to Implementing Computational Intelligence-Based Agriculture System." In *Computational Intelligence in Internet of Agricultural Things*, 193–219. Springer. https://doi.org/10.1007/978-3-031-67450-1_8.

Shafik, Wasswa. 2024b. "Biosensor-Based Drones Anomaly Detection Integration for Sustainable Agriculture Development." In *Achieving Food Security Through Sustainable Agriculture*, 283–316. IGI Global. https://doi.org/10.4018/979-8-3693-4240-4.ch012.

Shafik, Wasswa. 2025. "Sustainable Agriculture and Diet for Urban Development." In *Sustainable Smart Cities and the Future of Urban Development*, 165–92. IGI Global Scientific Publishing.

Shafik, Wasswa, and Kassim Kalinaki. 2024. "Societal and Ethical Implications of Technology-Enhanced Agriculture and Healthcare: An African Context." In *2024 IST-Africa Conference (IST-Africa)*, 1–11. IEEE. https://doi.org/10.23919/IST-Africa63983.2024.10569306.

Singh, Amandeep, Naser Nawayseh, Yash Kumar Dhabi, Siby Samuel, and Harwinder Singh. 2024. "Transforming Farming with Intelligence: Smart Vibration Monitoring and Alert System." *Journal of Engineering Research (Kuwait)* 12 (2). https://doi.org/10.1016/j.jer.2023.08.025.

Symeonaki, Eleni, Chrysanthos Maraveas, and Konstantinos G. Arvanitis. 2024. "Recent Advances in Digital Twins for Agriculture 5.0: Applications and Open Issues in Livestock Production Systems." *Applied Sciences (Switzerland)*. https://doi.org/10.3390/app14020686.

Thotho, Doreen, and Paul Macheso. 2024. "Comprehensive Survey on Applications of Internet of Things, Machine Learning and Artificial Intelligence in Precision Agriculture." *Tanzania Journal of Engineering and Technology* 42 (4). https://doi.org/10.52339/tjet.v42i4.922.

Tirrell, Ben, Neville Millar, and Bruno Basso. 2024. "Rules Get in the Way of Farmer Cooperation: How Digital Agriculture Could Change the Nature of the Game in the Food-Energy-Water-Climate Nexus." *PLOS Sustainability and Transformation* 3 (1). https://doi.org/10.1371/journal.pstr.0000092.

Wang, Jingyi, Qingning Lin, and Xuebiao Zhang. 2024. "How Does Digital Economy Promote Agricultural Development? Evidence from Sub-Saharan Africa." *Agriculture (Switzerland)* 14 (1). https://doi.org/10.3390/agriculture14010063.

Wang, Wei, and Tingting Mei. 2024. "Research on the Effect of Digital Economy Development on the Carbon Emission Intensity of Agriculture." *Sustainability (Switzerland)* 16 (4). https://doi.org/10.3390/su16041457.

Wang, Yi Jia, Naihui Wang, Mo Li, Heng Li, and George Q. Huang. 2024. "End-Users' Acceptance of Intelligent Decision-Making: A Case Study in Digital Agriculture." *Advanced Engineering Informatics* 60. https://doi.org/10.1016/j.aei.2024.102387.

Zhang, Jian, Dawn Trautman, Yingnan Liu, Chunguang Bi, Wei Chen, Lijun Ou, and Randy Goebel. 2024. "Achieving the Rewards of Smart Agriculture." *Agronomy* 14 (3). https://doi.org/10.3390/agronomy14030452.

Zhang, Yanhong. 2024. "Innovation in Thinking and Practice Path of Digital Economy Based on Big Data Analysis to Promote High-Quality Development of Agricultural Service Trade." *Applied Mathematics and Nonlinear Sciences* 9 (1). https://doi.org/10.2478/amns.2023.2.00600.

PART IV

CASE STUDIES AND FUTURE PERSPECTIVES

CHAPTER THIRTEEN

AI-ENHANCED SOIL MOISTURE PREDICTION USING SENSOR NETWORKS AND MACHINE LEARNING

Abstract

Modern agriculture requires accurate prediction of soil moisture variability in order to optimize irrigation, improve water use efficiency, and maximize crop yields. Driven by changes in ambient climate and plant physiology, accurate soil moisture forecasts can be used in resource management practices such as scheduling irrigation to maximize absolute transpiration or prevent water stress. For large-scale irrigation systems such as those used in arid and semi-arid regions and commercial crop farming, the availability of workers and water is highest in the early evening when it is cooler and the sun has already set. Thus, accurate prediction of soil moisture carried out late each day provides a predictive window for operations. Furthermore, many farms are small to medium scale and do not utilize expensive and centralized resource management software tools, thus necessitating cost-effective and accessible predictive tools. A lack of predictive systems for soil moisture tailored for such small scales exists, in part due to a lack of continuously updated model performance parameters derived as output statistics. Machine learning tools, particularly Convolutional Neural Networks, are capable of directly predicting soil moisture using image data provided by ground sensors, either individually or in a multi-sensor network configuration. The benefits of using genetic algorithms include the ability to handle complex, non-linear functions where input/output relationships are initially unknown. Furthermore, the use of hybrid, synergistic machine learning, and sensor tools are considered to be inherently more efficient and easier to adapt to a variety of soil and environmental conditions than traditional theoretical methods. Consequently, relatively few studies directly predict site-specific field scale soil moisture synergistically via machine learning and sensor networks, utilizing genetic algorithms as a further predictive parameter integrator. The scarce practice of such models motivates the value of what follows.

Keywords: AI-Driven Agriculture, Climate-Responsive Farming, Machine Learning Models, Precision Irrigation, Real-Time Data Analysis, Sensor Networks, Soil Moisture Prediction, Sustainable Water Management

13.1 Introduction

Global warming, rapid urbanization, population growth, shifting climates, and land use changes serve as the origins of severe changes in the natural environmental systems, which in turn restrict human existence and health. Among these constraints, an acute shortage of water is better classified as an invitation to both acute and long-term adverse effects on freshwater resources, which is closely linked to our food supplies and causes severe implications for the fundamental criteria of alleviating malnutrition and fulfilling food requirements. The global agricultural system currently demands a significant portion of the total annual flows. Thus, there is a great deal of need for agricultural productivity and water savings. The traditional agricultural system requires an immense amount of water, soil, and human resources (Al-Sager et al. 2024). However, changes in consumption patterns, climate variability, increasing environmental risk, globalization, and high production costs have increased the interest in precision and intensive agriculture.

Conventional methods are not enough for soil control and irrigation systems. Therefore, a sensor-based field system automatically observes soil and plants, receives data from the sensors, records it, and transmits it to a relevant computer system where it can be used online or offline to implement irrigation control programs, such as moisture rate regulation. Soil is key to the sustainable management of irrigation, closely linked to yield and the proportion of water used by the plants. One of the operational objectives of this system in the present study is the construction of an artificial intelligence system to anticipate soil moisture in the root zones and the quantity of water needed to establish the observation of hydrological and meteorological conditions (Kara et al. 2024).

The key aim of this research is to investigate the synergy between the Internet of Things sensor network and machine learning for the prediction of soil moisture content. Sensors located in the orchard send data about the environmental conditions to the central data storage servers automatically according to the threshold levels. The enormous data will need predictive algorithms for the prediction of soil moisture content for efficient irrigation and nutrient supply, in order to maximize fruit yield and minimize environmental degradation. The prediction accuracy and computational complexity are two main motivations for the investigation concerning the development of predictive models. The scalability, efficiency, and accuracy of emerging models are not well studied (G. Wang et al. 2024). The use of the central inferential machine learning model is planned rather than the existing manual models in order to address the challenges in modern smart precision agriculture systems. Neither assumed nor statistical inference automatically creates and distributes edge inferential models suggesting relevant benchmarks. Against this background, several objectives of this study are listed below, including the design of the prediction models for soil moisture content. This indeed helps decision-makers to minimize the separation of the orchard area and increase crop yield. The research outlines many benefits, such as efficient water requirements and nutrient supply, that, in turn, cost-effectively maximize fruit production. Thus, the smart precision agricultural system is energy-efficient, where the user accepts the deep learning algorithm for soil moisture prediction (Soylu and Bras 2024). Moreover, the environment is not to be affected by the IoT sensor life scenario in the orchard field.

13.2 Soil Moisture Prediction

Recent advances in prediction techniques have enabled real-time soil moisture content predictions with increasing accuracy. At the same time, breakthroughs in sensor technologies have led to the rise of a wide variety of sensor networks, which are increasingly adopted in soil moisture prediction. Exciting novel applications are being explored, contributing to closing the gap between research and real-world use. In this section, we provide a review of the prediction techniques developed in recent years and the sensor networks employed in the field. Rather than an exhaustive list of existing methodologies, we opt to show a variety of techniques so that overall trends can be understood. We will proceed with the state-of-the-art methodologies that have preceded this effort as follows. Soil moisture prediction techniques can be roughly categorized into two categories: physical model-based predictions and data-driven models. Sensor networks enable the collection of soil moisture data with high spatiotemporal resolution, covering small plots or individual fields (Yundong Wang et al. 2024). It is very important to obtain accurate soil moisture data to improve the accuracy of agricultural activities. Thus, combining different prediction models with sensor networks will be essential for realizing more accurate predictions. The extent of advances shown in a large number of publications demonstrates the increasing demand for collaborative work between sensor and machine learning technologies to improve prediction accuracies at various levels of the scientific research community (R. Wang et al. 2024). These facts lead us to the motivation and objective of this study, which lies in the gap between the two branches of research.

13.2.1 Soil Moisture Prediction Techniques

Soil moisture prediction has been the focus of many studies over the years. A number of different methods have been used to predict soil moisture content. One of the most commonly employed methods to evaluate changes in ground moisture content is gravimetric measurements. The rise of remote sensing technology has precipitated further advancements in the prediction of soil moisture that is now employed in developed countries. Satellite remote sensing, in particular, has recently and repetitively provided large quantities of soil moisture data, leading to its increased usage in soil moisture prediction. Moreover, with the advent of technology, increasingly complex methods are used to predict soil moisture (Kara et al. 2024). With the emergence and growth of the IoT, forgotten on-the-go information obtained from sensors is widely used for the prediction and monitoring of soil moisture content. Soil moisture monitoring is extremely helpful to the agricultural sector. Many traditional systems are also based on soil moisture retrieval models, which are limited in terms of their capacity to forecast soil moisture precisely. Nevertheless, the same systems indicating the good behavior of PMLs also suggest that some problems are linked to their implementation in an actual environment connected to real-world agricultural organizations. Both storage and mineralization of carbon are found in the soil, and moisture affects these protecting processes (Yanling Wang et al. 2024). The moisture content of the soil controls the amount of water expended and affects the yield.

13.2.2 Sensor Networks in Agriculture

Sensor networks have gained significant attention in different areas of human life, including healthcare, the environment, and precision agriculture, to monitor the spatial and temporal behavior of the area of interest. Sensor networks in agriculture enable highly precise and specific data collection that is used for monitoring soil and weather conditions. The collected information, in turn, is used to optimize growing conditions for crops depending on the current environment in the field. Employing sensor networks in agriculture drives the dimensions of smart farming and links to big data, IoT, and predictive analytics that offer business intelligence, including farm traceability and environmental analysis (Yanling Wang et al. 2024). The increasing use of sensor networks means increased demands in various aspects: scientific, technical, socio-economical, and ethical. Importance of Sensor Networks Applications of sensor networks ranges from benign monitoring of patient data in a hospital to measuring air pollutants in a city or estimating soil and weather conditions in an agricultural field. They enable the capture of vast and reliable data underlying systems, objects, or events. In agriculture, they can inform about optimum conditions for planting and soil nutrition through hyper-specified meteorological data in micro-climates of vineyards for generating selective heatmaps between high-performing and low-performing parcels in different crop and growth stages (Yundong Wang et al. 2024).

13.3 Sensor Technologies

Tensiometers are designed to measure suction, not volumetric water content. Nevertheless, it can be related to volumetric water content after the installation of the tensiometer. The time domain transmission sensor measures the dielectric property of the soils using an oscillating frequency; the reduced frequency and the amplitude are proportional to the water content of the soils. In agriculture, the usable frequency is around 1500 or 1800 MHz. The other sensor technologies are frequency domain reflection, a capacitance sensor, and fast transmission decay. These technologies measure volumetric water content, which can be fully used in the agricultural field. These sensors can be used for control applications. Various mechanical solutions are also used for soil moisture detection, especially in field experimental devices (Al-Sager et al. 2024). These sensors are set to measure the moisture of materials and are often used in experimental devices. The up-to-date situation of the sensor technologies was discussed. The most promising solution is represented today by capacitive measurement principles, such as the well-known frequency domain reflection and its most recent developments. The bullet point list summarizes the state of the art of the existing methods in the following applications: soil moisture detection by sensors and agricultural water management on-site irrigation control through sensors. It can also be shown that the digital generation makes it possible to use soil moisture detectors digitally to save energy and lower costs. There are many sensor technologies in soil moisture measurement. Most of them have good resolutions. The trade-off is either costs or maintenance. The sensors are mainly capacitance sensors. In terms of the technology for soil moisture detection, the accuracy of the soil tendency and time domain transmission sensors are the best (Pan, Li, and Li 2024). A tensiometer can provide the most accurate suction, though these technologies are impacted by ethical AI, as illustrated in Fig 13.1.

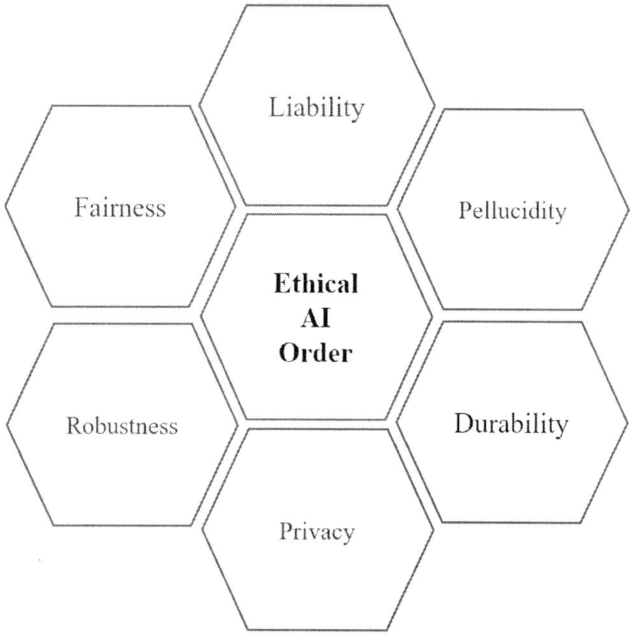

Fig. 13-1 Ethical Artificial Intelligence Order

13.3.1 Types of Sensors

Capacitive sensors are the most widely used volumetric moisture sensors and work by measuring the change in the dielectric constant of soil, which is directly connected to the soil water content; hence, the dielectric constant is directly proportional to the soil moisture level. Resistors use the same principle; when the water content increases, the electrical conductivity increases, which is directly proportional to water content. Other sensors, such as time domain reflectometry sensors and neutron moisture gauges, are available to measure the moisture content. The application of the different types of sensors in the field is significantly dependent on their effectiveness, accuracy, and limitations. For example, resistive sensors are relatively lower in cost. However, due to chemical reactions with the metallic rods of the sensor, the transit of water between the sensor prongs is decreased, often resulting in unreliability of the measurements (Dolaptsis et al. 2024). With the use of remote sensing, satellite-based microwave sensors are used to gauge the amount of moisture in the soil up to one meter deep.

The combined reflective, emissive, and transmitted values from microwave sensors measure the soil moisture content, where the sensitivity of this technique can reach up to 5% volumetric water content or even lower if the linearization of the model is included. In fact, soil moisture content at different timescales can be estimated using active and passive microwave sensors due to their sensitivity and specificity of such. The most appropriate sensor type is particularly relevant to the applied agricultural sector. In orchards or vineyards, for instance, where pivot irrigation is used, the most appropriate sensor type is a capacitive one. On the other hand, when using linear irrigation or in the presence of sandy soil with low water holding capacity, a tensiometer or time domain transducer will likely be more reliable. The capacitance sensor type is also used in the southern Australian region to control land forming and guide decisions regarding the installation of tiling systems to drain water out of the soil in case of high water content. Today, wireless, long battery life, and remote monitoring are key components included in the design of advanced sensors in order to facilitate data collection (Hughes et al. 2024). The agricultural sector is quite diverse, and therefore, the development of a range of sensors is necessary.

13.3.2 Deployment Strategies

Due to soil variability and changes in the environment during the cropping season, a simple soil water sensor alone cannot deliver soil moisture conditions for a whole field. Typically, the crop root zone and soil moisture spatial-temporal patterns will depend on the variations in topography, geology, soil, land use, and management practices. Variations are typically high in areas near field boundaries, and they need to be well-represented by the sensors. However, the detailed variability in soil properties and water content in these fine-textured surficial soils will always exist, even when the soil profile is relatively uniform and deep. Soil sensors must be strategically placed in these areas accordingly. Moreover, the main objectives of the experiments also have to determine where the sensors need to be deployed (Thomas, Yadav, and Šimůnek 2024). For all cases, communication features must be considered, along with sensor selections for agricultural soil moisture applications. Multiple studies inter-comparing the strategies and planning of networks are limited. However, a variation of experimental field studies exists in each application domain. Various data collection intervals from a day to a few minutes, and spatial sensor distributions were applied in these experiments. Case studies in this subsection are mostly with networks of sensors. In a few studies, the data from additional sensors were also applied. The analyses are associated with effective soil moisture software tools to demonstrate the roles of data collection intervals and the sensor density in establishing the soil moisture conditions for the watershed and fields (Dafalla and Al-Mahbashi 2024).

13.4 Machine Learning Algorithms

Different algorithms can be employed for prediction in machine learning. Overall, it can be divided into two categories: supervised and unsupervised learning. For large datasets, supervised learning-based algorithms are preferred due to their high prediction accuracy. On the other hand, unsupervised learning-based algorithms are preferred for local simulations of soil moisture as they can capture the complex patterns of soil moisture.

Supervised learning models can predict future soil moisture dynamics by using current and past soil moisture, linear regression combined with the maximum likelihood method, and ensemble methods. Some unsupervised models can solve complex and high-dimensional clustering problems (Kňažková and Hrbáček 2024). These models are able to learn and model complex patterns in large data representations not separated into independent components and can reconstruct all levels of input systems. Some recent studies have shown that machine learning outperformed physical models in terms of processing sensor-based data. Various approaches to models have also been used to improve physical modeling, including spatiotemporal models in areas of a water-limited environment. Selecting the best algorithm is not straightforward and is highly case-dependent. Moreover, proper training before the application of the model is needed to improve predictive power. Despite recent advances related to the use of machine learning, the synergies between sensor data and machine learning have a positive impact on data-driven soil moisture predictions. Collected sensor information will be incorporated as measurement data in the training or identification stages, calibration of a model, or learning of model parameters (Guo et al. 2024; Banda et al. 2024). Despite the minor work conducted in agricultural research and practices related to soil moisture management, it is expected to improve soil moisture management and contribute to large-area policy development where sensor application is limited and influenced by ethical factors, as demonstrated in Fig 13.2.

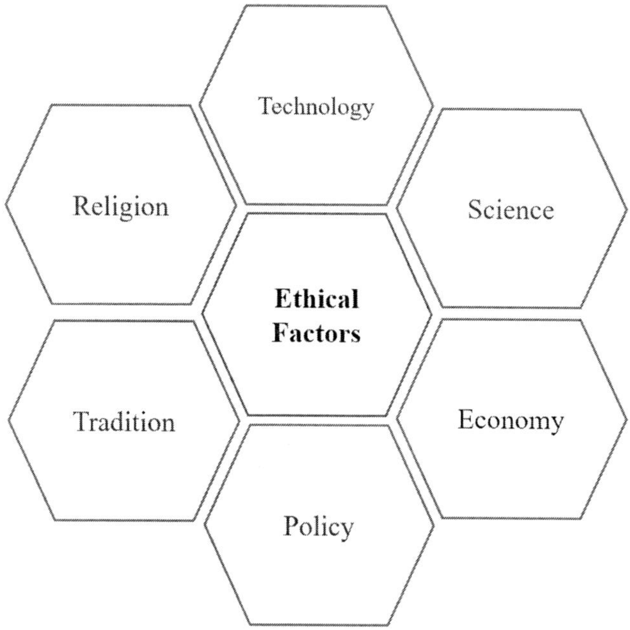

Fig. 13-2 Ethical factors

13.4.1 Supervised Learning

Supervised learning algorithms are a group of algorithms devised to learn patterns that distinguish the inputs and form desired outputs based on these patterns. These patterns are learned within the training process using input data and known output data to train the model. This paired input dataset is referred to as labeled data. In the context of soil moisture prediction, a labeled dataset comprises historical observed soil moisture and the actual amount of water used by the root zones. The primary objective of the regression model is to recognize the relationship between the two variables or even among multiple inputs and outputs and then use this relationship to make predictions (Tájmel, Cruz-Paredes, and Rousk 2024). There are several models used in scientific research for precise soil moisture update prediction, including ridge and Lasso regression, regression trees, support vector machines for regression, XGBoost, neural networks, and LSTM models for predicting temporal changes in soil moisture. Additionally, the ability to predict soil moisture using a regression model is closely connected to the model used to approximate input/output statistical nonlinearity. Some of the major techniques of regression used for soil moisture prediction include linear regression, support vector machines for regression, artificial neural networks, and LSTM (Hu et al. 2024).

Their strength has been demonstrated as a practical model that can be used for numerous modeling claims. The limitation of such models is directly related to the quality and size of the initial dataset: the more observations you have, the more precise your solution would be in a purely supervised way, i.e., accurately in the sense of the complexity of the prediction function you can model. It should be remarked that, in some cases, minor increases in dataset sizes have been found to enhance results significantly. Such a representation of reality in supervised learning depends on the acquisition of adequate datasets with thousands of examples of well-chosen and accurate learning material. In agricultural settings, several applications using supervised learning have indicated improved soil water predictions achieved by the learning stage of soil models (Ye et al. 2024). During the learning process, algorithms are trained to generalize from input-output pairs; the potential challenge of overfitting can arise where the model possesses an excessive number of training parameters, and so becomes too flexible. Model learning is then potentially driven by large amounts of noise rather than by generalizing from the true relationship between independent and dependent variables. Consequently, models fail to perform well at predicting new or unobserved examples; this is also graphically displayed following the train/test phase or within learning curves created by plotting the error on the training and testing datasets as a function of sample size. Although overfitting is a model-tuning problem, that is, an intrinsic feature of learning processes that can be handled, it is an important aspect to consider from an integrative point of view. The major technique employed to counter the risk of overfitting is cross-validation, a process that can be employed during the phase of model input coefficients and in the process of tuning the final model values (Zheng et al. 2024).

13.4.2 Unsupervised Learning

In contrast to supervised techniques, where the underlying mechanisms have already been trained and validated using labeled data, unsupervised learning can be applied to unstructured and unlabeled data. Unsupervised methods analyze the given data to extract hidden patterns and relationships. Clustering methods group similar sensor data that reflect the same soil moisture dynamic patterns. Clustering validates the individual state representation and recognizes the number of states within the unlabeled data. Dimensionality reduction methods can reduce data redundancy and compress the important features, which are used further for soil moisture prediction (Yang et al. 2024).

In particular, principal component analysis can identify sensors from the large sensor network with very similar information, which reduces redundancy in the prediction of soil moisture, changes, and behavior. For a large sensor network, clustering and dimensionality reduction are very powerful tools that will significantly reduce the computational cost and make the soil moisture prediction model practical. In addition, unsupervised methods also have some of the standard issues and challenges that are faced in the analysis. Finding an optimal number of clusters in the case of a clustering method is quite a challenging task, as there exists a plethora of ways using which these can be identified (Yu Wang et al. 2024; Szabó et al. 2024).

Moreover, the validation results are also confusing to interpret, and the results may vary based on the input parameters and the algorithm being used. It is clear from this that unsupervised learning and clustering techniques will improve the descriptive capabilities in expanding the soil moisture dynamics. Furthermore, supervised learning with the relevant clustering-state-to-label conversion method shows great verifiable results.

Applications and Case Studies In agriculture, where sensor networks are of primary interest, clustering methods are extensively used for soil moisture dynamic analysis. In particular, clustering methods have been used to cluster sensor data based on the fluctuation patterns of non-vegetation spectroscopy, which are the reflectance and temperature data of the sensors that are placed near the soil surface (Al-Sager et al. 2024). The analysis of the hydrological characteristics of the subsurface parameters of the node network has been explored by applying various methods. Daily fluctuation patterns in the sensor data, which are strongly related to the soil moisture content fluctuations, have been identified. The optimum number of clusters is one of the main issues that arise with clustering techniques. Fuzzy C-means and the k-means clustering methods, which are two main clustering methods, require the input of the desired number of clusters. It is quite challenging to find this optimal value, as different input values lead to different results, based on the assessment (Dolaptsis et al. 2024).

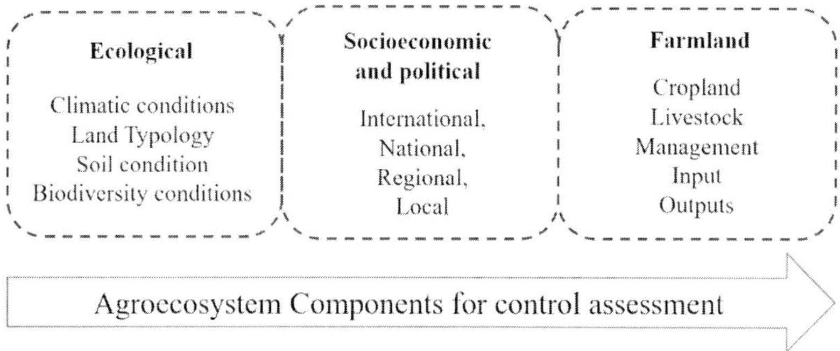

Fig. 13-3 Agroecosystem components for control assessment

13.5 Integration of Sensor Networks and Machine Learning

The present study pursues the integration of sensor networks with advanced data analytics, mostly machine learning techniques, to develop a holistic approach to soil moisture prediction. It is only the synergy between advanced analytics and real-time data collection, accomplished within smart agriculture, that can address technology issues affecting soil moisture prediction based on sensor networks and successfully implement the prototype of the crop management system. Both statistical data preprocessing and biotechnical processing are essential backbone sub-processes that permute and complement digital machine learning methods. Machine learning offers novel methodologies in soil moisture research, facilitating the challenge related to human expert discrimination relying on conventional sensors. It can comprehensively cover training to results obtained after sensor data interpretation accompanied by suggested actions based on analytical results (Banda et al. 2024). The data that comes from these sources may not necessarily be adapted to what is expected by the data analytical method to yield sound outcomes. It requires effective data preprocessing from raw data precleaning, cleaning, transformation, and post-processing activities for best practice and overall performance. The future tech trend suggests convergent usage of machine learning and agronomic sensors with AI and cloud services, and will continue to provide continuous improvements in the development of smart soil moisture management (Tájmel, Cruz-Paredes, and Rousk 2024). The physical synergy of cyber-physical systems may offer a one-size-fits-all and scale-independent solution in future soil moisture research based on compounding precipitation and drought forecasting methodologies and associated soil crop production dynamics in extended farmlands and its associated benefits, as presented in Fig.13.4.

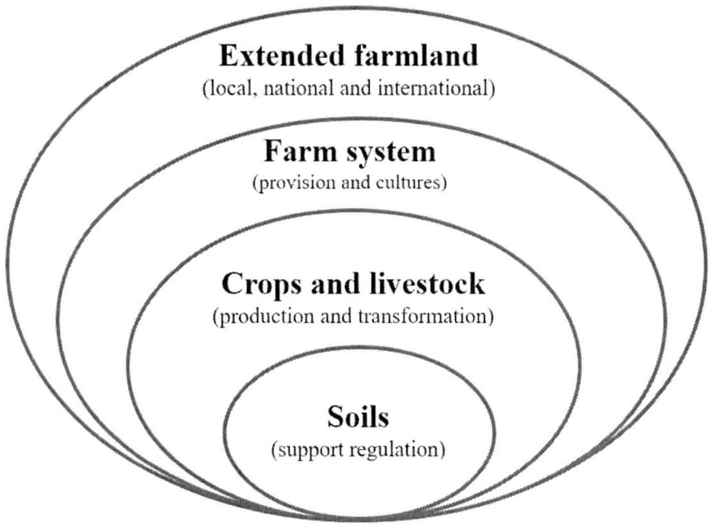

Fig. 13-4 Extended Approaches in sustainable farming

13.5.1 Data Preprocessing

In the integration of WSN and ML, data preprocessing is considered a crucial phase. The task of preprocessing the raw data is especially important for sensor data, as the data recorded and collected by the sensors may consist of missing data, outliers, or noise. As a result, data preprocessing methods are used to prepare the collected raw data for the analysis phase. After that, the enhanced data is input into the ML model or learning algorithm. This enhances the efficiency and performance of the ML model. Missing value and outlier treatment are the initial important steps in data preprocessing. A number of techniques are employed to address missing values and eliminate outliers. These include handling methods, deep learning imputation techniques, interpolation, label methods, and cost-sensitive learning methods (Hu et al. 2024).

After missing value and outlier treatment, the dataset can be prepared using preprocessing techniques such as normalization, scaling, and reformatting to maximize data quality, thus minimizing any discrepancy in the prediction or approximation of the model generated by the learning algorithm. The improvement in the quality of raw data directly impacts the effectiveness of learning algorithms. Several tools and technologies can be exploited in data preprocessing to guarantee optimum performance in soil moisture prediction applications. The use of datasets utilizing preprocessing techniques is a recommended approach in order to maximize the prediction's accuracy. Even without a preprocessed data requirement, some ML libraries accept unprocessed datasets. The challenge is the huge number of computational resources required to preprocess the data prior to being used as input in the predictive models and validations, especially for large-scale projects (Yang et al. 2024).

13.5.2 Feature Selection

Selecting the smallest number of relevant features is necessary to develop AI models efficiently. Feature selection is important to improve the robustness and generalizability of machine learning models by reducing overfitting, increasing model accuracy, and decreasing model complexity by extracting useful information that can lead to finding causal relationships. It is useful for evaluating model performance because reducing the number of features and only retaining the quantitatively high-quality features can avoid the deterioration of model performance caused by irrelevant high-dimensional feature data. Additionally, it is helpful for further reducing the computational costs for both training and prediction because operations performed on every feature can be time-consuming. On the other hand, including irrelevant and low-quality features can also slow down the learning process (Szabó et al. 2024). Data dimensionality is a concern because too many features require large storage and relevant resources.

Statistical and optimization methods, as well as filter and wrapper approaches, are among the popular feature selection techniques. Feature importance from input data can be selected on the basis of statistical measures such as mean, standard deviation, variance, chi-square, correlation, or those indexed in scoring and ranking. Additionally, feature engineering should also be applied to develop high correlation metrics. A host of machine learning algorithms such as Naïve Bayes, decision trees, linear regression, and linear discriminant analysis is specifically designed for agricultural areas in the predictor, e.g., large-scale agricultural datasets predict based on stochastic weather changes and apply to real-world disease case studies to ensure feature stability. Furthermore, feature selection should apply different methods fitting to the model and consider whether inclusion is important to achieve high quality (Yu Wang et al. 2024). We should remember that feature selection can overfit the model when it does not split into training, testing, and validation depending on sustainable agricultural aspects, as sampled in Fig 13.5.

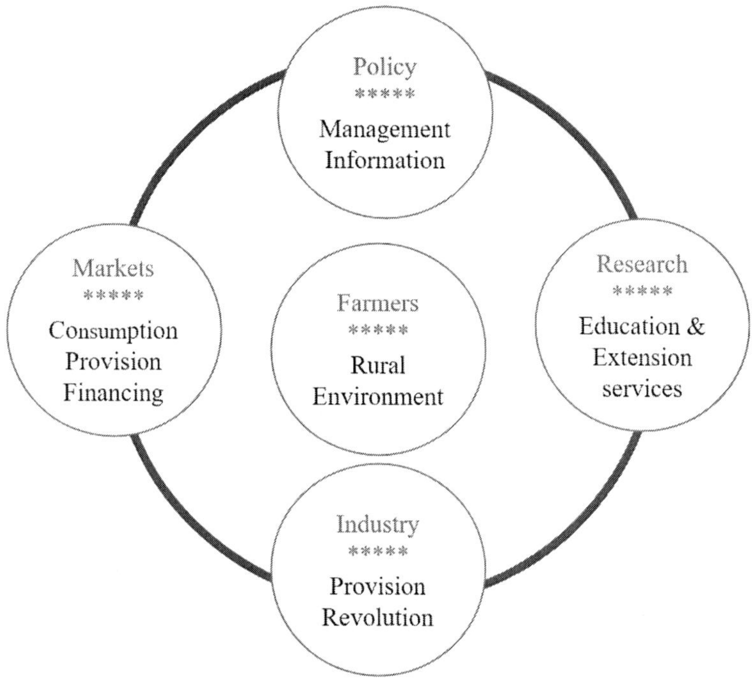

Fig. 13-5 Sustainable agricultural aspects

13.6 Case Studies

In this section, a series of case studies of the application of the AI-enhanced soil moisture prediction technique in sensor networks are presented. For that, an initial brief description is made of each case and its implementation settings, as well as the results obtained in the experiments and their relevance, which show improved performance metrics and area representation in agricultural scenarios. Additionally, the difficulties faced and the lessons learned in each case are discussed to emphasize the essential requirements for the adoption of these proposals and the results possible to be achieved in the field, both in terms of research and agriculture (Lu and Wang 2024). We can highlight that the diversity of the cases ensures that the results and conclusions can be comprehensive enough to encourage site-specific adoption in various locations or to enable further investigation in various regions.

13.6.1 Gro Drought 4.0

The implementation of the first case has partners Zamorano and Gro Drought 4.0. The purpose of the partnership was to develop a monitoring and prediction platform as a service aiming at supporting Gro Drought 4.0 consulting in their advising proposals in the management of a farm located at the Escuela Agricola Panamericana El Zamorano trading company. The chosen site, which is the pilot project for the Mexico location, is situated almost in a steppe climate zone and had recently suffered from extreme drought in the period from fall 2019 to spring 2020. The approach of precision irrigation by means of AI-enhanced soil moisture prediction using sensor networks and machine learning has shown successful results that created an impact in realistic agricultural settings. In the literature, several such projects have been carried out (Greiser et al. 2024). Yields could be increased by 14%, and water use efficiency was improved by 36% in Indian sweet cherry orchards. Socio-economically, bridge loans with terms matching mean farmers' cash flow cycles were provided in order to introduce more financially secure integrated development mechanisms. Indian apple orchards, too, were shown to benefit; by saving 17 MWh of electricity for pump operation per year, a decrease of €1,700 per year in electricity bill costs was noticed, given that annual energy requirements totaled some 1 to 1.5 MWh per hectare (Suresh et al. 2024).

Within the specific type of orchard, the data of the weather forecast were fed into the training model with an aim to prepare for hailstorm periods. The sensor nodes need to be equipped with mobile functionality to ensure scalability. This requirement was fulfilled in numerous other projects. A project in India is on the way to being scaled up from 1 to 20 stations (covering 30 ha) in local food crop assurance while incorporating it into crop insurance policy. Scalability can be achieved as assessed based on the considerably larger planting area quantities desired by farmers. The editor of the commercially maintained dataset of a project applicable to a mass population had an aim of using a 15,000 to 20,000 sensor node cellular Internet of Things (IoT) network in African countries (Atanasov 2024).

Major headliners for exchanging water inputs for agricultural outputs are reduced risks for stakeholders, water-saving, and improved production. They all faced challenges of false premises about the status quo of irrigation technology and constraints preventing workers from efficiently using all necessary information from science, giving a chance to come up with innovative exception rules that, when met, help keep user empowerment in decision-making. These success stories serve as inspiration that there are simple solutions that are just a few mouse clicks away. Who better to describe these stakeholder-backed achievements than the success stories themselves? The overall innovation project needs to know what is out there and against which benchmark it is expected to deliver in the developments in the field of AI-enhanced soil moisture prediction. Increased crop yield with precision irrigation: Irrigation Service delivers monthly site-specific irrigation advice and 5-day and 10-day irrigation forecasts. The Irrigation Service integrates hydrological, meteorological, and satellite data, climate forecasts, and high-resolution soil-plant data into a decision support system. By using machine learning models, it has been possible to make use of the variability of the fields, the local water movements within the fields, and the heterogeneity of the root depth in the use of water and mineral fertilization due to precipitation (Vorobevskii, Luong, and Kronenberg 2024).

13.7 Challenges and Future Directions

The future of soil moisture prediction with AI applications appears promising. However, to translate the potential into operational on-the-ground applications, we need to address several research challenges and future research directions:

(i) Data security and privacy: The demand for agricultural data has increased since the inception of sensor networks. However, the data are sensitive because they are also associated with the location and cost of agricultural properties, which may put farmers at a disadvantage. Therefore, future research may need industry policies to improve user data security, privacy, legal, and compliance (Choi et al. 2024).

(ii) Scalability: Interpretable models are data-intensive, and handling vast volumes of data represents computational, management, and storage hurdles. To address this challenge, we need to develop robust mathematics and algorithms for big data integrations.

(iii) Deployment cost and sensor preparation: Sensor network deployment, which involves installing remote sensors, needs close attention. The cost of sensor deployment, calibration, and power supply is high. Hence, there is a need to advance field data collection as part of portable sensors. In research where existing farm sensor networks are used, agricultural industry partners are required to minimize the downside of sensor placement (Lee et al. 2024). For the future development of sensor networks, there is a need to develop portable sensors that only measure soil moisture and are not physically attached to crop roots to expand the deployment.

(iv) Current connected tools: AI may offer better use in future agriculture by developing new models or advancing current concepts. We will attract sufficient academic, industry, and government collaborations to establish best practices in the practical implementation of AI in agriculture to convert AI into applied outcomes. For on-farm applications and changing environmental conditions, research into new AI techniques must be adapted and, at the same time, be straightforward and simple to use. More applications are likely to be relevant to tri-state areas, which predict soil water status as one part of land management decisions (Li, Shen, and Liu 2024).

13.7.1 Data Security and Privacy

Increasingly, sensor networks are deployed not only in research and precision agriculture but also to optimize farm operations. The need for real-time distributed data collected by these sensor networks raises serious concerns regarding data security and privacy (Shafik 2024b). These sensor networks are often subjected to low-layer DoS attacks with adversaries intent on malicious data modification, thus providing inaccurate data to control and decision support systems. The generation of high-fidelity, real-time data with inherent trust is also critical to facilitate the uptake of this technology and demonstrate adherence to regulatory and compliance frameworks (Vorobevskii, Luong, and Kronenberg 2024). Ensuring robust privacy security in an interoperable manner presents a variety of data sharing paradigms and user identification and access control paradigms more complex than simply ensuring secure and private data exchanges; thus, encryption may not be a satisfactory methodology in the context of sensitive agricultural information. One promising research direction could be the investigation of secure multi-party computing and differential privacy techniques in the context of sensor networks and distributed data logic so that data privacy can be ensured regardless of data storage location (Greiser et al. 2024). The synergistic effects of differential privacy and homomorphic encryption with expert systems-based decision-making also represent a promising research direction. More generally, priorities for future work consist of the formalization and implementation of appropriate real-time security and privacy-preserving frameworks that can interact with users to ensure a robust, trust-based process and to ensure the continued usage and benefits of this technology in improved resources for the broader agricultural demographic (R. Wang et al. 2024).

13.7.2 Scalability Issues

The widespread availability of AI-enhanced soil moisture prediction systems could increase the marginal benefit for farmers' investments in soil water sensor networks for precision irrigation practices. Taking the implementation of powerful sensor networks over whole crop areas into consideration, many challenges for the scalability of such a prediction system emerge (Shafik 2025). First, soil properties such as texture, soil structure, porosity, aeration status, and pH interact with the local climate and surface cover conditions on subfield scales. Consequently, local machine learning prediction models will possibly perform optimally, but communication technologies might fail to transfer data between the diverse scales and conditions in larger-scale systems. In particular, transferability is a greater concern when operating with high-resolution irrigation schemes for very localized strategies in contrast to large-scale zoning approaches (Yanling Wang et al. 2024). Second, the roll-out of sensor networks over multiple acres can have a substantially higher price tag. In addition to largely increased investment costs, maintenance, and management efforts for handling sensor nodes and their low-power capabilities, management interfaces have to be improved. Upfront investments in network size may also translate into upfront investments in maintenance costs. The management of different CMUs has always been identified as a key bottleneck, especially when running a larger-scale implementation with a combination of different communication media. Here, the diversity of cross-media configurations translates into different sets of configuration parameters that are generally handled per time slot for subsequent simulation runs. However, these configurations depend on the media and its specifications. A change of media implies thorough technical planning at all levels and may result in the need for different configurations to adapt it to novel medium conditions, which are generally difficult to measure or anticipate (Liu et al. 2024). However, further commercial solutions and even the use of legacy technologies can result in an even greater upfront monetary spending.

Modularity refers to the ability of a system to function with variable capacities and specifications. In this specific case, the possible candidate modularity levels range from sensors' ground truth validation to the subsequent machine learning data processing. Data analysis and machine learning-based processors need to harness possibly diverse, uncertain, and error-polluted valuable information exits. Adaptable technologies are seen as an important sector for enablers of local collaborations, research, and adoption. They can help improve local solutions at the farm level, population management, and market access, and at the same time, help drive adoption in wider landscapes. They can support research, development, and application for site-specific and landscape-scale processes and products based on the integration of communication, modeling, predictive analytics, and visualizations (Fjeld et al. 2024). The deployment of adaptable systems at agricultural landscapes needs to consider the preferred choices of farmers and practitioners and facilitate easy access, interpretability, and actionable insights. Pending issues need collaborative efforts between the scientific community and the stakeholder populations to be solved. As with many experiments, outreach, technology adoption, implementation adoption, development of undervalued market integration, and knowledge of the population needs, as well as in perceptual science, are experiencing new approaches. The identification of codes and meta-codes of purpose and usage in land-related data, digital twins, and systems in adaptation was addressed with experiences and dialogue (Asadollah et al. 2024; Wu et al. 2024). These make clear that failure mechanisms still need to be understood and addressed for the successful scaling of hardware, software, and human implementations.

13.8 Implications for Future Research

At the outset of this thesis, one of the objectives was to evaluate whether high-density sensor networks and artificial intelligence machine learning algorithms can be integrated to enhance agricultural field soil moisture prediction. This work has demonstrated in three incisive ways it can: the recent deployment of high-density sensor networks, their detailed, up-to-date examination, and the practicality of combining several state-of-the-art AI machine learning algorithms in a tightly focused manner while avoiding algorithm over-configuration. It was shown that this choice has increased predictive power without increasing the complexity of the prediction effort. Importantly, evaluating the new predictive models across real and diverse farming landscapes increased the application of the research for the agricultural stakeholders. Here, a thorough investigation and discussion of the accuracy and other metrics associated with a 4-week ahead daily forecast was the chief focus (Tájmel, Cruz-Paredes, and Rousk 2024). Trends and discrepancies were interpreted in the light of daily data quality, and a robust data pre-processing strategy was created and examined.

Findings primarily support that sensor networks can be deployed alongside AI machine learning algorithms to predict soil moisture, thereby allowing busy farmers to utilize ready at hand tracks for irrigation management. Despite the promising applications, the installation of sensor networks at this density is a costly affair for the typical farmer (Shafik 2024a). Furthermore, while the sensors are easy to install, data quality and systemic error are very important. By deploying a predictive model that adjusts for real-time data quality, it is reasonable to forecast soil moisture for most crops. However, the use of ensemble algorithms may be complicated by such implementation in a large-scale environment. As such, research that focuses on the development of new algorithms suitable for this application is an area of interest for future research, as is also the scalability and data security for both the soil moisture interfaces used in this case study (Hu et al. 2024; Ye et al. 2024). Future research should continue to focus on advancements in ML techniques for soil water prediction, with special attention paid to data-adaptive techniques to keep pace with evolving agricultural landscapes.

Research should be strengthened through interdisciplinary collaborative environments, allowing the development of methodologies that will collectively tackle current and future challenges in the adoption of these techniques. Research is also required to facilitate the development of methods and to evaluate existing practices for securing data and ensuring privacy in modern agriculture (Shafik et al. 2024). With the increased use of AI analytics in agriculture, the security of databases containing yield, soil, environmental, and even farm practices data becomes a key concern. At the same time, preventive measures are also required to avoid planting strategies, irrigation regimes, and nitrogen fertilizer applications being exposed to misuse or theft of data. There is still scope to identify innovations that will assist farmers in generating benefits to increase confidence and scale soil moisture monitoring and prediction approaches across a range of farming systems, including technology adoption forecasting tools, informed business consulting services, and linking satellite sensing with on-farm moisture measurements (Sumarsono, Afiatna, and Muflihah 2024; Jiao 2024). Finally, these research findings will impact the economic feasibility of these approaches in the medium to long term but will also impact agricultural systems' sustainability over the long term. By integrating AI learning into existing tools, researchers should also pay closer attention to potential advancements in technology and methods. In the context of the findings of this review, a future research approach should aim to adopt a proactive stance in dealing with evolving research areas and significantly intensify innovation in anticipation of future research challenges related to the uptake of developed tools and strategies (Letcher et al. 2024; Steger et al. 2024).

13.9 Conclusion

This study presents an approach integrating AI with sensor networks for enhancing soil moisture prediction. Key findings suggest that integrating sensor and forecast information increased soil moisture predictions at all stations and improved or had no change on irrigation triggers. There is a shift in agriculture from extensive to precision farming, where much research and development revolves around the proper management of key inputs such as water. The integration of recent modern and advanced technologies in the field of agriculture is essential; these include AI, IoT, and sensor-based networks. The implementation of AI in precision agriculture would pave new thinking to devise resource management practices and offer necessary amendments in predictions and alerts systems. In this study, an AI-enhanced soil moisture prediction model is proposed for lower reproducibility and applicability of a machine learning model using realistic datasets. The coupling of sensor and forecast information in a predictive model resulted in the improvement of results with respect to a model forecast by just image data. Therefore, the main findings of the study were: (i) the coupling of sensor and forecast information for a machine learning model improves soil moisture predictions; (ii) sensor and forecasts were useful to trigger the irrigation need for most stations; (iii) no improvement in irrigation operations was quantified at some stations when forecasts were coupled with sensors; and (iv) rainfall estimates had a major improvement if forecasts were based on radar rather than the model. Continuously improving and adapting these models and integrating multiple ICT and IT-based technologies is important in the emerging concept of an AI-enriched future agriculture. These strategies could help address the hurdles that may affect the responsiveness of stakeholders. One effort is the continuation and promotion of research and development studies that have an AI orientation and innovative structure of models, forecasts, and applications.

References

Al-Sager, Saleh M., Saad S. Almady, Samy A. Marey, Saad A. Al-Hamed, and Abdulwahed M. Aboukarima. 2024. "Prediction of Specific Fuel Consumption of a Tractor during the Tillage Process Using an Artificial Neural Network Method." *Agronomy* 14 (3). https://doi.org/10.3390/agronomy14030492.

Asadollah, Seyed Babak Haji Seyed, Ahmad Sharafati, Mohammad Saeedi, and Shamsuddin Shahid. 2024. "Estimation of Soil Moisture from Remote Sensing Products Using an Ensemble Machine Learning Model:

A Case Study of Lake Urmia Basin, Iran." *Earth Science Informatics* 17 (1). https://doi.org/10.1007/s12145-023-01172-8.

Atanasov, Svetoslav. 2024. "Automated Remote Sensing System for Crops Monitoring and Irrigation Management, Based on Leaf Color Change and Piecewise Linear Regression Models for Soil Moisture Content Predicting." *Scientific Horizons* 27 (1). https://doi.org/10.48077/scihor1.2024.127.

Banda, Enid, Vahid Rafiei, Josué Kpodo, A. Pouyan Nejadhashemi, Gurjeet Singh, Narendra N. Das, Rabin Kc, and Amadiane Diallo. 2024. "Millet Yield Estimations in Senegal: Unveiling the Power of Regional Water Stress Analysis and Advanced Predictive Modeling." *Agricultural Water Management* 291. https://doi.org/10.1016/j.agwat.2023.108618.

Choi, Su Hoon, Sang Hyun Lee, Ung Yang, and Min Soo Kim. 2024. "Data Mining Approach to Predicting Soil Moisture Based on Meteorological Factors and Flow Rates." *Horticultural Science and Technology* 42 (1). https://doi.org/10.7235/HORT.20240001.

Dafalla, Muawia, and Ahmed M. Al-Mahbashi. 2024. "Identifying Problematic Soils Using Compressibility and Suction Characteristics." *Buildings* 14 (2). https://doi.org/10.3390/buildings14020521.

Dolaptsis, Konstantinos, Xanthoula Eirini Pantazi, Charalampos Paraskevas, Selçuk Arslan, Yücel Tekin, Bere Benjamin Bantchina, Yahya Ulusoy, et al. 2024. "A Hybrid LSTM Approach for Irrigation Scheduling in Maize Crop." *Agriculture (Switzerland)* 14 (2). https://doi.org/10.3390/agriculture14020210.

Fjeld, Dag, Magnus Persson, Johan E.S. Fransson, Jan Bjerketvedt, and Martin Bråthen. 2024. "Modelling Forest Road Trafficability with Satellite-Based Soil Moisture Variables." *International Journal of Forest Engineering* 35 (1). https://doi.org/10.1080/14942119.2023.2276628.

Greiser, Caroline, Lucia Hederová, Giulia Vico, Jan Wild, Martin Macek, and Martin Kopecký. 2024. "Higher Soil Moisture Increases Microclimate Temperature Buffering in Temperate Broadleaf Forests." *Agricultural and Forest Meteorology* 345. https://doi.org/10.1016/j.agrformet.2023.109828.

Guo, Jingjing, Martin Ziegler, Niko Wanders, Mike Vreeken, Qiuzhen Yin, Hao Lu, Louise Fuchs, Jibao Dong, Youbin Sun, and Francien Peterse. 2024. "Robust Land Surface Temperature Record for North China over the Past 21,000 Years." *Science Advances* 10 (8). https://doi.org/10.1126/sciadv.adj4800.

Hu, Feichi, Qinli Yang, Junran Yang, Zhengming Luo, Junming Shao, and Guoqing Wang. 2024. "Incorporating Multiple Grid-Based Data in CNN-LSTM Hybrid Model for Daily Runoff Prediction in the Source

Region of the Yellow River Basin." *Journal of Hydrology: Regional Studies* 51. https://doi.org/10.1016/j.ejrh.2023.101652.

Hughes, M., D. L. Jackson, D. Unruh, H. Wang, M. Hobbins, F. L. Ogden, R. Cifelli, et al. 2024. "Evaluation of Retrospective National Water Model Soil Moisture and Streamflow for Drought-Monitoring Applications." *Journal of Geophysical Research: Atmospheres* 129 (6). https://doi.org/10.1029/2023JD038522.

Jiao, Zehua. 2024. "The Application of Remote Sensing Techniques in Ecological Environment Monitoring." *Highlights in Science, Engineering and Technology* 81. https://doi.org/10.54097/7dqegz64.

Kara, Ahmet, Engin Pekel, Erdener Ozcetin, and Gazi Bilal Yıldız. 2024. "Genetic Algorithm Optimized a Deep Learning Method with Attention Mechanism for Soil Moisture Prediction." *Neural Computing and Applications* 36 (4). https://doi.org/10.1007/s00521-023-09168-7.

Kňažková, Michaela, and Filip Hrbáček. 2024. "Interannual Variability of Soil Thermal Conductivity and Moisture on the Abernethy Flats (James Ross Island) during Thawing Seasons 2015–2023." *Catena* 234. https://doi.org/10.1016/j.catena.2023.107640.

Lee, Jared A., Pedro A. Jiménez, Rajesh Kumar, and Cenlin He. 2024. "Impact of Direct Insertion of SMAP Soil Moisture Retrievals in WRF-Chem for Dust Storm Events in the Western U.S." *Atmospheric Environment* 321. https://doi.org/10.1016/j.atmosenv.2024.120349.

Letcher, Theodore, John Eylander, Sally Shoop, and Susan Frankenstein. 2024. "Are the Noah and Noah-MP Land Surface Models Accurate for Frozen Soil Conditions?" *Cold Regions Science and Technology* 220. https://doi.org/10.1016/j.coldregions.2024.104149.

Li, Binnan, Lixia Shen, and Shuhui Liu. 2024. "A Comparison of Soil Water Infiltration Models of Moistube Irrigation." *Italian Journal of Agronomy* 19 (1). https://doi.org/10.4081/ija.2024.2216.

Liu, Dandan, Yaowen Chang, Lei Sun, Yunpeng Wang, Jiayu Guo, Luyue Xu, Xia Liu, and Zhaofei Fan. 2024. "Multi-Scale Periodic Variations in Soil Moisture in the Desert Steppe Environment of Inner Mongolia, China." *Water (Switzerland)* 16 (1). https://doi.org/10.3390/w16010123.

Lu, Yuming, and Wei Wang. 2024. "Elastoplastic Responses of Unsaturated Highly Expansive Clay Stabilized by MOC-Based Multiphase Agent under Static and Cyclic Loadings." *Case Studies in Construction Materials* 20. https://doi.org/10.1016/j.cscm.2024.e03005.

Pan, Yongjie, Xia Li, and Suosuo Li. 2024. "Effects of Different Soil Thermal Conductivity Schemes on the Simulation of Permafrost on the Tibetan Plateau." *Geoderma* 442.

https://doi.org/10.1016/j.geoderma.2024.116789.

Shafik, Wasswa. 2024a. "Barriers to Implementing Computational Intelligence-Based Agriculture System." In *Computational Intelligence in Internet of Agricultural Things*, 193–219. Springer. https://doi.org/10.1007/978-3-031-67450-1_8.

Shafik, Wasswa. 2024b. "Biosensor-Based Drones Anomaly Detection Integration for Sustainable Agriculture Development." In *Achieving Food Security Through Sustainable Agriculture*, 283–316. IGI Global. https://doi.org/10.4018/979-8-3693-4240-4.ch012.

Shafik, Wasswa. 2025. "Sustainable Agriculture and Diet for Urban Development." In *Sustainable Smart Cities and the Future of Urban Development*, 165–92. IGI Global Scientific Publishing.

Shafik, Wasswa, Ali Tufail, Rosyzie Anna Awg Haji Mohd Apong, and Liyanage Chandratilak De Silva. 2024. "Internet of Things for Smart Agricultural Practices." In *Internet of Things Applications and Technology*, 190–217. Boca Raton: Auerbach Publications. https://doi.org/10.1201/9781003458401-13.

Soylu, Mehmet Evren, and Rafael L. Bras. 2024. "Quantifying and Valuing Irrigation in Energy and Water Limited Agroecosystems." *Journal of Hydrology X* 22. https://doi.org/10.1016/j.hydroa.2023.100169.

Steger, David N., Richard L. Peters, Theresa Blume, Alexander G. Hurley, Daniel Balanzategui, Daniel F. Balting, and Ingo Heinrich. 2024. "Site Matters - Canopy Conductance Regulation in Mature Temperate Trees Diverges at Two Sites with Contrasting Soil Water Availability." *Agricultural and Forest Meteorology* 345. https://doi.org/10.1016/j.agrformet.2023.109850.

Sumarsono, Fatma Ayu Nuning Farida Afiatna, and Nur Muflihah. 2024. "The Monitoring System of Soil PH Factor Using IoT-Webserver-Android and Machine Learning: A Case Study." *International Journal on Advanced Science, Engineering and Information Technology* 14 (1). https://doi.org/10.18517/ijaseit.14.1.18745.

Suresh, P., R. H. Aswathy, Veena Divya Krishnappa, and P. M. Rajasree. 2024. "Efficient Iot -Machine Learning Based Smart Irrigation Using Support Tree Algae Algorithm." *IETE Journal of Research*. https://doi.org/10.1080/03772063.2024.2319707.

Szabó, Brigitta, János Mészáros, Annamária Laborczi, Katalin Takács, Gábor Szatmári, Zsófia Bakacsi, András Makó, and László Pásztor. 2024. "From EU-SoilHydroGrids to HU-SoilHydroGrids: A Leap Forward in Soil Hydraulic Mapping." *Science of the Total Environment* 921. https://doi.org/10.1016/j.scitotenv.2024.171258.

Tájmel, Dániel, Carla Cruz-Paredes, and Johannes Rousk. 2024. "Heat Wave-Induced Microbial Thermal Trait Adaptation and Its Reversal in the Subarctic." *Global Change Biology* 30 (1). https://doi.org/10.1111/gcb.17032.

Thomas, Anooja, Brijesh Kumar Yadav, and Jiří Šimůnek. 2024. "Water Uptake by Plants under Nonuniform Soil Moisture Conditions: A Comprehensive Numerical and Experimental Analysis." *Agricultural Water Management* 292. https://doi.org/10.1016/j.agwat.2024.108668.

Vorobevskii, Ivan, Thi Thanh Luong, and Rico Kronenberg. 2024. "Seasonal Forecasting of Local-Scale Soil Moisture Droughts with Global BROOK90: A Case Study of the European Drought of 2018." *Natural Hazards and Earth System Sciences* 24 (2). https://doi.org/10.5194/nhess-24-681-2024.

Wang, Guowei, Chunying Wei, Li Yan, and Jian Li. 2024. "Soil Moisture Prediction Model Based on Improved GRU Recurrent Neural Network." *Strategic Planning for Energy and the Environment* 43 (2). https://doi.org/10.13052/spee1048-5236.4329.

Wang, Ruizhen, Siyang Wan, Weitao Chen, Xuwen Qin, Guo Zhang, and Lizhe Wang. 2024. "A Novel Finer Soil Strength Mapping Framework Based on Machine Learning and Remote Sensing Images." *Computers and Geosciences* 182. https://doi.org/10.1016/j.cageo.2023.105479.

Wang, Yanling, Liangsheng Shi, Yaan Hu, Xiaolong Hu, Wenxiang Song, and Lijun Wang. 2024. "A Comprehensive Study of Deep Learning for Soil Moisture Prediction." *Hydrology and Earth System Sciences* 28 (4). https://doi.org/10.5194/hess-28-917-2024.

Wang, Yu, Bifeng Hu, Yongsheng Hong, Songchao Chen, Chengyi Zhao, and Jie Peng. 2024. "Minimize of Moisture Effects from Laboratory Simulations of In-Situ Vis-NIR Spectral for the Prediction of Soil Salinity." *Infrared Physics and Technology* 137. https://doi.org/10.1016/j.infrared.2024.105194.

Wang, Yundong, Yujie Wei, Yingni Du, Zhaoxia Li, and Tianwei Wang. 2024. "Estimation of Spatial Distribution of Soil Moisture on Steep Hillslopes by State-Space Approach (SSA)." *Science of the Total Environment* 914. https://doi.org/10.1016/j.scitotenv.2024.169973.

Wu, Zongjun, Ningbo Cui, Wenjiang Zhang, Daozhi Gong, Chunwei Liu, Quanshan Liu, Shunsheng Zheng, Zhihui Wang, Lu Zhao, and Yenan Yang. 2024. "Inversion of Large-Scale Citrus Soil Moisture Using Multi-Temporal Sentinel-1 and Landsat-8 Data." *Agricultural Water Management* 294. https://doi.org/10.1016/j.agwat.2024.108718.

Yang, Chuan, Yue Yin, Jiantong Zhang, Penghui Ding, and Jian Liu. 2024. "A Graph Deep Learning Method for Landslide Displacement Prediction Based on Global Navigation Satellite System Positioning." *Geoscience Frontiers* 15 (1). https://doi.org/10.1016/j.gsf.2023.101690.

Ye, Yuchen, Haishan Chen, Siguang Zhu, and Yinshuo Dong. 2024. "Machine Learning-Based Prediction of Summer Extended-Range Precipitation and Possible Contribution of Soil Moisture over China." *Plateau Meteorology* 43 (1). https://doi.org/10.7522/j.issn.1000-0534.2023.00025.

Zheng, Wengang, Kai Zheng, Lutao Gao, Lili Zhangzhong, Renping Lan, Linlin Xu, and Jingxin Yu. 2024. "GRU–Transformer: A Novel Hybrid Model for Predicting Soil Moisture Content in Root Zones." *Agronomy* 14 (3). https://doi.org/10.3390/agronomy14030432.

CHAPTER FOURTEEN

AI-BASED TOOLS FOR OPTIMIZING FERTILIZER USE AND MINIMIZING ENVIRONMENTAL IMPACT IN DEVELOPING NATIONS

Abstract

Modern agriculture is expected to satisfy the nutritional demands of an additional 2 billion people by 2050, and this must be achieved using fewer resources and by generating less pollution. The presence of excessive nutrients or pesticides in water can damage aquatic ecosystems and the biosphere. Inefficient application of mineral fertilizers can decrease ecosystem services and deprive the economy, both directly through a fall in productivity and indirectly through increased disease, environmental, and climate change vulnerability. Potassium, phosphorus, and nitrogen deficiency ranges around 30-40% in African soil, which is a major concern. Most crops uptake nutrients in their yield, and if under or over-fertilized, the economy will pay the price through yield reduction. Artificial intelligence and machine learning techniques can be used to optimize fertilizer use to minimize any negative impacts, either on the environment or on the economy. This discusses the potential use of AI for the management of fertilizer applications for smallholder farmers in developing nations and potentially in developed nations. Biodiversity provides ecosystem services at both local and global scales, making ecological integration into agricultural management of intellectual and practical interest, particularly when trying to meet the increasing need for agricultural production. Approximately 75% of people living in the developing world depend on farming for their livelihoods, and this livelihood resource is generally sustained by them growing their food. In-situ crop production for home consumption plays a crucial role in ensuring a healthy diet endowed with bioavailable nutrients. Moreover, healthy soils, responsible agricultural practices, and passive income can play an essential role in ensuring food security and sustainable livelihoods. Given these global and regional challenges, this discusses the potential for the use of AI techniques to build robust and user-friendly models that provide rational fertilizer recommendations for smallholder farmers living in agro-biodiversity hotspots in developing nations.

Keywords: Agricultural Innovation, AI-Driven Fertilizer Optimization, Data-Driven Decision Making, Developing Nations, Environmental Impact Reduction, Machine Learning, Smart Agriculture, Precision Nutrient Management, Soil Health Monitoring, Sustainable Agriculture Practices

14.1 Introduction

There is a pressing need to optimize the use of nitrogen fertilizer in modern agriculture. Farmers in the world's most productive regions often apply between one-third and one-fifth more fertilizer than necessary in the hope of hedging against annual weather variability. If the effectiveness of currently available weather-based yield prediction tools could be improved through the application of AI-based fertilization optimization tools, such improvements could enhance the economic bottom line of most of the world's farmers (Kara et al. 2024). More importantly, we must optimize the amount and timing of nitrogen application to minimize environmental and ecological costs associated with excessive fertilization, including GHG emissions and eutrophication of aquifers and rivers. Among other things, excessive fertilization causes algal blooms that can create ocean dead zones. AI-based tools could fundamentally alter the global nitrogen balance equation, allowing low-income countries to more efficiently use the nitrogen fertilizers they import and improve environmental sustainability in the process (Yanling Wang et al. 2024).

Agricultural production, chiefly food production, navigates a tricky balance. To keep agriculture on a productive path in the extended fight against hunger, relying solely on the organic nutrient resources available at the farm site, such as cow or pig dung and plant residues, or borrowing animal manure from collective resources, would be inadequate in the shorter to medium term and lead to widespread ecosystem disturbances. In addition, many developing nations would likely be unable to sustain satisfactory agricultural production, which could lead to continued food insecurity. Fertilizer use in the form of essential plant nutrients will be almost indispensable for achieving significant and sustainable food production. Nevertheless, excessive use of chemical fertilizers, together with poor timing and spatial management, creates environmental risks. Runoff and leaching losses of excess nitrogen and partly phosphorus contribute to food and water quality issues, including smothering algal blooms and dead zones in coastal and fresh surface waters, and these losses also contribute to the greenhouse gas effect, climate change, and disturbance of nutrient cycling (Pan, Li, and Li 2024). Our motivation is to review the role of agriculture in contributing to this balance and to identify the ways crop management decisions could manage the balance more effectively.

Several studies have shown that traditional farming practices are currently unsustainable and that the nitrogen use efficiency of maize and wheat can be improved through smarter nitrogen scheduling. Several reports have provided ample evidence suggesting that 21st-century agricultural practices using the latest advances in crop genetic improvements, fertilization, disease management, and agricultural mechanization will either have to produce an additional 40-60% output by 2050, relative to 2013 levels or will have to bring millions of deforested hectares back into agricultural production to avoid massive global food shortages associated with population growth. However, economic and educational disparities mean that a disproportionate burden will rapidly fall on low-income countries (Dolaptsis et al. 2024). In the absence of new and innovative approaches to increasing productivity, new forests will likely be cleared.

While AI systems for various components of food production are growing in significance, little AI has been deployed to optimize fertilizer use. We focus on this as the context for a review of AI interventions to increase the productivity of agricultural inputs specific to the socio-economic conditions of developing nations. Over the years, many areas of the world have seen significant benefits from the adoption of AI technologies. As a result, the majority of world nations are gaining interest in the deployment of AI as a means of economically and culturally transforming these nations from a traditional agricultural approach to a more sophisticated yet still sustainable approach (Thomas, Yadav, and Šimůnek 2024; Al-Sager et al. 2024). The importance of this is that agriculture accounts for as little as 30% of least industrialized economies to as much as 57% of economies in transition. When many individual nations enter economies in transition status, involving the diversification of skills and promoting the development of these alternative work opportunities, it is often the case that a large number of people are engaged in smallholder agriculture, concentrating on food crops like maize, often with traditional tools and methods. The central question is, thus, how to reduce these people's subsistence and alleviate their poverty while promoting economic growth in their societies. The AI intervention field in developing countries is explored in the literature in a comparatively limited way for agricultural inputs. This chapter attempts to fill that gap by reviewing AI interventions developed for agricultural inputs, specifically fertilizer, in developing nations (Dolaptsis et al. 2024; Hughes et al. 2024). It also provides an initial synthetic understanding of how these different interventions work according to the evidence. After this introductory part, we provide a top-down general discussion of the problem in relation to traditional agricultural fertilizer application in newly industrializing and least industrialized economies. We demonstrate the ambiguity in the frequent claim that the absence of chemical energy from fertilizers explains agricultural productivity. We discuss the socio-economic evidence about what people in developing nations do, particularly the empirical evidence surrounding winter cropping during the dry season. Regional information also supports this (Dafalla and Al-Mahbashi 2024).

14.2 Current Practices in Fertilizer Use

Fertilizer application rates and practices in developing countries differ depending on land size, technology adoption, and farm management practices. In South Asia, the average nutrient application rates are 102 kg N, 21 kg P2O5, 66 kg K2O, and 34 kg S per hectare per year. Conversely, in Sub-Saharan Africa, the addition rates are much lower at 14 kg N, 6 kg P2O5, 13 kg K2O, and 4.4 kg S. This discrepancy can be attributed to various factors, some of which have external impacts such as the differences in technology adoption, smallholder farm management, and access to credit. One of the several challenges of low application rates is inefficient fertilizer application. The rest of this subsection discusses the general practices of fertilizer use and the agronomic efficiency of fertilizers in maize cultivation in the developing world. Despite fertilizer applications below soil nutrient removals, more than 60% of Asian smallholder farmers believe that the average maize yield (2.2 t/ha) is good. In maize-cultivating countries in Asia, farmers apply a well-developed package containing organic manure (1.4 t/ha), urea (109 kg/ha), DAP (31 kg/ha), MOP (28 kg/ha), and Zn (15 g/ha). In Sub-Saharan Africa, farmers rely heavily on organic and inorganic sources for nutrient maintenance. Essentially, fertilizers used by African smallholders over 10 years include limestone (350–1000 kg/ha), DAP (11–38 kg/ha), urea (14–80 kg/ha), and NPS-Zn + organic manure (11–33 kg/ha) (Banda et al. 2024).

14.2.1 Traditional Methods

Agriculture in developing countries is often undertaken with limited financial resources and without access to significant education. As a result, farmers frequently use methods that are indicative of significant traditional knowledge but are non-optimal regarding plant nutrition. Farmers typically rely on traditional practices that have been used for generations, usually related to existing practices in the area, without necessarily following the scientific research that has proven these practices optimal. For example, some rural farmers often apply nitrogen fertilizer at a random stage of plant development, possibly causing nutrient losses and environmental damage. Low literacy rates and limited access to information sources, combined with uncertain results and high costs of chemical analysis, prevent making informed decisions about the time, rate, and type of nutrients required (Pan, Li, and Li 2024; Al-Sager et al. 2024). Consequently, farmers are susceptible to the poor quality of chemical fertilizers; they can be from contaminated or diluted sources because they rely on distributors instead of direct and controlled sources. Therefore, custom-made chemical fertilizers may not reflect plant needs on a specific farm according to the law of the minimum, which indicates that yield is proportional to the nutrient whose minimum concentration is available in the soil. Fertilizer use may be excessive in many applications, as quantities are primarily based on the size of the farm, without consideration of plant requirements. This causes nutrient losses to the subsoil, mainly in the form of nitrates, which can be washed away by excessive rains and end up in the air, contaminating both the soil and the environment (Thomas, Yadav, and Šimůnek 2024; Hughes et al. 2024). Compliance with some technical recommendations could enhance plant health and crop tolerance to stress and thereby could increase farm yield, farmer income, and farmer quality of life, in addition to additional benefits for soil management and the environment through the "5Rs" as presented in Fig 14.1.

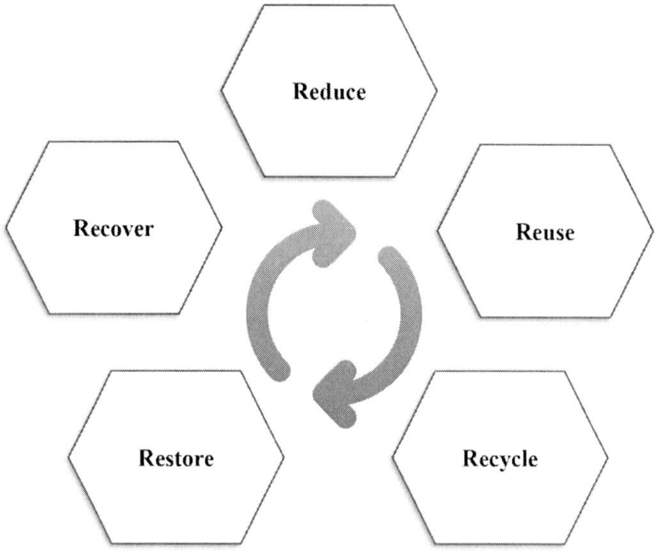

Fig. 14-1 The 5Rs

14.2.2 Existing Technologies

Soil and plant tissue tests are traditional tools that have been interacting very closely since the second half of the 19th century when the first tests and methods of interpretation were developed. They provide the most essential data in programs of soil and crop fertility. Stability in soil fertility programs cannot exist without the accurate and speedy determination of the fertility level of soil and the development of fertilizer recommendations that help producers achieve the desired crop response and reduce the impact on the environment (Dafalla and Al-Mahbashi 2024). The relative efficacy of different soil test methods is important to establish the most appropriate for a certain purpose, such as evaluation of the fertility of an area with several management systems, calibration of crop nutrient uptake or response data to fertilizer addition, and comparison of new crop or soil management systems. Rapid methods of monitoring soil fertility are available, but generally, commercial analyses are of low cost and have high precision and a broad calibration range for the various nutrients tested, making them the most commonly used methods (Guo et al. 2024).

Nevertheless, while currently implemented, they have several disadvantages, including high costs for developing countries, measurement of fertility upon chemical extractants that alter soil structure by removing cations from the exchange sites, time-consuming pickup of the samples, and no real-time response. Suppose we want to fulfill the need for quick on-site soil tests to monitor soil fertility and maintain crop productivity at an affordable cost. In that case, we need soil and plant tests to provide accurate quantitative information about chemical and biological soil properties, which are rapid, affordable, simple, and suitable to use in the field. The development of one diagnostic alternative, an optical sensor test system for the prediction of extractable soil nutrients, P, K, and soil organic matter specific and total nitrogen, was made using chemometric techniques to build calibration models based on mid-infrared spectral terms obtained with the instrument, nutrient concentrations determined from chemical analyses with flame photometry, and colorimetric techniques (Hu et al. 2024; Tájmel, Cruz-Paredes, and Rousk 2024).

14.3 Current Challenges in Fertilizer Use

One of the biggest challenges of fertilizers in agriculture is environmental pollution due to soil degradation resulting from the overuse of some fertilizers and leaching, vapor loss, and loss of food for humans due to the irrigation of polluted water. Various agricultural sources are available to use less water, protect the environment, and reduce the cost of agricultural inputs for sustainable agriculture free of heavy applications. This issue gets less attention. In general, fertilizers account for 40-60 percent of the total costs of farming operations in developing countries, which is a significant cost factor in these countries. Developed countries spend between 10-20% of their total expenses on fertilizers, despite significant advances in farm technology that increase crop yields. The costs of preparing land and planting are higher expenses and labor input is vulnerable across Africa. Given the current world food crisis caused by higher food prices, countries in Africa must redesign the way they produce food (Ye et al. 2024). Many farmers in these areas usually can't afford to plant at the right time; they often sow cotton seeds and spend time-saving fertilizer. The distributed fertilizer results in non-uniform growth, fewer crops, and ultimately lower yields. The agricultural waste in these countries has many political and economic implications, namely employment and GDP growth associated with overseas development and food imports through agriculture. Fertilizers are used for some diseases in crops but not for others. Sometimes, plants don't get enough nutrients, leading to slower growth and reduced plant production. There is disproportionate access to modern fertilizers compared to existing low-technology fertilizers that meet rural farmers' needs, given their financial conditions (Zheng et al. 2024).

14.3.1 Environmental Impact

Soil fertility improvements have historically been the key to increasing agricultural yields. From the Paleolithic man who planted the first seeds from wild grasses to the present day, where megacities demand greater food production, all require the use of mineral fertilizers. Crop farmers, in particular, rely significantly on chemical fertilizers. Sub-Saharan Africa has the lowest use of mineral fertilizers per hectare, which is attributed to its low purchasing power. However, the main environmental impact of these unreasonable fertilization technologies lies in the soil. In addition to global warming, greenhouse gas emissions, and air pollution, the progressive toxicity of soils has been linked back to proportionally high yield and specific use of chemicals. Cultivating biological life is a main concern for developing nations, where ecosystems are under a protectionist umbrella. Neglected over time, ESC was at first an academic focus dedicated to the study of their adaptive capacities (Yang et al. 2024). Today, concern centers on the consequences of excessive use of mineral fertilizers over a long period. The buildup of chemical ions promotes the characteristic depletion of mobile chemical elements such as Ca, K, and Na and toxic elements such as Al, Cu, Zn, and B, which may ultimately lead to the infertility of the soil. The decline of soil fertility translates to structural and functional damage to the landscapes, including global water resources. Water pollution that arises from the profuse use of fertilizers is as old as the industrialized fertilization of agriculture itself and has translated into water pollution events. These water crises have had serious consequences for organisms and the biodiversity of the water ecosystem, including fish, which have received worldwide attention. There is an urgent need to innovate and for farming to adopt optimized protocols for responsible fertilization, especially in the south, where it is the backbone of food and security; an illustration of water application is presented in Fig 14.2. After all, most waterborne diseases of poverty are associated with the ecosystem and biodiversity of developing countries (Sumarsono, Afiatna, and Muflihah 2024; Kiran Pandiri, Murugan, and Goel 2024).

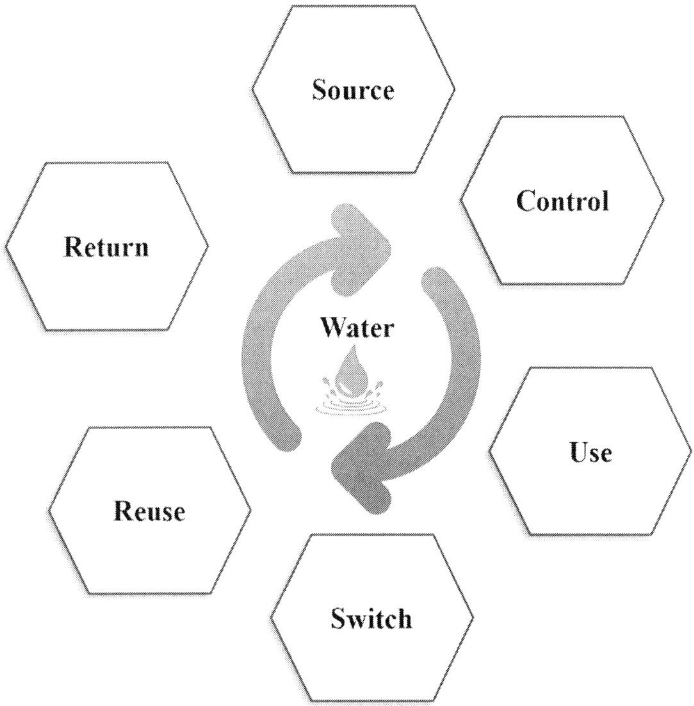

Fig. 14-2 Water and sustainable management

14.3.2 Economic Considerations

Fertilizers have application costs, which are not insignificant. For example, using tractor and machine time for pre-plant fertilizer applications is more costly than spreading fertilizer after the seeds are already sown. Fertilizers also have a cost, and as a result, farmers sometimes apply more of them than is sufficient. Any excess fertilizer is ultimately wasted. There is a 'front-loaded' payment for most types of fertilizer—inputs must be paid for, purchased, and in hand months in advance of being used. Thus, a lack of trust in inputs sold at local farm stores can force a farmer to travel long distances or make more expensive purchases than is strictly optimal (Lu and Wang 2024). For smallholder farmers in the developing world, this is a significant financial burden since they are often credit-constrained, cannot afford significant investments, and are unable to shoulder a large loss if a crop fails. In response, many attempt to skimp on fertilizer purchases and rather use a smaller quantity of fertilizer in the traditional way, which is to mix it into the crop bed uniformly. However, the inefficiency of the applied fertilizer decreases the overall amount of fertilizer that 'sticks' onto the crop, ultimately leading to a reduction in the quantity of food produced and, as a result, farmers' income. In short, existing policies and practices 'penalize' farmers who operate at a low economic scale, even though they make up over 90% of all farmers in most African countries. Meanwhile, fertilizer pricing and availability in the developing world highlight significant low-hanging fruit that can be improved with AI-assisted applications. Indicative studies suggest that the additional marginal increase in crop yields made possible by using AI-assisted solutions can help pay for and make them profitable (Zhang et al. 2024). Such estimates are, however, likely to be highly context-specific.

14.3.3 Data Accessibility

An important factor remarked in various sections of this literature review is that data visualization, regardless of its level of detail, must be as user-friendly as possible in order to cater to potentially dozens of fertilizer management specialists and hundreds of extension agents in several different countries, preferably operating in multiple languages. Additionally, both the hourly updated weather data and the visualized crop simulation outputs must rely on the internet so they can be accessed by any device regardless of its processing abilities. To reiterate, the design goal of the interface was accessibility. These design elements might seem less than state-of-the-art in technology in very advanced nations, but in the context of the spread of technology use, the need for these accessibility factors becomes of paramount importance (Jiao 2024; dos Santos, Pinheiro, and Specht 2024). The spread of information technology and the relevance of using information and communication technologies to keep farmers informed and to ensure efficiency and effectiveness in improving agricultural programs and policies have already been established. However, in addressing a framework regarding agriculture and rural development, special attention must be given to the levels of digital information access, distribution, and usage. The deep digital divide between regions, countries, and different social and demographic groups must be reduced, keeping in mind the new concept of the information society. Therefore, efforts must be made to establish the catalysts for the inclusion of all social sectors in the so-called Information Society (Letcher et al. 2024).

14.3.4 Infrastructure Constraints

By taking a more systems-oriented view of the problem, we can begin to understand mechanisms that may resolve some of these barriers. Providing meaningful advice on complex farming problems is computationally difficult. Despite important progress in simulation-oriented technologies in the physical sciences, the ability to model crop yields in spatially varied and dynamic environments is less certain. Limitations extend to our ability to measure the features of crop production systems that we would like to build predictive simulations around. But even if we had dynamic crop models and systems-to-model information, the ability of farmers to utilize such technology would still be circumscribed (Steger et al. 2024; Challa et al. 2024). Because of a lack of infrastructure, the information already available is not fully used. Even the most conservative investments in quality infrastructure set the stage for more advanced data sharing and AI applications. In Africa, only a small percentage of arable land is already equipped for irrigation. However, the potential markets for such technologies in developing nations are especially large. Furthermore, additional capabilities for precision agriculture are within easy reach of most African countries simply by exploiting more satellite imagery. Infestations of birds could become a worse problem in the age of biofuels, and attention has to be paid to the way that different land uses can make these problems worse (Iaccheri et al. 2024; Ran et al. 2024).

14.4 Role of AI in Agriculture

Artificial intelligence (AI) and its suite of powerful technologies, such as predictive analytics and pattern recognition, have great potential for adoption across all facets of the agricultural sector. The Internet of Things (IoT), drones, satellites, sensors, blockchain, machine learning (ML), and remote sensing can gather a range of data across socio-demographic, economic, agricultural, and environmental spheres for many global regions. Moreover, AI technologies are developed to be generic, automated, adaptable, and easily scalable. These technologies are programmed and trained to recognize a wide range of parameters and nested response relationships (Lv et al. 2024). Consequently, a new technology can be trained to understand specific scenarios with a modest increment to the initial cost of adapting a particular device for another application. AI technologies can integrate multiple data streams, processing huge datasets of complex linked and non-linear data relationships to make decisions on any number of points of interest in farm-to-fork value chains. Artificial intelligence is championed as a means of addressing many of the key challenges hindering agriculture over decades (Z. Li, Yuan, and Su 2024).

Parameters preventing agriculture from making the most of AI include limited agricultural data, technology development driven by the producers of technology rather than the clients and technology development largely performed independently of the agricultural production system. The application of AI technologies can deliver increased efficiency and sustainability to agriculture and food processing from paddock to plate. AI-based software can undertake analyses covering supply routes, machinery use, predictive crop development, and yield and market financial risk management. These applications are in addition to making decisions related to monitoring and targeting strategies for natural resource capital investment when managing large tracts of land (Chen, Wei, and Huang 2024). Most of these benefits occur by increasing how effectively we use water, maintaining healthy soils, suppressing impacts on the global environment, and appropriately using fertilizers and agrochemicals.

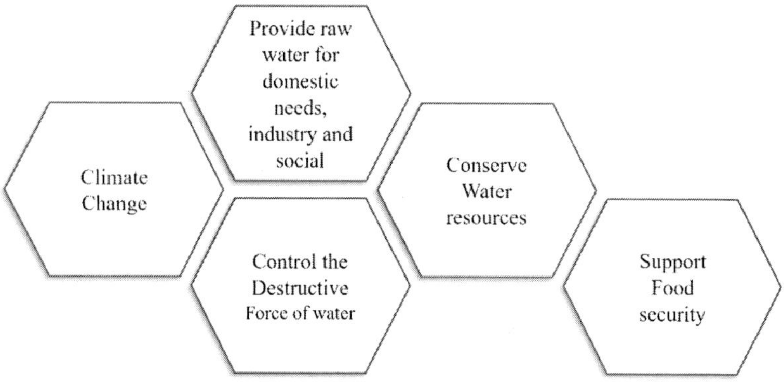

Fig. 14-3 Climate change control measures

Over the past few decades, we have seen remarkable growth in the field of artificial intelligence mapping with human intelligence. AI-based technologies may offer potential solutions to reform agricultural practices through various methods based on machine learning tools. In practice, AI technologies not only can monitor the fields regularly, but can also play a pivotal role in the precision management of farms. Various AI tools can revolutionize agricultural practices based on basic tools of AI such as remote sensing, machine learning, and data analytics. The AI-based technologies enforce remote sensing that is integrated with physical tools, such as drones and slow-moving robots or smart machines. The widespread adaptability of AI provides the confidence to implement AI-based tools to solve and address agricultural issues in diverse contexts around the world (Wu et al. 2024). Many researchers and professionals are demonstrating that in agricultural practice, AI will help users minimize human intervention and control the destructive forces of water, as illustrated in Fig 14.3.

AI-driven tools have the potential to integrate data from various sensors monitoring the conditions of crops, soil, weather, climate, equipment, etc., for treating the crop fields with genetic varieties having different agricultural inputs to adapt to different site conditions. The AI process mainly consists of two sub-processes called Data Transfer and Support Systems. The Data Transfer process acts as a unit trainer where it trains the transfer genetic algorithm. This algorithm reads the data from the Genetic Direct Weather Generator, Genetic Crop Growth Builder, Genetic Yield Predictor, Genetic Harvest Loss Builder, Genetic Economic Evaluator, and inputs to GA (W. Li et al. 2024). Then, the algorithm reads the GA output and transfers the population members to the Genetic Direct Weather Generator, purifies the genetic inputs by interacting with the output GA member by ecstatic technique, and evolves the GA populations based on the purified climate inputs. The main function of the Support Systems process is to manipulate GA within the transfer genetic algorithm cycle and make the final decision on the agroecological inputs and techniques to use during the next cropping season (Fjeld et al. 2024).

Precise agricultural technologies are becoming increasingly important, and AI can play a large part in optimizing inputs for individual plants or animals. For instance, AI has been used for precision irrigation or to control the nutrient supply required for optimal plant growth. Due to the complexity of food production systems, AI can also inform decisions necessary at the farm level, such as crop or yield predictions, disease monitoring, or selection of seeds. AI tools for crop management either exploit monitoring data from satellites to provide actionable intelligence or rely on simulations based on climate and soil attributes. The tool illustrates how AI and digitization can offer solutions that focus both on the productivity and sustainability of crops in order to mitigate food insecurity and climate change (G. Wang et al. 2024). The app advises decision-makers by allowing local context to be taken into account. For instance, the threshold for collating soil samples is set at 20 ha because this is the maximum area a farmer will want to work by hand after collating too few samples. These new technological opportunities are becoming increasingly relevant for resource-poor farmers as they offer a cheap, live connection to vast sources of information. AI and digitization can provide advice and decision-making tools tailored to local information, which can be readily updated and tested on evidence where they are to be applied. The technology does not replace fieldwork; instead, it transforms the nature of fieldwork so it can be based on continuous data collection in the field rather than targeting fieldwork to data collected previously (Soylu and Bras 2024).

14.5 AI-Based Solutions for Fertilizer Optimization

Some of the most promising uses of AI in agriculture have emerged in the pursuit of effective and efficient fertilizer applications. Currently, a highly inelastic function between nutrient application and crop yield is highly undesirable. For farmers, in particular those in developing countries, the technical and financial access to the data and time-consuming methods prescribed by traditional agronomy and agrology play the most inhibiting roles. Methods and technologies in development are targeted at cost-effective AI analysis of high-throughput soil, water, and agricultural data, with the goal of determining the amount of fertilizer each square meter of agricultural land needs as well as when to apply different nutrients. Based on these findings, two classes of technologies are under development: precision agriculture, which delivers locally informed fertilizer supply needs or solutions, and smart fertilization systems, with different variations of AI-driven automatic day-to-day decisions (Yundong Wang et al. 2024). The efficiency improvement brought by the above-outlined technologies is expected to justify the proposed investment on the grounds of reducing greenhouse gas emissions and limiting water and soil pollution to the benefit of the long-term availability of healthy soils.

Precision agriculture is an aggregate of different strategies aimed at increasing the efficiency of nutrient supplies. In developed countries where most are wealthy enough to afford it, using precision agriculture for their farms translates to lower economic and environmental costs and higher yields. Artificial neural networks were simulated for the potential use of artificial intelligence in supplementing the chemical soil test and recommendation system. The best-fit model was observed with a multilayer feed-forward network. Results showed that the trained network modeled the complex relationships among nutrients, data sets, and fertility levels accurately and was able to make reasonable predictions of soil nutrient supply and crop demands. The use of artificial intelligence in supplementing the chemical soil test and its recommendation in predictive soil fertility testing is a tool for predicting nutrient supplies and subsequent assessment of a potential supply of nutrients to a crop plant system (R. Wang et al. 2024).

The concept of precision agriculture involves tailoring the amount of fertilizer applied to the field based on the right time and place. Precision agriculture requires using data, either from a local hardware system or from remote sensing data such as land registry, soil type, crop type, and weather forecast, which is updated in real-time, such as with the use of harvested parameters, GPS measurements, and crop performance data. Whether a combination of different sources of data or not, this data needs to be interpreted by various algorithms to model the environment where the crop is grown. Precision agriculture can be defined as producing and marketing products according to the most appropriate spatial area, such as using the best-suited variety for the specific conditions in the field or the technological possibilities of the day. Resources are utilized in order to produce the yield and quality that is demanded by the market while using any management strategy (Suresh et al. 2024). Therefore, it is considered that site-specific approaches are for fulfilling the demands of the markets and increasing the profit of the farmer. Moreover, these technical possibilities are due to technological advancements in geospatial data determination, since there are important technological advances in technologies such as GPS, as well as in electronic card machinery and even in GIS. The information that can be captured from these advancements relates to soil, crop, and economic management for a specific site in real-time. Most importantly, it is essential to locate exactly the required information in site-specific technologies that increase yield, reduce excessive input use, and minimize or eliminate environmental pollution. In this manner, by manipulating the factors that affect the sole crop using precise farming, we can move agriculture toward a sophisticated solution (Atanasov 2024; Vorobevskii, Luong, and Kronenberg 2024).

In the field of AI for agriculture, one of the most prominent application elements is fertilizer management services supported by AI, often referred to as smart fertilization. One consequence of the digitalization trend in the agricultural sector is that vast amounts of data have been generated over the last decade that can be used in smart fertilization systems like this. The primary purpose of these systems is to decide on the rate and timing of nutrient delivery in a way that conforms closely to the needs of the growing crop without providing any more nutrients than needed. This is usually known as "right source, right rate, right time" (Greiser et al. 2024).Typically, smart systems use algorithms accessing real-time information about the nutrient status of a specific field, local weather for the short term and forecasts, and a set of algorithms and databases that determine the requirements of the crop and the amount of nutrients likely to be provided by the crop, soil, and any pre-season fertilizer application. The concept is that the substantial variation in the climate can be mitigated by using above and below-ground sensors together with local short-term weather forecast data and AI techniques. Accurate matching of nutrient supply with crop demand should have a better impact than using annual weather norms or historical averages like in standard good agricultural practice in many cases (Kara et al. 2024; Yanling Wang et al. 2024). By optimizing the delivery of the input, the smart systems should save on the input cost and have a significant environmental impact that can be derived from two factors: reduced nutrient losses and reduced energy usage.

14.6 Case Studies and Success Stories

Several case studies exemplify the success of AI-driven approaches for optimizing fertilizer use. These case studies reveal the tangible benefits of AI-driven approaches implemented in actual developing nation conditions. Several guiding principles emerged from these case studies that are often crucial for such an implementation, though their relevance and importance may vary across the four scenarios. For instance, in large-scale commercial vegetable greenhouses, having site-specific protocols is crucial. Yet general country- or continent-tailored approaches are more appropriate for widespread traditional vegetable cropping systems. Lessons learned from agriculture research and extension activities include the importance of scale and the need for tools to operate across a variety of farming systems. A single tool might not be useful for targeting fertilizer recommendations on small-scale farms if the needs of the farming systems are sufficiently different. However, an online tool could increase the impact of AI-driven recommendations. In southern India, a program demonstrated that one-size-fits-one recommendations could be sustainable, with initial traction leading to a domino effect of farmers training one another and forming communities. This site-specific approach to unirrigated dryland farmers is reminiscent of the field-level approach using satellite data to map field boundaries and crops to provide AI-based inputs. Lastly, the innovative partnerships described above speak to the importance of developing multiple collaborations with the private sector for programs to be successful. Success stories such as these bolster the case for further investment in the development of similar AI-based tools for smallholder farmers in developing countries. The high cost of fertilizer use optimization in smallholder contexts is a key driver for the success of the interventions described.

14.6.1 Implementations in Developing Nations

AI fertilization predictions were carried out at a trial in Kenya for the developing-world workshop. The trials on the ground used the Optimized Soil and Plant Health Management model, which provides a mobile app to select an optimized input regime for smallholders. Communities have successfully adopted mobile technologies, and smallholder farmers have adopted recommendations provided by that model, leading to improved productivity and reduced environmental impacts. The applications have been delivered in collaboration with local associations, which facilitated communication and co-sponsored recruitment of farmer participants under the support of national agricultural organizations. AI-driven fertilization trials have also been carried out, again with positive results, in Zambia and Ethiopia (Sospedra-Alfonso et al. 2024). In each country, the app was tailor-made for the specific local agricultural context and to the requirements and practices of local farmers under the guidance of national agriculture departments. Also, in Ethiopia, NGOs were closely involved in aid trials. In the Zambian case, a development aid agency was involved. In both countries, the projects have also involved collaboration with seed companies, which may help to push the adoption of the apps. All the trials showed positive increases in farm productivity and profitability. All projects also worked intensively with national government bodies in agriculture, with public extension officers in each country trained to use the apps; in Ethiopia in particular, there was also the education of university staff and the latest education of their students (Lv et al. 2024).

14.6.2 Benefits and Potential Impact

It is typically estimated that the use of AI tools could double crop yields as profitably as the use of external human experts to guide crop and fertilizer practices. Thus, farmers in developed countries are also potential end-users of our app. Our app allows farmers to be more productive, efficient, and sustainable. As much as 50% of the cost of farming in developed nations is the purchase and application of fertilizers. Anything that can make that fertilizer work better will have significant economic benefits, especially for farm owners. Because 90 to 97% of the farms in most developing nations are smallholders, producing the food and economic resources they need to live, the same wealth-creating potential exists for them as for farms in developed nations that sell what they produce to other consumers (Pan, Li, and Li 2024). The main users of this product or service will be smallholders or individual farmers who will be able to afford the cost. The bottom line improvement in income via increased production and greater ability to fine-tune crop nitrogen requirements to actual crop growing conditions will pay more than the annual cost of the license. The economic benefits are, in the short term, reduced input costs of nitrogen fertilizers, possible increased yields, and thus more continuity to give assurance to the farmer that their annual yielding capacity will be less dependent upon climatic problems (which are becoming more and more prevalent and extreme). What benefits and services will this product provide to a local economy? The more productive and capable a farming unit or business is, the more it supports small rural or urban communities because it can pay more in wages to employees, provide more stable employment throughout the year, and generate more tax for the support of locally determined services and infrastructural developments. The ability to service loans to help improve the farm by increasing semi-permanent capital can help improve the productivity, profitability, and asset worth of many family farm businesses (Szabó et al. 2024). Corporate farm businesses will employ more staff, whether seasonal or permanent.

14.6.3 Economic Benefits

Improved nutrient management can have significant economic advantages; this section explores the potential cost savings and alterations in cost structure when using appropriate amounts of fertilizers. Throughout this paper, it is argued that this is particularly relevant in the case of Sub-Saharan Africa and other developing nations. Quick and relatively straightforward estimations of what these costs could be due to suboptimal fertilizer use are detailed, based on a combination of high-level data and experience. Consequently, it is shown that in these countries, appropriate levels of investment in this new technology could be made significantly quicker than in the developed world, where not doing so would lead to a long-term reduction in profit and potential for significant economic, environmental, and social impacts that are often hard to quantify (Yu Wang et al. 2024). Attributions of economic performance as a result of interventions are difficult to quantify and require a deep understanding of baseline data, which, in many cases, is unlikely to exist. However, through the development of sectoral and regional agricultural models, country data, and top-down assessments of the fertilizer market, the economic performance of the fertilizer application method can be made to support the appropriateness of investment. This tool would allow estimation of farmer net benefit, often based on changes in yields and maximal cost savings. These scenarios could then be applied to numerous farmers across several growing seasons in order to derive an approximate net benefit for the country of concern (Kiran Pandiri, Murugan, and Goel 2024).

14.6.4 Environmental Benefits

Environmental benefits from the use of these tools include reduced expenditure of resources with minimal productivity losses. Precision application of fertilizers means that inputs are limited to exactly what is required by the plants and not a drop more. Any small environmental loss minimizes actual losses. Fertilizer that is needed by the plants but is not used can be wasted financially, staying in the soil until later crops have need of it, causing crop failures, or simply staying in the soil for the next use. Smart systems can further minimize resource wastage. Any of these precision agricultural systems hold great potential as strategic tools to optimize fertilizer application (Challa et al. 2024). The environmental benefits from any of these precision agricultural systems are diverse and numerous. The environment can also benefit through decreased chemical runoff, resulting from more crops taking up more nutrients when they are needed, reducing pollutants. Consequently, less soil and land are degraded or destroyed by the salinization and eutrophication that typically follow agricultural runoff. Such reduced soil degradation and pollution can improve regional land and ecosystem quality and, as such, improve resource efficiency (Asadollah et al. 2024).

Native plant and animal communities can also show potential for improvement, especially in *eutrophied* environments, where higher nitrogen concentrations favor some species over others. These precision agricultural systems are a promising direction for agri-environmental sustainability and provide methods for agriculture that cope not only with sustainability in its narrow sense but also with minimal impacts as interventions occur across entire agroecosystems, which are integral for protection against climate change. There are also reductions in carbon footprints. Sustainable, increased adaptation capabilities can prevent future famines. The local, regional, and global ecological systems have built-in systems of checks and balances to prevent overpopulation and sustain balance (W. Li et al. 2024; Fjeld et al. 2024).

14.7 Challenges and Limitations

The use of data to optimize farm management practices is now an essential part of any modern agribusiness. Frequently, this can be achieved through the use of machine learning solutions. In this context, we are particularly interested in using such systems to optimize the level of fertilizer application. This is of great interest not just from an economic standpoint, in terms of ensuring that fertilizer rates match crop requirements, but also from an environmental perspective, since over-application of fertilizers is a significant source of environmental pollution. While there are numerous reasons and potential benefits for the use of AI in fertilizer optimization, the adoption of these technologies may be hindered for various reasons. Firstly, while high-input agriculture and industrial food efficiency have reduced the number of people with first-hand experience in agriculture, many of the people with access to and interest in adopting precision agriculture technologies globally reside in developing nations where these technologies stand to have the most substantial impact (Liu et al. 2024; Sospedra-Alfonso et al. 2024). Adoption of and access to AI tools for agriculture will be heavily influenced by the socioeconomic and political landscape as well as the cultural and philosophical positions of the stakeholders. Also of concern is the lack of general knowledge of AI and low computational literacy in many agricultural sectors. The adoption of these technologies is likely to be influenced by the level of support that the system and data custodians provide in terms of access, training, and application of the developed tools. A further potential barrier is the level of technological integration and automation in many developing nations. Many nations of the Asia-Pacific region, in particular, are characterized by a contraction of the rural workforce due to rapid industrial growth. Consequently, the uptake of such AI-based services by farmers might be limited by a lack of economic imperative for a replacement workforce (Choi et al. 2024; Vorobevskii, Luong, and Kronenberg 2024). Closely related is the availability of infrastructure for the implementation of these systems.

14.7.1 Technological Challenges

Nations in the global South face many challenges that affect the use and impact of AI. This includes poor infrastructure in rural areas, low literacy levels, and the lack of reliable internet. Agricultural AI tools are still in development and are yet to reach many parts of the world. For many of these applications, access to the internet and to digital records from which data can be extrapolated is required. This allows AI applications access to large data sets and the ability to 'learn' and generate predictions and recommendations. This is a challenge in the context of developing nations, as databases may not exist outside government departments, and those that do exist may be fragmented, out of date, or insecure (Hu et al. 2024). These data are also not the property of the current farmers, and therefore, there are significant issues related to ownership, long-term storage, and accessibility via the Internet. It may be more realistic to develop solutions that help to gather that data while still providing useful insights for the farmers.

In some countries, technological solutions that have been developed in the past for use in agriculture can often no longer be used or replicated due to a lack of support and familiarity with the original technology, changes in regional environmental conditions, or due to changing patterns of pest dynamics. New technology for agriculture is therefore likely to be most successful if it is not incompatible with existing practices. This has diverse implications, including not only the ability to integrate with what is already used to grow food and the specification of the tools used, but also the words used to describe the technology (Ye et al. 2024). It is of no use identifying problems and offering fields of data if farmers cannot relate this data back to an existing problem. The use of these applications also needs to be straightforward and tooltip-free. Overloading the user with lists of choices may cause decision-making overload and reduce user experience. There are significant differences in the level of technology that different farmers are familiar with. Factors that can impact this include access to the relevant tools or handling previously obtained technology. Tailoring technology and also additional technical support will be necessary for farmers without much previous experience or digital literacy to make the use of new technology accessible to more people (Yang et al. 2024; Zheng et al. 2024).

14.7.2 Access and Adoption in Developing Nations

Access and adoption within the context of AI technologies in developing nations are at the heart of any research effort aimed at enhancing the use of and efficiency in fertilizer management. As of 2019, well over 600 million people living in sub-Saharan Africa do not have access to electricity to power cell phones. Providing these technologies to such people requires much more than technological advancements. Currently, many farmers depend on family labor to do much of the work near their homestead, so disruptions in good health become a serious concern. To stay better connected in the future, farmers will make technology and infrastructure near them adapt, in addition to the adaptation and adoption of new practices by the farmer (Sumarsono, Afiatna, and Muflihah 2024). Education and Training: As with most new technologies or interventions, the level of success in the adoption of AI technologies, in terms of either market penetration or how it enhances agricultural practices, is never fully realized. Considering these insights can be employed to make great progress on the societal adoption of

AI tools and many of the new technologies that will come out of this research in this fast-evolving digital age. Innovation related to these technologies requires a greatly supportive ecosystem that incorporates experts in AI, agriculture, fertilizer management, the environment, users of such technologies, government programs or policies, NGOs, and for-profit industry (Z. Li, Yuan, and Su 2024; Lv et al. 2024).

14.8 Ethical and Social Implications

Ethical issues are brought up when AI interventions are introduced into the domain of agriculture. A question that arises is how egalitarian this AI tool is. Equity issues are often reported in the adoption of agricultural technology, including digital tools. Access to new technology could sometimes escalate perceived and actual socioeconomic disparities for large and small farmers (Shafik et al. 2024). The social difference in the use of AI tools is often noticed in the demographics of the farmers who adopt them. Marginal and small farmers, often due to issues such as language barriers, access to the internet, the digital divide, digital literacy, and trust, do not get to use the available AI tools for agriculture. Some possible risks related to AI in general include informational, economic, social, and psychological risks to specific stakeholders (Szabó et al. 2024). Overreliance on digital and AI tools could lead to the loss of tech-free or organic tags, reduce the diversity of plants, and alter traditional knowledge about sustainable farming. When not adopted properly, the impacts could be quite distressing, including malnutrition, land loss, and economic consequences. Data privacy and data security concerns, especially in geographies where data containing farmer-level recommended insights or outputs could affect the physical security of the farmers if known, can further prevent the adoption of fertilizer tools. Open communication and community engagement play a crucial role in fostering acceptance of the AI tool (Ye et al. 2024; Kalinaki et al. 2023). Inclusion and participation of the direct and potential beneficiaries of the AI interventions make them more locally meaningful, appropriate, and sustainable. The tool should be open-ended enough to cater to the needs and social contexts of different farmers. This can also raise questions about the tool's efficacy and ethics. If a recommendation is so general that it is relevant for a wide range of geographies, then questions could arise regarding the efficacy of information given the extreme range of climates that Indian geography and agriculture cover. Given evolving contexts, NGOs should be looking for AI interventions being developed within the target countries that have the potential to lead to tools that could also be mass-scaled, eventually enforcing the fortification of the government's role in AI for good itself (Shafik et al. 2023).

14.8.1 Equity and Inclusivity

Innovations in agriculture demand access to technologies such as AI and algorithms; thus, if only larger farmers can invest in adopting these technologies, there is concern that smaller farmers will be marginalized further. The only pathway to build fairness is to ensure that the benefits of AI are distributed across all sizes of farms, which has been a goal of many governments and agencies involved in agricultural development. Many AI models and algorithms are currently in development, so there is an opportunity to build equity into AI technologies for agriculture, especially if there are deliberate policies focused on the inclusion of AI to include all producers. Many farmers have access to or can afford to procure traditional precision agriculture technologies, including AI components. Given these access issues, the disparities in wealth and power, country and community income levels, and broader levels of human development and education indicate that many of the existing AI technologies used on farms are not accessible to the world's poorest farmers, who could potentially benefit the most (Steger et al. 2024).

With the development of new digital agriculture opportunities, it is essential to consider equity implications. The development of new AI technologies will become a tool applied to the world's decisions regarding investments, research and development, policy, and intervention. To bring the most benefit, AI strategies seek to provide fair and equal access to AI opportunities (Shafik 2025). A focus on inclusive data and learning strategies for those currently marginalized in data may mitigate problems in some social systems, although care is needed to manage issues of power and prejudice. Equity is about fairness; inclusivity is about intentionally including all people in a social system. Inclusivity has been a target in some sectors, with success in improving the effectiveness of interventions. Legal policy also involves showing responsibility and due diligence in participating with and listening to disadvantaged or suppressed groups in deciding potential future actions in the sector. The capacity for inclusivity is present. It involves education and persuasion about participating in their environment, but also understanding what their participation would look like (Banda et al. 2024; Thomas, Yadav, and Šimůnek 2024). Successful technology innovation policies have promoted awareness among target groups of the tools and support available. Awareness activities are often part of broader extension services and must consider differential abilities to respond to media, creating equitable mirrors of access (Shafik and Kalinaki 2024).

14.9 Future Directions and Emerging Trends

Experimental prototypes to integrate AI-based tools with off-the-shelf soil and leaf nutrition sensors are currently under construction. Increased miniaturization of technologies such as X-ray fluorescence technology for nutrient sensing and cameras that can identify crop diseases and pests continues to inspire the development of new tools and technologies (Shafik 2024b). Also, there is a growing trend toward more automation in the application of agricultural inputs. This technology can be made more accessible to resource-limited farmers in the developing world as additional technological advances continue to improve its scalability. With the rise of data-driven farming, a new focus has been put on industry partnerships and their potential impact. Specifically, data collection and analysis are particularly important for companies specializing in seed, pesticide, and fertilizer products (Wu et al. 2024).

Although stakeholder inclusion is mentioned as critical to the success of a tool, there currently appears to be a tendency to consult the stakeholders only at the research stage rather than allowing sufficient time and resources for two-way feedback in order to ensure appropriateness fully. Those quoted characterize participatory consultations with key stakeholders as critical to the development of tools as well as an opportunity to share some of our techniques. The development of further stakeholder relationships is necessary to expand the global reach of AI for Good in Agriculture. For instance, professionals in digital research methods should seek to develop partnerships in sub-Saharan Africa, Southeast Asia, and South America. This would have the dual effect of increasing the global focus of such research and reducing the perception that the blog covers Western agricultural practices. Finally, there is the pursuit of new technological developments that will drive greater collaboration with stakeholders (Dolaptsis et al. 2024; Steger et al. 2024). Investigating combinatory design projects or other such research would help to explore how ideas can continue to be developed to suit farmer input and expert opinion better.

Stakeholders in a Digital Research Network have a responsibility to continue searching for new techniques and building upon findings should they thereby surface. The network, composed of researchers, industry professionals, government employees from around the world, and scholarship recipients, was characterized as a network that should continue the dialogue. The esteemed network recently held a webinar that made clear that the future of many industries needs conversations about where we are today and where we are headed. The webinar concluded that grants should push beyond the country-specific focus that has been common in academia and work across borders to find universal outcomes that drive the future. Agriculture is no exception (Wu et al. 2024). The pursuit of such grant initiatives also helps to give researchers ideas to guide their grant applications.

The development of better machine learning algorithms in recent decades, in addition to increases in compute power, has led to advancements in AI with a broad range of applications. Specifically, new ML model architectures, such as deep learning models, have shown potential for more accurate model predictions with a variety of geospatial and environmental research applications, including information extraction, change detection, data fusion, multitemporal land-cover change detection, and crop classification (Shafik 2024a). Furthermore, the development of advanced deep learning models for image segmentation has the potential for identifying and quantifying soil and crop characteristics for more precise fertilizer application practices. The application of high-resolution in-situ sensor network technology has potential for informing new policies and enhancing tools for producers to change their fertilization practices in real-time (Szabó et al. 2024). Better Predictive Analytics and Interoperability AI technologies offer the potential for improved integration between multiple data sources and improved ability to provide producers with science-based knowledge and electrical tools to effectively manage risk associated with nitrogen fertilizer application. Tools that allow producers to predict crop performance will relate these predictions to relevant soil sealing rates and economic performance. In addition, these tools need to be easy to operate and provide a feedback mechanism that captures user feedback in real-time for future improvement and development. AI tools will enable broad deployment across various agriculturally driven economic platforms and policymakers in developing nations. The ability to use chronic anthropogenic nitrogen as a key input to transmitted risk provides promise for broader application in developing countries to optimize fertilizer use (Sumarsono, Afiatna, and Muflihah 2024).

14.10 Conclusion

Adoption of AI can play a role in addressing the barriers to successful large-scale fertilizer optimization programs for emerging markets by significantly reducing the combined time, cost, and risk involved in the compilation of relevant data and decision tools to extend access to smallholders. As soil health rather than crop yield has become the key focus of fertilizer use in many systems, it is important to consider optimization solutions oriented at addressing the multitude of sustainable agriculture indicators necessitated by this change in focus. The need for solutions beyond those of a singular economic context render parser based AI an important tool for design of microdosing based approaches that seek to maximize returns from investment as well as minimizing environmental impact. The AI-based approaches described will help governments, agroindustry, and development organizations navigate these changes in fertilizer advice delivery mechanisms using input and context-independent data and are potentially appropriate for use in every world region where fertilizer is used to fuel agricultural development and manage plant nutrient stocks. It is rare to encounter a new technology that can so rapidly and significantly rectify chronic environmental, social, and economic problems, along with the basic building capacity for addressing those issues. Public sector funders, and large international foundations, meeting their obligations to benefit the people and environments of developing countries directly and without hesitation, should fund further improvements of these tools. Care must be taken to explicitly include potential constraints and ethical use issues in the evolutionary development of this technology. Policymakers must consider how these cutting-edge tools can be used to the greatest positive effect in the areas of greatest need, and that they are used to the benefit rather than the exploitation of marginalized people. International forums can also foster consensus amongst national actors to include development of these tools and early commitment to supporting their use in national policy agreements, targeting a significant reduction in fertilizer over-/misuse rates as measured by surveys of such use in all major agronomic zones around the world.

References

Al-Sager, Saleh M., Saad S. Almady, Samy A. Marey, Saad A. Al-Hamed, and Abdulwahed M. Aboukarima. 2024. "Prediction of Specific Fuel Consumption of a Tractor during the Tillage Process Using an Artificial Neural Network Method." *Agronomy* 14 (3). https://doi.org/10.3390/agronomy14030492.

Asadollah, Seyed Babak Haji Seyed, Ahmad Sharafati, Mohammad Saeedi, and Shamsuddin Shahid. 2024. "Estimation of Soil Moisture from Remote Sensing Products Using an Ensemble Machine Learning Model: A Case Study of Lake Urmia Basin, Iran." *Earth Science Informatics* 17 (1). https://doi.org/10.1007/s12145-023-01172-8.

Atanasov, Svetoslav. 2024. "Automated Remote Sensing System for Crops Monitoring and Irrigation Management, Based on Leaf Color Change and Piecewise Linear Regression Models for Soil Moisture Content Predicting." *Scientific Horizons* 27 (1). https://doi.org/10.48077/scihor1.2024.127.

Banda, Enid, Vahid Rafiei, Josué Kpodo, A. Pouyan Nejadhashemi, Gurjeet Singh, Narendra N. Das, Rabin Kc, and Amadiane Diallo. 2024. "Millet Yield Estimations in Senegal: Unveiling the Power of Regional Water Stress Analysis and Advanced Predictive Modeling." *Agricultural Water Management* 291. https://doi.org/10.1016/j.agwat.2023.108618.

Challa, Lakshmi Poojitha, Chandra Deep Singh, Kondapalli Venkata Ramana Rao, Anakkallan Subeesh, and Mandru Srilakshmi. 2024. "Prediction of Soil Moisture Using Machine Learning Techniques: A Case Study of an IoT-Based Irrigation System in a Naturally Ventilated Polyhouse." *Irrigation and Drainage* 73 (3). https://doi.org/10.1002/ird.2933.

Chen, Shengnan, Wei Wei, and Yue Huang. 2024. "Biophysical Controls on Canopy Transpiration of Pinus Tabulaeformis under Different Soil Moisture Conditions in the Loess Plateau of China." *Journal of Hydrology* 631. https://doi.org/10.1016/j.jhydrol.2024.130799.

Choi, Su Hoon, Sang Hyun Lee, Ung Yang, and Min Soo Kim. 2024. "Data Mining Approach to Predicting Soil Moisture Based on Meteorological Factors and Flow Rates." *Horticultural Science and Technology* 42 (1). https://doi.org/10.7235/HORT.20240001.

Dafalla, Muawia, and Ahmed M. Al-Mahbashi. 2024. "Identifying Problematic Soils Using Compressibility and Suction Characteristics." *Buildings* 14 (2). https://doi.org/10.3390/buildings14020521.

Dolaptsis, Konstantinos, Xanthoula Eirini Pantazi, Charalampos Paraskevas, Selçuk Arslan, Yücel Tekin, Bere Benjamin Bantchina, Yahya Ulusoy, et al. 2024. "A Hybrid LSTM Approach for Irrigation Scheduling in Maize Crop." *Agriculture (Switzerland)* 14 (2). https://doi.org/10.3390/agriculture14020210.

Fjeld, Dag, Magnus Persson, Johan E.S. Fransson, Jan Bjerketvedt, and Martin Bråthen. 2024. "Modelling Forest Road Trafficability with Satellite-Based Soil Moisture Variables." *International Journal of Forest Engineering* 35 (1). https://doi.org/10.1080/14942119.2023.2276628.

Greiser, Caroline, Lucia Hederová, Giulia Vico, Jan Wild, Martin Macek, and Martin Kopecký. 2024. "Higher Soil Moisture Increases Microclimate Temperature Buffering in Temperate Broadleaf Forests." *Agricultural and Forest Meteorology* 345. https://doi.org/10.1016/j.agrformet.2023.109828.

Guo, Jingjing, Martin Ziegler, Niko Wanders, Mike Vreeken, Qiuzhen Yin, Hao Lu, Louise Fuchs, Jibao Dong, Youbin Sun, and Francien Peterse. 2024. "Robust Land Surface Temperature Record for North China over the Past 21,000 Years." *Science Advances* 10 (8). https://doi.org/10.1126/sciadv.adj4800.

Hu, Feichi, Qinli Yang, Junran Yang, Zhengming Luo, Junming Shao, and Guoqing Wang. 2024. "Incorporating Multiple Grid-Based Data in CNN-LSTM Hybrid Model for Daily Runoff Prediction in the Source Region of the Yellow River Basin." *Journal of Hydrology: Regional Studies* 51. https://doi.org/10.1016/j.ejrh.2023.101652.

Hughes, M., D. L. Jackson, D. Unruh, H. Wang, M. Hobbins, F. L. Ogden, R. Cifelli, et al. 2024. "Evaluation of Retrospective National Water Model Soil Moisture and Streamflow for Drought-Monitoring Applications." *Journal of Geophysical Research: Atmospheres* 129 (6). https://doi.org/10.1029/2023JD038522.

Iaccheri, Eleonora, Annachiara Berardinelli, Marco Tartagni, and Luigi Ragni. 2024. "Affordable Microwave Soil Moisture Detector." *IEEE Sensors Journal* 24 (6). https://doi.org/10.1109/JSEN.2024.3355762.

Jiao, Zehua. 2024. "The Application of Remote Sensing Techniques in Ecological Environment Monitoring." *Highlights in Science, Engineering and Technology* 81. https://doi.org/10.54097/7dqegz64.

Kalinaki, Kassim, Wasswa Shafik, Tar J. L. Gutu, and Owais Ahmed Malik. 2023. "Computer Vision and Machine Learning for Smart Farming and Agriculture Practices." In . https://doi.org/10.4018/978-1-6684-8516-3.ch005.

Kara, Ahmet, Engin Pekel, Erdener Ozcetin, and Gazi Bilal Yıldız. 2024. "Genetic Algorithm Optimized a Deep Learning Method with Attention Mechanism for Soil Moisture Prediction." *Neural Computing and Applications* 36 (4). https://doi.org/10.1007/s00521-023-09168-7.

Kiran Pandiri, D. N., R. Murugan, and Tripti Goel. 2024. "Smart Soil Image Classification System Using Lightweight Convolutional Neural Network." *Expert Systems with Applications* 238. https://doi.org/10.1016/j.eswa.2023.122185.

Letcher, Theodore, John Eylander, Sally Shoop, and Susan Frankenstein. 2024. "Are the Noah and Noah-MP Land Surface Models Accurate for Frozen Soil Conditions?" *Cold Regions Science and Technology* 220. https://doi.org/10.1016/j.coldregions.2024.104149.

Li, Wangyang, Youzhen Xiang, Xiaochi Liu, Zijun Tang, Xin Wang, Xiangyang Huang, Hongzhao Shi, et al. 2024. "Estimation of Soil Moisture Content Based on Fractional Differential and Optimal Spectral Index." *Agronomy* 14 (1). https://doi.org/10.3390/agronomy14010184.

Li, Zhenghao, Qiangqiang Yuan, and Xin Su. 2024. "High-Spatial-Resolution Surface Soil Moisture Retrieval Using the Deep Forest Model in the Cloud Environment over the Tibetan Plateau." *Geo-Spatial Information Science*. https://doi.org/10.1080/10095020.2024.2307931.

Liu, Dandan, Yaowen Chang, Lei Sun, Yunpeng Wang, Jiayu Guo, Luyue Xu, Xia Liu, and Zhaofei Fan. 2024. "Multi-Scale Periodic Variations in Soil Moisture in the Desert Steppe Environment of Inner Mongolia, China." *Water (Switzerland)* 16 (1). https://doi.org/10.3390/w16010123.

Lu, Yuming, and Wei Wang. 2024. "Elastoplastic Responses of Unsaturated Highly Expansive Clay Stabilized by MOC-Based Multiphase Agent under Static and Cyclic Loadings." *Case Studies in Construction Materials* 20. https://doi.org/10.1016/j.cscm.2024.e03005.

Lv, Yijin, Jialiang Kuang, Zecong Ding, Rong Li, and Zhenqing Shi. 2024. "Soil Moisture Dynamics Regulates the Release Rates and Lability of Copper in Contaminated Paddy Soils." *Science of the Total Environment* 908. https://doi.org/10.1016/j.scitotenv.2023.168525.

Pan, Yongjie, Xia Li, and Suosuo Li. 2024. "Effects of Different Soil Thermal Conductivity Schemes on the Simulation of Permafrost on the Tibetan Plateau." *Geoderma* 442. https://doi.org/10.1016/j.geoderma.2024.116789.

Ran, Yuling, Wei Bai, Lingwei Kong, Henghui Fan, Xiujuan Yang, and Xuemei Li. 2024. "Prediction of Soil Degree of Compaction Based on Machine Learning: A Case Study of Two Fine-Grained Soils." *Engineering Computations (Swansea, Wales)* 41 (1). https://doi.org/10.1108/EC-06-2023-0304.

Santos, Thaís Aquino dos, Rinaldo José Barbosa Pinheiro, and Luciano Pivoto Specht. 2024. "Effects of Moisture Content and Soil Suction on the Permanent Deformation of Tropical Subgrade Soils: Experimental Investigations and Modelling." *Road Materials and Pavement Design*. https://doi.org/10.1080/14680629.2024.2310823.

Shafik, Wasswa. 2024a. "Barriers to Implementing Computational Intelligence-Based Agriculture System." In *Computational Intelligence in Internet of Agricultural Things*, 193–219. Springer. https://doi.org/10.1007/978-3-031-67450-1_8.

Shafik, Wasswa. 2024b. "Biosensor-Based Drones Anomaly Detection Integration for Sustainable Agriculture Development." In *Achieving Food Security Through Sustainable Agriculture*, 283–316. IGI Global. https://doi.org/10.4018/979-8-3693-4240-4.ch012.

Shafik, Wasswa. 2025. "Sustainable Agriculture and Diet for Urban Development." In *Sustainable Smart Cities and the Future of Urban Development*, 165–92. IGI Global Scientific Publishing.

Shafik, Wasswa, and Kassim Kalinaki. 2024. "Societal and Ethical Implications of Technology-Enhanced Agriculture and Healthcare: An African Context." In *2024 IST-Africa Conference (IST-Africa)*, 1–11. IEEE. https://doi.org/10.23919/IST-Africa63983.2024.10569306.

Shafik, Wasswa, Ali Tufail, Rosyzie Anna Awg Haji Mohd Apong, and Liyanage Chandratilak De Silva. 2024. "Internet of Things for Smart Agricultural Practices." In *Internet of Things Applications and Technology*, 190–217. Boca Raton: Auerbach Publications. https://doi.org/10.1201/9781003458401-13.

Shafik, Wasswa, Ali Tufail, Chandratilak De Silva Liyanage, and Rosyzie Anna Awg Haji Mohd Apong. 2023. "Using a Novel Convolutional Neural Network for Plant Pests Detection and Disease Classification." *Journal of the Science of Food and Agriculture*. https://doi.org/10.1002/jsfa.12700.

Sospedra-Alfonso, Reinel, William J. Merryfield, Viatsheslav V. Kharin, Woo Sung Lee, Hai Lin, Gulilat T. Diro, and Ryan Muncaster. 2024. "Evaluation of Soil Moisture in the Canadian Seasonal to Interannual Prediction System, Version 2.1 (CanSIPSv2.1)." *Journal of Applied Meteorology and Climatology* 63 (1). https://doi.org/10.1175/JAMC-D-23-0034.1.

Soylu, Mehmet Evren, and Rafael L. Bras. 2024. "Quantifying and Valuing Irrigation in Energy and Water Limited Agroecosystems." *Journal of Hydrology X* 22. https://doi.org/10.1016/j.hydroa.2023.100169.

Steger, David N., Richard L. Peters, Theresa Blume, Alexander G. Hurley, Daniel Balanzategui, Daniel F. Balting, and Ingo Heinrich. 2024. "Site Matters - Canopy Conductance Regulation in Mature Temperate Trees Diverges at Two Sites with Contrasting Soil Water Availability." *Agricultural and Forest Meteorology* 345. https://doi.org/10.1016/j.agrformet.2023.109850.

Sumarsono, Fatma Ayu Nuning Farida Afiatna, and Nur Muflihah. 2024. "The Monitoring System of Soil PH Factor Using IoT-Webserver-Android and Machine Learning: A Case Study." *International Journal on Advanced Science, Engineering and Information Technology* 14 (1). https://doi.org/10.18517/ijaseit.14.1.18745.

Suresh, P., R. H. Aswathy, Veena Divya Krishnappa, and P. M. Rajasree. 2024. "Efficient Iot -Machine Learning Based Smart Irrigation Using Support Tree Algae Algorithm." *IETE Journal of Research*. https://doi.org/10.1080/03772063.2024.2319707.

Szabó, Brigitta, János Mészáros, Annamária Laborczi, Katalin Takács, Gábor Szatmári, Zsófia Bakacsi, András Makó, and László Pásztor.

2024. "From EU-SoilHydroGrids to HU-SoilHydroGrids: A Leap Forward in Soil Hydraulic Mapping." *Science of the Total Environment* 921. https://doi.org/10.1016/j.scitotenv.2024.171258.

Tájmel, Dániel, Carla Cruz-Paredes, and Johannes Rousk. 2024. "Heat Wave-Induced Microbial Thermal Trait Adaptation and Its Reversal in the Subarctic." *Global Change Biology* 30 (1). https://doi.org/10.1111/gcb.17032.

Thomas, Anooja, Brijesh Kumar Yadav, and Jiří Šimůnek. 2024. "Water Uptake by Plants under Nonuniform Soil Moisture Conditions: A Comprehensive Numerical and Experimental Analysis." *Agricultural Water Management* 292. https://doi.org/10.1016/j.agwat.2024.108668.

Vorobevskii, Ivan, Thi Thanh Luong, and Rico Kronenberg. 2024. "Seasonal Forecasting of Local-Scale Soil Moisture Droughts with Global BROOK90: A Case Study of the European Drought of 2018." *Natural Hazards and Earth System Sciences* 24 (2). https://doi.org/10.5194/nhess-24-681-2024.

Wang, Guowei, Chunying Wei, Li Yan, and Jian Li. 2024. "Soil Moisture Prediction Model Based on Improved GRU Recurrent Neural Network." *Strategic Planning for Energy and the Environment* 43 (2). https://doi.org/10.13052/spee1048-5236.4329.

Wang, Ruizhen, Siyang Wan, Weitao Chen, Xuwen Qin, Guo Zhang, and Lizhe Wang. 2024. "A Novel Finer Soil Strength Mapping Framework Based on Machine Learning and Remote Sensing Images." *Computers and Geosciences* 182. https://doi.org/10.1016/j.cageo.2023.105479.

Wang, Yanling, Liangsheng Shi, Yaan Hu, Xiaolong Hu, Wenxiang Song, and Lijun Wang. 2024. "A Comprehensive Study of Deep Learning for Soil Moisture Prediction." *Hydrology and Earth System Sciences* 28 (4). https://doi.org/10.5194/hess-28-917-2024.

Wang, Yu, Bifeng Hu, Yongsheng Hong, Songchao Chen, Chengyi Zhao, and Jie Peng. 2024. "Minimize of Moisture Effects from Laboratory Simulations of In-Situ Vis-NIR Spectral for the Prediction of Soil Salinity." *Infrared Physics and Technology* 137. https://doi.org/10.1016/j.infrared.2024.105194.

Wang, Yundong, Yujie Wei, Yingni Du, Zhaoxia Li, and Tianwei Wang. 2024. "Estimation of Spatial Distribution of Soil Moisture on Steep Hillslopes by State-Space Approach (SSA)." *Science of the Total Environment* 914. https://doi.org/10.1016/j.scitotenv.2024.169973.

Wu, Zongjun, Ningbo Cui, Wenjiang Zhang, Daozhi Gong, Chunwei Liu, Quanshan Liu, Shunsheng Zheng, Zhihui Wang, Lu Zhao, and Yenan Yang. 2024. "Inversion of Large-Scale Citrus Soil Moisture Using

Multi-Temporal Sentinel-1 and Landsat-8 Data." *Agricultural Water Management* 294. https://doi.org/10.1016/j.agwat.2024.108718.

Yang, Chuan, Yue Yin, Jiantong Zhang, Penghui Ding, and Jian Liu. 2024. "A Graph Deep Learning Method for Landslide Displacement Prediction Based on Global Navigation Satellite System Positioning." *Geoscience Frontiers* 15 (1). https://doi.org/10.1016/j.gsf.2023.101690.

Ye, Yuchen, Haishan Chen, Siguang Zhu, and Yinshuo Dong. 2024. "Machine Learning-Based Prediction of Summer Extended-Range Precipitation and Possible Contribution of Soil Moisture over China." *Plateau Meteorology* 43 (1). https://doi.org/10.7522/j.issn.1000-0534.2023.00025.

Zhang, Jun, Sheng Wang, Zhiyong Fu, Fa Wang, Kelin Wang, and Hongsong Chen. 2024. "Soil Thickness Influences the Control Effect of Micro-Topography on Subsurface Runoff Generation in the Karst Hillslope Critical Zone." *Catena* 239. https://doi.org/10.1016/j.catena.2024.107957.

Zheng, Wengang, Kai Zheng, Lutao Gao, Lili Zhangzhong, Renping Lan, Linlin Xu, and Jingxin Yu. 2024. "GRU–Transformer: A Novel Hybrid Model for Predicting Soil Moisture Content in Root Zones." *Agronomy* 14 (3). https://doi.org/10.3390/agronomy14030432.

CHAPTER FIFTEEN

ETHICAL AND ENVIRONMENTAL CONSIDERATIONS IN ENABLING TECHNOLOGIES FOR SUSTAINABLE AGRICULTURE

Abstract

Enabling technologies and precision farming strategies are considered key drivers to move towards more sustainable agriculture. Indeed, the 20th century's 'green revolution' increased capital from land, thus reducing the occurrence of famines, particularly in Asia, but at the cost of environmental problems, such as pollution, soil erosion, loss of biodiversity, landscape degradation, infringement of animal welfare, reduction of crops' organoleptic properties, as well as the degradation of farmers' social perception by the public. The recent economic and environmental burden of agriculture has amplified the expectation for innovations that meet society's demand for sustainability, the major frontiers of which are ethical and environmental concerns that are challenging the technological sector. In the last decades, research has approached huge knowledge in new technologies. Most of these technologies have been focused on human health; hence, it is reasonable to think that these studies, devices, principles, and knowledge can represent the frontier of influencing innovation toward 'sustainable agriculture.' Moreover, for these reasons, ethical and environmental considerations in enabling technologies for sustainable agriculture can represent a priority for policymakers as well as society. This ambition cannot ignore that in the ancient agricultural world, there has always been geographical and historical know-how for the most efficient possible exploitation of natural resources.

Keywords: Agricultural Technology Governance, Climate-Smart Technologies, Environmental Sustainability, Ethical Agricultural Technologies, Green Agricultural Practices, Precision Farming Ethics, Responsible Innovation, Socioeconomic Impact Assessment, Sustainable Resource Management

15.1 Introduction

Sustainable agriculture is essential for attaining food security, safeguarding environmental health, and achieving a degree of economic viability in agriculture and rural areas. When implemented, sustainable farming practices aim to ensure that the natural resource base that supports production remains productive indefinitely and can be used by future generations. By contrast, unsustainable practices of agriculture and resource use can contribute to a range of socio-economic issues, including resource degradation, increased poverty, and the accumulation of social deprivation and vulnerability. If production, procurement, trade, or consumption practices are modified to decrease environmental and social burdens, food systems become more sustainable. Sustainable agriculture is necessary to enable agriculture to maintain the ecological balance of a site, which is required to guarantee that biodiversity and ecosystems are preserved. Agriculture and ecosystems are intimately linked, with each being heavily reliant on the other (Ferdouse et al. 2024). Farmers must thus look at issues from an organic or combined basis since no element exists independently. The number of people on Earth is growing, and as a result, the climate is changing. Without the protection of farmers' health, biodiversity, agricultural resources, and the environment that sustains them, people will be more exposed to risks. The preservation of organic resources at landscape and planetary levels is important for people to live decent, healthy lives in 2050. Because preservation is an important farming issue, it must be combined with the creation process. By providing enough healthy food, the agricultural industry is required to protect biodiversity and develop regular solutions that allow farmers to handle threats. In doing this, the farming industry is obliged to keep operating in a way that preserves natural resources (Pradha, More, and Prasad 2024).

Technologies enabling agriculture to meet the sustainable development goals of zero carbon emissions, zero hunger, and zero soil erosion, to mention just a few, are rapidly being developed and are capable of being cost-effectively deployed. This chapter will explore a number of these technologies. Firstly, we will explore precision agriculture techniques that streamline agronomic research and can lead to better decisions for optimizing inputs, monitoring and enhancing yields, and ultimately boosting farm efficiency. Secondly, we will investigate the potential of smart irrigation systems that can automatically switch off when it is raining, sense moisture in the soil, and only water when needed, and in the future, potentially be connected to management systems for inputs, fertilizer, and crop protection products, providing a one-stop shop for all farm needs. Finally, we will explore some biotechnology developments that could see plants that are more resistant to disease and drought, as briefly presented (Manville, Levitt, and Lai 2024).

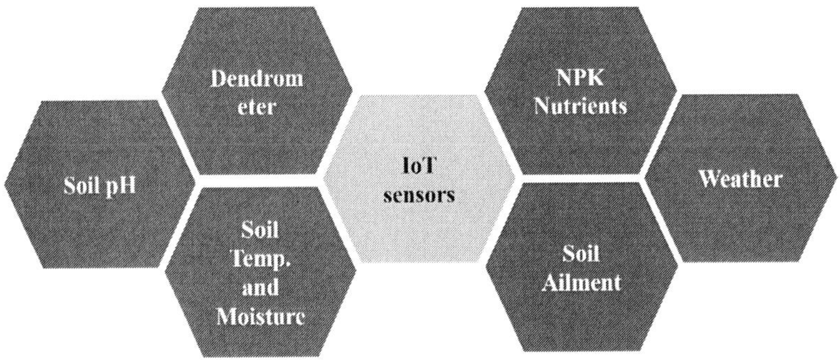

Fig. 15-1 Internet of Things-based sensors

Enabling this technological revolution, meanwhile, is the advent of big data, greater interconnectedness, the industrial internet of things, exponential technological progress in sensors and electronic systems, as well as co-developments in machine learning, connected devices, robotics, and biotechnology. A number of challenges are in place limiting the development of these technologies that must be looked at. They include the lack of digital agronomists to support the processing of this data, along with the complexity and uncertainty of knowledge and technologies themselves (Martey 2024). Proceeding through correct ethical pathways before knowing the possible consequences of research and technology deployment has meant that many genes with potential anti-infective or anti-disease qualities are patent-protected and are not able to be utilized due to prohibitive costs. There are also issues surrounding universal access to new technologies that will allow zero-interference meat cultivation, clean water supplies, and localized control of the genetics of locally grown crops based on different IoT sensors, as illustrated in Fig 15.1 (Keleş 2024; Pfeiffer 2024).

15.1.1 Precision Agriculture

Precision agriculture is an innovative and technology-driven approach, mostly data-driven, to manage spatial and temporal variability in agricultural fields. Timely and appropriate information at the right scale is needed because it is impractical to conduct empirical physical or chemical analysis at each point within a field in order to make an accurate assessment of the distribution of resources. Various sensors and associated data collection can be used to develop irrigation scheduling prescriptions, define land-leveling needs, detect pests, diseases, weeds, and vegetative stress, and manage resources. The technology for precision agriculture is the readiness of tools for farmers that allow data from their fields and assets to be collected, directly analyzed through software, and then recommendations returned to the farmer (Chidiogo Uzoamaka Akpuokwe et al. 2024).

Many integrated technologies can be applied for precision agriculture include Global Navigation Satellite Systems, Geographic Information Systems, Global Positioning Systems with real-time kinematic enhancement, Differential Global Positioning Systems, Geographic Positioning Systems, Global Systems for Mobile Communication or Code Division Multiple Access communication systems, Farm Management Information Systems, Yield Monitors, Telemetry, and Internet of Things, remote sensing, Geographic Information Systems, and Global Positioning Systems for higher profit and maximum yield to help minimize waste, protect the environment, and save energy (Arianto et al. 2024). This area is widely accepted as representing the future of the new farming business for sustainable development combined with environmental health. However, due to the multi-aspect nature of precision farming technologies, they must be technically understood, and their broader ethical, environmental, socioeconomic, and policy issues must also be considered, along with their business roles (Saeudy and Hussainey 2024; Sadowski et al. 2024). Different IoT Applications are demonstrated in Fig 15.2.

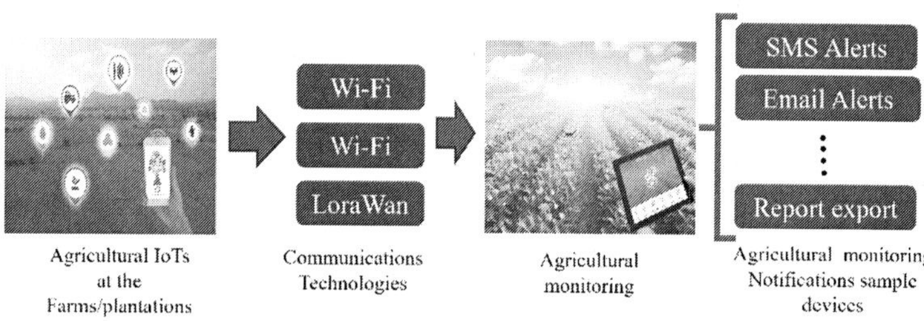

Fig. 15-2 Application of agricultural technologies

15.1.2 Smart Irrigation Systems

When enabled to use advanced technologies, irrigation systems are often described as smart because they manage water delivery in a way that can help ensure both water use efficiency and water productivity, whether in the short, medium, or long term. A smart irrigation system can sense how much water a crop needs, and it applies that amount of water at the right time in the right irrigation method, from standard surface irrigation to the most advanced subsurface drip systems. This presents advanced irrigation control technologies, including the challenges to their adoption and the possible integration within sustainable practices. Smart technologies can help to sense the right amount of water for a crop, while standard sensors can also sense soil moisture for a desired precision in irrigation. Finally, smart technologies can choose the right crop coefficient or the right soil moisture setting according to the soil and the available water in terms of moisture or through a dynamic scheduling system that takes into account meteorological forecasts and crop growth stages (Iozzo et al. 2024). Therefore, according to a general overview, smart irrigation systems use various combinations of sensors and automated controls to optimize water delivery in agricultural systems. There are numerous sensors and systems available, usually measuring different characteristics of the plant, soil, or atmosphere. These instruments have high initial costs, but they offer advantages such as ease of use and accurate and fast readings. Providing usable, meaningful information and interpreting these inputs can be challenging, particularly in cases where the operator does not have technical expertise. Then, the adoption of these technologies is limited due to a lack of training, extension services, cost recovery, and access to advanced technologies by some farmers or irrigation practitioners (Khamisu, Paluri, and Sonwaney 2024). The use of smart technologies in a smart irrigation system may also have environmental concerns, but sustainably managed, they represent some of the most practical and innovative technological solutions available today to face the actual challenges that modern agriculture has in arid and non-arid regions. Therefore, the bottom line connecting smart irrigation in terms of sustainable agriculture with both economic and environmental sustainability is water-using factors influencing disruptive technologies, as presented in Fig 15.3.

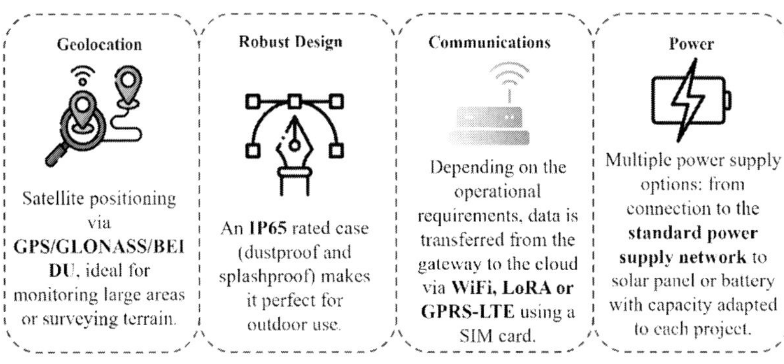

Fig. 15-3 Factors influencing agricultural disruptive technologies

15.1.3 Biotechnology and Genetic Engineering

Biotechnology and genetic engineering are considered two enabling technologies in sustainable agriculture. Much effort has been made to modify crops using genetic engineering techniques to allow them to express traits not naturally found. This can be achieved by these two main methods currently used by researchers: classical transgenesis and gene editing. The main purpose of researchers using these methods is focused on improving traits related to growth conditions, such as biosynthetic pathways, enhancing resistance to pests and diseases, and tolerance against abiotic stress. Moreover, some scientists are developing a new platform known as vertical farming, and next to this new form of agriculture, gene editing techniques would be beneficial as a method to modify food crops to grow food with a higher nutritional value (Sánchez-García et al. 2024). Most research in genetic engineering has focused on the economic benefits rather than considering ethical and environmental factors. However, these new enabling technologies for sustainable agriculture have a considerable impact on raising ethical questions as biotechnologies are considered not natural. There are some positive outcomes of producing GM crops, which can counter some of the ethical arguments against using this technology. Genetic modification presents varied regulatory frameworks in different parts of the world. This technology has attempted to save more food production, reduce insecticide application, and adjust agroecosystems to adverse production constraints, combined with the possibility of future co-tolerance to abiotic stresses. It makes GM crops important strategies for crop improvement in the future. This might result in higher yields, reduced chemical application, and increased income for the producers in certain areas of the world (Khoshkbarchi and Shahriari 2024). There are many examples of successful deployment of biotechnologies to produce crops that reduce the need for agrochemicals. The Bt genes have been deployed to protect a number of crops against caterpillar pests. By making these modifications in crops, farmers were able to reduce their pesticide applications. Because of the development of pests resistant to normally applied pesticides, Bt crops are now considered important in the control of crop destruction. Growers in the United States have had a strong reduction in chemical insecticide use. Biological pest control has increased because Bt crops are non-toxic to many non-target organisms, replacing alternative control over the surrounding agroecosystems (de la Torre and Briones 2024; Starke and Butkus 2024). Additionally, a substantial reduction of fungicide applications in soybeans has been achieved using GM soy with resistance to soybean rust, based on the different pathogens, as illustrated in Fig 15.4.

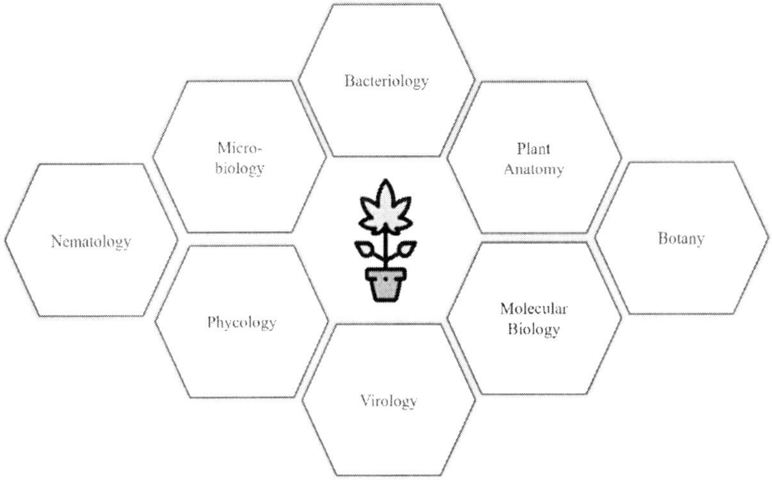

Fig. 15-4 Pathogens

15.2 Ethical Considerations in Sustainable Agriculture Technologies

A crucial dimension of the discussion on the adoption of enabling technologies for sustainable and regenerative agriculture involves ethical considerations. In this area, many enabling technologies are data-intensive, which is a concern in terms of privacy and data security. In addition, advanced technologies such as decision support systems should be built considering ethical implications since they are data-intensive, and their use might raise ethical concerns.

15.2.1 Data, Security, and Privacy

This becomes increasingly important when enabling technologies are considered. Future decision support systems will exchange and merge data about farmers for decision-making. With this in mind, will the adoption of these technologies bring new ethical challenges? Such systems could be only available to farmers who produce a large amount of data or who can afford to pay for them. This bears potential problems related to data stratification and the digital divide if a proportion of farmers is left out of small- or large-scale digital agriculture due to initial structural issues linked to digital gaps or to individual choices not to share personal data. Moreover, access and use of information can generate power platforms that could result in the worsening of problems of inequity and marginalization. Intellectual property issues should also be critically examined (Ajayi et al. 2024). Who owns the data traced by the system? As discussed above in this and other sections, ethical policy dimensions are of paramount importance related to the adoption and diffusion of enabling technologies. People's perceptions and belief systems must be met. Benefits are likely to be not perceived unless equity, ethical, and social issues are taken into account, including acts of directly and indirectly involved people. However, these issues are hardly addressed, promoting in science and technology development a "modernity-rationality" that is purely technocratic and instrumental (Jansen et al. 2024). The main critique of this approach is the general refusal of the existence of myriads of possible mechanisms for the improvement of the food and agriculture systems and the environment in developing an all-embracing system based on the "hard ground" of "scientific truth." Some scientists and stakeholders have pointed out several times that in the domain of the assessment of all components (socio-economic entities, ethical considerations, environmental, systemic aspects) surrounding one specific technology or innovation, there is time and necessity for the joint reflection of all involved (Pastorino, Prearo, and Barceló 2024). From this perspective, it is important to find a new framework for assessing, ideally in a shared exercise of Participatory Technology Assessment and Oriented Research. This approach is based on the strong belief that ethical and social issues are not external or independent but are part of the social and ethical structuration of techniques. If applied in agricultural technologies, then the acceptability of this technology will depend on a social and technical network of association. In this perspective, some actors are given credit for getting to act, whether it be involved in the preparation of legal narrow design, in scientific research programs, or in the enforcement and control of these technologies (Raza, Khan, and Qamar 2024; Ejuma Martha Adaga et al. 2024).

15.2.2 Technological Miss Application

The collection of an increasing number of different types of data seems to be one of the key features and drivers of future sustainable and smart agriculture. From farm machinery to soil, water, and plant sensors connected to the Internet of Things to satellite data in the sky, modern agricultural environments generate enormous and highly diverse data sets. Across countries with differences in data protection approaches from regulation to self-regulation, how farmers' data privacy is protected and what kind of data can be collected and used is a crucial ethical and legal consideration. It remains unclear if or which data from precision agriculture technologies directly represent sensitive personal data. However, this does not preempt the considerable potential for privacy intrusion of big agricultural data by combining datasets from different sources. Potential associated risks, if such data were no longer purely anonymized, comprise personal location information, family finances, and any authorized pattern of behavior such as rest time from work when the farm would be unoccupied (Ajayi et al. 2024).

The consideration of security and data protection on the farm involves cybersecurity risks of unauthorized access to personal data and dangers to property, such as hacking of machinery. Ethical criticism of data usage in agriculture is also linked to property rights and fairness considerations. Data portability and potential use of cloud services interfere with national property laws. Furthermore, by claiming data for research, new knowledge could be produced at cheaper prices since the data producers are not paid for the data value (Patil, Jadhav, and Nimbagal 2024). Smallholders in livestock farms might lose bargaining power vis-à-vis large agricultural technology providers, holding leverage for the required digital services. In particular, this problem might pose an obstacle to the enforcement of competition law, as large technology providers could exchange information to distort competition. Therefore, keeping in mind differences in the data protection legal frameworks around the world and the limited effectiveness of data protection regulations outside Europe, data localization measures should be deemed a serious lever for retaining access to trade with high-quality data. In setting such limits, the European Union could contribute to its regional digital autonomy using computer vision (Conneely and Coates 2024).

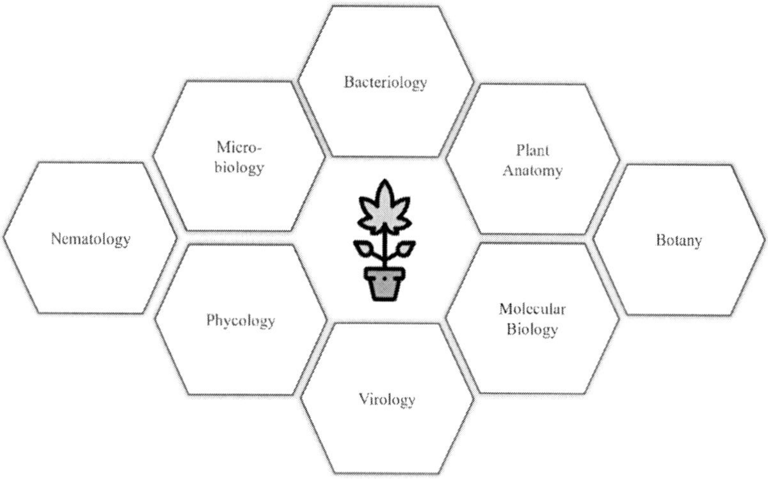

Fig. 15-5 Application process of computer vision in sustainable pest detection and classification

15.2.3 Equity and Access

While populations in developed nations often discuss sustainable and enabling technologies as a universal concept, there is growing evidence showing a disparity in the adoption rates of these technologies by different farming communities. One of the largest concerns is that these enabling technologies, which may offer opportunities for stewardship and sustainable agriculture, are most available to, and thereby useful for, farmers who are already financially prosperous. There is reason to be concerned about the impacts of these technologies if this tool is only available to and useful for one of three primary categories of producers. This would mean that a large percentage of the people living in industrial and rural communities are left behind and, if they have an interest in continuing to farm, are forced into a remaining category with potentially detrimental social, environmental, economic, and personal health impacts (Hariyanti et al. 2024).
Socioeconomically, not all farmers have equal access to the same resources, including enabling technologies and the financial support to be able to implement these innovations. Policy and decision-makers are continually seeking the most financially feasible and socially desired solutions to incentivize broad-scale adoption of sustainable agricultural systems and regain ecosystem functions lost during increasing intensification. In order to ensure

that these future agricultural systems are as equitable and functional as possible, it is the role of academics, policy professionals, industry, and the public to move beyond the implementation of regulatory policies, certifications, and discounted ecosystem service markets to bolster market-supporting solutions. A priority is to support the dissemination of enabling technologies to farmers who are currently unable to weather or bank the economic savings of investing in these systems (Davidavičienė, Abou Fayad, and Gergess 2024). This includes opening up the process for families currently excluded by increasing farm sizes.

Start-up incubators and other private and public sector business support groups should be encouraged to broaden their services to focus on societal goals of inclusive, smart agricultural practices for any start-up operation seeking to innovate or expand in novel ways. Renewed and new partnerships between municipal and regional government, industry, nonprofit trades, and community governance organizations could construct mentoring and mutually developing resources for potential new entrant farmer colleagues. It is not the role of the production market to decrease liability and financial risk to consumers, neighbors, and the biosphere (Saeudy and Hussainey 2024). Currently, however, existing infrastructure is rapidly being developed to shift smart farm management to align with those social and capital goals. It is here that the municipal governing body can begin to shed a more practical role in environmental oversight: in helping new entrants or transitioning operations to adhere to modern equitable socio-ecological practices confidently and quickly. Farmers documented as having human and ecological respect ingrained in their business model will be more attractive to a variety of markets, including the urban-rural relationships seeking to become more resilient and socially equitable in light of climate change (Akpan and Oluwagbade 2024).

15.2.4 Intellectual Property Rights

When it comes to technology, intellectual property rights play a vital role. Intellectual property law provides creators with ownership over their creations, which spurs them to innovate. On the other hand, people's access to that technology may be limited as a result of intellectual property laws that privilege the rights of the creator. This can result in something important or necessary becoming a vehicle for the rights of the first owner. This section focuses specifically on the types of intellectual property rights in the landscape for sustainable agriculture technologies (Khamisu, Paluri, and Sonwaney 2024).

Many people believe that intellectual property plays a significant role in the agriculture field. This has consequences for how farmers can develop their innovations over time, specifically whether that right is vested or licensed. Farmers face new possibilities as a result of new technology. Some patented tools make it difficult for farmers to practice their traditional farming methods, which they claim can be a crucial influence on their decision to adopt new technology. Part of the explanation for this could be the idea that farming is a way of life that is often passed down through generations. Different national perspectives vary on this issue. Farmers face legal discrepancies about plant genetics and the use of genetically modified technology due to the increased controversy surrounding patenting plant innovations and the roles of the genetically modified industry, particularly those involving technology. This has important implications for food sovereignty due to the difficulty of seeds increasingly controlled by a few companies and the risk of declining agricultural biological diversity. Many argue that technology is a basic human right that must be shared with farmers all over the world. By contrast, proponents of intellectual property rights posit that such privileges provide the motivations needed for vital research and development. According to this legislation, the roles of the courts and administrative court judges, as well as issues of national sovereignty and equitable treatment for international nationals, must be clarified (Narkhede et al. 2024; Pezzullo 2024). In order to assess the environmental impacts of artificial intelligence applied to agriculture, there are some international attempts, as presented in Table 15.1.

Table 15-1 Ethical Considerations in Enabling Technologies for Sustainable Agriculture

Category	Consideration	Description	Impact on Agriculture	Application	Reference
Global Responsibility	Avoiding exploitation in developing regions	Ensures fair global partnerships	Reduces technology disparity	Reduces technology disparity	(de la Torre and Briones 2024)
Ethical Financing	Supporting green and responsible funding	Encourages sustainable technology growth	Promotes ethical investments	Microloans for sustainable tools	(Pastorino, Prearo, and Barceló 2024)
Fair Compensation	Ensuring fair returns for technology use	Protects farmer income	Improves livelihoods	Revenue-sharing models for AI services	(Raza, Khan, and Qamar 2024)
Ownership Models	Supporting cooperative ownership	Distributes technology benefits equitably	Encourages community investment	Shared machinery models	(Zacher et al. 2024)
Data Privacy and Security	Ensuring farmers' data control	Protects personal and farm data	Builds trust in digital agriculture	Secure data-sharing platforms	(Nzubechukwu Chukwudum Ohalete et al. 2024)
Transparency in AI Decisions	Clear AI decision-making processes	Ensures understandable outcomes	Increases trust in AI predictions	Increases trust in AI predictions	(Sánchez-García et al. 2024)

Cultural Sensitivity	Respecting local farming traditions	Avoids disrupting cultural practices	Enhances adoption of suitable tech	Customized solutions for Indigenous farms	(Akpan and Oluwagbade 2024)
Employment Implications	Addressing job displacement concerns	Mitigates potential automation job losses	Supports workforce transition	Training programs for AI usage	(Iozzo et al. 2024)
Ethical AI Use	Preventing bias in algorithm design	Reduces unequal outcomes	Reduces unequal outcomes	Bias monitoring in AI systems	(Sánchez-García et al. 2024)
Intellectual Property Rights	Fair access to innovations	Avoids monopolization of key technologies	Encourages innovation sharing	Open-source agri-tech tools	(Jain et al. 2024)
Stakeholder Involvement	Engaging local communities	Promotes collaborative technology adoption	Fosters community-driven solutions	Farmer cooperatives	(Keleş 2024)
Informed Consent	Ensuring user consent in data usage	Protects farmer autonomy	Enhances ethical data collection	User agreements in precision farming	(Chakraborty 2024)
Open-source agri-tech tools	Addressing job displacement concerns	Mitigates potential automation job losses	Supports workforce transition	Training programs for AI usage	(Pradha, More, and Prasad 2024)

15.3 Environmental Considerations in Sustainable Agriculture Technologies

Agriculture is an essential part of human activities, and therefore, we have to move towards the certainty of sustainable practices. But we also need to consider advances that investigate and implement philosophical views that take into account a comprehensive understanding of the reported problems, consequences, and scientific data. One of these views is about environmental issues, and it focuses on the knowledge of all elements that interact mutually in our soil. When it comes to agriculture, one of the principal parts is soil and water. Agriculture is responsible for most of the soil being degraded and also for watercourses being polluted (Keleş 2024). Integrating environmental considerations into new technologies for sustainable agriculture would result in not just a partial alleviation of environmental problems but guarantee their long-term sustainability as well. Technological advancements in agriculture that have become widespread until now are predominantly the product of engineering thinking, which has been focused on solving individual problems in a given environment. Agrotechnical research has to take into account the interactions among all elements in agroecosystems and should focus on developing intelligent technologies to improve efficiency. As a result, we have to deal with shifts in perspectives in order to initiate the responsible development of interventions. Implementing our proposed technologies and teaching them have to be done based on the same principle of environmental respect. The outcome of this moral action—teaching, intervening, and teaching for—has to target not just a general increase in annual yield but a sustainable movement in agricultural systems toward environmental resilience as well (Nguyen-Viet and Nguyen 2024; Salameh and Al-Alami 2024).

15.3.1 Soil Health and Fertility

Soil is 'the collection of natural bodies on the earth's surface, containing living matter and supporting or capable of supporting plants outdoors.' It is the essential extractive, exchange, and transductive medium for both primary production and essential ecosystem services. Soil health is the continued capacity of the soil to function as a living system within ecosystem and land-use boundaries to sustain plant and animal productivity and biodiversity. Soil fertility is a central component or indicator of soil health. With long-recognized agricultural practices and goals for maintenance and improvement, and a history of policy measures, soil health, quality, and function are coming back to the forefront of sustainable agricultural and land-use policy. Enabling technologies for soil improvement date back to the voluntary development of the plow and the smelting of metals and can be seen as part of agronomy, even though the chemical-oriented artificial fertilizers created a better, more detailed agronomic scientific background (Chidiogo Uzoamaka Akpuokwe et al. 2024; Ahmad Hamdan et al. 2024).

Cultural eutrophication, compaction, and formation of water-repellent soils are evidence of serious soil degradation. Soil erosion, or sediments and nutrients ending up in surface water or the atmosphere, is the most acute problem in this context, for which structural funds can be used to invest in soil reparation. It is also the subject of complex agreements among countries in river basins, lakes, seas, or even oceans into the respective conventions. Different scenarios prepare for technological developments in this direction spurred by an awareness of the limits between a healthy living soil and a sub-product of its technical use. The discussion also refers to a sustainable consciousness-ethical approach to sustaining a living soil. There is an increasing need for healthy and fertile soil 'climate-smart' technologies for precision or metis agriculture (at foliar spray level) as well as resilience to non-technology irrigated farming. Democratic participation of individual farmers, food companies, the concentration of tobacco/food, and conservation of tobacco/food and natural resources water/soil dependent consumers is thus a necessity to fostering this shift. Enabling technologies can be an inventive-driven solution, and more attention shall be devoted to attitudes and ethical behaviors that spontaneously defend soil mores and, more generally, natural resources (Atutxa et al. 2024). In the past, industry 'produced,' society and the environment 'assumed costs and received the residuals.' In the agenda of the fiftieth year of environmental ethics, there is also a need to go beyond the loophole of the 'duty-to-perform obligations and the kindness no-obligation' and open the door to other options that can generate a greater commitment. We must also be careful, also in the use of enabling technologies in agriculture, to the prohibition ethics, that is, to avoid practices. Scientific innovation reflects human goodwill, but to align with an ethic based on prudence, traceability from biological foods and agriculture is useful. Soil mores must be reconstructed in the mind even if it is also important to reconstitute them socially (and successfully achieved in the economics of pleasure). 'Implementation Systems' also involve ethical repercussions, and it is ethical, in our view, to find solutions to adjust the economic benefit of national GDP increases to reduce the ecological footprint of the planet within the boundaries of the carrying capacity of the planet (Piunno and Shahmuradyan 2024).

15.3.2 Water Conservation and Quality

Water is a premium resource and needs to be managed efficiently, both in terms of consumption and the reuse of waste. There is a growing concern about global water scarcity and increased pollution of rivers, leading to the accumulation of heavy metals in agricultural fields. Therefore, smart irrigation to grow more with less is the need of the day. Around 3.6 billion people (50% of the world's population) are currently living in potentially water-scarce areas at least one month per year. Existing irrigation technologies are inefficient in water use and have low water productivity. Agricultural land management significantly affects watershed health in any region. A number of watershed-quality parameters are directly influenced by agricultural practices, and spatial variability of these indicators helps determine the impacts of various land use practices on a watershed (Al-Swidi, Al-Hakimi, and Al-Hattami 2024). The application of wireless devices in the development of precision agricultural systems is spreading rapidly for improving and monitoring both water quality and its use in agricultural lands. Most of the chemicals were found at higher concentrations at inputs relative to the outputs, and significant statistical differences were found for major chemicals between inputs and outputs. Nonpoint source pollution is diffuse runoff water that is contaminated with excess fertilizers and pesticides. Shallow tile flow at the field's edge is considered to be nonpoint source pollution in agricultural lands. There is a variety of other potential causes of poor water quality, including erosion, contamination, and runoff. These include microbial and parasitic pathogens, oxygen-demanding organic wastes, plant nutrients, pesticides, and other potentially harmful chemicals. Contaminants that affect the environment are complex and interconnected (Khamisu, Paluri, and Sonwaney 2024; Sánchez-García et al. 2024). Hence, a breakdown in access to any given sanitary resource affects agricultural productivity at each level, which, in turn, impacts public health and welfare.

Heavy rainfall can lead to surface water flooding and loss of soil from the fields. Managing and overcoming ground accumulation represent viable ways of securing high biomass productivity and overcoming food scarcity. Water reuse is an attractive option, and some land application components may require treatment, but that treatment might not need to be as high as required for potable water. The key issue concerning water use is sustainability over the long term. Thus, management for environmental cleanliness is essential, but should still allow productivity in agriculture. The primary objective of water management on and off the farm should be sustainability and the protection and rebuilding of natural resources above everything else (Khoshkbarchi and Shahriari 2024). There are already several technologies that have been developed to help protect surface waters from runoff contamination, with the most important of these being the plan. Mulching and no-till practices can have the best effect on water conservation, improve infiltration, reduce surface runoff, increase soil water storage, and conserve energy. Overall optimization in water management can have the largest impact on maintaining a quality supply of water for a productive environment and simultaneously deliver enough water supply for agriculture. Rapid depletion of clean water sources and poor quality of existing water bodies have replaced it as an essential measure to ensure sustainable water supplies both in industrialized and developing nations (Madjozi 2024).

15.3.3 Biodiversity Preservation

Biodiversity and the genetic diversity of both crops and livestock allow various crop species and varieties, livestock breeds, and non-crop organisms to co-adapt to the effects of climate change, and so to maintain and even increase food production when faced with biotic and abiotic stresses. Presently, small farms and diverse agricultural ecosystems are found to harbor abundant biodiversity and to sustain food production and populations in developing countries. Henceforth, agriculture must move towards more diversified crop and livestock systems guided by the fostering of healthy agroecosystems and not by the extraction of biodiversity for market exploitation (Zacher et al. 2024).

Enabling technologies aim to lead humanity and agriculture in that direction. However, technologies also present a risk that biodiversity is decreased further unless closely monitored. Recent history shows that, in many places, the introduction of new high-yielding crop varieties went alongside the loss of local traditional crop varieties. The resulting constant appearance of large areas planted to near-identical, high-yielding crop monocultures favors the abnormal build-up of injurious crop pests. This then fuels further production of crop protection chemicals. Maintenance or even enhancement of the overall agricultural biodiversity in every technological transition remains very important in working towards overall sustainable farming. This is especially true since the benefits of conservation tillage and agroforestry are also being appreciated; these farming systems can be enhanced through the planting of diverse crop varieties and maintaining a web of interacting organismal biodiversity (Ajayi et al. 2024). Some technologies are able to support this aim of conserving agriculture's biodiversity while achieving productivity.

Inversely, biodiversity has its own role in making crop production sustainable. Biodiversity, including genetic diversity, is an important aspect of good farm management through its close links with agroecosystem functions and cultivated agricultural systems. Diverse agroecosystems harbor a range of ecosystem functions as a response to their complexity, which usually sustain high levels of productivity and resilience. Those functions favor water conservation and recycling, erosion reduction and surface water quality enhancement, promotion of free environmental control of pests and pathogens, generation and recycling of plant nutrients, pest damage dilution, and the dispersal and control of disease and pests (Hariyanti et al. 2024). These command a high reliance by farmers and large costs. Agriculture also serves as a gene bank for crop and related genetic resources. This means that not only are valuable genes for crop improvement held by traditional farming populations, but farming wild and locally adapted crop varieties continues a sophisticated in situ breeding program aimed at producing high-quality food plants adapted to local conditions. Thus, at the farm level, a degree of biodiversity maximizes the prospects for attaining maximum sustainable performance levels in all these roles (Sánchez-García et al. 2024).

Biodiversity conservation thus represents an equally important goal and driver for future agricultural research and development. Since almost one-fifth of all organisms in Europe are considered to be living with a serious risk of extinction, there is ample justification for actions that consider not only economic land uses but also uphold the interests of the Indigenous biological systems and ecological functions in any agricultural land management system (Shafik 2024a). Unlike in-field biodiversity, which can be both managed and enhanced in cropping management, many wild and semi-natural habitats also support, and indeed depend upon, a range of rare and valuable species and neglected plants. For these many intertwined reasons, it is thus important that the planning and execution of research and new technology developments in agriculture be guided by consideration for biodiversity conservation (Starke and Butkus 2024; de la Torre and Briones 2024). This ensures that future agriculture is developed within policies that draw inspiration from biodiversity and foster the integration of conservation as an objective in multifunctional landscapes, as summarized in Table 15.2.

Table 15-2 Environmental Considerations in Enabling Technologies for Sustainable Agriculture

Category	Consideration	Description	Impact on Agriculture	Application	Reference
Precision Agriculture	Real-time monitoring for precise actions	Optimizes resource allocation	Enhances productivity and sustainability	Enhances productivity and sustainability	(Ferdouse et al. 2024)
Renewable Energy Integration	Utilizing renewable energy sources	Lowers reliance on fossil fuels	Promotes green energy use	Wind and solar energy on farms	(Pradha, More, and Prasad 2024)
Waste Management	Wind and solar energy on farms	Reduces environmental pollution	Promotes circular economy practices	Composting systems	(Manville, Levitt, and Lai 2024)
Sustainable Livestock Systems	Minimizing environmental impact of livestock	Reduces methane emissions	Supports greener animal farming	Enhanced livestock feed technologies	(Ahmad Hamdan et al. 2024; Chidiogo Uzoamaka Akpuokwe et al. 2024)
Pollution Control	Limiting chemical runoff	Prevents contamination of ecosystems	Protects water bodies	Controlled fertilizer application	(Piunno and Shahmuradyan 2024)
Climate Adaptation	Enhancing resilience to climate changes	Supports adaptation to extreme weather	Improves crop survival rates	Climate-smart agriculture models	(Ferdouse et al. 2024)
Energy Efficiency	Reducing energy consumption	Increases energy use efficiency	Cuts operational costs	Solar-powered agricultural machinery	(Keleş 2024; Jain et al. 2024)

Pest and Disease Management	Environmentally safe pest control	Reduces chemical pesticide usage	Promotes safer food production	Biocontrol agents and AI monitoring	
Biodiversity Protection	Safeguarding ecosystem diversity	Promotes ecological balance	Enhances resilience to pests	Agroecological farming practices	(Salameh and Al-Alami 2024)
Carbon Footprint Reduction	Lowering emissions from farming	Supports climate change mitigation	Reduces agricultural emissions	Use of renewable energy on farms	(Chakraborty 2024)
Water Management	Sustainable water use	Reduces overuse of water resources	Improves water availability	IoT-enabled water sensors	(Piunno and Shahmuradyan 2024)
Soil Health Conservation	Preventing soil degradation	Preventing soil degradation	Ensures future productivity	Soil nutrient monitoring tools	(Salameh and Al-Alami 2024)
Resource Efficiency	Optimizing input use (water, nutrients)	Minimizes waste and maximizes yield	Minimizes waste and maximizes yield	Precision irrigation systems	(Martey 2024)

15.4 Case Studies of Ethical and Environmental Issues in Sustainable Agriculture Technologies

The case studies deal with different enabling technologies and vary in scope, structure, and length, but they all discuss in detail the ethical and environmental issues that were raised during the projects (Shafik 2025). Through the documents presented here, the authors help illustrate some of the potential challenges in balancing the drive for innovation and sustainable agriculture with the need to consider local socio-political, ethical, and environmental contexts, and ensure that any particular enabling technology does not marginalize or disadvantage marginalized groups of people in both the global north and global south (Shafik 2024b). The use of drones in agricultural development, also known as precision agriculture, is a key example of a potentially enabling technology for sustainable agriculture, but one that raises numerous ethical issues. Precision agriculture has the capability to improve production, minimize the use of fertilizers and pesticides, and make smallholder farmers more resilient to shocks. Precision agriculture allows for a focus on individual problem fields and patches or degraded wastelands, increases understanding about soil fertility, pests and diseases, availability of agronomy inputs, climate, and water resources. Drones have the potential to revolutionize the way farmers, communities, governments, and development agencies approach sustainable agriculture research and extension. However, as with many enabling technologies, drones and precision agriculture have an array of potential negative environmental and ethical implications, including the exclusion of the poorest or those who are otherwise most marginalized (Shafik et al. 2024; 2023). The ethical implications of using GMOs for invasive fish and impacting precursor research on GM plants in Mozambique are two quite different, if related, issues. Unraveling the main strands of each of the case mentioned above studies will offer different perspectives on technology impacts, on small farm peoples' participation decision areas, and the industry creation processes that emerged. The studies illustrate that technology adoption is not clearly a positive or negative experience. In both cases, different values and stakeholder terminology intersect and entwine with local socio-political dynamics, creating potential exclusion opportunities or barriers to different technology uses in different settings (Anguiano-Santos and Rodríguez-Entrena 2024).

15.4.1 The Use of Drones in Precision Agriculture

Drone technology has robust potential to enhance the collection of big data that underpins precision agriculture. Ongoing developments will help improve the effective and accurate monitoring of plant stress, water use efficiency, and available resources in plant production landscapes, particularly in horticultural systems. Patrons of this technology promote its benefits to more effectively apply inputs (such as water and nutrients), improve productivity and crop quality, reduce labor and business costs, and reduce potential environmental harm and our overall environmental footprint. Furthermore, drones and other robotic machinery can be applied to a range of other on-farm activities beyond the management of plants, further expanding the efficiency and safety of other on-farm activities. But what are the ethical and other considerations when applying the precision farming principles of drones in agricultural contexts (Davidavičienė, Abou Fayad, and Gergess 2024)? In particular, how should we apply what we can do, when does the application of technology end or potentially compromise the social relationship with farming and farmers, what are the social implications of drones in agriculture, and who gets the data they generate and how can it be used?

Today, drones are being used (or have been used) for a variety of purposes in agriculture. Initial experiences in using drones for precision agriculture purposes in non-research contexts were for a mix of crop monitoring and also to assess the overall impact of heavy rainfall events on on-farm landscapes (Shafik and Kalinaki 2024). For agronomy purposes, small 'off-the-shelf' drones have been used for yield estimates of fodder beet crops. Larger scale 'research grade' drones have been used for a variety of sensors to be attached and flown over different agricultural production landscapes including apples, pears, vineyards (both commercial and experimental), and sunflower crops. In all cases, the primary focus has been looking at (near) plant system response and how the drones can be used in the development of plant health monitoring or plant stress assessment. Drones are being combined routinely on-farm with other precision agriculture tools, such as EM plate reading and ground-based sensors. These drones have NDVI and other cameras to assess plant stress. Furthermore, other available sensors include loss of phenological stages using a high-end RedEdge multi-spectral sensor unit and multi and hyper-spectral cameras, either purchased as built units or available built-in on mini UAS drones (Pradha, More, and Prasad 2024). Preliminary outputs of this research in which drones are involved are anticipated pilot projects for data collection and processing of data relating to:

- Plant biophysical, biotic, and abiotic health and yield production.
- Precise historical and present-day soil classifications and bed topographic designs.
- The effectiveness of different inputs, such as soil amendments and beneficials in canopy research programs.

15.4.2 GMOs and Ethics in Agriculture

The second case study engages with a technology that is highly debated and, in some instances, highly contested – genetically modified organisms. Inputs were gathered from several stakeholders, including two regulatory bodies, a private-public grouping, and three non-governmental organizations. The organization's work in Kenya focuses on the eastern side of the country and so, when they convened their workshop, the three activists, drawing as they do from 'the ground', represented a broad range of interests and concerns. Again, the narrative includes lengthy quotations that reflect the range of viewpoints discussed at the meeting. GMOs have the potential to support the achievement of sustainable agriculture. For example, crops can be engineered to reduce yield loss under conditions of drought and waterlogging, to increase the nutritional value of food, or to provide resistance to insect pests, reducing the need for chemical intervention (Manville, Levitt, and Lai 2024). The potential application of this technology is vast, and the choices about how and where GMOs may contribute to sustainable agriculture must include choices about their ethical use. The possibility of investment making these choices financially unviable is perhaps the greatest threat to the link between ethical rationales and the research and development necessary to make such technologies available. As a minimum, greater transparency about how GM animals, plants, and microorganisms are expected to contribute to sustainable agriculture could only be achieved through an open and robust ethical debate (Ahmad Hamdan et al. 2024).

15.5 Regulatory Frameworks and Policies

We focus on the regulatory framework and policies that have shaped and will continue to shape the products of sustainable intensification. Important to discuss in this context are the roles of various regulatory bodies in protecting the public from food that harms them or is not fit for purpose. It is the role of the government to regulate against products that may harm the population and to support innovation that has the potential to provide benefits for all or some of the population. Of equal importance is the legislative framework that supports innovations, allowing entrepreneurs and innovators to develop products that are assessed for the purpose that such an innovation seeks (Sadowski et al. 2024). Important areas of regulatory consideration in ensuring the safe and ethical application of genetic technology include the extent to which the outputs of this approach may have unexpected secondary effects, what the principles of imposed innovations will be, and the broader regulatory environmental standards, such as the climate change impacts of any new technology. These issues are currently considered within the framework of the European Common Agricultural Policy. The regulatory portfolio should include the protection of intellectual property essential to encourage and maintain the new user-led innovations, not just the enforcement of genetic patents, which is currently the preoccupation of regulatory discussions in many forums (Khamisu, Paluri, and Sonwaney 2024).

Collaboration in regulatory and legislative development is required. This is a responsibility of technologists; hence, the co-development approaches and the major investment in dialogue between farmers, technologists, and policymakers embodied by the Farm Platforms of the UK Sustainable Intensification Programmes. Regulatory frameworks exist to reassure the public that products are safe and robust against the impacts from the products. Ongoing challenges include how to integrate new products within the broader policies of land use and the rapidly evolving relationship between policy frameworks, technological developments, and public perceptions. Successful and ineffective models of regulatory frameworks were reviewed through a review of scholarly articles covering GM innovation and regulation (Pezzullo 2024; Narkhede et al. 2024). We selected innovative aspects of regulations because innovation is considered a key driver of responsible research and innovation. Further aspects of the innovations and governance of the resulting framework were analyzed by reviewing five selected European research projects carried out over the years 2000, conducted at an EU level, while the rest were carried out at a national level. The selection of the projects mostly covers physical sciences and some parts of life sciences, for which various terms such as novel foods, GM, GMO assessment, and applications of science and technology were used. All five projects used the EU as a test arena selected for reasons regarding the panel of stakeholders, the quality of debate, and the quality of research that they offered (Kostyra et al. 2024).

15.6 Future Directions and Emerging Technologies

The global population is projected to exceed 9 billion people by 2050, which is expected to drive a substantial increase in demand for agricultural products. Climate change is predicted to expose the world's most vulnerable populations to increasingly variable growing conditions and new production challenges, such as decreased nutrient density and increased pest and disease pressures. We forecast that trends in enabling technologies for sustainable agriculture will center around our adaptive capacity to address these challenges. One of these technologies is artificial intelligence, which has enormous potential to revolutionize agricultural practices. Another is big data, which enables us to combine meteorological, environmental, and whole-genome data to more accurately predict the incidence, severity, and distribution of pests and diseases, offering new opportunities for early mitigation efforts (Hariyanti et al. 2024). Furthermore, we anticipate increased development of robotics, in terms of both field robots for seeding, weeding, and harvesting and greenhouses for transplanting and automated pollination, which may be valuable to support growers with labor shortages as changing climates shift production locations. As we progress toward more innovative farming practices, it will be critical to ensure the alignment of developed technologies with critical sustainability goals such as decreasing greenhouse gas emissions, reducing water consumption, conserving habitats, improving water quality, ensuring resilient seed systems, and guaranteeing access to healthful foodstuffs. Additionally, the ethical considerations raised in the transition from smaller farm operations to larger, more corporate models should be taken into account. Most developed countries agree on the importance of supporting innovative new technologies in order to meet food security goals (Saeudy and Hussainey 2024). Collaboration between academia, industry, and federal government has resulted in national research and development strategies that will help reduce greenhouse gas emissions, secure more sustainable supply chains, and enhance agricultural productivity. Still, we will fail to meet global food security goals if we do not also invest in the development of smallholder agriculture, ensure capacity-building and education for relevant populations, and actually transfer innovations to farmers working in diverse and varying environments. Climate-resilient agriculture requires education and training for extension services and farmers, awareness-raising in rural populations, closer collaboration between scientists seeking innovative alternatives and traditional breeders, and new regional collaborations to further relevant policy and research efforts. Given increasing global market and security threats, it is imperative that our policymakers assess the security and sustainability of national food systems on a continual basis, engaging appropriate scientific and technical experts as well as broader agricultural stakeholders (Sadowski et al. 2024). Executing this vision requires a concerted effort by the global research community, greater investment in the agricultural sector, and committed partnerships between government, industry, and farmers. In addition, changes in consumer acceptance of sustainably produced foods will incentivize large companies to make substantial commitments to increasing sustainable agricultural technologies while continuing to decrease their carbon and environmental footprints. Addressing climate change will also require increased global focus on global-scale problems such as the sustainability of fish supplies and research on carbon sequestration methods. In the end, advanced agricultural methods must continue to align with the goals of developing the most abundant, nutritious, and safest foodstuffs (Kostyra et al. 2024; Carvalho, Mota, and Ramos 2024). The world is not so much at risk of running out of food as it is of running out of food that is safe to eat.

15.7 Conclusion

This collection of chapter has delved deeply into the myriad aspects of enabling technologies for agriculture and how, interconnectedly, they require close scrutiny of the ethical, social, and environmental dimensions. At the crux of a majority of this collection is an evolutionary narrative: technological capabilities are advancing at a staggering rate; in no area is this more profound than in the agricultural biotechnology landscape. We are rapidly moving beyond the static, purely genetic economy to an economy that leverages various sources of information and harnesses this knowledge to make better food and environmental products. Given the challenges of climate change, the exodus of laborers from food production, the spread of food insecurity, and the loss of arable land, agriculture must evolve. Regenerative agriculture, incorporating technologies that boost choices and accelerate and ensure the safe design of many pathways, is in some respects our only hope. We have made it clear throughout this discourse that there are ethical and environmental concerns that need to be interrogated and addressed. Technological advances do have a significant role to play in underpinning and making possible a sustainable future. On the horizon, we must work to ensure that these systems push in the direction of promoting not only wariness but ownership. When considering growing populations, shifting climates, and finite resources, delicate indeed is the balance of possible regards. The solutions when considering technological innovation and agriculture are not black and white; they exist in shades of grey. Reconciling the dichotomy between embracing the clear-eyed acknowledgment that technological evolution is essential in meeting sustainability goals while also being aware of potential dangers is a complex and vital nexus that we as a global society must clarify. An inclusive policy must be established to aid farmers in the adoption of enabling technologies, regardless of scale, and accessibility must be ensured. An ongoing dialogue must exist in the regulatory sphere and among stakeholders in order to address new challenges and correct misplaced expectations. Ongoing research and innovation are inescapably needed, and expanding upon collaborations in research and development will be essential in securing and harnessing the great potential of sustainability-oriented enabling technologies for agriculture.

References

Ahmad Hamdan, Kenneth Ifeanyi Ibekwe, Emmanuel Augustine Etukudoh, Aniekan Akpan Umoh, and Valentine Ikenna Ilojianya. 2024. "AI and Machine Learning in Climate Change Research: A Review of Predictive Models and Environmental Impact." *World Journal of Advanced Research and Reviews* 21 (1). https://doi.org/10.30574/wjarr.2024.21.1.0257.

Ajayi, Samson Abidemi, Oluwaseun Oladeji Olaniyi, Tunbosun Oyewale Oladoyinbo, Nneka Damola Ajayi, and Folashade Gloria Olaniyi. 2024. "Sustainable Sourcing of Organic Skincare Ingredients: A Critical Analysis of Ethical Concerns and Environmental Implications." *Asian Journal of Advanced Research and Reports* 18 (1). https://doi.org/10.9734/ajarr/2024/v18i1598.

Akpan, James U., and Oluyinka I. Oluwagbade. 2024. "Social and Environmental Responsibility in Accounting: Beyond Financial Metrics." *INTERNATIONAL JOURNAL OF SOCIAL SCIENCES AND MANAGEMENT RESEARCH* 9 (9). https://doi.org/10.56201/ijssmr.v9.no9.2023.pg163.188.

Al-Swidi, Abdullah Kaid, Mohammed A. Al-Hakimi, and Hamood Mohammed Al-Hattami. 2024. "Fostering Environmental Preservation: Exploring the Synergy of Green Human Resource Management and Corporate Environmental Ethics." *Bottom Line* 37 (1). https://doi.org/10.1108/BL-06-2023-0191.

Anguiano-Santos, Carlos, and Macario Rodríguez-Entrena. 2024. "Sustainability Reporting in Focus: Analysing Spanish Transposition of the Non-Financial Reporting European Directive in the Agri-Food Sector." *Agricultural and Food Economics* 12 (1). https://doi.org/10.1186/s40100-024-00303-z.

Arianto, Tomi, Adrias, Septriani, and Febri Yulika. 2024. "The Ethnoecology of the Malay Sacred Myth through Narratives of Kampung Tua Nongsa in Batam." *Studies in English Language and Education* 11 (1). https://doi.org/10.24815/siele.v11i1.19006.

Atutxa, Ekhi, Sofia Garcia-Torres, Charalampos Kyfonidis, Dimitrios Karanassos, Evangelos Kopsacheilis, Christina Tsita, Diego Casado-

Mansilla, et al. 2024. "Engagement and Accessibility Tools for Pro-Environmental Action on Air Quality: The SOCIO-BEE Paradigm." *Universal Access in the Information Society*. https://doi.org/10.1007/s10209-023-01072-0.

Carvalho, Luísa, Carlos Mota, and Patrícia Ramos. 2024. "Socially Responsible Investment Funds—An Analysis Applied to Funds Domiciled in the Portuguese and Spanish Markets." *Risks* 12 (1). https://doi.org/10.3390/risks12010009.

Chakraborty, Srabanti. 2024. "Bovine Milk Alternative and Its Potential in Indian Market- A Critical Study." *INTERANTIONAL JOURNAL OF SCIENTIFIC RESEARCH IN ENGINEERING AND MANAGEMENT* 08 (03). https://doi.org/10.55041/ijsrem29225.

Chidiogo Uzoamaka Akpuokwe, Seun Solomon Bakare, Nkechi Emmanuella Eneh, and Adekunle Oyeyemi Adeniyi. 2024. "CORPORATE LAW IN THE ERA OF GLOBALIZATION: A REVIEW OF ETHICAL IMPLICATIONS AND GLOBAL IMPACTS." *Finance & Accounting Research Journal* 6 (3). https://doi.org/10.51594/farj.v6i3.857.

Conneely, Ellie Ann, and Christopher J. Coates. 2024. "Meta-Analytic Assessment of Physiological Markers for Decapod Crustacean Welfare." *Fish and Fisheries* 25 (1). https://doi.org/10.1111/faf.12798.

Davidavičienė, Vida, Cendrella Abou Fayad, and Marc Gergess. 2024. "INTEGRATING SUSTAINABLE DEVELOPMENT INTO THEIR MARKETING STRATEGY: PRACTICES OF LEBANESE AGRI-FOOD INDUSTRIES." *Business: Theory and Practice* 25 (1). https://doi.org/10.3846/btp.2024.19457.

Ejuma Martha Adaga, Zainab Efe Egieya, Sarah Kuzankah Ewuga, Adekunle Abiola Abdul, and Temitayo Oluwaseun Abrahams. 2024. "PHILOSOPHY IN BUSINESS ANALYTICS: A REVIEW OF SUSTAINABLE AND ETHICAL APPROACHES." *International Journal of Management & Entrepreneurship Research* 6 (1). https://doi.org/10.51594/ijmer.v6i1.710.

Ferdouse, Jannatul, Beatriz Q. Silva, Marie Christin Baune, Nino Terjung, and Sergiy Smetana. 2024. "Life Cycle Assessment of Hemp-Based Milk Alternative Production in Lower Saxony, Germany, Based on a Material Flow Analysis of a Pilot Scale." *International Journal of Life Cycle Assessment* 29 (4). https://doi.org/10.1007/s11367-023-02264-9.

Hariyanti, Fitri, Almasdi Syahza, Zulkarnain, and Nofrizal. 2024. "Economic Transformation Based on Leading Commodities through Sustainable Development of the Oil Palm Industry." *Heliyon* 10 (4). https://doi.org/10.1016/j.heliyon.2024.e25674.

Iozzo, Ivano, Magda Belmontesi, Carlo Di Gregorio, Matteo Tretti Clementoni, and Valentina Angela Antonucci. 2024. "Revolutionizing Neck Rejuvenation: ABO Botulinum Toxin A Solution in Multipoint Technique and NASHA Gel Skinbooster." *Journal of Drugs in Dermatology : JDD* 23 (1). https://doi.org/10.36849/JDD.8095.

Jain, Shubham, G. Ranganna, Priyanka Shivaji Jadhav, Ujjwal Shrivastava, Ankita Sood, Lalu Prasad, Suraj Luthra, and Anushi. 2024. "Cultivating Tomorrow: Harnessing Gene Editing and Silencing for Precision Horticulture." *Journal of Advances in Biology & Biotechnology* 27 (1). https://doi.org/10.9734/jabb/2024/v27i1688.

Jansen, Sammie N.G., Bart A. Kamphorst, Bob C. Mulder, Irene van Kamp, Sandra Boekhold, Peter van den Hazel, and Marcel F. Verweij. 2024. "Ethics of Early Detection of Disease Risk Factors: A Scoping Review." *BMC Medical Ethics* 25 (1). https://doi.org/10.1186/s12910-024-01012-4.

Keleş, Ruşen. 2024. "Harmonious Human Settlements in the Light of the Principles of Ekistics." *Ekistics and the New Habitat* 84 (January 2024). https://doi.org/10.53910/26531313-e2024841678.

Khamisu, Muhammad Sani, Ratna Achuta Paluri, and Vandana Sonwaney. 2024. "Environmental Social and Governance (ESG) Disclosure Motives for Environmentally Sensitive Industry: An Emerging Economy Perspective." *Cogent Business and Management* 11 (1). https://doi.org/10.1080/23311975.2024.2322027.

Khoshkbarchi, Amir, and Hamid Reza Shahriari. 2024. "Improving Agents Trust in Service-Oriented Environment Based on Entropy Structure and Information Ethics Principles." *International Journal of Human-Computer Interaction* 40 (2). https://doi.org/10.1080/10447318.2022.2115639.

Kostyra, Eliza, Klaudia Gawlińska, Sylwia Żakowska-Biemans, Anna Piotrowska, Magdalena Gantner, Klaudia Kulik, and Katarzyna Świąder. 2024. "Exploring the Sensory Characteristics of the Soybean Spreads Enhanced with Vegetables and Spices." *Applied Sciences (Switzerland)* 14 (3). https://doi.org/10.3390/app14031096.

la Torre, Ana de, and Víctor Briones. 2024. "Ethical-Philosophical Considerations. Environmental Health." *Sistema* 2024 (269–270).

Madjozi, Monica. 2024. "Nanotechnology and Its Impact on Physical Sciences." *Journal of Physical Sciences* 5 (1). https://doi.org/10.47941/jps.1630.

Manville, Albert M., B. Blake Levitt, and Henry C. Lai. 2024. "Health and Environmental Effects to Wildlife from Radio Telemetry and Tracking

Devices—State of the Science and Best Management Practices." *Frontiers in Veterinary Science*. https://doi.org/10.3389/fvets.2024.1283709.

Martey, Edward Markwei. 2024. "Customer's Attitude and Purchasing Behavior of Green Food in Ghana: The Moderating Role of Environmental Views and Green Trust." *Electronic Green Journal* 2024 (49). https://doi.org/10.5070/G314960819.

Narkhede, Ganesh, Satish Chinchanikar, Rupesh Narkhede, and Tansen Chaudhari. 2024. "Role of Industry 5.0 for Driving Sustainability in the Manufacturing Sector: An Emerging Research Agenda." *Journal of Strategy and Management*. https://doi.org/10.1108/JSMA-06-2023-0144.

Nguyen-Viet, Bang, and Anh Thi Lan Nguyen. 2024. "Vietnamese Consumer's Perspective on Green Beauty Care Products: Exploring the Impact of Animal Welfare Concerns and Skepticism toward Green Advertising." *Acta Psychologica* 244. https://doi.org/10.1016/j.actpsy.2024.104210.

Nzubechukwu Chukwudum Ohalete, Oluwatoyin Ayo-Farai, Tolulope O Olorunsogo, Paschal Maduka, and Temidayo Olorunsogo. 2024. "AI-DRIVEN ENVIRONMENTAL HEALTH DISEASE MODELING: A REVIEW OF TECHNIQUES AND THEIR IMPACT ON PUBLIC HEALTH IN THE USA AND AFRICAN CONTEXTS." *International Medical Science Research Journal* 4 (1). https://doi.org/10.51594/imsrj.v4i1.737.

Pastorino, Paolo, Marino Prearo, and Damià Barceló. 2024. "Ethical Principles and Scientific Advancements: In Vitro, in Silico, and Non-Vertebrate Animal Approaches for a Green Ecotoxicology." *Green Analytical Chemistry* 8. https://doi.org/10.1016/j.greeac.2024.100096.

Patil, Shrinivas R., Smita Nadiger Jadhav, and Suvarna Nimbagal. 2024. "A Study on Ethical Implications of Using Technology in ESG Investing and Ensuring Unbiased Decision Making." *Multidisciplinary Science Journal* 6 (8). https://doi.org/10.31893/multiscience.2024153.

Pezzullo, Phaedra C. 2024. "On Environmental Communication as a Care Discipline." *Environmental Communication* 18 (1–2). https://doi.org/10.1080/17524032.2023.2300361.

Pfeiffer, Katie. 2024. "Homemaking for More-Than-Humans: Biodesign and the Redesign of Homes as Spaces for Multispecies Care." *Home Cultures*. https://doi.org/10.1080/17406315.2024.2335444.

Piunno, Paul A.E., and Anna Shahmuradyan. 2024. "A Science, Technology, Society, and Environmental Impact Assignment for Senior-Level

Analytical Chemistry Students." *Journal of Chemical Education* 101 (3). https://doi.org/10.1021/acs.jchemed.3c00883.

Pradha, Adhira, Taarush More, and Maandavid Prasad. 2024. "Models and Strategies for Implementation of Social Responsibility in Modern Enterprises." *Journal of Economics, Innovative Management and Entrepreneurship* 1 (1). https://doi.org/10.59652/jeime.v1i1.119.

Raza, Syed Ali, Komal Akram Khan, and Bushra Qamar. 2024. "Understanding the Influence of Environmental Triggers on Tourists' pro-Environmental Behaviors in the Pakistan's Tourism Industry." *Journal of Tourism Futures* 10 (1). https://doi.org/10.1108/JTF-12-2021-0269.

Sadowski, Norah, Resham Talwar, Edward F. Fischer, and Rowena Merritt. 2024. "Generating Demand for Alternative Protein in Low- and Middle-Income Countries: Opportunities and Experiences from Nutritious and Sustainable Market Solutions." *Current Developments in Nutrition* 8. https://doi.org/10.1016/j.cdnut.2023.101996.

Saeudy, Mohamed, and Khaled Hussainey. 2024. "An Institutional Perspective on the Shifts in Banking and Capitalist Ideology: Sustainability, Social and Environmental Insights." *International Journal of Bank Marketing* 42 (4). https://doi.org/10.1108/IJBM-02-2023-0125.

Salameh, Elias, and Hakam Al-Alami. 2024. "Hydrogeological, Hydrochemical and Environmental Consequences of the Extraction of Nonrenewable Groundwater in Jordan." *Hydrogeology Journal* 32 (1). https://doi.org/10.1007/s10040-023-02739-x.

Sánchez-García, Eduardo, Javier Martínez-Falcó, Bartolomé Marco-Lajara, and Encarnación Manresa-Marhuenda. 2024. "Revolutionizing the Circular Economy through New Technologies: A New Era of Sustainable Progress." *Environmental Technology and Innovation* 33. https://doi.org/10.1016/j.eti.2023.103509.

Shafik, Wasswa. 2024a. "Barriers to Implementing Computational Intelligence-Based Agriculture System." In *Computational Intelligence in Internet of Agricultural Things*, 193–219. Springer. https://doi.org/10.1007/978-3-031-67450-1_8.

Shafik, Wasswa. 2024b. "Biosensor-Based Drones Anomaly Detection Integration for Sustainable Agriculture Development." In *Achieving Food Security Through Sustainable Agriculture*, 283–316. IGI Global. https://doi.org/10.4018/979-8-3693-4240-4.ch012.

Shafik, Wasswa. 2025. "Sustainable Agriculture and Diet for Urban Development." In *Sustainable Smart Cities and the Future of Urban Development*, 165–92. IGI Global Scientific Publishing.

Shafik, Wasswa, and Kassim Kalinaki. 2024. "Societal and Ethical Implications of Technology-Enhanced Agriculture and Healthcare: An African Context." In *2024 IST-Africa Conference (IST-Africa)*, 1–11. IEEE. https://doi.org/10.23919/IST-Africa63983.2024.10569306.

Shafik, Wasswa, Ali Tufail, Chandratilak De Silva Liyanage, and Rosyzie Anna Awg Haji Mohd Apong. 2023. "Using a Novel Convolutional Neural Network for Plant Pests Detection and Disease Classification." *Journal of the Science of Food and Agriculture*. https://doi.org/10.1002/jsfa.12700.

Shafik, Wasswa, Ali Tufail, Chandratilak De Silva Liyanage, and Rosyzie Anna Awg Haji Mohd Apong. 2024. "Using Transfer Learning-Based Plant Disease Classification and Detection for Sustainable Agriculture." *BMC Plant Biology* 24 (1): 136. https://doi.org/10.1186/s12870-024-04825-y.

Starke, Jeffrey A., and Michael A. Butkus. 2024. "Tell Us What You Think: Exploring Environmental Ethical Considerations in the Problem-Solving Studio." *Journal of Civil Engineering Education* 150 (1). https://doi.org/10.1061/jceecd.eieng-1893.

Zacher, Hannes, Clara Kühner, Ian M. Katz, and Cort W. Rudolph. 2024. "Leadership and Environmental Sustainability: An Integrative Conceptual Model of Multilevel Antecedents and Consequences of Leader Green Behavior." *Group and Organization Management* 49 (2). https://doi.org/10.1177/10596011241229891.

CHAPTER SIXTEEN

TOWARDS A SUSTAINABLE AND TECH-DRIVEN AGRONOMY FUTURE

Abstract

Past breakthroughs in agriculture and their adoption have propelled mankind into the world's largest geometric progression, as discussed in earlier chapters. Ever-increasing food needs made technology a central element of accelerating growth. We must, however, always bear in mind the inalienable fact that agriculture is the very foundation of development and growth. Sustainable development in agriculture, therefore, is a supreme imperative. The meaning of sustainable agriculture in real terms is far broader than caring only for the health of the soil. Meeting the target of food for all is, of course, one of the primary elements of sustainable agriculture. Critical water resources must be sustainably managed. Valley by valley, the world is dramatically finding that it has been taking too much water for granted. Sustainable agriculture must strive to integrate pure agriculture with forestry, fisheries, and animal production all important sources of food, employment, and other benefits for rural development. Due to the era of knowledge, continuous interaction, and synchronized functioning between essentially all economic endeavors, a highly productive knowledge-based agriculture is a prerequisite for conforming to the goals of social and economic development. The increasing demand for technology and the market for rural development and agriculture is shaping the economy from mainly agricultural to one in which universally accessible knowledge and technologically flexible practices have success. The cellular, economic, and technological imbalances of the community can only be solved by broadly inclusive extended technical, environmental, and social educational strategies. As global resources, particularly for food production, decrease, their management becomes increasingly crucial. Hyperproduction has left many lands depleted and eroded, overused water reserves, and polluted the environment. Agricultural development must now adapt to be a factor in the nourishment of the earth and not a cause of its problems. Increased agricultural productivity, to be economically and biologically effective over the long term, must be based on more approvable use of resources and advanced strategies of harvest control.

Keywords: Agricultural Technology Integration, Climate-Resilient Agriculture, Data-Driven Decision Making, Eco-Friendly Agricultural Techniques, Precision Farming Solutions, Renewable Resource Management, Smart Farming Innovations, Sustainable Agronomy Practices

16.1 Introduction

In recent years, the field of agronomy has witnessed tremendous technological advancements that transformed the way farming is done, making it more productive, efficient, and ensuring sustainability. Agronomy now embraces the application of various integrated systems, such as cutting-edge information technology, global positioning systems, and geographical information systems. In the current scenario, digital agriculture is referred to as precision agriculture, which involves the digitization of field practices focusing on optimizing the use of water, pesticides, and fertilizers to achieve a balance between economic benefits and environmental impact. The integration of IoT and sensors in the field of agronomy has given new dimensions to gather real-time data on crop and soil (Sara et al., 2024). Sensors monitor minute-by-minute changes in weather variables, crop development, insect occurrence, disease prevalence, and the condition of the crop and soil. The enormous data generated from various sources are transformed into valuable information using big data analytics, making decision-making processes more reliable and user-friendly. Big data is the act of collecting large data sets that may be analyzed to predict certain trends and behaviors. Big data analyses can prove to be significant for the prediction of crop yield and improving decisions for sustainable resource management. Additionally, more recently proposed innovative technologies, such as blockchain and cryptocurrency, are now being utilized for managing and funding sustainable agronomic practices (Weyler et al., 2024).

The technological revolution has significantly improved the yield and conservation of natural resources and eventually redefined modern agronomic practices. Subsequently, the continuation of advanced research in the field of agronomy can help hasten the pace of a sustainable and tech-driven future in agronomic practices. To recap, agronomy can be defined as the science and technology of producing and using plants in agriculture for food, fuel, fiber, and land restoration. It includes plant genetics, breeding, biotechnology, phytopathology, crop production, management, soil management and conservation, and quality. Agronomists often specialize in areas such as crop rotation, irrigation and drainage, plant breeding, soil, crop, or seed technology, eco-physiology, and disease management. Agronomists work to integrate information from various disciplines, including ecology, chemistry, microbiology, molecular biology, genetics, and geology, to develop new practices and technologies to improve the quality and productivity of agricultural produce and to harvest them more efficiently while increasing the sustainability of the land (Кузнецов et al., 2024).

Agronomy is about understanding how plants grow and the management skills to produce healthy and abundant crops with minimum impact on the environment. Agronomists help manage to feed people all around the world, with many other aspects of the career, including conservation, urban soil management, and even some work to save our planet from global climate change. Without agronomy, we wouldn't have sufficient quantities of food, fuel, or fiber to sustain life on Earth, wouldn't have food able to be adapted to specific environments or resistant to diseases and pests, and wouldn't be taking such good care of the land. Agronomy has always been about technology, from the invention of the plough to irrigation (Lopes et al., 2024). But now agronomy is more about cutting-edge science and information technology to manage more efficient, high-quality production to feed an expanding world population.

The ability to produce enough food by 2050 will depend on various interacting factors. The increasing food demand that growing populations will bring near at hand, the finite nature of land, water, and greenhouse gas mitigation options in the face of climate change, and the necessary improvement in plant and animal enterprise adaptation to other rising constraints are among the most important. However, the high cost of many agronomic interventions means that, without appropriate attention to the distribution of the gains from technology use, many agronomic benefits will not generate broad sufficient incentives for relevant stakeholders to bear the risks and costs associated with testing, investment, and learning (Sara et al., 2024; Weyler et al., 2024).

First, training levels among key stakeholders are inadequate to ensure the sustainable delivery of known beneficial agronomic practices. In many developing regions, less than 10% of farmers utilize basic soil management tools that would increase productivity. Soil testing rates also remain very low, typically under 5% in Sub-Saharan Africa (SSA), while the main input suppliers have difficulties accessing the credit needed for long-term infrastructure investments. Second, the models that underlie efforts to remediate and sequester the emissions generated outside food production do not aim to ensure benefits that reflect actual land funds. This is problematic. Where the design of agronomic interventions to benefit the climate does not match encouragingly with productivity co-benefits, there may be a limited take-up of environmentally friendly, feasible mitigation outcomes (Bantjes, 2024; Wang et al., 2024). Such disparity may also discourage the development and diffusion of potentially important financial tools implemented to help the protection of the SSA environment (Moalosi et al., 2024).

The carbon price, which is inevitable in carbon emission control, could revive agriculture to become a carbon sink, but land use and reforestation should be suggested and evaluated. Agriculture, though initiated by domestication and the Green Revolution, receives exacerbated criticism instead of acclaim in carbon contribution, with inevitability and ligament. From 1961 to 2005, the repeal of carbon emissions in land use conversion exposed that the average emission ratio is 0.47×10^3 gC/m²/yr, and America is responsible for 19 to 23%. During 1990–2000, the severest emission propensity was transformed to cultivation in Latin America and Asia (12.7 and 15.7×10^3 gC/m²/yr). By 1999, China and America rose to 20% and 18%, ranking second and third; 6% and 7% contributions were achieved by India and Indonesia, respectively. Optimistic long-term emission suppression is facilitated by carbon sequestration for tropical countries, and developed countries migrate to emerging markets during the short term (Lopes et al., 2024). The potential market will interstitially elevate tropical countries from frying pans to fires and conserve foreign exchange supplements with less imported fuel.

Agriculture's partial credit and complete blame deserve much examination. During the Holocene, the Land Use/Cover Change repository reveals that the carbon sink had been germinating for 10,000 years in mid-latitudes, accumulating a total of 46 Gt C, and decreasing by 31% of the pre-industrial range. In total, due to the repositories of domestication and the Green Revolution, human society refrains from safeguarding 34% CO2, excluding the ingrowth effect. Contradictory suggestions exist. On about one third of the arable area, more than 70% of colonists produce wealth from industry and societal modernity. Why should agriculture accept a carbon subsidy? This work reveres the suggestions for energy cultivation: lush green with no plow case, one-time clearing, and soil protection (Joshua Kyle Beltran et al., 2024). Visible prosperity should be transformed into sustainable land use. Carbon emission containment and no plow management provide across-the-spectrum benefits for human bio-resources, domesticated species, and indigenous species. Vegetation is attracted, spreading from reclamation areas, recently burnt areas, and strips of dangling saplings. In addition to soil reforestation for algal blooms, no flooding accompanies warm events. Establishing vegetation cover to generate spring flooding eliminates sediment heralds adjacent to cities' drinking water supplies. The forest protects itself by disrupting the water yield for land reservations, sequestering carbon, mitigating ingestion management and ecosystems, and rescuing the evaporation of their microcosms on land they have lived for the global hydrological cycle (Shen et al., 2024).

Water and arable land scarcity have emerged as one of society's greatest challenges. The use of traditional soil for the cultivation of plants in agricultural activities is a practice as old as the history of agriculture. Soil nutrition in the right manner has been ignored and sometimes given low interest because it leads to higher fertilizer usage. These addressing techniques are complex, with high costs and consequences due to their impact on the environment. Watering through rainfall, surface, and drip irrigation is usual for plants, especially for many flowers, trunks, and leaves, but these processes require considerable amounts of water. A sustainable agricultural future completely depends on soil and water. Using current technology and the huge amount of data related to the environment in space, a strategy has been charted to use it to build a tool to control and optimize the water applied to plants and monitor this aspect flow (Asiminari et al., 2024; Kahya et al., 2024). Current field sensors do not always provide the necessary data on the root plants throughout their growth. AI may not have the necessary accuracy but can be, in the future, a great ally for solving the two mentioned restrictions. In the right direction, the possible development of radars and image processing capable of providing the right information and the greatest potential advantage: the measurement by the noninvasive root soil. Hydrated plants offer less resistance to light microwaves, using little energy while reflecting radio and electromagnetic waves in frequencies outside the disturbance band of the plants (Beisekenov & Hasegawa, 2024). With this data, it is possible to assess the needs of the plants.

16.1.1 Practicability of Sustainable Agronomy in the Information Era

In order to create a more sustainable agronomy future, it is necessary to contribute to the development of a knowledgeable society in agriculture, improve the management and control of food resources, and optimize the production processes that provide food resources. The term agronomy is defined as science applied to the existing and future agricultural production systems because it is based on the joint application of mathematical and experimental sciences to the fundamental aspects, engineering sciences to the transformation and management aspects of the production system (Li & Chen, 2024). Precision agriculture has introduced sensor systems where dedicated software tools manage the recorded data to guide the agriculture operation planners to configure a set of data-driven information needs for a specific production system. The result is the centralized and sustainable management of the entire food production system landscape for different crop cycles. Among the different digital concepts of the agronomy sector, decision designers can contribute to the sustainable and environmentally friendly improvement of the sector. The emphasis of this narrative review is on territory information technology and the characterization of the technology used in precision agriculture for the design of sustainable agronomy production system models (Armanini et al., 2024).

A redefinition of feeding strategies with digital precision feeding index technical rules has been proposed based on the usage of innovative technology to monitor and control the eating behavior and health status of different animal categories. Sectors of the sustainable agronomy field have already demonstrated their potential in processing information driven through sensor technology. Their application has proven valuable both in the animal and in the crop management decision processes (Yagnik, 2024). It has been recognized that the application of the concepts of sustainable agronomy, which featured the precision agriculture technology boom, can also be extended to enhance the performance of agricultural production systems based on controlled and optimized animal-rearing practices. Animal dietary strategies redefining the feeding guidelines for sustainable specialty crop production have been proposed. In particular, precision crop systems can be operated, considering every single managed standalone decision point applying the precision agronomy concept. Digital agronomy production system models have effectively contributed to enhancing the results of existing agriculture and have redirected decision-making processes to improve and ensure a food supply chain (Bertolini & Limongelli, 2024).

16.1.2 Precision Agronomy

In addressing the food demand in 2050, new operating models need to be established in food production and agribusiness. Digital tools such as satellite monitoring and global positioning system tractor systems are used to create high-resolution maps so that farmers can use inputs such as water, pesticides, and fertilizers only when and where needed. Farmers had already started embracing better information technology tools a few years ago, summarized under the term precision farming, to optimize operations and minimize the environmental impact of their activity. In this way, biodiverse ecosystems that help maintain healthy cropland ecosystems can coexist with food production (Sakya et al., 2024). In recent years, the technological evolution in this field has been significant, but considerable challenges remain, and the speed and success of adoption are still unsatisfactory. Precision agriculture enables the monitoring of environmental parameters for subsequent data analysis. Precision agriculture is one of the various digital initiatives that are open and multidisciplinary and could stand as innovative solutions. Achieving precision agriculture is not simple. The implementation of applied digital technologies is highly technical, and these digital technologies are complex systems. Moreover, the digital transformation of agriculture is not exempt from the risk of entry by multinational companies in formulas of digital plantation systems on which farmers' survival may depend. It is up to the countryside population to fight to position digital transformation as an innovation that re-empowers farmers and technological sovereignty as an important public policy pillar of its democracies (Pesonen et al., 2024).

16.1.3 Remote Sensing and Digital Agronomy

Remote sensing approaches, particularly satellite sensors, have long been used in agriculture. Some of the satellite-based remote sensing current applications in agriculture include Mapping cropped areas, including the estimation of planted areas of different crops and their mixture within pixels, extraction of crop rotations and fallows, and changes in the use of agricultural land. Estimating crop growth and yield: including the assessment of crop health, such as vegetative vigor, canopy density and development, leaf area, nitrogen status, chlorophyll content, photosynthesis, biomass, and crop maturity (Raja, 2024). Monitoring crop and soil area-specific characteristics and soil-plant nitrogen feedback, including the tracking of actual growth stage, water uptake, and drought conditions of crops. Mapping and quantifying biophysical conditions and associated changes, both those that condition or affect the natural vegetation productivity. Detecting pests and diseases: given the different visible and shortwave infrared hyperspectral characteristics of healthy and non-healthy leaves, particularly those caused by biotic factors. This variability also affects the biochemical content of cell structures and the leaf's water content and affects their reflectance properties; the technology-driven farm prospects are presented in Fig 16.1 (Deepak et al., 2024; Sharma et al., 2024).

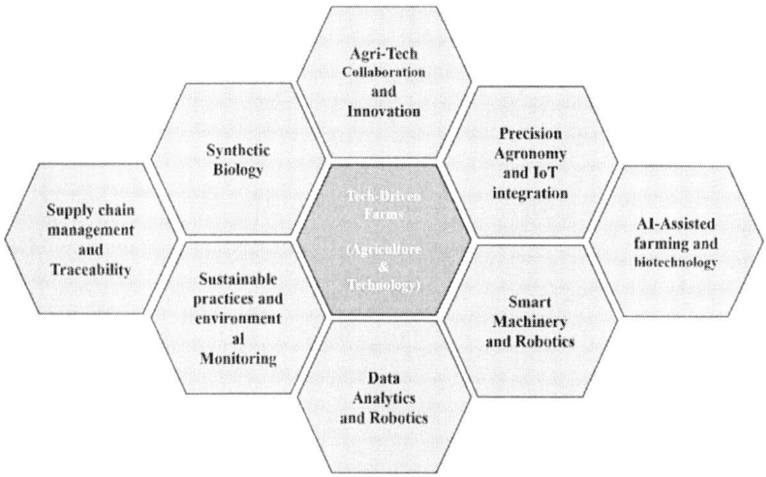

Fig. 16-1 The future of agriculture and software integration

16.2 Agronomy and Sustainable Practices

Sustainable Practices Sustainability on the farm is not just an end goal but a continuous process of improvement through sharing best practices, learning, and innovation. Using digital technologies is critical for the adaptation of sustainable agronomy practices. There exists a continuum of agronomy practices moving from conventional methods to agro-ecological practices. Conventional agriculture, relying on mechanization and chemical inputs, uses modern crop varieties but has a high nitrogen footprint, consumes large amounts of water, and poses risks to biodiversity (Loukatos et al., 2024). To create sustainable agriculture, agroecological practices must be followed. Integrated Pest Management (IPM) is one such agroecological approach, attracting and increasing natural pollinating species and biological control agents. Digital technologies like drones could be used for precise decision-making in the specific areas of the field where chemical applications are actually required, hence reducing the overuse of pesticides and helping the transition towards agroecological practices (Lowenberg-DeBoer, 2024).

16.2.1 Organic Farming

The present agricultural system contains inherent flaws. These include high energy consumption, environmental damage, product quality, and food safety issues. They create considerable pressure to reach food security and healthy life goals. Solving these problems will require the use of innovative techniques and technologies to change the current scenario. In fact, societies have always been looking for ways to enhance agricultural productivity. Organic farming excludes the use of synthetic chemicals, agrochemicals, and modified organisms in the agricultural ecosystem. Organic farming depends heavily on traditional methods and materials from non-scientific disciplines. Lower crop yields reduce their potential to meet food security requirements, although it leads to better tasting, longer-lasting, and more authentic crops (Shafik, 2024a; Shafik, Tufail, et al., 2024). Organic farming contributes to the slowing of some major causes of pollution and environmental destruction, such as agrochemical residue concentration and ecological process depletion. Consequently, some components of the agricultural community prefer it. However, low productivity and a lack of robust selection marker systems remain major obstacles to its development. Such a complicated scenario forces breeders to explore strategic solutions that can help organic farming reach food quality and safety supply and environmental goals and move closer to the development direction of sustainably driven, tech-supported agriculture (Shafik, 2024b).

Agroecology

When agriculture claims to restore or promote "natural processes," it should be remembered that agricultural species usually have naturally remained outside the system. What most distinguishes a system is the number of viable links, species, and processes within it; no system can go beyond its weakest link. However, it does not ultimately reduce to a matter of "making the best of the current situation." Existing forms of high-input agriculture are seen to contribute to a degree of environmental collapse that is integrally linked to the structure of national and global economies and to ways in which nations are linked economically to one another. To a large extent, agricultural intensification would not be possible without the support of economic structures, especially trade and finance arrangements that distribute comparative advantage and patterns of externalization (Shafik, Tufail, et al., 2024). Agroecology mobilizes a wide range of social, environmental, cultural, and economic changes, and the degree of change required to shift from our current mode of meal production to one based largely on agroecology may seem excessively or even impossibly high, while direct negative impacts on farm income are foreseeable. These matters pose a major challenge to decision-makers at every scale, especially at the international scale, and force a high degree of discomfort; theoretical work on agroecology and the results from systems in various regions offer particularly valuable guidance.

However, like other systems, agroecology will only be achievable and stable if built on a strong, desirable structure, which highlights both the extent of change necessary and the key roles of externalization and the institutions of economics (Shafik, 2025).

16.3 Innovative Solutions for Sustainable Agriculture

The attention given to sustainability allowed the development of innovative solutions to ensure and improve the efficiency of more sustainable production practices, such as those contributing to minimizing the carbon and water footprints of different agricultural systems. The research for sustainable development goals has opened the way for different public research institutions, startups, and companies to investigate, develop, and present different solutions in various areas, including those related to carbon footprints and technology. These innovative solutions, tools, and insights use information technology (IT) to integrate, analyze, measure, monitor, and simulate all the dynamics that involve the use of natural and human resources and the conditions of the different natural resources in agricultural systems and chains (Shafik et al., 2025; Shafik & Kalinaki, 2024). The work consists of reviewing many of these new solutions with the inherent limitations of a work of this kind. Agriculture is responsible for a significant proportion of land use, and consequently, carbon emissions mitigating measures involve the adoption of both precision farming strategies and knowledge of the effects of soil management. Some practices are intended to improve the environment and the users' health or to have positive long-term environmental impacts. The challenge for society in the current context is to increase agricultural production more efficiently and yet guarantee less environmental impact, i.e., obtain the "sustainable intensification" of agriculture. To do this, we propose a focus on agronomic issues and incorporate solutions drawn from IT (Wang et al., 2024).

One of the concepts that has gained a lot of attention over the recent years to ensure food security and sustainability of food production is the vertical farming concept. In terms of business or industrial applications, vertical farming can be considered as the use of a controlled environment in which plants are grown within a facility. Many other controlled environment agriculture systems are in operation today, such as greenhouses and indoor containers. However, the significant distinction with vertical farming is the vertical arrangement of crop production inside the facility. This is achieved through a construction with several levels of crop production, and the application of controlled environment technologies is planned and implemented in a way to grow crops at each level successfully (Beisekenov & Hasegawa, 2024).

At present, large and small-scale vertical farming facilities are being operated worldwide. They either operate as farming facilities providing harvested products to consumers or as platform providers that host and facilitate the farming of others. Along with vertical farming, citizens are exposed to and get familiar with other sustainable agriculture concepts like rooftop farming and urban agriculture. These operations are smaller in scale and rely on various degrees of technology inputs to achieve their sustainable agriculture goals. Starting from the early years of the concept formation in the 1980s, there has been significant interest in vertical farming in the scientific literature. However, the number of operational farms in the world was limited by 2010. The recent years of the 2010s showed a remarkable growth rate in operational farms and in the scale of operations (Pesonen et al., 2024; Sakya et al., 2024).

Many companies are operating by earning income through farm produce sales, but farm implements, and technology companies are also making significant technology investments. These technologies are directed to enable the use of LEDs, sensors, irrigation systems, thermal cooling, robotics, machine learning, and artificial intelligence to be operated at scale and efficiency for vertical farms. Suppose the market and policy incentives thrive in the direction of further market growth. In that case, it should be stated that both existing companies and new companies can become subjects of cooperation and coordination with each other and with public organizations. This is because for large-scale vertical farming to operate and to be sustainable, a technology push and market pull would require significant investment costs at each of the stages from design to operation. The brand-new facilities with technological and infrastructure capabilities also require substantial project management effort (Sharma et al., 2024). This could be a single-use development, or it could encompass and catalyze urban infrastructure and logistics networks.

Biotechnology applications represent the markets in which significant use of biotechnology methods, raw materials, and processes is being made. Nanotechnology in agriculture promises precision farming through nanosensors and nano-controlled release systems. It streamlines agricultural practices when confronted with global challenges at the crossroads of increasing population, agricultural productivity, and environmental sustainability. It has facilitated the genetic improvement of crops and microorganisms that can be used in an efficient and emergency-resistant way in farming. Using custom-designed proteins facilitates stress-resilient transgenic crops (Smitt et al., 2024). Genetically modified crops expressing tolerance to various stresses can be produced using various transformation methods. These methods have also been used for the genetic transformation of various orphan crops, as well as for the transfer of genes needed for various traits in pulses and legumes. In Brassica species, Agrobacterium-mediated transformation has been widely used not only for the introduction of genes for traits such as herbicidal tolerance, salt tolerance, and insect resistance but also for the production of low- and high-erucic-acid-free canola/rapeseed cultivars and high-lauric acid cultivars (Wenkai et al., 2024). The vectors specifically constructed have been tested and found to provide transgenic events at a good frequency.

16.4 Policy and Regulatory Frameworks

Without an enabling policy, regulatory and institutional environment, sustainable intensification of cropping systems at the global level will not be possible, as a wide range of issues must be addressed, including natural resource use efficiency, pollution issues, conservation and recreation values of ecosystems, loss of genetic resources, and marketing characteristics. Regulatory frameworks and market surroundings have to be seen as essential elements of policy instruments needed for the exclusion of farmers from access to polluting practices that are inefficient. This is difficult to achieve within a policy strictly defined by a standard tax or subsidy approach and might need special regulation. All these free-riding situations suppose the assurance of public goods as essential elements that are costly to guarantee and the necessity to internalize costs that are externalized concerning any transaction to be performed (Lopes et al., 2024; Кузнецов et al., 2024). When dealing with policy and regulatory frameworks and the enabling environment in their specific countries, Country Teams now have four main alternatives to consider concerning national scale by checking:

(i) to what extent the country's six conditions are met;
(ii) to what extent a higher reliance on market forces and well-functioning markets may solve remaining problems;
(iii) to what extent local governments are in a position to deal with this themselves by the provision of well-elaborated customized microscale enabling environment measurements; and
(iv) available technical support and financial assistance. Important policy development can be noted on national environmental and natural resource management policies, including agro-environmental

> and food safety issues, and this rapidly enlarging set of issues dealing with a broader range of public services that agricultural producers can provide.

This prompted substantial support and assistance for countries. It is stated that territorial differentiation will be a persistent issue in regional policy for a very long time, so countries will further increase their assistance and national or regional development programs and spend investments in public goods realization for increasing policy co-financing possibilities, as derived from the reformulated rules in this respect (Kahya et al., 2024).

16.4.1 Government Incentives

Government incentives on sustainable and tech-driven agronomy. The government should show a strong commitment to supporting sustainable and tech-driven agronomy by providing incentives in the form of credits, tax exemptions, or even subsidies to the stakeholders in the sector, particularly smallholders and private sectors. The incentives are given to promote technology adoption in agricultural production. The strong government commitment can give good signals to the industry stakeholders, especially private sectors such as banks, to support agricultural technology needs. These private sectors can facilitate the technology through facilitating credit access. The credit makes the technology available to farmers who are capital-constrained (Armanini et al., 2024). The technology providers will increase their research and development activities related to sustainable agriculture. The increased activities provide more alternative technologies to choose the best and the cheapest one. The sustainable agricultural technology costs will be more affordable for farmers. In addition, research and development also contribute to the improvement of overall agriculture. The increased productivity then leads to increased revenue for communities in the potential economic regions. It remains to be seen what possible consequences this will have regarding immigration to the cities. To foster technology adoption, governments also have to increase legal rights guarantees for the farmers. Legal guarantees such as property rights will encourage farmers to invest more in the technology (Bertolini & Limongelli, 2024; Yagnik, 2024). They are more interested in long-term benefits if they are assured that they will have the right to exploit it in the future.

16.4.2 International Agreements

Such an evolution towards a more sustainable and tech-driven future will require various adjustments to the way we currently produce our food, as well as a collective capacity for technological innovation. An effective enabler for this transformation is international cooperation, both regarding the changes needed in the agri-food sector and the research that underpins technological progress. Currently, several international agreements are setting the standards for future agriculture. At the first level, important global and regional institutions are crucial for providing suitable conditions for technological breakthroughs in agriculture. Several of the most successful scientific breakthroughs of recent decades have been made possible by setting up international research organizations. This was true for photonic research, which led to the first optical computers and fiber optics (Pesonen et al., 2024; Raja, 2024). Various international cooperation organizations have made great investments in helping to create conditions where important technological breakthroughs can take place. Agriculture is particularly sensitive to global and regional conditions, including implications regarding the climate, clean water, and healthy soil. In this respect, recently changing patterns in international cooperation are troubling. Despite important progress in the last millennium, institution building is currently less prominent on the public policy agendas across the world than twenty years ago. However, discontinuing these efforts could pose substantial risks. Future developments will involve significant challenges, particularly in terms of climate, preservation of biodiversity, and the efficient management of limited resources, such as nitrogen, phosphorus, and water. Handling these challenges will require specific agreements between countries if negative externalities are to be avoided (Bogue, 2024; Smitt et al., 2024). At a second level, adaptation of international agreements to new requirements in the primary sector is needed. Several international agreements are not well adapted to new agricultural practices, particularly those aimed at decreasing harmful impacts on the ecosystem pathways, as illustrated in Fig. 16.2.

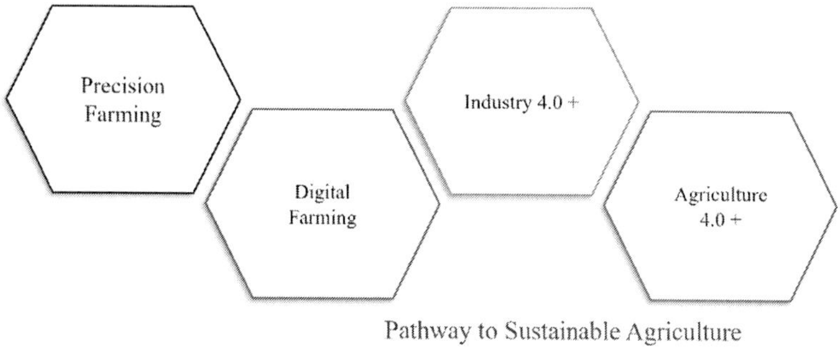

Fig. 16-2 The pathway to sustainable agricultural ecosystems

16.4.3 Compliance and Enforcement

Compliance and enforcement are critical components of agricultural policy designed to uphold the principle of rules and laws that are adhered to, and that also ensure enforcement measures for those who do not adhere to these rules and regulations concerning technology use and sustainable agricultural practices. Compliance ensures that the laws and policies are both workable and required, resulting in consequences for those who do not adhere to these rules. While producers may find it challenging to develop a common technology usage policy, there are no enforcement issues associated with this instance. However, when policies become more prescriptive in defining what can and cannot be done, compliance becomes more difficult to monitor and enforce (Wenkai et al., 2024). Given the uncertainty surrounding compliance and the challenges of monitoring and enforcing compliance with general agricultural policy and regulatory frameworks, can the effectiveness of technology policies and regulations designed to protect human health and the environment be effectively monitored and enforced? It is vitally important to monitor and enforce the various pesticide rules and regulations. A three-tiered approach to compliance assurance is adopted: education and outreach, compliance assistance, and regulation (Zhang, 2024).

Monitoring and enforcement are necessary for several reasons. Some of these include: The effectiveness of government risk management policy may be compromised if non-compliant production practices lead to higher pest thresholds. The cost-effectiveness of resistance management is reduced if there are low levels of compliance. For example, several case studies of successful biotechnology compliance and monitoring enforcement programs have been presented. The Nexus program follows the release of genetically modified products into the environment and has shown that while proactive compliance programs can assist industry, technology stewardship must always be used. Accountability has been reported as one effective way to enforce organic standards within a country. In Australia, the enforcement of organic standards was effectively left in the hands of individual consumers to hold others accountable for the products they produce. Norms and traditions are the main enforcers of social order, and it is no different in the context of technical progress in agriculture. Enforcement will provide structure, transparency, and an understanding among stakeholders, particularly where there are gaps, to ensure that a level of agreement or understanding exists concerning the role and responsibility of stakeholders regarding technological advancements (Ayris et al., 2024; Raja, 2024). Enforcement should see regulatory flexibility in response to knowledge and understanding of issues, thus exploiting advantages while enabling the greatest full scope of impacts to be realized, particularly positive environmental outcomes.

16.4.4 Economic Implications of Sustainable Agronomy

Agronomists and their stakeholders in different countries aim at a triple win of the sustainability agenda: doing more with fewer resources, preserving ecosystem services, and improving farmer economies and the livelihoods of farming communities. Agricultural research and smart agronomy must participate in fostering this shared process for the benefit of all. The three targets are not only reconcilable but interdependent. Improvements in technical intensity and positive externalities in agriculture act as both substitutes and synergies among land, labor, and capital. In addition, technical advances can unleash more interest in agricultural conservation and develop new financial markets financing investments in natural fixed capital and ecosystem services (Sara et al., 2024). When we evaluate the aggregate implications, however, variations in economic performance become evident. The flexibility to technological change—in the short and longer run—differs and influences the ability of actors to reap the full benefits, as does the flexibility of biological systems to new technologies and services. As a result, national economic performances and the changes experienced by many groups of agricultural stakeholders adapt differently to the land- and resource-using characteristics of the sustainable scenarios (Kahya et al., 2024). These variations promote technological change in terms of inputs and outputs, i.e., how inputs and outputs respond to changes in asset-ready technologies. This outcome impacts farmer behavior and actual parcel profitability. There are both winners and losers based on current cost and revenue technical gradients, ratifying that the rules governing techno-systems can generate long-lasting interest in protecting and fostering a development-friendly environment (Li & Chen, 2024).

16.4.5 Cost-Benefit Analysis

Various types of technological platforms can be used to define a cost-benefit analysis of crop protection research and development (R&D), including emerging technology, genetics, other sustainability strategies, biological control methods, and agrochemicals. The advantages of these platforms may vary according to specific systems, but a serious limitation to this comparative analysis is relevant data availability, costs of data management, and provision of reagents for ever-growing bioprospecting needs. Economic evaluations provide important insights into the welfare-economic implications of R&D in the field of plant protection. They permit the analysis of several important policy issues, including the effect of intellectual property and so-called "public good" aspects of R&D investment in basic and applied areas. Numerous previous studies have examined the economic costs and benefits of R&D expenditure in various fields. Given the concern of decision-makers and scientists about a fairly general decrease in R&D expenditures, it becomes apparent that a number of well-controlled empirical investigations adopted a more disaggregated approach by concentrating on specific factors promising high social returns (Lowenberg-DeBoer, 2024; Sharma et al., 2024). In the field of agricultural plant protection, economic literature is still limited, despite a consensus that new mutually compatible plant protection methods and cropping systems are needed.

16.4.6 Market Trends

Digital tools are penetrating powerfully into the traditional agriculture market. In 2019, precision agronomy revenues in Spain reached 245 million euros, with a growth rate of 11% with respect to 2018. Based on this 11% growth rate in 2019, a growth of 13% in 2020 to a total of 277 million euros was predicted, and on the back of this strong 2020 level, an average growth of 13.52% over the period 2020-2023 was anticipated. It is now thought that this commercial use of precision agronomy tools is going to be greater and that the growth of the sector is going to be accelerated due to the growing interest in sustainable food production. Therefore, a higher growth rate of big-data precision services that use cutting-edge artificial intelligence models in two different fields is predicted: the optimization of agronomic businesses and the maximization of the ecological impact of agricultural activities. Precision agriculture does not only reduce costs and optimize yield; it also contributes to a substantial extension of the natural and rural landscape. Some experts agree and add that with machine learning strategies, motorized interventions are facilitated using tractors and combines (Seitov, 2024). In addition, when automatically controlled by AI algorithms, these motorized interventions drastically reduce the use of fertilizers and agrochemicals in general. Some use cases are presented in Fig. 16.3.

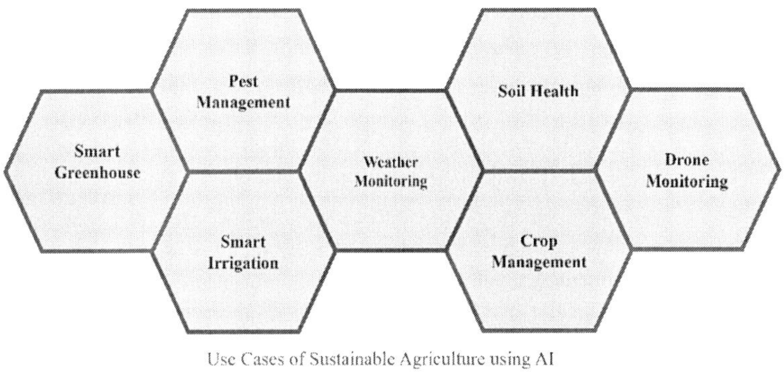

Fig. 16-3 Use cases of Sustainable agriculture using artificial intelligence.

16.5 Community Engagement and Education

Community participation in data collection, processing, and analysis is seen as engagement, which refers to the nature and degree to which information is shared between the providers and resource users. This type of engagement allows community members to have greater control over the data they collect and learn necessary data collection methodologies, including proper handling and maintenance of equipment, sensors, benchmarking, and data use on-farm. Technology will be a major pillar in achieving this concept; hence, many existing solutions need to adapt from mere data collection tools to multi-faceted platforms used for education, engagement, sensitization, and providing good data that can be easily interpreted by both the developed and developing worlds. Such platforms' advisory services include information on suitable practices, helping farmers make sustainable alternatives for improving crop productivity while ensuring quality, and helping smallholder farmers collect farm data by providing them with reliable sensors and weather measuring stations (Zhang, 2024). At the end of the production period, the platforms will help the farmers fetch better prices for their produce. In this manner, we can empower them with superior information and real-time solutions to shared challenges faced in the quest to attain food security. Key to this is the fact that we are helping smallholder farmers gain knowledge and decision-making capabilities, empowering all parties through data-driven technologies that can best alleviate poverty (Ayris et al., 2024).

16.6 Farmer Training Programs

Each decision a farmer makes impacts yield, from timing crop treatments in season and equipment in use to which seed variety to plant before the season. There is often a wealth of data available to influence these decisions, but farmers often do not know how to access it. Technology-centric training programs and solution-focused specialists can actively teach farmers by combining extensive product knowledge and hands-on training experience with a diversity of technology solutions, utilizing and interpreting these tools themselves for a greater end-to-end understanding. Farmers are then more effectively capable of making better-informed decisions and improving their agricultural businesses without being at the mercy of increasing multinational vertical integration within the agricultural industry (Joshua Kyle Beltran et al., 2024). Many farmer-focused agronomy programs partner with local land-grant universities, farmer co-ops, and other related organizations with a vested interest in tailoring training programs by growing regions, incorporating region-specific curriculum, resistance management plans, personalized ratios and types of services, and by leveraging local precision farming consultants. Some of these programs have modeled professional sales development and multi-week training programs to teach farmers. Instead of focusing on technology's value for the manufacturer, agronomy programs put the farmer consultant in the right light by focusing on the ability to advise and guide farmers in all elements of precision planning, nearly eliminating the need for endless support calls to and from farmers and manufacturers (Sakya et al., 2024).

16.6.1 Public Awareness Campaigns

Public awareness campaigns should become a norm, creating regional materials explaining the benefits of certain sustainable technologies or good practices and spreading this information through all possible channels used by farming communities. Unlike other markets, the farming sector also includes a variety of consumers with lower digital knowledge, thus highlighting the importance of regional events and personalized communication (Deepak et al., 2024). These campaigns will inspire trust and provide a golden opportunity for farmers to address concerns, learn implementation techniques, and secure their practices. Ultimately, these campaigns will contribute to dynamic market performance and the digital presence needed for good performers to initiate self-growth mechanisms. Recently, open field events organized by seed companies have been overshadowed by digital packages presented at farmers' homes. Each farming location had a list of planned visits, allowing all interested farmers, regardless of their internet connection speed, to stay up-to-date in real-time on all developments, ask necessary questions, and check implementation techniques. If a stagnant farming market, not only for certain commodities, can be identified somewhere in the world, this technique will provide options for use through agronomy extension services during the implementation period, although most likely on a limited territory (Lowenberg-DeBoer, 2024).

16.7 Future Trends and Emerging Technologies

The future of agronomy is expected to be highly influenced by new sensors and big data technologies. To support the farm-level implementation of sustainable agronomic practices, tools are very much needed to enable more mechanized, interconnected, soil- and plant-sensing practices that will then help in decision support. New and up-to-date technology is needed to collect spatially variable farm nutrient data and the identification of land characteristics and best practices that influence this variability. We expect that with the drive towards more sustainable agronomic practices, there will be a corresponding increase in the use of sensor technology on farms. This will enable precision placement of fertilizers and chemicals and will help to avoid fertilizer application overlap or misses (Bogue, 2024). The near-universal implementation of technology has unlocked the potential and enjoyment of working with high technology and data on the farm.

In large measure, the appeal of cyber-agriculture technologies across domains in agronomy lies in the use of technology to increase efficiency. Mechanical tools enable timely planting and harvest, reduce human labor, and cut crop loss at each stage, from pre-planting to transportation to market. Social appeal is also a strong factor to consider. The growing popularity of sensor technologies and the general completeness of new agricultural machinery offerings with digital data connection capabilities means high technology is not only for the wealthy but a serious consideration for any agricultural enterprise (Wenkai et al., 2024). People like to work with technology; there is joy in watching a vehicle driving hands-free along a straight path, collecting data, and using that data to inform decision-making on how best to solve the numerous problems that are of interest to agronomy scholars and practitioners. The feasibility of wireless communication channels applying data to solve problems that are specific to a given plant or farm and returning requested information to the farmer or farm equipment applying resources to a specific location in real-time opens opportunities in a way that is new and inherently exciting (Weyler et al., 2024).

Agriculture plays a key role in achieving the United Nations Sustainable Development Goals in eliminating poverty, guaranteeing food security, and dealing with climate change. The Sixth World Congress of Agronomists presented a unique opportunity to gather and publish a comprehensive set of articles that examine, inform, speculate, and provide novel visions for the future of agriculture and the scientific and technological needs. A total of 127 women and men agronomists and professionals active in interdisciplinary areas of applied research and technological development from 17 countries and 4 continents came together with the support of UNESCO (Wang et al., 2024). Artificial Intelligence (AI) in agriculture refers to the use of complex algorithms designed to perform specific tasks in an 'intelligent' way similar to human cognitive understanding. AI is already being used by many large agricultural businesses that require real-time decision-making and analytics to enhance crop yields. Through the application of AI and Machine Learning (ML) algorithms in robotics, particularly combined with hyper-spectral camera systems, precision and granular use and deployment of input resources have been shown to reduce and minimize chemical application in the agricultural ecosystem significantly. The implementation of AI practices in agriculture is also supported by the rapidly advancing Internet of Things (IoT) environment, remote sensing, unmanned aerial vehicles, satellite image acquisition, and Global Positioning System (GPS) data capture trends. As a result, all of this data capture and analysis can lead to a sustainable farming outcome with high-yield production. Consequently, AI can steer agriculture in the desired direction to meet ambitious UN Sustainable Development Goals in relation to agriculture (Ayris et al., 2024; Zhang, 2024).

Supply chain convergence through blockchain as an emerging technology in the domain of agriculture is currently in the limelight for creating traceability of toxic elements and components in commodities and contributing to an important problem in trade, namely food fraud. It is commonly accepted that while efficient traces may enjoy private benefits that typically help reduce or externalize operational or regulatory costs, they suffer when other industries, especially social problems, are externalized. This research aims to expand our perspective on existing traceability system designs, as well as bring in a new element to enrich our conversation around the association between traceability and transparency in today's information and its effect on holding enterprises and starting to move away from combined economic field crops to more climate change-mitigable field crops in agriculture. Spending solutions will be difficult, though a range of standards face us in the coming years, with compliance concerns (Sakya et al., 2024). These concerns for agri-complex businesses could be different and more stringent regarding sustainability and environmental friendliness. Blockchain technologies in the form of smart contracts in the food and agri-complex present a tantalizing opportunity to not just trace individual transactions but to lock into the blockchain descriptions of complex patterns of transactions, from planting through harvesting to transportation, processing, packaging, delivery, and public recognition. What is holding back this simple solution? We believe blockchain technological capabilities for smart contracts are a huge plus for agri-complex until we address climate risk in scenarios people around agriculture face regarding economic fear that is certainly coming but not mentioned in mitigation (Bertolini & Limongelli, 2024).

Blockchain is actually a new technology that is good for solving the problems related to the supply chain. Within the food industry, applying blockchain in the supply chain will enhance transparency and traceability throughout the chain. All parties involved, including the end user, can track the journey through the supply chain due to the ability to keep records of data transactions. As a result, concerns about food sources, food safety, fraudulent meals, and forced labor are better addressed. All these advantages will further create customer trust. That's why they not only rely on certifications but they also consider blockchain technology. There are already some international companies that have embraced blockchain technology (Pesonen et al., 2024; Shafik, 2024c; Shafik, Kalinaki, et al., 2024). For example, Walmart has joined IBM's blockchain to enhance its food supply chain traceability and transparency in China. Farm2fork is one of the blockchain platforms that are used to connect farmers with the finances they need and suppliers. This platform will show that the foods they are ordering are sourced and considered to bring benefits to society, animals, and the environment. Namely, blockchain has a concept that can carry out the process of tracing agricultural products based on identification using agriculture maps, climate characteristics, geographical features, seasons of production, the name of the place, and biosafety certificates (Wang et al., 2024). However, blockchain is not free from risks and challenges because it is not yet known by society and is not regulated by the government, so regulation is needed to protect customer transactions.

16.8 Summary of Key Findings

In summary, key literature study findings and thoughts leading to the conceptual foundation are highlighted here under key subheadings – trends, challenges, opportunities, best practices, and choices for the way forward, recognizing that the above are mutually interdependent. These will precede the main elements and conceptual framework of a technology-driven agronomy research and industry development model towards a sustainable future. The methodology used included a comprehensive review of literature related to agricultural development, agronomy, technology deployment, and their impact on society and the environment (Asiminari et al., 2024; Kahya et al., 2024). Such key findings will underpin a future agronomy model, recognizing that:

(1) the global agronomy community and their societal connectedness should further increase by being part of smart commodity-focused value nets;
(2) addressing the societal challenges such as resource constraints and the expectation to double food supply in the next 30 years;
(3) leveraging digital infrastructure exponentially and IT technologies incrementally;
(4) engaging in large policy experiments;
(5) creating valuable real options while making clear and prudent initial choices;
(6) averting potential backlash with strategies for critical self-reflection, educating society, building adaptive capacity, and enabling ongoing upgrading of technology; and
(7) clinical studies to further support the model building.

16.9 Conclusion

Agronomists have always tried, and still, they try their best to incorporate all available tools and knowledge into Agriculture. Those tools have been evolving for centuries, and they do it now at a fast pace. Precision Agriculture promises to reduce the inputs that are being used beyond an optimal level, thus becoming an essential tool for Sustainability. Besides, the knowledge that Agronomists afford is not just available to a bunch of people. The evidence pointing towards a decreased number of food crises, together with an increased population, is clear. Only by embracing knowledge availability and harnessing tools and methods do we get a sustainable view of the future. The view presented here is that SDGs opened a well-paved road towards Sustainability, enhancing the importance and the role of those working within its development. This view is not only pragmatic but is well supported by the technological advances that modern societies are open to. Additional efforts must be sustained for society to not just move into the roadmap that these Goals are but also to bring them close to reality. Agronomy, in particular crop production, plays a pivotal role in some of the SDGs. An abundance of tools is available for the science that suits this specific crop production area, thus deserving an effort to let it go further into the nice intersection of science, technology, and society.

References

Armanini, C., Junge, K., Johnson, P., Whitfield, C., Renda, F., Calisti, M., & Hughes, J. (2024). Soft robotics for farm to fork: applications in agriculture & farming. In *Bioinspiration and Biomimetics* (Vol. 19, Issue 2). https://doi.org/10.1088/1748-3190/ad2084

Asiminari, G., Moysiadis, V., Kateris, D., Busato, P., Wu, C., Achillas, C., Sørensen, C. G., Pearson, S., & Bochtis, D. (2024). Integrated Route-Planning System for Agricultural Robots. *AgriEngineering*, *6*(1). https://doi.org/10.3390/agriengineering6010039

Ayris, K., Jackman, A., Mauchline, A., & Rose, D. C. (2024). Exploring inclusion in UK agricultural robotics development: who, how, and why? *Agriculture and Human Values*. https://doi.org/10.1007/s10460-024-10555-6

Bantjes, J. (2024). Global knowledge-power asymmetries and student mental health in sub-Saharan Africa: A case study from South Africa. *Journal of International Students*, *14*(2). https://doi.org/10.32674/jis.v14i2.6390

Beisekenov, N., & Hasegawa, H. (2024). Advanced Preprocessing Technique for Tomato Imagery in Gravimetric Analysis Applied to Robotic Harvesting. *Applied Sciences (Switzerland)*, *14*(2). https://doi.org/10.3390/app14020511

Bertolini, A., & Limongelli, R. (2024). *Regulatory frameworks and standards for agricultural robotics in the European Union*. https://doi.org/10.19103/as.2023.0124.18

Bogue, R. (2024). Robots addressing agricultural labour shortages and environmental issues. *Industrial Robot*, *51*(1). https://doi.org/10.1108/IR-10-2023-0245

Deepak, A., Sharma, K., Naik, G. R., Shah, D. U., Modi, G., Poonguzhali, S., & Singh, D. P. (2024). Image Processing based Robotic Car for Agricultural Ploughing using Machine Learning Approach. *International Journal of Intelligent Systems and Applications in Engineering*, *12*(2s).

Joshua Kyle Beltran, Dominee Kyle Ibarlin, Mark Ivan Mapa, & Edwin Romeroso Arboleda. (2024). Exploring computer vision, machine learning, and robotics applications in banana grading: A review. *International Journal of Science and Research Archive, 11*(1). https://doi.org/10.30574/ijsra.2024.11.1.0180

Kahya, E., Ozduven, F. F., & Aslan, Y. (2024). YOLOv5 Model Application in Real-Time Robotic Eggplant Harvesting. *Journal of Agricultural Science, 16*(2). https://doi.org/10.5539/jas.v16n2p9

Li, Y., & Chen, Y. (2024). Robotics, environmental regulation, and agricultural carbon emissions: an examination of the environmental Kuznets curve theory and moderating effects. *Frontiers in Sustainable Food Systems, 8*. https://doi.org/10.3389/fsufs.2024.1336877

Lopes, M. S., Moreira, A. P., Silva, M. F., & Santos, F. (2024). Robotic Arm Development for a Quadruped Robot. *Lecture Notes in Networks and Systems, 811*. https://doi.org/10.1007/978-3-031-47272-5_6

Loukatos, D., Arapostathis, V., Karavas, C. S., Arvanitis, K. G., & Papadakis, G. (2024). Power Consumption Analysis of a Prototype Lightweight Autonomous Electric Cargo Robot in Agricultural Field Operation Scenarios. *Energies, 17*(5). https://doi.org/10.3390/en17051244

Lowenberg-DeBoer, J. (2024). *Economics of agricultural robotics*. https://doi.org/10.19103/as.2023.0124.19

Moalosi, R., Rapitsenyane, Y., & Sealetsa, O. J. (2024). Africa's transition towards net zero in the transport industry. In *Research Journeys to Net Zero: Current and Future Leaders*. https://doi.org/10.4324/9781003380566-17

Pesonen, L., Calvo Alonso, D., Backman, J., Kaivosoja, J., Recio Martinez, J., & Soininen, J.-P. (2024). *Advances in connectivity and distributed intelligence in agricultural robotics*. https://doi.org/10.19103/as.2023.0124.14

Raja, R. (2024). Software Architecture for Agricultural Robots: Systems, Requirements, Challenges, Case Studies, and Future Perspectives. *IEEE Transactions on AgriFood Electronics, 2*(1). https://doi.org/10.1109/tafe.2024.3366335

Sakya, G., Shivam, Rani, N., Tripathi, T., & Vats, S. (2024). Cloud-Based Swarm Robotics for Modern Agriculture. *Lecture Notes in Networks and Systems, 786*. https://doi.org/10.1007/978-981-99-6547-2_13

Sara, G., Todde, G., Pinna, D., Waked, J., & Caria, M. (2024). Implementation and Assessment of an Autonomous Ground Vehicle (AGV) for On-Field Agricultural Operations. *Lecture Notes in Civil Engineering, 458 LNCE*. https://doi.org/10.1007/978-3-031-51579-8_30

Seitov, S. K. (2024). Advantages of aluminum as a structural material for an agricultural robotic truck. *Horticulture and Viticulture*, *1*. https://doi.org/10.31676/0235-2591-2024-1-51-58

Shafik, W. (2024a). Barriers to Implementing Computational Intelligence-Based Agriculture System. In *Computational Intelligence in Internet of Agricultural Things* (pp. 193–219). Springer. https://doi.org/10.1007/978-3-031-67450-1_8

Shafik, W. (2024b). Biosensor-Based Drones Anomaly Detection Integration for Sustainable Agriculture Development. In *Achieving Food Security Through Sustainable Agriculture* (pp. 283–316). IGI Global. https://doi.org/10.4018/979-8-3693-4240-4.ch012

Shafik, W. (2024c). Blockchain-Based Internet of Things (B-IoT): Challenges, Solutions, Opportunities, Open Research Questions, and Future Trends. In *Blockchain-based Internet of Things* (pp. 35–58). Chapman and Hall/CRC. https://doi.org/10.1201/9781003407096-3

Shafik, W. (2025). Sustainable Agriculture and Diet for Urban Development. In *Sustainable Smart Cities and the Future of Urban Development* (pp. 165–192). IGI Global Scientific Publishing.

Shafik, W., & Kalinaki, K. (2024). Societal and Ethical Implications of Technology-Enhanced Agriculture and Healthcare: An African Context. *2024 IST-Africa Conference (IST-Africa)*, 1–11. https://doi.org/10.23919/IST-Africa63983.2024.10569306

Shafik, W., Kalinaki, K., & Zakari, R. Y. (2024). Blockchain's Motivation for IoT-Enabled Smart City. In *Secure and Intelligent IoT-Enabled Smart Cities* (pp. 195–221). IGI Global. https://doi.org/10.4018/979-8-3693-2373-1.ch010

Shafik, W., Tufail, A., Apong, R. A. A. H. M., & De Silva, L. C. (2024). Internet of Things for Smart Agricultural Practices. In *Internet of Things Applications and Technology* (pp. 190–217). Auerbach Publications. https://doi.org/10.1201/9781003458401-13

Shafik, W., Tufail, A., De Silva, C. L., Haji, R. A. A., & Apong, M. (2025). The Role, Application, and Impact of Artificial Intelligence in the Agriculture Industry. In *Future Tech Startups and Innovation in the Age of AI* (pp. 36–60). CRC Press.

Sharma, A., Kumar, S., Singh, A., Kumar, S., Saurabh, Yadav, H. C., Hazarika, S., & Hasan, R. (2024). Exploring the Role of Robotic Automation in Climate Vulnerability Mitigation: Towards Sustainable Horticulture. *International Journal of Environment and Climate Change*, *14*(2). https://doi.org/10.9734/ijecc/2024/v14i23914

Shen, J., Tang, S., Ariffin, M. K. A. M., Arry, A. A., & Zhao, R. (2024). Prospects and challenges of novel agricultural robots in vertical farming: A review. *Journal of Biotech Research, 16.*

Smitt, C., Halstead, M., Zimmer, P., Labe, T., Guclu, E., Stachniss, C., & Mccool, C. (2024). PAg-NeRF: Towards Fast and Efficient End-to-End Panoptic 3D Representations for Agricultural Robotics. *IEEE Robotics and Automation Letters*, *9*(1). https://doi.org/10.1109/LRA.2023.3338515

Wang, D., Zhang, B., Zhou, J., Xiong, Y., Liu, L., & Tan, Q. (2024). Three-dimensional mapping and immersive human–robot interfacing utilize Kinect-style depth cameras and virtual reality for agricultural mobile robots. *Journal of Field Robotics*. https://doi.org/10.1002/rob.22294

Wenkai, Y., Ruihang, J., Yiran, Y., Zhonghan, G., Wanyang, S., & Shuzhi, S. G. (2024). Agricultural Robotic System: The Automation of Detection and Speech Control. *Lecture Notes in Computer Science (Including Subseries Lecture Notes in Artificial Intelligence and Lecture Notes in Bioinformatics)*, *14454 LNAI*. https://doi.org/10.1007/978-981-99-8718-4_23

Weyler, J., Labe, T., Behley, J., & Stachniss, C. (2024). Panoptic Segmentation With Partial Annotations for Agricultural Robots. *IEEE Robotics and Automation Letters*, *9*(2). https://doi.org/10.1109/LRA.2023.3346760

Yagnik, N. S. (2024). COMPREHENSIVE REVIEW IN AGRICULTURAL ROBOTICS: POST-COVID-19 PERSPECTIVE OF ADVANCED ROBOTICS WITH SMART FARM. In *Precision Agriculture for Sustainability: Use of Smart Sensors, Actuators, and Decision Support Systems*. https://doi.org/10.1201/9781003435228-8

Zhang, Q. (2024). Opinion: Agricultural Robotics, issues worthy to study. In *Computers and Electronics in Agriculture* (Vol. 216). https://doi.org/10.1016/j.compag.2023.108528

Кузнецов, В. В., Шароватова, Т. И., & Холодов, О. А. (2024). Features of the transition of the russian agricultural sector to a new technological way of life. *Учет и Статистика*, *20*(4). https://doi.org/10.54220/2605.2023.45.93.005

List of Abbreviations

3D	Three Dimension
AFLPs	Amplified Fragment Length Polymorphisms
AgTech	Agricultural Technology
AI	Artificial Intelligence
AIoT	Agricultural Internet of Things
Al	Aluminum
AL	Automation and Logistics
AM	Additive Manufacturing
AR	Augmented Reality
ATP	Adenosine Triphosphate
B	Boron
BLE	Bluetooth Low Energy
BNF	Biological Nitrogen Fixation
Ca	Calcium
CAFO	Concentrated Animal Feeding Operation
Cas	CRISPR-associated Protein
cDNA	Complementary DNA
CMUs	Central Management Units
CNN	Convolutional Neural Network
COVID-19	Coronavirus disease 2019
CPS	Cyber-Physical Systems
CRISPR	Clustered Regularly Interspaced Short Palindromic Repeats
CSA	Community Supported Agriculture
CSAg	Climate-smart agriculture
Cu	Copper
DBMS	Data Base Management System
DL	Deep Learning
DNA	Deoxyribonucleic Acid
DoS	Denial of Service Attacks
DSBs	Double-Strand Breaks
DSS	Decision Support Systems
EDTs	Enterprise Digital Technology Services
ESG	Environmental, Social and Governance
FAO	Food and Agriculture Organization

FMIS	Farm Management Information System
GDP	Gross Domestic Product
GEV	Generalized Extreme Value
GIS	Geographic Information Systems
GMC	Genetically Modified Crops
GMOs	Genetically Modified Organisms
GNSS	Global Navigation Satellite System
GPS	Global Positioning System
GPU	Graphics Processing Unit
HGM	Heat Shock Gene Modulator
HGS1	Hypergravity Sensor 1
Hsp70	Heat Shock Protein 70 kilodaltons (kDa)
ICTs	Information and Communication Technologies
IoE	Internet of Everything
IoTs	Internet of Things
IPDM	Integrated Pest and Disease Management
IPM	Integrated Pest Management
IT	Information Technology
K	Potassium
LED	Light Emitting Diode
LiDAR	Light Detection and Ranging
LPIS	Land Parcel Identification System
LPR	Laryngopharyngeal Reflux
LSTM	Long Short-Term Memory
M2M	Machine-to-Machine
MAS	Marker-Assisted Selection
ML	Machine Learning
NDVI	Normalized Difference Vegetation Index
NFC	Near-Field Communication
NGO	Non-Governmental Organization
NICAM	Nicotinamide Adenine Dinucleotide Biosynthetic Pathway
OsSAP17-1	Oryza sativa Stress-Associated Protein 17-1
PA	Precision Agriculture
PAM	Polyacrylamide
PCR	Polymerase chain reaction
pDE-CAS9	Plasmid Deactivated Enhanced CRISPR-associated Protein 9
pH	Potential of Hydrogen
PMLs	Plant Membrane Lipids
QTL	Quantitative Trait Loci

R&D	Research and Development
RFID	Radio Frequency Identification
RGB	Red, Green and Blue
RNA	Ribonucleic acid
ROI	Return on Investment
SAMs	Site-Specific Agricultural Management
SAR	Synthetic Aperture Radar
SAR	Synthetic Aperture Radar
SCM	Supply Chain Management
SDG	Sustainable Development Goals
SDI	Subsurface Drip Irrigation
SDS	Sudden Death Syndrome
SEEA	System of Environmental-Economic Accounting
siRNA	Small Interfering RNA
SMEs	Small and Midsize Enterprises
SMS	Short Message Service
SNP	Single Nucleotide Polymorphism)
SRAPs	Sequence-Related Amplified Polymorphisms)
SSA	Sub-Saharan Africa
S-SAPs	Sequence-Specific Amplified Polymorphisms)
SSL	Secure Sockets Layer
SSRs	Simple Sequence Repeats
TALE	Transcription activator-like effector
TALENs	Transcription Activator-Like Effector Nuclease
UAV	Unmanned Aerial Vehicles
UV	Ultraviolet Radiation
VR	Virtual Reality
VRT	Variable Rate Technology
VTOL	Vertical Take-Off and Landing
Zn	Zinc

ABOUT THE AUTHOR

Wasswa Shafik (Member, IEEE) is a Computer Scientist, an Information Technologist, and a Research Director at the Dig Connectivity Research Laboratory (DCRLab), Kampala, Uganda. He received his Bachelor in Information Technology at Ndejje University, Luweero, Uganda, a Master in Information Technology Engineering (Communication and Computer Networks Option) at Yazd University, Yazd, Islamic Republic of Iran. He pursued his PhD in Digital Science (Computer Science) at the School of Digital Science, Universiti Brunei Darussalam, Brunei Darussalam. His research broadly examines, integrates, and focuses on developing computationally and statistically efficient models and algorithms to address complex questions in artificial intelligence and machine learning problems. His specific research interests include Smart Agriculture, Computer Vision, Ecological Informatics, Applied Artificial Intelligence, the Internet of Things, Cybersecurity and Privacy, and Smart Computing. He has authored, edited, and published hundreds of peer-reviewed books, technical papers, book sections, and numerous IEEE International Conferences and prestigious international journals. He has served as a reviewer of several international journals, Scopus, Compendex (Elsevier Engineering Index), and WoS International Journals. He further served in different capacities as department support for Mathematics for Data Science, Advanced Topics in Computing, Advanced Algorithms, and System Performance and Evaluation. Physical Computing for Intelligent Systems: Prior to this, as a department fellow, he served as a researcher associate at the Intelligent Network Laboratory in Iran. He served in different capacities as a Community Data Officer at the Programme for Accessible Health, Communication and Education, Research Associate and Data Manager at Population Services International, Data Manager and Research Assistant at the Socio-Economic Data Center, Research Lead at TechnoServe and Ag. Chief Executive Officer at Asmaah Charity Organisation.